CONTINUUM MODELS OF DISCRETE SYSTEMS 4

CONTINUUM MODELS OF DISCRETE SYSTEMS 4

Proceedings of the Fourth International Conference
on Continuum Models of Discrete Systems
Stockholm, Sweden, June 29 - July 3, 1981

Edited by

O. BRULIN and R. K. T. HSIEH

Department of Mechanics
Royal Institute of Technology
S-10044 Stockholm, Sweden

50
1931 N·H 1981
P C

NORTH-HOLLAND PUBLISHING COMPANY – AMSTERDAM • NEW YORK • OXFORD

ISBN: 0 444 86309 5

Published by:
North-Holland Publishing Company
Amsterdam • New York • Oxford

Sole distributors for the U.S.A. and Canada:
Elsevier North-Holland, Inc.
52 Vanderbilt Avenue
New York, N.Y. 10017

Library of Congress Cataloging in Publication Data

International Conference on Continuum Models of Discrete Systems (4th : 1981 : Stockholm, Sweden)
Continuum models of discrete systems IV.

Bibliography: p.
Includes index.
1. Continuum mechanics--Congresses. 2. System analysis--Congresses. I. Brulin O. (Olof), 1918- . II. Hsieh, R. K. T. (Richard Kin Tchang), 1950- . III. Title.
QA808.2.I565 1981 530.1'5 81-16826
ISBN 0-444-86309-5 AACR2

PRINTED IN THE NETHERLANDS

Editors' Preface

This Conference on Continuum Models of Discrete Systems (CMDS 4) is the fourth in the biennial series after the very successful conferences held at Jodlowy Dwor, Poland in June 1975; at Mont Gabriel, Canada in June 1977 and at Freudenstadt, Federal Republic of Germany in June 1979. CMDS 4 was held in the beautiful surroundings at the Grand Hotel Saltsjöbaden, Stockholm, Sweden from June 29th to July 3rd, 1981.

The purpose of these conferences is to bring together internationally renowned researchers active in continuum modelling of discrete systems within various fields so that topics of common interest may be discussed, and recent developments be exposed to critical review. Judging from the attendance record at the sessions, this objective was met by the Conference.

During the week, forty-eight invited contributions, general and sessional, were presented. All of them are recorded in these Proceedings together with six articles by individuals that were not able to attend the meeting in person.

The Organizing and Scientific Committees for the Fourth Conference consisted of:

Chairman,	O. Brulin	Sweden
	C. P. Enz	Switzerland
	W. Fiszdon	Poland
	G. Herrmann	USA
Past-Chairman,	E. Kröner	FR Germany
	I. A. Kunin	USA
	J. W. Provan	Canada
	A. J. M. Spencer	United Kingdom
	H. Zorski	Poland
Conference Secretary,	R. K. T. Hsieh	Sweden

A fifth conference is being organized in the United Kingdom with G. H. A. Cole and A. J. M. Spencer as organizers. Additional members of the Organizing Committee are O. Brulin, Past-Chairman, and G. Duvaut, G. Herrmann, E. Kröner, I. A. Kunin, H. Zorski.

The publication of the Proceedings concludes the work of the conference and we hereby would like to take the opportunity to thank the members of the Scientific Committee for valuable help in proposing a number of highly qualified contributors, as well as advising on the program of the CMDS 4. Merit is due to those who agreed to act as Chairmen of the various sessions, and we wish to acknowledge their help: K. H. Anthony, M. Beran, G. H. A. Cole, S. C. Cowin, G. Duvaut, C. P. Enz, K. Kondo, E. Kröner, I. A. Kunin, M. T. Landahl, R. E. Rosensweig. Thanks are also due to the participants, their contributions making the meeting worthwhile. We want to acknowledge the administrative help of Mrs. Christine Söderström as well as the work of our wives, Ingrid and Eva, who organized the social program. We also wish to thank Mathematics Editor Dr. A. Sevenster and Technical Editor Mrs. T. Kraaij, of the North Holland Publishing Company for their good cooperation.

Finally, we acknowledge with gratitude the financial support of the Department of Mechanics, and the Materials Centre of The Royal Institute of Technology; The State Department of Education, The Royal Swedish Academy of Engineering Sciences, The Royal Swedish Academy of Sciences, The Swedish Institute, The British Council at Stockholm, The Goethe Institute at Stockholm. Without their sponsorship it would not have been possible to arrange the Conference.

In conclusion we express our sincere wishes of success for the CMDS 5 in 1983.

O. Brulin and R.K.T. Hsieh
Editors
The Royal Institute of Technology

CONTENTS

PART 3 - FLUIDS, COUPLED FIELDS

PART 4 - BIOLOGICAL SYSTEMS, CONTINUUM MODELS OF DISCRETE SYSTEMS

PART 5 - MICROSTRUCTURE, FRACTURE, FATIGUE

LIST OF ATTENDEES

ANTHONY, K. H. — Fachbereich 6-Naturwissenschaften I-Physik
Universität Gesamthochschule-Paderborn
Warburger Strasse 100, Postfach 1621
D-4790 Paderborn, FR Germany

ARRIDGE, R. G. C. — H.H. Wills Physics Laboratory
University of Bristol
Royal Fort, Tyndall Avenue
Bristol BS8 1TL, United Kingdom

BARGMANN, H. W. — Swiss Federal Institute for Reactor Research
CH-5303 Würenlingen, Switzerland

BERAN, M. — School of Engineering
Tel-Aviv University
Ramat-Aviv, Tel-Aviv 69978, Israel

BRODIN, G. — Department of Electrical Measurements
Royal Institute of Technology
S-10044 Stockholm 70, Sweden

BRULIN, O. — Department of Mechanics
Royal Institute of Technology
S-10044 Stockholm 70, Sweden

COLE, G. H. A. — Department of Physics
University of Hull
Hull HU6 7RX, North Humberside, United Kingdom

COWIN, S. C. — Department of Biomedical Engineering
Tulane University
6823 St. Charles Avenue, New Orleans
Louisiana 70118, USA

DUVAUT, G. — Laboratory of Theoretical Mechanics
University of Paris VI
Tour 66, 4 Place Jussieu
F-75230 Paris 05, France

EDELEN, D. G. B. — Center for the Application of Mathematics
Lehigh University
4 West 4th Street, Bethlehem
Pennsylvania 18015, USA

ENZ, C. P. — Department of Theoretical Physics
University of Genève
32 Boulevard d'Yvoy
CH-1211 Genève 4, Switzerland

FELDERHOF, B. U. — Institut für Theoretische Physik A
Rheinisch-Westfälische Technische
Hochschule Aachen
Sommerfeldstrasse
D-5100 Aachen, FR Germany

FISCHER- HJALMARS, I.
Department of Theoretical Physics
University of Stockholm
Vanadisvägen 9, S-11346 Stockholm, Sweden

GAIROLA, B. K. D.
Institut für Theoretische & Angewandte Physik
Universität Stuttgart
Pfaffenwaldring 57/VI
D-7000 Stuttgart 80, FR Germany

GUO, Z.
Lehrstuhl für Mechanik I
Ruhr Universität Bochum
Postfach 2148, D-4630 Bochum 1, FR Germany
and
Department of Mathematics
Peking University
Peking, China

HAKEN, H.
Institut für Theoretische Physik
Universität Stuttgart
Pfaffenwaldring 57/IV
D-7000 Stuttgart 80, FR Germany

HINCH, E. J.
Department of Applied Mathematics &
Theoretical Physics
University of Cambridge
Silver Street, Cambridge CB3 9 EW, United Kingdom

HJALMARS, S.
Department of Mechanics
Royal Institute of Technology
S-10044 Stockholm 70, Sweden

HSIEH, R. K. T.
Department of Mechanics
Royal Institute of Technology
S-10044 Stockholm 70, Sweden

JEFFREY, D. J.
Department of Applied Mathematics &
Theoretical Physics
University of Cambridge
Silver Street, Cambridge CB3 9EW, United Kingdom

JENKINS, J. T.
Department of Theoretical & Applied Mechanics
Cornell University, Thurston Hall
Ithaca, New York 14853, USA

KLEMAN, M.
Centre d'Orsay
University Paris-Sud
Bâtiment 510, F-91405 Orsay, France

KONDO, K.
University of Tokyo
C.P.N.P.
1570 Yotsukaido, Yotsukaido City
Chiba-Ken, 284 Japan

KOVACS, I.
Institute for General Physics
Eötvös University
Muzeum Krt. 6-8, H-1088 Budapest, Hungary

KRATOCHVIL, J.
Institute of Physics
Czechoslovak Academy of Sciences
Na Slovance 2, 18040 Praha-8, Czechoslovakia

KRÖNER, E.
Institut für Theoretische & Angewandte Physik
Universität Stuttgart
Pfaffenwaldring 57/VI
D-7000 Stuttgart 80, FR Germany

KUNIN, I. A.
Department of Mechanical Engineering
University of Houston
Houston, Texas 77004, USA

LAMBERMONT, J.
Institute for Mechanical Constructions
TNO
P.O. Box 29, NL-2600 AA Delft, The Netherlands

LANCHON, H.
Laboratoire d'Energétique & de Mécanique
Théorique & Appliquée
2, Rue de la Citadelle, B.P. 850
F-54011 Nancy Cedex, France

LANDAHL, M. T.
Department of Mechanics
Royal Institute of Technology
S-10044 Stockholm 70, Sweden
and
Department of Aeronautics
Massachusetts Institute of Technology
Cambridge, Massachusetts 021139, USA

LINDGREN, R. E.
Department of Mechanics
Royal Institute of Technology
S-10044 Stockholm 70, Sweden
and
Department of Aeronautics
University of Florida
Gainesville, Florida 32611, USA

LUDWIG, W.
Institut für Theoretische Physik II
Westfälische Wilhelms-Universität
Domagkstrasse/Corrensstrasse
D-4400 Münster, FR Germany

MARINOV, P. A.
Institute of Mechanics & Biomechanics
Bulgarian Academy of Sciences
P.O. Box 373, 1090 Sofia, Bulgaria

MARKENSCOFF, X
Department of Mechanical & Environmental
Engineering
University of California
Santa Barbara, California 93106, USA

MARKOV, K. Z.
Faculty of Mathematics & Mechanics
University of Sofia
P.O. Box 373, 1090 Sofia, Bulgaria

MAUGIN, G. A.
Laboratory of Theoretical Mechanics
University of Paris VI
Tour 66, 4 Place Jussieu
F-75230 Paris 05, France

MAZILU, P.	Institut für Mechanik Ruhr-Universität Bochum Postfach 102148, Universitätstrasse 150 D-4630 Bochum 1, FR Germany
McCARTHY, M. F.	Faculty of Science National University of Ireland University College, Galway, Ireland
MUGHRABI, H.	Institut für Physik Max-Planck-Institut für Metallforschung Heisenbergstrasse 1 D-7000 Stuttgart 80, FR Germany
MULLER, I.	Institut für Thermo & Fluiddynamik Technische Universität Berlin Sekr. HF1-Strasse des 17 Juni 135 D-1000 Berlin 12, FR Germany
MUSCHIK, W.	Institut für Theoretische Physik HA D1 Technische Universität Berlin Hardenbergstrasse 4-5 D-1000 Berlin 12, FR Germany
NUR, A. M.	Department of Geophysics Stanford University Stanford, California 94305, USA
PARRY, G.	School of Mathematics University of Bath Claverton Down Bath BA2 7AY, Avon, United Kingdom
PROVAN, J. W.	Department of Mechanical Engineering McGill University 817 Sherbrooke Street West Montreal H3A 2K6, Quebec, Canada
RASKY, D. J.	Department of Mechanical & Environmental Engineering University of California Santa Barbara, California 93106, USA
ROSENSWEIG, R. E.	Corporate Research Science Laboratories Exxon Research & Engineering Company P.O. Box 45, Linden New Jersey 07036, USA
SÄLTZER, W. D.	Institut für Material und Festkörperforschung Universität und Kernforschungszentrum Karlsruhe Postfach 3640-IMF, D-7500 Karlsruhe 1, FR Germany
SAVAGE, S. B.	Department of Civil Engineering & Applied Mechanics McGill University 817 Sherbrooke Street West Montreal H3A 2K6, Quebec, Canada

SCHULZ, B.
Institut für Material und Festkörperforschung
Universität und Kernforschungszentrum Karlrune
Postfach 3640-IMF
D-7500 Karlsruhe 1, FR Germany

SINGH, M.
Department of Mathematics
Simon Fraser University
Burnaby V5A 1S6
British Columbia, Canada

SJÖLANDER, A.
Department of Theoretical Physics
Chalmers University of Technology
S-40220 Gothenburg 5, Sweden

SKALAK, R.
Bioengineering Institute
Columbia University
New York, New York 10027, USA

SÖDERHOLM, L.
Department of Mechanics
Royal Institute of Technology
S-10044 Stockholm 70, Sweden
and
Department of Theoretical Physics
University of Stockholm
Vanadisvägen 9, S-11346 Stockholm, Sweden

STEFANIAK, J.
Institute of Technical Mechanics
Technical University of Poznan
Ul. Piotrowo 3
60-965 Poznan, Poland

TALBOT, D. R. S.
School of Mathematics
University of Bath
Claverton Down
Bath BA2 7AY, Avon, United Kingdom

VERMA, P. D. S.
Department of Mathematics
Regional Engineering College
Kurukshetra-132119, India

WALPOLE, L. J.
School of Mathematics & Physical Science
University of East Anglia
Wilberforce Road
Norwich NR4 7TJ, United Kingdom

WILLIS, J. R.
School of Mathematics
University of Bath
Claverton Down
Bath BA2 7AY, Avon, United Kingdom

Additional Contributions to the Proceedings

DIENER, G. and WEISSBARTH, J.	Sektion Physik Technische Universität Dresden Mommsenstrasse 13, DDR-8027 Dresden German Democratic Republic
ESTRIN, Y. and KUBIN, L. P.	Institut für Metallkunde & Metallphysik Rheinisch-Westfälische Technische Hochschule Aachen Kopernikusstrasse 14 D-5100 Aachen, FR Germany
GAUTHIER, R. D.	Basic Engineering Department Colorado School of Mines Golden, Colorado 80401, USA
JAHSMAN, W. E.	Department 256 Intel Corporation Santa Clara, California 93106, USA
KLUGE, G.	Sektion Physik Friedrich-Schiller-Universität Jena Max-Wien-Platz 1, DDR-69 Jena German Democratic Republic
LIONS, J. L.	College de France 11, Place Marcelin Berthelot F-75231 Paris 05, France
TEODOSIU, C.	Department of Solid Mechanics Institute for Physics and Technology of Materials Str. Constantin Mille 15 70701 Bucharest, Romania

Part 1
SYNERGETICS, STATISTICAL MECHANICS, PLASTICITY

Continuum Models of Discrete Systems 4
eds. O. Brulin and R.K.T. Hsieh

FIELD FLUCTUATIONS IN A RANDOM MEDIUM

M.J. Beran

School of Engineering
Tel-Aviv University
Tel-Aviv, Israel

In this paper we begin by reviewing lowest order perturbation solutions for the field fluctuations in an infinite random medium. We then present some new results that specifically apply to a two-phase random medium and show their relation to bounds presently available. Finally we present a discussion of the next-order in the perturbation solution.

INTRODUCTION

In this paper we consider a random dielectric medium in which the electric and displacement fields (respectively $\underset{\sim}{E}$ and $\underset{\sim}{D}$) are given by the equations

$$\nabla \cdot \varepsilon \underset{\sim}{E} = 0 \tag{1}$$

$$\nabla \times \underset{\sim}{E} = 0 \tag{2}$$

$$\underset{\sim}{D} = \varepsilon \underset{\sim}{E} \tag{3}$$

Here ε is considered to be a random function of position and we shall choose the statistics of the ε field to be homogeneous and isotropic. The same equations apply to heat conductivity, electrical conductivity etc. and most of the results in this paper can easily be extended to the elastic problem.

We may define an effective dielectric constant, ε^*, by the equation

$$\langle \underset{\sim}{D} \rangle = \varepsilon^* \langle \underset{\sim}{E} \rangle \tag{4}$$

where the brackets denote an ensemble average[1]. (We note that the definition $\langle \underset{\sim}{D} \cdot \underset{\sim}{E} \rangle = \varepsilon^* \langle \underset{\sim}{E} \rangle \cdot \langle \underset{\sim}{E} \rangle$ is an equivalent definition). The functions $\langle \underset{\sim}{E} \rangle$ and $\langle \underset{\sim}{D} \rangle$ are constant vectors. References 2-8 are a few useful references for the dielectric and elastic problems.

The field fluctuations, $\underset{\sim}{E}'$ and $\underset{\sim}{D}'$ are defined by the equations

$$\underset{\sim}{E} = \langle\underset{\sim}{E}\rangle + \underset{\sim}{E}' \tag{5}$$

$$\underset{\sim}{D} = \langle\underset{\sim}{D}\rangle + \underset{\sim}{D}' \tag{6}$$

The most readily available statistical quantity associated with $\underset{\sim}{E}$ is the variance defined as

$$\sigma_E^2 = \frac{\langle\underset{\sim}{E}' \cdot \underset{\sim}{E}'\rangle}{\langle\underset{\sim}{E}\rangle \cdot \langle\underset{\sim}{E}\rangle} \tag{7}$$

The variance may be obtained from the two point correlation function

$$\mathcal{E}_{ij}(\underset{\sim}{r}) = \langle E_i'(\underset{\sim}{x}_1)\, E_j'(\underset{\sim}{x}_2)\rangle \tag{8}$$

$$(\underset{\sim}{r} = \underset{\sim}{x}_2 - \underset{\sim}{x}_1)$$

by the relation

$$\sigma_E^2 = \mathcal{E}_{ii}\ (0)/(\langle\underset{\sim}{E}\rangle \cdot \langle\underset{\sim}{E}\rangle) \tag{9}$$

Similar quantities may be defined for $\underset{\sim}{D}$.

Perturbation solutions may be obtained using an integral equation iteration procedure or more directly from the differential equations. We use the latter procedure here since it is easier to see how general results may be applied to the two phase case.

Following Beran & Molyneux[2] we define the function

$$L_i(\underset{\sim}{r}) = \langle E_i'(\underset{\sim}{x}_1)\ \varepsilon'(\underset{\sim}{x}_2)\rangle \tag{10}$$

where

$$\varepsilon(\underset{\sim}{x}) = \langle\varepsilon\rangle + \varepsilon'(\underset{\sim}{x}) \tag{11}$$

Equation (1) may be written in the form

$$<\varepsilon> \frac{\partial E_i'(\underset{\sim}{x})}{\partial x_i} + <E_3> \frac{\partial \varepsilon'(\underset{\sim}{x})}{\partial x_3} + \frac{\partial}{\partial x_i}[E_i(\underset{\sim}{x})\varepsilon'(\underset{\sim}{x})] = 0 \qquad (12)$$

where for convenience we have taken $<E_i> = <E_3>\,\delta_{i3}$.

Multiplying both sides of Equation (12) by $\varepsilon'(\underset{\sim}{x}_2)$ and ensemble averaging we find (using the $\underset{\sim}{x}_1$ in Equation (12))

$$<\varepsilon> \frac{\partial}{\partial x_{1i}} L_i(\underset{\sim}{x}_1,\underset{\sim}{x}_2) + <E_3> \frac{\partial}{\partial x_{13}} C_0(r) + \frac{\partial}{\partial x_{1i}} <E_i'(\underset{\sim}{x}_1)\varepsilon'(\underset{\sim}{x}_1)\varepsilon'(\underset{\sim}{x}_2)> = 0 \qquad (13)$$

where

$$C_0(r) = <\varepsilon'(\underset{\sim}{x}_1)\varepsilon'(\underset{\sim}{x}_2)> \qquad (14)$$

Assuming $\varepsilon'(x)$ is small the lowest order perturbation solution is obtained by neglecting the triple correlation function.

From Equation (2) we find

$$\delta_{ijk} \frac{\partial}{\partial x_{ij}} L_k(\underset{\sim}{r}) = 0 \qquad (15)$$

where δ_{ijk} is the Levi Civita tensor. In terms of $\underset{\sim}{r}$ the lowest order perturbation equations for $L_i(\underset{\sim}{r})$ are

$$\frac{\partial}{\partial r_i} L_i(\underset{\sim}{r}) = -\frac{<E_3>}{<\varepsilon>} \frac{\partial}{\partial r_3} C_0(r) \qquad (16)$$

$$\delta_{ijk} \frac{\partial}{\partial r_j} L_k(\underset{\sim}{r}) = 0 \qquad (17)$$

Similar manipulation for $\mathcal{E}_{ij}(\underset{\sim}{r})$ leads to

$$\frac{\partial}{\partial r_i} \mathcal{E}_{ij} (\underset{\sim}{r}) = - \frac{<E_3>}{<\varepsilon>} \frac{\partial}{\partial r_3} L_i (-\underset{\sim}{r}) \qquad (18)$$

$$\delta_{ijk} \frac{\partial}{\partial r_j} \mathcal{E}_{km} (\underset{\sim}{r}) = 0 \qquad (19)$$

The equations for $L_i(r)$ and $\mathcal{E}_{km}(r)$ may be solved by expressing these functions in the form of an axially symmetric vector and tensor respectively[9,2]. We find that both functions are dependent on $C_0(r)$. However both $L_3(0)$ and $\mathcal{E}_{ij}(0)$ are independent of the r dependence of $C_0(r)$ and depend only on the value of $C_0(0) = <\varepsilon'^2>$ i.e. on the strength of the fluctuations in ε'. To lowest order the results are

$$L_3(0) = - \frac{1}{3} \frac{E_3}{<\varepsilon>} <\varepsilon'^2> \qquad (20)$$

$$\mathcal{E}_{11}(0) = \mathcal{E}_{22}(0) = \frac{1}{3}\mathcal{E}_{33} (0) \qquad (21,a)$$

$$\mathcal{E}_{33} (0) = \frac{1}{5} \frac{<E_3>^2}{<\varepsilon>^2} <\varepsilon'^2> \qquad (21,b)$$

$$\mathcal{E}_{ii} (0) = \frac{1}{3} \frac{<E_3>^2}{<\varepsilon>^2} <\varepsilon'^2> \qquad (21,c)$$

$$\mathcal{E}_{ij} (0) = 0 \qquad i \neq j \qquad (21,d)$$

We note that the effective constant ε^* is given by

$$\varepsilon^* = <\varepsilon> + L_3(0) = <\varepsilon> - \frac{1}{3} \frac{<\varepsilon'^2>}{<\varepsilon>} \qquad (22)$$

In the next section we restrict our attention to a two-phase medium and will find expressions for the variance of fluctuations in each phase. These expressions will depend on the volume fractions v_1 and v_2 and the respective dielectric constants ε_1 and ε_2. The results will be compared to bounds previously obtained. In the final section we shall discuss the next order in the perturbation.

TWO PHASE MEDIUM

Calculation of Variances

In a two phase medium it is of interest to consider the variances in each phase separately. Thus in place of Equation (7) we define the two quantities

$$\tilde{\sigma}^2_{E_\alpha} = \frac{<(\underset{\sim}{E}_\alpha - <\underset{\sim}{E}_\alpha>) \cdot (\underset{\sim}{E}_\alpha - <\underset{\sim}{E}_\alpha>)>}{<\underset{\sim}{E}_\alpha> \cdot <\underset{\sim}{E}_\alpha>} \tag{23}$$

where $\alpha = 1,2$. (no sum on α)

The relations between $<\underset{\sim}{E}_\alpha>$ and $<\underset{\sim}{E}>$ are easily derived in terms of ε_1, ε_2 and ε^*. Choosing $\varepsilon_1 > \varepsilon_2$ they are

$$<\underset{\sim}{E}_1> = \frac{(\varepsilon^* - \varepsilon_2)}{v_1(\varepsilon_1 - \varepsilon_2)} <\underset{\sim}{E}> \tag{24,a}$$

$$<\underset{\sim}{E}_2> = \frac{(\varepsilon_1 - \varepsilon^*)}{v_2(\varepsilon_1 - \varepsilon_2)} <\underset{\sim}{E}> \tag{24,b}$$

To find the numerator in Equation (23) we return to Equations (16) - (19). In a general medium $\underset{\sim}{r}$ is defined as $\underset{\sim}{x}_2 - \underset{\sim}{x}_1$ where the location of both points is arbitrary. In order to find the quantity

$$\tilde{\mathcal{E}}_{ij\alpha}(\underset{\sim}{r}) = <(E_{i\alpha} - <E_{i\alpha}>)(E_j - <E_j>)> \tag{25}$$

we first consider the quantity

$$\mathcal{E}_{ij\alpha}(\underset{\sim}{r}) = <(E_{i\alpha} - <E_i>)(E_j - <E_j>)> \tag{26}$$

Here we require that the point $\underset{\sim}{x}_1$ lie in the phase α. The equations (16) - (19) remain the same but now $C_o(r)$ must be replaced by $C_{o\alpha}(r)$. The functions $C_{o1}(r)$, $C_{o2}(r)$ and $C_o(r)$ are in general not equal and

$$C_{o1}(0) = <(\varepsilon_1 - <\varepsilon>)^2> = v_2^{\,2}\ (\Delta\varepsilon)^2 \tag{27,a}$$

$$C_{o2}(0) = <(\varepsilon_2 - <\varepsilon>)^2> = v_1{}^2(\Delta\varepsilon)^2 \tag{27,b}$$

$$C_o(0) = <(\varepsilon - <\varepsilon>)^2> = v_1 v_2(\Delta\varepsilon)^2 \tag{27,c}$$

For a two phase medium we thus find for example

$$\sigma_E^2 = \frac{1}{3} v_1 v_2 (\frac{\Delta\varepsilon}{<\varepsilon>})^2 \tag{28,a}$$

$$\sigma_{E_1}^2 = \frac{1}{3} v_2{}^2 (\frac{\Delta\varepsilon}{<\varepsilon>})^2 \tag{28,b}$$

$$\sigma_{E_2}^2 = \frac{1}{3} v_1{}^2 (\frac{\Delta\varepsilon}{<\varepsilon>})^2 \tag{28,c}$$

To find $\tilde{\sigma}_\alpha^2$ we must replace $<\tilde{E}>$ in Equation (26) by $<\tilde{E}_\alpha>$(Equation (25)). This is easily done using Equations (24) and (21). The final result is

$$\tilde{\sigma}_{E_1}^2 = \frac{2}{9} v_2{}^2 (\frac{\Delta\varepsilon}{<\varepsilon>})^2 \tag{29,a}$$

$$\tilde{\sigma}_{E_2}^2 = \frac{2}{9} v_1{}^2 (\frac{\Delta\varepsilon}{<\varepsilon>})^2 \tag{29,b}$$

Similar expressions may be derived for $\tilde{\sigma}_{D_\alpha}^2$.

The constant $<\varepsilon>$ is given by

$$<\varepsilon> = v_1\varepsilon_1 + v_2\varepsilon_2 \tag{30}$$

and thus $\sigma_{E_1}^2$ and $\sigma_{E_2}^2$ are easily calculated from a knowledge of $v_1 v_2$, ε_1 and ε_2.

From Equations (29,a) and (29,b) we see that to order $(\frac{\Delta\varepsilon}{<\varepsilon>})^2$ the ratio of fluctuations in the phases depends only upon the volume fractions. If the volume fraction of phase '1' is high (i.e. $v_1 >> v_2$) then the fluctuations are low and vice versa. If we think of the ε_1 material being the 'matrix' phase and ε_2 the 'inclusion' phase then, as expected, fluctuations are weaker in the 'matrix' phase.

In addition since $\langle\varepsilon\rangle$ may be written as

$$\langle\varepsilon\rangle = v_1\varepsilon_1 + v_2\varepsilon_2 = \varepsilon_2 + v_1\Delta\varepsilon \approx \varepsilon_2 \approx \varepsilon_1 \qquad (31)$$

then to order $(\Delta\varepsilon)^2$ the results are symmetrical. That is, although $\varepsilon_1 > \varepsilon_2$ the ratio $(\Delta\varepsilon/\langle\varepsilon\rangle)^2$ does not change if v_1 and v_2 are interchanged.

Bounds

In a recent paper [10] we showed that

$$\frac{v_1\varepsilon_1}{\langle\varepsilon\rangle}\left[\frac{\langle \underset{\sim}{E}_1\rangle\cdot\langle \underset{\sim}{E}_1\rangle}{\langle E\rangle\cdot\langle E\rangle}\,\tilde{\sigma}^2_{E_1}\right] + \frac{v_2\varepsilon_2}{\langle\varepsilon\rangle}\left[\frac{\langle \underset{\sim}{E}_2\rangle\cdot\langle \underset{\sim}{E}_2\rangle}{\langle E\rangle\cdot\langle E\rangle}\,\tilde{\sigma}^2_{E_2}\right] =$$

$$\frac{\varepsilon_1\varepsilon_2}{v_1v_2(\Delta\varepsilon)^2}\left(1 - \frac{\varepsilon^*}{\langle\varepsilon\rangle}\right)\left(\varepsilon^*\left\langle\frac{1}{\varepsilon}\right\rangle - 1\right) \qquad (32)$$

From this expression it is possible to find bounds for $\tilde{\sigma}^2_{E_1}$ and $\tilde{\sigma}^2_{E_2}$. There is a similar expression for the $\underset{\sim}{D}$ field variances.
See also Mendelson[11].

In the lowest order perturbation we find from Equation (32) the expression

$$v_1\,\tilde{\sigma}^2_{E_2} + v_2\,\tilde{\sigma}^2_{E_2} = \frac{2}{9}\,v_1v_2\,\frac{(\Delta\varepsilon)^2}{\langle\varepsilon\rangle^2} \qquad (33)$$

and the bounds

$$\tilde{\sigma}^2_{E_2} \leqslant \frac{2}{9}\,v_2\,\frac{(\Delta\varepsilon)^2}{\langle\varepsilon\rangle^2} \qquad (34,a)$$

$$\tilde{\sigma}_{E_2} \leqslant \frac{2}{9}\,v_1\,\frac{(\Delta\varepsilon)^2}{\langle\varepsilon\rangle^2} \qquad (34,b)$$

Comparing Equations (34,a) and (34,b) to Equations (29,a) and (29,b) we see that the bounds are too high by a factor of v_2 and v_1 respectively. Thus the bounds derived from Equation (32) are clearly not the best bounds. Our main interest in using Equation (32) is, however, in the case when $\varepsilon_1/\varepsilon_2 >> 1$ and it remains an open question as to how good the bounds are in this limit.

HIGHER ORDER PERTURBATION CALCULATIONS

Equation (12) may be used to calculate higher order perturbations. Molyneux[12,1] showed that to next order.

$$\sigma_E^2 = \frac{1}{3}\frac{\langle\varepsilon'^2\rangle}{\langle\varepsilon\rangle^2} - \frac{2}{9}\frac{\langle\varepsilon'^3\rangle}{\langle\varepsilon\rangle^3} + \frac{1}{\langle\varepsilon\rangle^3}\int_0^\infty dr \int_0^\infty\int_{-1}^{1} u\,\frac{(u^2-1)}{\rho r}\,\frac{\partial C_{10}(\rho,r,u)}{\partial u}\,du\,d\rho \tag{35}$$

where $C_{10}(\ ,r,u)$ is the three point correlation function for three points on a triangle with sides of length ρ and r and included angle $\cos^{-1} u$.

The effective constant is given by a similar integral and we find the relation

$$\sigma_E^2 = \left[2\left(1 - \frac{\varepsilon^*}{\langle\varepsilon\rangle}\right) - \frac{1}{3}\frac{\langle\varepsilon'^2\rangle}{\langle\varepsilon\rangle^2}\right] \tag{36}$$

Since ε^* may be easily measured we can easily find σ_E^2 to the next order in the perturbation. To my knowledge further orders in the perturbation have not been calculated and thus we do not know if there is a relationship between σ_E^2, ε^* and the moments of ε' to higher orders.

In the two phase case we cannot, however, even find a relationship between $\tilde{\sigma}_{E_\alpha}^2$ and ε^*. The reason for this is that $\tilde{\sigma}_{E_\alpha}^2$ is found from an expression similar to Equation (35) with $C_{10}(\rho,r,u)$ replaced by $C_{1\alpha}(\rho,r,u)$, the later function being determined by the requirement that one of the points must lie in phase α. On the other hand ε^* is determined by $C_{10}(\rho,r,u)$. The relationship between C_{10} and $C_{1\alpha}$ depends on the statistics of the medium and hence in general no simple relationship exists between $\tilde{\sigma}_{E_\alpha}^2$ and ε^*.

If we wish to determine $\tilde{\sigma}_{E_\alpha}^2$ beyond the lowest order perturbation we are thus forced to consider the geometry of the medium. In place of this the best we can do is determine bounds for $\tilde{\sigma}_{E_\alpha}^2$. Similar statements may be made for the $\underset{\sim}{D}$ field.

REFERENCES:

[1] Beran, M., Statistical Continuum Theories, Wiley, New York, 1968.

[2] Beran, M. & Molyneux, J., Nuovo Cimento 30, 1406, 1963.

[3] Brown, W.F., J. Chem. Phys., 23, 1514, 1955.

[4] Hori, M., J. Math. Phys., 18, 487, 1977 and six previous articles.

[5] Kroner, E., J. Mech. Phys. Solids, 15, 319, 1967.

[6] Kerner, E., Proc. Phys. Soc., 69B, 802, 1956.

[7] McCoy, J., Nuovo Cimento, 57B, 139, 1968.

[8] Molyneux, J., J. Math. Phys., 10, 912, 1969.

[9] Batchelor, G., Theory of Homogeneous Turbulence, Cambridge Univ. Press, 1953.

[10] Beran, M., J. Math. Phys., 21, 2583, 1980.

[11] Mendelson, K., J. Appl. Phys., 48, 25, 1977.

[12] Molyneux, J., Ph.D. Thesis, Dept. of Mechanical Eng., Univ. of Penn., Phila. PA., 1964.

Continuum Models of Discrete Systems 4
eds. O. Brulin and R.K.T. Hsieh

THERMOMECHANICAL INSTABILITIES OF LOW TEMPERATURE PLASTIC FLOW

Y.ESTRIN* and L.P.KUBIN**

* Institut für Allgemeine Metallkunde und Metallphysik, RWTH Aachen, W.-Germany

**Laboratoire de Métallurgie Physique, Faculté des Sciences, Poitiers, France

A phenomenological macroscopic description of low temperature plastic deformation accompanied by heat release has been proposed. The assumption of homogeneous heat production associated with plastic deformation has been shown to be fulfilled under realistic experimental conditions, regardless of specific deformation patterns. Computer simulation of the temporal behaviour of the flow stress has been carried out. Instabilities of plastic flow have been shown to occur in a certain range of thermal, mechanical, and geometrical parameters, in good agreement with experiment. Analytical criteria for the onset of instabilities have been formulated. The effect of extreme slip localization (e.g. necking) on the occurrence of instabilities has been investigated.

INTRODUCTION

One of the most striking features of low temperature plasticity of metals and alloys is the discontinuous character of plastic flow at cryogenic (liquid helium) temperatures. Fig. 1 illustrates this phenomenon for the case of Nb. Though the occurrence of instabilities appears erratic rather than systematic, being sensitive even to slight variation of mechanical, geometrical, or thermal parameters, the phenomenon seems to be inherent in the plasticity at cryogenic temperatures. Discontinuous plastic flow was observed for a wide variety of metals and alloys characterized by different dislocation mechanisms of plasticity and deformation patterns, as manifested in the dislocation structure and the morphology of slip traces. It can be thus concluded that, despite the discrete nature of carriers as well as of elementary events of plasticity, the description of discontinuous flow should be based on a macroscopic model which need not consider a discrete deformation pattern. One such model, going back to Basinski /1/, was outlined in a paper /2/ presented at the preceding CMDS Conference. Discontinuous plastic flow was attributed there to the effect of heat release associated with thermally activated plastic deformation on the plastic strain rate (thermal instability of plastic flow). In close analogy with the combustion theory /3,4/, a criterion for such a thermal instability to occur was obtained by studying the stability of steady state solutions of the heat conduction equation containing a source term whose temperature dependence is given by an Arrhenius-type formula. This approach was pursued further by taking into account the coupling between the equation of heat conduction and the

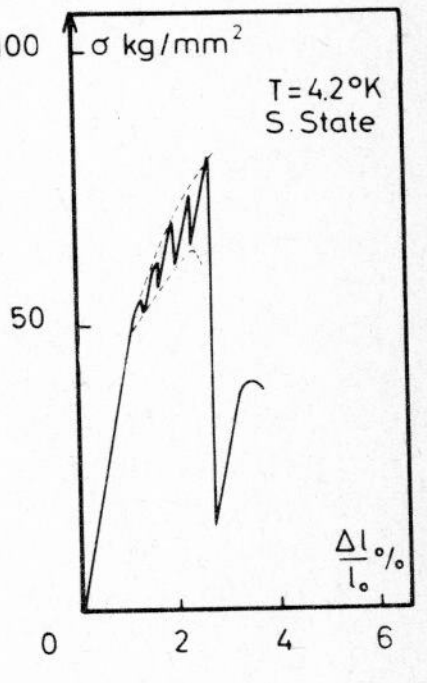

FIG. 1
Serrated stress-strain curve for tensile deformation of a Nb single crystal at 4.2 K (strain rate 10^{-4} s^{-1})
(Standard notation: σ normal stress, $\Delta\ell/\ell$ relative elongation)

mechanical equation for a deforming specimen. Analytical criteria for the occurrence of instability of plastic flow (referred to as thermomechanical instability (TMI)) were derived /5,6/. The temporal behaviour of the applied stress as well as the detailed shape of the stress drop associated with TMI were examined by means of computer simulation /7/. The present report gives an account of this investigation together with a discussion of experimental possibilities of verifying the theoretical description propounded.

EQUATION OF HEAT CONDUCTION

The analysis /5-7/ has shown two important physical features of the thermal effects accompanying the plastic deformation at cryogenic temperatures which make it possible to considerably simplify phenomenological description of these effects. First, the rate of heat removal turns out to be controlled by heat exchange with the thermal bath (given by the coefficient of surface heat transfer, h /8/) rather than by heat conduction within the specimen (given by the thermal conductivity, K). As a result, the temperature can be considered homogeneous throughout every cross-section of the specimen normal to the (tensile or compressive) axis (apart from negligible near-surface regions). The formal condition for this is represented as

$$hR/2K \ll 1 \tag{1}$$

where R is the specimen radius.

Second, for typical values of parameters, the width of the temperature profile around an active slip band, $\lambda_0 \cong (KR/2h)^{1/2}$, is much larger than the average distance between active slip bands, ℓ/n (where ℓ is the specimen length and n the number of active slip bands). In fact, this condition, represented as

$$n \gg \frac{\ell}{R}(2hR/K)^{1/2}, \text{ i.e. } n \gg 2(\ell/R), \tag{2}$$

obviously can only fail in case of extreme slip localization when one or a few slip bands are operative. Apart from this exceptional case which will be treated separately below, heat generation can be considered homogeneous along the strain axis and thus throughout the entire specimen. Hence, in spite of the discrete nature of plastic deformation, the problem under consideration proves to be homogeneous in the thermal sense.

The equation of heat conduction assumes in this homogeneous case (referred to as the H-case) a simple form /5/:

$$c\dot{T} = -(2h/R)(T-T_0) + r\,\tau\dot{\gamma}_p \tag{3}$$

where c is the heat capacity per unit volume, T the specimen temperature, T_0 that of the thermal bath, τ the resolved shear stress, and $\dot{\gamma}_p$ the plastic strain rate; the fraction r of the mechanical work dissipated as heat is, in the physical situation under consideration, close to unity so that r can be omitted.

For the case of extremely localized plastic deformation, e.g. slip confined to a single active site, like in the case of necking (illustrated by Fig. 2 and referred to as the N-case), the equation of heat conduction is no longer homogeneous along the strain axis x, and Eq.(3)

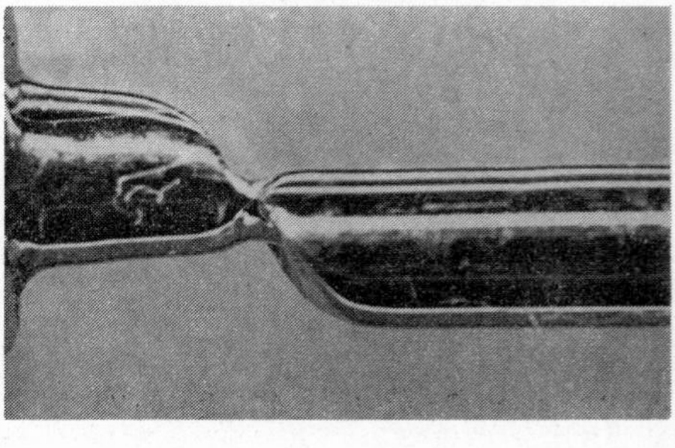

FIG. 2

Catastrophic shear of a Nb single crystal corresponding to the N-case (T_0=4.2 K, strain rate 10^{-4} s^{-1}).

must be replaced by

$$c\dot{T} = \partial^2 T/\partial x^2 - (2h/R)(T-T_o) + \tau\dot{\gamma}_p d\delta(x) \qquad (4)$$

where d is the width of the active plastic zone and $\delta(x)$ is the delta-function indicating the location of the active slip zone at x=0. The use of the delta-function representation of the source term implies that $d \ll \lambda_o$.

MECHANICAL EQUATION

In tension (or compression) at a constant strain rate $\dot{\gamma}_a$ the mechanical behaviour of the specimen is described by the equation for the plastic strain rate γ_p in the active slip bands:

$$\dot{\gamma}_p = (\ell/nd)(\dot{\gamma}_a - \dot{\tau}/E) \qquad (5)$$

with E denoting the combined elastic modulus of the specimen and the testing machine. (In the N-case, n should be put equal to unity.) On the other hand, the plastic strain rate can be represented by the Arrhenius law for thermally activated plastic deformation:

$$\dot{\gamma}_p = \dot{\gamma}_o \exp(-\Delta G/kT) \qquad (6)$$

where the activation enthalpy $\Delta G = \Delta G(\tau)$ is determined by the specific mechanism underlying the process of plastic deformation, kT has the usual meaning, and the pre-exponential factor γ_o is considered constant. By substituting Eq. (6) into Eq. (5) we obtain the desired mechanical equation.

It is instructive to represent this equation in a dimensionless form. Following the procedure of Refs. /5/-/7/ we introduce a reference state $\tau = \tau_o$, $T=T_o$ describing steady state ($\dot{\tau} = \dot{T} = 0$) in the absence of heat release. Using a linear expansion of the activation enthalpy near the reference stress τ_o,

$$\Delta G(\tau) = \Delta G(\tau_o) - v(\tau - \tau_o)$$

(where $v = -(\partial \Delta G/\partial \tau)_T$ is the activation volume which is considered constant), and introducing the dimensionless stress $\sigma = \tau/\tau_o$, temperature $\theta = (T-T_o)/T_o$, and time $z = (2h/Rc_o)t$ (where $c_o = c(T_o)$), we obtain an equation describing in the dimensionless form the temporal variation of the flow stress

$$d\sigma/dz = \dot{\Sigma}_o\{1 - \exp[\frac{\beta\theta + \alpha(\sigma-1)}{1+\theta}]\}. \qquad (7)$$

Here a normalized elastic slope $\dot{\Sigma}_o = R\dot{\gamma}_a Ec_o/2h\tau_o$ as well as the constants $\alpha = v\tau_o/kT_o$ and $\beta = \ln(\dot{\gamma}_o/\dot{\gamma}_a)$ (the latter being a constant typically equal to 20-30) have been introduced.

STEADY STATE SOLUTIONS

The above consideration shows the existence of a coupling between the thermal and the mechanical equations. Any change in the thermal state of the specimen will be thus reflected in its mechanical behaviour. In particular, thermal fluctuations bringing the system out of steady state may, under certain conditions to be specified below, result in a "catastrophic" growth of the deviation from steady state exhibited as a serration on the stress vs. time (or on the stress-strain) curve. Our aim in this Section is to find first steady state solutions of the above sets of coupled equations. Stability of these solutions will be investigated in the next Section.

We proceed from representing also the thermal equation, Eq. (3) in the H-case or Eq. (4) in the N-case, in a dimensionless form:

$$\psi d\theta/dz = -\theta + \mu\sigma(\dot{\Sigma}_o - d\sigma/dz), \qquad (3')$$

$$\psi d\theta/dz = \partial^2\theta/\partial\xi^2 - \theta + 2\mu^*\sigma(\dot{\Sigma}_o - d\sigma/dz)\delta(\xi) \qquad (4')$$

where the dimensionless parameter $\mu = \tau_o^2/Ec_oT_o$ represents the ratio of the elastic and caloric energies and $\psi = c/c_o$ is a function of temperature defined by the temperature dependence of the heat capacity. The parameter μ^* in Eq. (4') corresponding to the N-case represents a re-normalized analogue of μ: $\mu^*=(\ell/2\lambda_0)\mu$; the dimensionless coordinate $\xi=x/\lambda_0$ has been introduced.

It should be noted that, in contrast to the mechanical equation whose dimensionless form, Eq. (7), does not bear any features of the deformation mode considered (the slip localization parameter ℓ/nd determining only the level of the reference stress τ_0), the form of the equation of heat conduction depends on whether the homogeneous or the localized deformation is considered (cf. Eqs. (3') and (4')).

The steady state solution ($d\sigma/dz=d\theta/dz=0$) for the H-case given by the set of Eqs. (3') and (7) is found readily:

$$\sigma_s=\alpha/(\alpha+\nu\dot{\Sigma}_0),\quad \theta_s=(\nu\dot{\Sigma}_0/\beta)\sigma_s \tag{8}$$

with $\nu=\beta\mu$ and the subscript "s" indicating steady state.

In the N-case described by the set of Eqs. (4') and (7), the steady state stress and the steady state temperature within the active plastic zone, $\theta_s(0)$, can be found using the following procedure (cf. Ref. /9/).

With the boundary condition that θ be finite at $\xi\to\pm\infty$, we obtain the steady state solution of Eq. (4') beyond the active zone ($\xi<0$, $\xi>0$):

$$\theta_s(\xi) = \theta_s(0)\exp(-|\xi|). \tag{9}$$

The "integration constant" $\theta_s(0)$ can be found from the equation

$$\theta_s'|_{\xi=+0} - \theta_s'|_{\xi=-0} + 2\mu^*\dot{\Sigma}_0\sigma_s = 0 \tag{10}$$

whose physical meaning is obvious: the heat fluxes from the active zone into the specimen bulk have been equated here to the rate of heat production within this zone. (Eq. (10) can be obtained formally by integrating Eq. (4') with respect to ξ in the limits $(-\Delta,+\Delta)$ and taking the limit $\Delta\to 0$.) The relationship

$$\theta_s(0) = \mu^*\dot{\Sigma}_0\sigma_s \tag{11}$$

follows immediately upon substitution of Eq. (9) into Eq. (10). Combining Eq. (11) with the steady state solution of the mechanical equation, Eq. (7), gives

$$\sigma_s = \alpha/(\alpha+\beta\mu^*\dot{\Sigma}_0). \tag{12}$$

It is seen that the form of the steady state solution of the problem in the N-case, Eqs. (11), (12), is identical with that for the H-case, Eq. (8), provided that a new meaning is given to the parameter ν denoting the product $\beta\mu$ in the H-case: now ν denotes $\beta\mu^* = \beta(\ell/2\lambda_0)\mu$. As can be easily estimated by using the typical values of parameters compiled in Ref. /7/, the modifying factor $(\ell/2\lambda_0)$ does not substantially differ from unity. Neither do the steady state solutions for the two cases differ noticeably. The tendency to instability in the N-case is, however, much greater, as will be shown in the next Section.

INSTABILITY CRITERIA

As seen from the results of the preceding Section, the steady state values of stress and temperature do not depend on the heat capacity, c, cancelling in the expression for ν. It is natural since c enters the basic set of equations only in combination with $\dot{T}$ disappearing in steady state. On the contrary, both the magnitude of c_0 and the temperature dependence of c reflected in ψ essentially determine instability criteria. For practical applications, a cubic temperature dependence of c, $c=c_0(T/T_0)^3$, characteristic of the low temperature region (T much below the Debye temperature) as well as the temperature independent heat capacity, $c=c_0$=const (T above the Debye temperature), are of most importance. These two possibilities will be considered both for the H-case (denoted Hc and Hc_0) and for the N-case (denoted Nc and Nc_0). The cases Hc and Hc_0 are described by the set of Eqs. (7), (3') with ψ equal to unity and to $(1+\theta)^3$, respectively; the cases Nc and Nc_0 by the set of

Eqs. (7), (4') with $\psi=1$ and $\psi=(1+\theta)^3$, respectively.

An instability criterion is formulated as the condition that small deviations from steady state represented conventionally as

$$\theta-\theta_s=\delta\theta\exp\lambda z, \quad \sigma_s-\sigma=\delta\sigma\exp\lambda z \tag{13}$$

increase with time, i.e. Re $\lambda > 0$. The threshold condition for the occurrence of instabilities, Re $\lambda = 0$, can be conveniently represented on the phase diagram in the coordinates ν/α vs. $\Sigma_0\alpha$ (Fig. 3) where the stability and instability ranges are separated by the lines corresponding to this condition /5,6/. Plastic flow is stable beyond the lines corresponding to Hc and Nc and unstable within the areas bounded by these lines. For Hc_0 and Nc_0, the instability range lies above the respective lines.

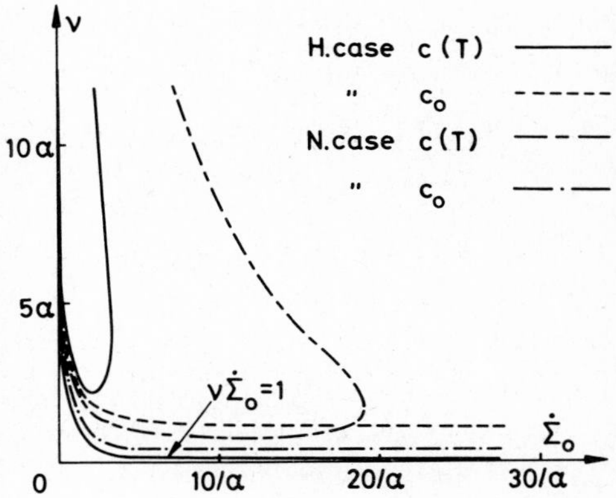

FIG. 3

Phase diagram representing the instability criteria.

$\nu=\beta\mu$ for the Hc and Hc_0 cases;
$\nu=\beta\mu^*$ for the Nc and Nc_0 cases;
$\beta=25$.

The analytical form of the TMI criteria for the homogeneous case,

$$\psi\alpha\dot{\Sigma}_0 = \frac{\beta-1}{\beta}\nu\dot{\Sigma}_0 - 1, \tag{14}$$

has been given a detailed treatment in a previous paper /5/. Here, to complete the "catalogue" of practically useful criteria, we add those corresponding to the case of extremely localized deformation (Nc and Nc_0).

Substituting the expressions $\theta(\xi,z)=\theta_s(\xi) + \delta\theta(\xi)\exp(\lambda z)$, $\sigma(z)=\sigma_s - \delta\sigma\exp(\lambda z)$ (where the steady state values $\theta_s(\xi)$ and σ_s are given by Eqs.(9),(11),(12)) into Eqs.(7),(4') and linearizing them with respect to the small deviations from steady state, we get:

$$\delta\theta''(\xi)-(\psi_s\lambda+1)\delta\theta(\xi)+2\mu^*(\sigma_s\lambda-\dot{\Sigma}_0)\delta\sigma\cdot\delta(\xi) = 0 \tag{15}$$

$$\delta\sigma = \beta\dot{\Sigma}_0\{\alpha\dot{\Sigma}_0+\lambda(1+\theta_s(0))\}^{-1}\cdot\delta\theta(0). \tag{16}$$

Here

$$\psi_s = \begin{cases} 1 & \text{for the } Nc_0\text{-case,} \\ (1+\theta_s(\xi))^3 & \text{for the Nc-case.} \end{cases}$$

From Eq. (15) the following equation is derived in the same way as Eq. (10) was derived from Eq. (4'):

$$\delta\theta'|_{\xi=+0} - \delta\theta'|_{\xi=-0} +2\mu^*(\sigma_s\lambda-\dot{\Sigma}_0)\delta\sigma = 0 . \tag{17}$$

Since only the solution of Eq. (15) in an infinitesimal vicinity of zero is needed to be substituted into Eq. (17), the expression for ψ_s in the Nc-case can be replaced by $(1+\theta_s(0))^3$. Putting into Eq. (17) the resulting solution of Eq. (15), $\delta\theta(\xi)=\delta\theta(0)\exp\{-(1+\lambda\psi_s)|\xi|\}$, as well as the expression (16) for $\delta\sigma$ we eventually obtain

$$(1+\lambda\psi_s)^{1/2} = \nu\dot{\Sigma}_o \frac{\sigma_s\lambda - \dot{\Sigma}_o}{\dot{\Sigma}_o\alpha+\lambda(1+\theta_s(0))} . \tag{18}$$

(Note that here $\nu=\beta\mu^*$.)

Eq. (18) can be represented as a cubic equation with respect to λ. Neglecting $\dot{\Sigma}_o\nu/\alpha$ as compared to unity (which is justified by a numerical estimate, cf. Ref. /7/) we have

$$\lambda^3 + \{[1-(\nu\dot{\Sigma}_o\sigma_s)^2\cdot(1+\theta_s(0))^{-2}]\psi_s^{-1} + 2\dot{\Sigma}_o\alpha(1+\theta_s(0))^{-1}\}\cdot\lambda^2 +$$
$$+ \dot{\Sigma}_o\alpha\{2(1+\theta_s(0))^{-1}\psi_s^{-1} + \dot{\Sigma}_o\alpha(1+\theta_s(0))^{-2}\}\cdot\lambda + (\dot{\Sigma}_o\alpha)^2(1+\theta_s(0))^{-2}\psi_s^{-1} = 0. \tag{19}$$

Looking for a threshold condition, i.e. for that of vanishing of the real part of λ, we represent λ as a purely imaginary quantity. Putting $\lambda=i\omega$ into Eq. (19) we obtain, after separating the real and the imaginary parts:

$$\omega^2\{1+2\dot{\Sigma}_o\alpha\psi_s(1+\theta_s(0))^{-1}-(\nu\dot{\Sigma}_o\sigma_s)^2(1+\theta_s(0))^{-2}\} = (\dot{\Sigma}_o\alpha)^2(1+\theta_s(0))^{-2} , \tag{20}$$

$$\omega^2 = \dot{\Sigma}_o\alpha\{2\psi_s^{-1}(1+\theta_s(0))^{-1} + \dot{\Sigma}_o\alpha(1+\theta_s(0))^{-2}\}. \tag{21}$$

Eqs. (20), (21) are consistent if

$$(\dot{\Sigma}_o\alpha)^2 + \{2(1+\theta_s(0))-(\nu\dot{\Sigma}_o\sigma_s)^2\cdot 2^{-1}(1+\theta_s(0))^{-1}\}\psi_s^{-1}\cdot(\dot{\Sigma}_o\alpha) +$$
$$+ \{(1+\theta_s(0))^2-(\nu\dot{\Sigma}_o\sigma_s)^2\}\psi_s^{-2} = 0 . \tag{22}$$

Eq. (22) is a quadratic equation with respect to $\dot{\Sigma}_o\alpha$. From it, $\dot{\Sigma}_o\alpha$ can be calculated as a function of $\dot{\Sigma}_o\nu$. The results are replotted on the phase diagram, Fig. 3. It can be recognized that the localization of plastic flow into a single plastic zone results in a significant spreading of the instability domains in the phase diagram.

The instability criteria presented in this Section cover all cases interesting for practical applications (both for low and medium temperature plasticity) and provide a means of predicting and interpreting TMI's of plastic flow of solids. An assessment of typical values of parameters entering the TMI criteria for the Hc-case ($\dot{\gamma}_a \sim 10^{-4}$ s^{-1}, $E \sim 10^{10}$ Pa, $\tau \sim 10^8$ Pa, $R \sim 10^{-2}$ m, $v \sim 10^{-27}$ m^3, $T \sim 4$K, $h \sim 10^3$ Jm^{-2}s^{-1}deg^{-1}, $c_o \sim 10^3$ Jm^{-3}deg^{-1}, all of these parameters being allowed to vary by one order of magnitude) shows that the representative point on the phase diagram typically lies within or close to the TMI range. This fact, together with strong experimental support for the predictions of the Hc-model of discontinuous plastic flow /7/ demonstrates that this model can realistically account for the phenomenon.

Another application of the theory proposed can be found in the room temperature (RT) deformation of polymers. Due to low heat conductivity (typically, $K \sim 10^{-2}$ Jm$^{-1}\cdot$s^{-1}deg^{-1} /10/), the effect of heat release may be appreciable here which is additionally enhanced by necking typical of polymer deformation. The instability criterion obtained for the Nc_o-case is thus relevant for the RT polymer deformation. (As at RT $h \cong 5$ Wm^{-2}deg^{-1} (the estimate following from the Stefan-Boltzmann law, cf. also Ref. /11/), inequality (1) underlying the present derivation holds for the RT polymer deformation.) It can be easily proved that with typical values of parameters ($\dot{\gamma}_a \sim 10^{-2}$ s^{-1}, $R \sim 10^{-2}$ m, $c_o \sim 10^5$ Jm^{-3}deg^{-1}) the RT deformation of polymers is adequately described by that part of the phase diagram of Fig. 3 (the Nc_o-case) which is characterized by the horizontal TMI boundary ($\dot{\Sigma}_o > 1/\alpha$ reducing to $kT_0/vE < 10$ which is obviously fulfilled). The occurrence of TMI requires here that $\nu=\beta\mu^*$ be larger than α. Taking v of the order of 10^{-28} m^3 and $\beta\cong25$ we see that this condition can be expressed roughly as

$$25(\tau_0/E)(\ell/2\lambda_0) > 1.$$

Remembering the definition of λ_0 which, for the values of parameters chosen, can be estimated to be $\cong 3\cdot10^{-3}$ m we conclude that the occurrence of TMI in the RT deformation of polymers appears fairly realistic.

COMPUTER CALCULATIONS

While the linearization procedure used made it possible to derive simple analytical criteria for the onset of TMI, an analysis of the detailed shape of a stress drop associated with TMI as well as of the evolution of the mechanical and thermal state of a specimen under straining requires, due to strongly non-linear character of the equations describing this evolution, computer calculations. The results of such calculations /7/, carried out for the Hc-case most adequately representing the deformation of metals and alloys at cryogenic temperatures, can be summarized as follows:

(i) Inside the solid line on Fig. 3 corresponding to the boundary of the TMI range for the Hc-case, discontinuous (serrated) stress vs. time curves were recorded;

(ii) Below the curve $\nu\Sigma_0=1$, stable plastic flow was established;

(iii) Above this curve but beyond the TMI range (i.e. outside the solid line), plastic flow is also stable, steady state being reached after a yield drop which occurs due to thermal effects.

Fig. 4 displays the simulated temporal variation of the flow stress during tensile deformation at various strain rates.

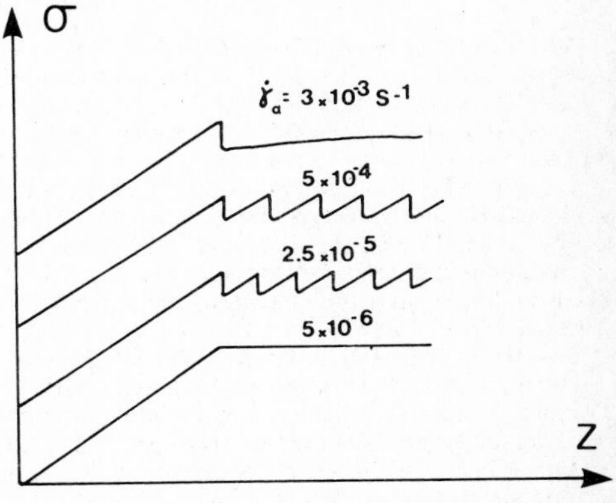

FIG. 4

Simulated stress vs. time curves recorded for various strain rates.

It is interesting to note that the strain rate interval in which TMI's take place has a lower and an upper bounds. This fact, confirmed by experiment /12/, can be understood by inspecting the shape of the TMI area on Fig. 3: by varying $\dot{\gamma}_a$ (and thus Σ_0), the boundary of this area is crossed twice by the representative point, in accordance with the result of computer simulation. Typical stress vs. time and temperature vs. time dependences corresponding to a single stress drop during TMI are presented on Fig. 5. The time scale (10^{-4} s) as well as the magnitude of the effect (stress drop of about 20 % and temperature increment of approximately 40 K at T_0=4.2 K) are in good agreement with experiment /1/, /13/. For further details about the effect of varying other parameters on the temporal behaviour of the flow stress and for comparison with experiment, we refer to Ref. /7/.

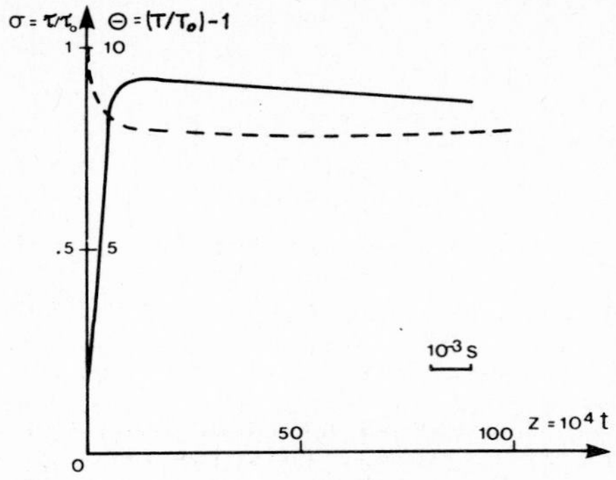

FIG. 5

Detailed shapes of the stress vs. time and temperature vs. time curves during a stress drop. (Time is reckoned from the moment the flow stress, $\sigma=1$, is reached)

—— θ (reduced temperature);
- - - σ (reduced stress).

REFERENCES

/1/ Basinski, Z.S., The Instability of Plastic Flow of Metals at Low Temperatures, Proc. Roy. Soc. A 240 (1957) 229-242.

/2/ Estrin, Y., and Mecking, H., Some Models for Instabilities of Plastic Flow, in: E.Kröner and K.-H.Anthony (Eds.) Continuum Models of Discrete Systems, University of Waterloo Press (1980) pp. 217-228.

/3/ Barenblatt, G.I., Methods of Combustion Theory in the Mechanics of Deformation, Flow, and Fracture of Polymers, in: H.Henning Kausch, John A.Hassel, and Robert J.Jaffee (Eds.) Deformation and Fracture of High Polymers (Plenum Press, New-York - London, 1973) pp. 91-108.

/4/ Zel'dovich, Ya.B., Barenblatt, G.I., Librovich, V.B., and Makhviladze, G.M., Mathematical Theory of Combustion and Explosion (Nauka, Moscow, 1980) Chap. 2.

/5/ Estrin, Y., and Kubin, L.P., Criterion for Thermomechanical Instability of Low Temperature Plastic Deformation, Scripta Met. 14 (1980) 1359-1364.

/6/ Estrin, Y., and Kubin, L.P., Characterization of Discontinuous Plastic Deformation at Low (Liquid Helium) Temperatures. Proc. of the Second Risø Intern. Symp. on Metallurgy and Materials Science, 14-18 September 1981 (in press).

/7/ Kubin, L.P., Spiesser, Ph., and Estrin, Y., Computer Simulation of the Low Temperature Instability of Plastic Flow, Acta Met. (to be published).

/8/ Carslaw, H.S., and Jaeger, J.C., Conduction of Heat in Solids (Oxford University Press, 1959), p. 19.

/9/ Mecking, H., and Estrin, Y., The Effect of Vacancy Generation on Plastic Deformation, Scripta Met. 14 (1980) 815-819.

/10/ Rehage, G., and Hirsch, G., On the Temperature Dependence of the Thermal Conductivity of Amorphous Polymers, in: R.W.Douglas and B.Ellis (Eds.) Amorphous Materials (Wiley-Interscience, 1971) pp. 151-157.

/11/ Rosenqvist, T., Principles of Extractive Metallurgy (McGraw-Hill Book Co., N.Y., 1974) p. 161.

/12/ Didenko, D.A., and Pustovalov, V.V., Velocity Dependence of the Discreteness of Slip and the Hardening Parameters in Aluminum at Low Temperatures, Strength of Materials 3 (1971) 1298-1304.

/13/ Schwarz, R.B., and Mitchell, J.W., Dynamic Dislocation Phenomena in Single Crystals of Cu-10.5 at. %-Al Alloys at 4.2 K, Phys. Rev. B 9 (1974) 3292-3299.

Continuum Models of Discrete Systems 4
eds. O. Brulin and R.K.T. Hsieh

SYNERGETICS - A STUDY OF COMPLEX SYSTEMS AT POINTS OF CRITICAL BEHAVIOR

H. Haken
Institut für theoretische Physik
Universität Stuttgart
W-Germany

A survey is given on some basic concepts and methods of synergetics. This interdisciplinary field deals with systems composed of many subsystems and studies how these systems change their macroscopic behavior at critical points of external parameters. It is shown that at such points discrete and continuous descriptions lead to the same results. Some further applications to metal physics are indicated.

§ 1 Introduction

The purpose of my contribution is twofold:
1) Because this conference is concerned with continuous models of discrete systems I want to show that there are discrete systems which behave in a way indistinguishable from continuous systems when they change their macroscopic properties.
2) I wish to outline some methods of how to cope with complex systems, i.e. systems of many subsystems with possibly complicated interactions.
To explain these points in more detail let us start with some phenomena wellknown since many years, namely phase transitions. Such phase transitions may be structural in which the structure of a lattice with one symmetry changes to another one. Other phase transitions involve melting or evaporation. It is wellknown that at a critical point, e.g. at a critical temperature, the behavior of the system is dominated by collective modes of long wave lengths, so that a microscopic description can be replaced by a continuous one. At such critical points usually long range order is established (depending on the direction of the transition). Critical fluctuations occur and symmetry can be broken. Such phenomena have been studied in the past in great detail but analysis was confined to systems in thermal equilibrium. Synergetics is dealing with systems which are driven away from thermal equilibrium, for instance by maintaining temperature gradients, by an in- and outflux of matter, by putting strain on solids or exposing them to random fluctuations or to oscillatory forces. Simultaneously one studies the dynamics of such systems, i.e. their temporal evolution.

It has turned out that such systems may change their macroscopic behavior at certain critical values of the driving forces dramatically and that at such points the systems behave very often quite similar to systems in thermal equilibrium, i.e. showing phase transition-like phenomena. In particular the discrete nature of the elements of the system and the individual kind of interaction between them do no more play a role. The behavior of such systems is rather governed by universal principles. The two main principles which we will explain below more mathematically are:

1) At critical points the dynamics of the system, be it even very complex, is governed by very few modes, the socalled order-parameters,
2) Close to these points all other degrees of freedom are governed by the slaving principle.

The laser was the first example in which this phase transition analogy was fully established[1], but since this system is not relevant to this present conference I shall not dwell on this example any longer. Furthermore since the number of examples has become rather large and the mathematical methods together with examples have been described in a book of mine [2] I shall confine my contribution to a brief outline. The reader, interested in more details, is referred to my book.

§ 2 Outline of the mathematical approach

Let us assume that a system is described by a set of variables which I shall describe by a vector q. When we deal with spatially distributed systems, $\underset{\sim}{q}$ may depend on the space variable $\underset{\sim}{x}$ as well as on time t. A good deal of processes in physics, chemistry and many other disciplines can be described by evolution equations of the form

$$\dot{\underset{\sim}{q}} = \underset{\sim}{N}(\underset{\sim}{q},\alpha,t) \tag{1}$$

$\underset{\sim}{N}$ is in general a nonlinear function of q and it depends on one or a set of control parameters α . It may or may not depend explicitly on time. In general a time dependence is introduced via fluctuations. When q depends on $\underset{\sim}{x}$, $\underset{\sim}{N}$ contains in general partial derivatives with respect to space. A general solution of these nonlinear stochastic partial equations is probably hopeless. However, there are classes of interesting phenomena in which these equations can be considerably simplified and reduced. As I have pointed out at various occasions, such a reduction becomes possible at critical points of control parameters α where the macroscopic behavior changes qualitatively. In the following I shall assume that the fluctuations do not shift the critical points appreciably. Therefore I shall assume that we can decompose $\underset{\sim}{N}$ according to

$$\underset{\sim}{N} = \underset{\sim}{N}_o + \underset{\sim}{F} \tag{2}$$

into a deterministic part, $\underset{\sim}{N}_o$, and a stochastic force, $\underset{\sim}{F}$, which I shall neglect in the first step of our following analysis but which can be taken into account in the following steps. Our procedure consists then of the following steps:

1) Let us assume that for the control parameter value $\alpha = \alpha_o$ an old state is realized which obeys the equation

$$\dot{\underset{\sim}{q}}^o = \underset{\sim}{N}(\underset{\sim}{q}^o, \alpha_o,t) \tag{3}$$

For this α_o, q^o is assumed stable. We now study what happens when we change the control parameter α.

2) Stability analysis
We assume that $\underset{\sim}{q}^{o}$ is still a solution of (1) but for a new value α. To check the stability of that solution we put

$$\underset{\sim}{q}(t) = \underset{\sim}{q}^{o} + \underset{\sim}{u}(t) \qquad (4)$$

and make a linear analysis by inserting (4) into (1) and linearizing $\underset{\sim}{N}$. Provided the system is autonomous, i.e. $\underset{\sim}{N}$ is independent of time t, the linearized equations read

$$\dot{\underset{\sim}{u}} = L(\underset{\sim}{q}^{o})\underset{\sim}{u} \quad . \qquad (5)$$

Let us discuss the following special cases:

a) q^{o} is time independent. Then L is also time independent. The solutions of (5) read

$$\underset{\sim}{u}(t) = \exp(\lambda t)\underset{\sim}{v} \qquad (6)$$

provided the characteristic exponents λ are nondegenerate. We shall call modes $\underset{\sim}{u}$ for which

$$\mathrm{Re}\lambda \geq 0 \qquad (7)$$

unstable and we shall write

$$\underset{\sim}{u}_{u}(t) = \exp(\lambda_{u} t)\underset{\sim}{v}_{u} \quad . \qquad (8)$$

Modes with

$$\mathrm{Re}\lambda < 0 \qquad (9)$$

are called stable and will be denoted by

$$\underset{\sim}{u}_{s}(t) = \exp(\lambda_{s} t)\underset{\sim}{v}_{s} \qquad (10)$$

b) the solution q^{o} is periodic
is L. By virtue of Floquet's theorem the solution of the non-degenerate case reads

$$\underset{\sim}{u}(t) = \exp(\lambda t)\underset{\sim}{v}(t) \qquad (11)$$

where $\underset{\sim}{v}(t)$ is again periodic.

3) The linear analysis of 2) gives us a set of solutions u or, when we split off the exponential factor, v. We now extend the wanted solution q(t) of the total nonlinear equations into the superposition of the form

$$q(t) = q^o + \sum_j \xi_j(t)\, v_j(t) \qquad (12)$$

When we deal with the bifurcation of limit cycles some care must be exercised with this hypothesis because of shifts of phase angles which in the course of time can become arbitrarily large. These shifts can be taken care of by introducing adequate phase angles into q^o and v as we have shown elsewhere. I shall not enter into the discussion because it is rather lengthy and I want just to give a sketch. In the next step of the analysis the hypothesis (12) is inserted into (1) and the equations are projected on the eigenvectors v. This leads us to equations for ξ which must be put into two classes namely the unstable modes ξ_u and the phases ϕ on the one hand and the stable mode amplitudes ξ_s on the other hand.

4) Slaving principle

As can be shown the resulting equations for ξ_s are of such a structure that for big enough damping i.e. big enough $|Re\lambda_s|$ and small enough amplitudes ξ_u the equations of ξ_s can be solved uniquely by explicit rules. This allows us to express ξ_s by means of

$$\xi_s(t) = F(\xi_u) \qquad (13)$$

Since I have discussed the slaving principle at various occasions I shall not dwell on it but just repeat the following feature. In many dissipative systems we have studied, the number of ξ_u is much smaller than that of ξ_s. Therefore we obtain an enormous reduction of the degrees of freedom. Inserting (13) into the equations for ξ_u we end up with a closed set of equations for ξ_u.

$$\dot{\xi}_u(t) = \tilde{N}(\xi_u(t')) \qquad (14)$$

Because the order parameters ξ_u refer to collective modes the detailed properties of the individual parts of the system and their mutual interaction do no more play a role explicitly so that the final equations for ξ_u have exactly the same appearence for discrete and continuous systems. The eigenvalues λ contain in both cases the dispersion law in the case that waves occur as a basis for the order parameters. Note, however, that the final equations are nonlinear which is necessary for the occurrecne of qualitative changes of a system.

§ 3 A simple example

A typical order parameter equation for the socalled unstable modes reads:

$$\dot{\xi}_u = \lambda\xi_u - b\xi_u^3 + F(t) \qquad (15)$$

This equation can be interpreted as the equation for the overdamped motion of a particle in the potential field V. When the control parameter α is changed the potential changes its form (compare Fig.1).

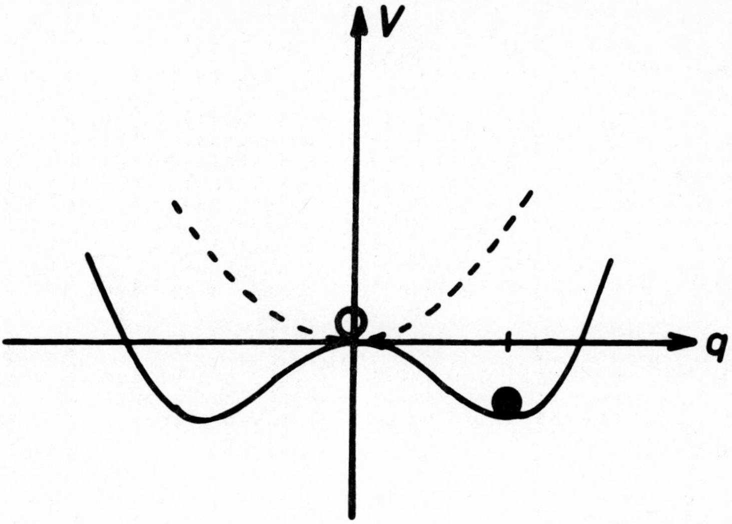

Fig. 1

This model reveals immediately that when α changes sign the particle relaxes more slowly to the equilibrium position (critical slowing down). The storing force becomes very small, the fluctuations are damped only weakly (critical fluctuations) and then the symmetry is broken. For further details of such considerations I must again refer the reader to my book.

§ 4 A recent application

Methods of synergetics were recently applied by Boffi, Bottani, Caglioti and Ossi to a study of the mechanisms of the thermomechanical instability of an elastically slowly strained solid. They start from the balance equations for entropy and linear momentum which they cast into the form

$$\frac{\partial\theta}{\partial t} - \chi \frac{\partial^2\theta}{\partial x^2} + \gamma T_o \frac{\partial\varepsilon}{\partial t} = -T_o \left.\frac{\partial\gamma}{\partial\varepsilon}\right|_o \varepsilon \frac{\partial\varepsilon}{\partial t} + \frac{\eta}{\partial C_\nu}\left(\frac{\partial\varepsilon}{\partial t}\right)^2 \qquad (16)$$

and

$$\frac{\partial^2 \varepsilon}{\partial t^2} - \nu_s^2 \frac{\partial^2 \varepsilon}{\partial x^2} - \frac{\eta}{\rho} \frac{\partial^3 \varepsilon}{\partial t \partial x^2} + \nu_s^2 \alpha \frac{\partial^2 \theta}{\partial x^2}$$

$$= - \frac{\nu^2 \alpha}{\gamma} \left. \frac{\partial \gamma}{\partial \varepsilon} \right|_o \left(\frac{\partial^2 \theta}{\partial x^2} \varepsilon + 2 \frac{\partial \theta}{\partial x} \frac{\partial \varepsilon}{\partial x} + \theta \frac{\partial^2 \varepsilon}{\partial x^2} \right) . \qquad (17)$$

The variables are temperature difference θ and strain ε . These authors could derive an order parameter equation by which the original problem could be solved.

For further details we refer the reader to the original paper. Such equations are a typical example of those to which our mathematical methods can be applied.

§ 5 Outlook

The methods of synergetics, of which only a very small section was presented here, have been applied to pattern formation and other processes in numerous systems, including fluid dynamics, chemical reactions leading to macroscopic patterns such as waves, processes of morphogenesis etc. I do hope that these methods will find also applications to problems discussed at this meeting.

REFERENCES:

Graham, R. and Haken, H., Z.Physik 213 (1968) 42o; 237 (197o) 31

Haken, H., Synergetics. An Introduction. 2nd enlarged edition, Springer Verlag, 1978

Boffi, S., Bottani, C.E., Caglioti, G. and Ossi, P.M., Strain Driven Thermoelastic Instability Toward Brittle Fracture, Z.Physik B - Condensed Matter 39, 135 (198o)

Continuum Models of Discrete Systems 4
eds. O. Brulin and R.K.T. Hsieh

PLASTICITY OF POLYCRYSTALS-UNUSUAL THERMOSTATIC BEHAVIOUR INDUCED BY DISLOCATIONS

J. Lambermont

T.N.O.
Leeghwaterstraat 5
Delft
HOLLAND

Material stability in conventional continuum theory requires that the second variation of the energy density is positive definite. This condition is, however, only true when the state variables can be varied arbitrarily. Plasticity is unusual in the sense that the dislocation state can be varied only in one direction, leading to the requirement that the first energy variation with respect to the dislocation state is positive. Whence the plastic affinity (partial energy derivative with respect to the dislocation state) is unequal to zero at equilibrium. Brief comments on the energy approach to plasticity are discussed.

Plasticity is characterized by conservative dislocation motion. During plasticity, dislocation segments become unstable and sweep over distances large compared with the lattice unit. An example is provided by a Frank-Read source which emits dislocation loops which pile up, say, against grain boundaries. In a work hardening solid, the back stress that an emitted loop exerts on the source raises the applied stress slightly while the solid deforms elastically, before the source segment again becomes unstable, giving rise to a small plastic strain increment. It is evident, therefore, that the stress-strain relation has a jogged (elastic-plastic steps) shape. Dislocation segments that are bowing out edge multipole bundles or dislocation walls do not usually have room to act as Frank-Read sources, so they create only one loop. With increasing plasticity, other segments, which are often observed to shorten with deformation, become unstable and sweep over a lattice plane. Clearly the resulting stress-strain relation is again jogged. Ponter, Bataille and Kestin [1] concentrated on describing this behaviour.

For a macroparticle of $1\,mm^3$, the number of jogs which appear in a plastic strain increment of 1 % is typically 10^{11}. The description of such steps requires microscopic knowledge of the exact length and the bow-out shape of the dislocation segments. We seek a macroscopic description of a work hardening solid in which the stress-strain relation had continuous derivatives. Therefore we adopt a continuous description of the dislocation state, as pile ups, which is often employed in dislocation theory. In [2] the thermodynamics of anelasticity due to the bow out of dislocation segments in their stable range has been discussed. Using this and the stability relations of thermostatics, which were modified to include the effect of fluctuation theory, the yield surface and condition have been derived in [3] where it was shown that stress assisted thermal fluctuations which aid the breakage of anchoring points of dislocation segments in long waiting times may give rise to a strong temperature dependence of the static yield and quasi-static flow stress. The continuous description implies that dislocation sources remain critically bowed out during plasticity and emit dislocations of infinitesimal Burgers vector strength which are added to the dislocation structure. Therefore it is necessary to introduce the auxiliary consistency condition to the Gibbs equation that the yield condition, which expresses the balance of Peach-Koehler force acting at critically bowed out sources, is satisfied during plasticity. Metallurgists such as Seeger went quite a bit further in the continuous description of plasticity when slip bands are formed. Experiments indicate that such bands are formed in a very small strain interval, and that they have about the same Burgers vector, but their geometry changes with plasticity. They introduce as a continuous dislocation variable essentially the number of slip bands. It is possible to define the dislocation state in such a way that this case is included in the analyses [4].

In the plasticity (conservative dislocation motion) of polycrystalline solids, the dislocations cannot leave the grains. It is therefore always possible to introduce the dislocation state, which is the internal or hidden thermodynamic variable and which must define the average dislocation structure, in such a way that, when it is made zero (by imaginary processes), the atoms are on the average shifted back to the relative positions they had in the undeformed state; the total strain tensor is then a state variable. To exhibit unusual thermodynamic behaviour, it suffices to consider only planar dislocation structures as (converted) pile ups, slip bands or edge multipole bundles (while for simplicity only one primary slip system in a representative element will be considered). Guided by observation it can be taken that those structures are continuously pinned down by dislocation reactions with secondary dislocations. They need not be considered explicitly in the theory because, although their density is often found to be of the same order as that of the primaries, their slip distance is so small compared with that of the primary dislocations that they hardly contribute to the plastic strain. They effect the theory therefore only in a two-fold indirect manner. Firstly they pin the structures down so that only local changes occur in them during plasticity, e.g., in converted pile-ups the dislocations already present do not move when new loops are added. This means that, although the fundamental energy relation is a functional, its derivative, the Gibbs equation, is an exact first-order differential equation. Secondly they influence the explicit form of the equations of state as they relax the internal stress field.

During plasticity, dislocation sources are bowed out critically in a definite direction and therefore can emit loops only of a definite kind. This implies that, when a solid is at the yield surface, the dislocation state can only be varied in one direction. From this it follows that the plastic affinity is unequal to zero at equilibrium, thus an energy-dislocation section of the energy surface terminates at the minimum in a corner point. This means that plasticity is always irreversible no matter how slow the deformation rate is. The theory [4] includes the case wherein the hardening is caused by the internal stress field (Seeger), by the change of the length of sources (Kuhlmann-Wilsdorf or by both effects occuring simultaneously (Hirsch). Furthermore, Drucker's relation is shown to follow whereby the unknown history dependence is replaced by the dislocation state. This bridges the gap between the metallurgical and phenomenological approaches.

REFERENCES

[1] Ponter, A., Bataille, J. and Kestin, J., "A Thermodynamic Model for the Time Independent Plastic Deformation of Solids", J.de Mécanique, 1980.
[2] Lambermont, J. "Dislocations as Internal Variables in the Thermodynamic Theory of Anelasticity" in Continuum Models of Discrete Systems (CMDS 3), Univ of Waterloo Press, 1980.
[3] Lambermont, J., "The Yield Condition of Crystalline Solids", in Physical Non-Linearities in Structural Analysis, Springer-Verlag, 1981.
[4] Lambermont, J., To be published.

Continuum Models of Discrete Systems 4
eds. O. Brulin and R.K.T. Hsieh

DISLOCATION DYNAMICS AND SOME FEATURES OF PLASTIC WAVE PROPAGATION

Xanthippi Markenscoff

Department of Mechanical & Environmental Engineering
University of California
Santa Barbara, California 93106

The decay of the amplitude of the leading wavefront of a plastic wave is considered in terms of the elastodynamics of the transient motion of straight dislocations and expanding dislocation loops. It is found in agreement with the results obtained from linear rate-dependent theories thus bridging continuum and dislocation approaches. There is a small correction accounting for different effects of dislocations moving in the direction of the pulse or opposite to it.

INTRODUCTION

Dynamic deformation in solids has been studied in the last 30 years by investigating the propagation of elastic-plastic stress pulses in wires, rods, tubes, and plates, in search of dynamic stress-strain relations to describe the response of structural materials. The propagation of plane pulses in plates of large lateral extent offers the advantage over bars of allowing true one-dimensional wave theory [1]. Here we are concerned with plate impact and one-dimensional incident waves. Recovery experiments of plate impact in single crystals offered the possibility of studying the microscopic mechanisms responsible for plastic flow by observation of the dislocations in the crystal before and after impact [1, 2, 3, 4]. This research has been instigated by such experiments [1, 3].

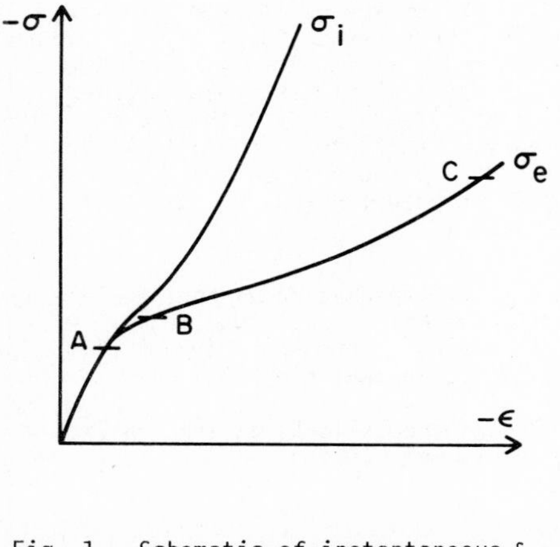

Fig. 1. Schematic of instantaneous & equilibrium (relaxed) stress-strain loading curves in uniaxial and finite strain.

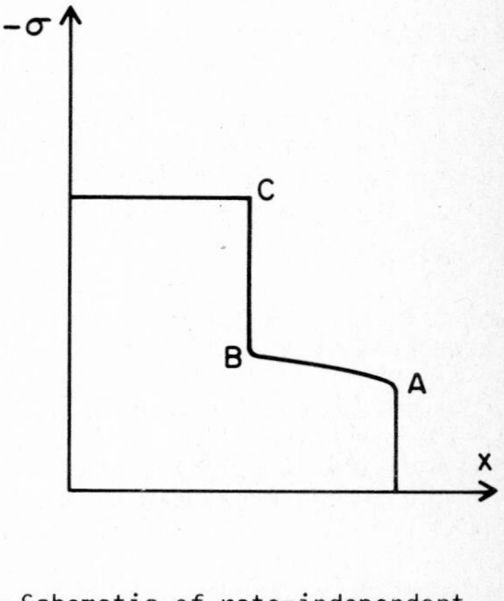

Fig. 2. Schematic of rate-independent (relaxed) wave profile vs. distance of propagation.

A rate-independent theory of elastoplastic wave propagation (with infinitesimal or finite strain) is adequate to describe the relaxed loading wave profile (Fig. 2). However, in order to describe the transition from the instantaneous stress at impact to the wave profile at later times, a rate dependent theory is needed (Fig. 1) [5].

The leading profile of a plastic wave travels with the elastic wave speed and is usually called elastic precursor. The amplitude of the elastic precursor decays progressively from the instantaneous impact stress to the equilibrium yield stress (Fig 3). Here we are concerned only with this particular feature of plastic wave propagation.

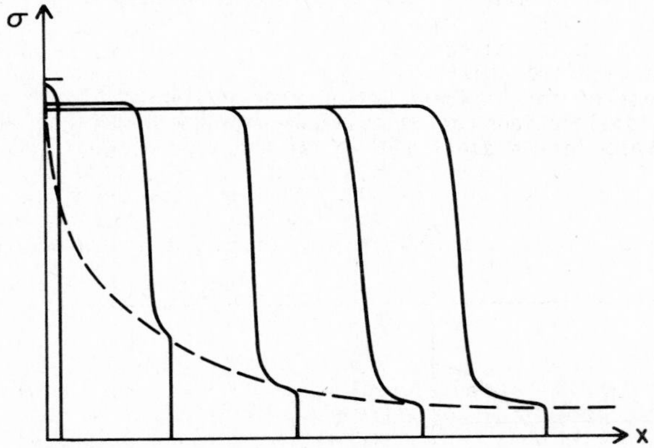

Fig. 3. Schematic of elastic precursor decay from linear rate-dependent theory.

Using a linear elastoviscoplastic constitutive relation, the rate of decay with distance of propagation x_3 of the jump $[v]$ in particle velocity at the wave-front is related to the rate of plastic strain $\dot{\gamma}^p$ [6] by

$$\frac{D[v]}{Dx_3} = \zeta \dot{\gamma}^p \tag{1}$$

where ζ is a dimensionless quantity depending on the slip directions. Moreover, the rate of plastic strain may be related to the dynamics of the dislocations at the wave-front according to Orowan's formula [7]

$$\dot{\gamma}^p = bNv_d \tag{2}$$

where N denotes the mobile dislocation density, b the Burgers vector, and v_d the average dislocation velocity at the wavefront.

Orowan's formula, however, does not account for the transient aspects of the dislocation motion. Here we describe a model developed in [8] and [9], which is based on the transient character of the motion of the dislocations that have been accelerated by the incident wave.

ELASTIC PRECURSOR DECAY FROM ELASTODYNAMICS OF DISLOCATIONS

We consider a plane wave propagating through an elastic solid containing a uniform distribution of dislocations. In the following we treat first the case of straight dislocations and subsequently of dislocation loops.

While both screw and edge dislocations are treated in [8], here we restrict attention to screw dislocations in order to make readily comparisons with the corresponding case in which the solid contains dislocation loops. Thus, we consider a plane shear wave propagating in an elastic solid containing uniformly distributed screw dislocations. The wave propagates along the $-x_3$ direction while the dislocation axis is parallel to the $-x_2$ direction. As the incident wave propagates it hits the dislocations that is meets on its way and sets them in motion along the x_1' direction (Fig. 4). The dislocations emanate wavelets traveling also with the shear wave speed c_2. Thus their wavefronts are tangent to the wavefront of the incident pulse. The cumulative effect of the superposition of many dislocations on the wavefront, which is obtained analytically, causes a decrease in the amplitude of the incident pulse, i.e. the elastic precursor decay.

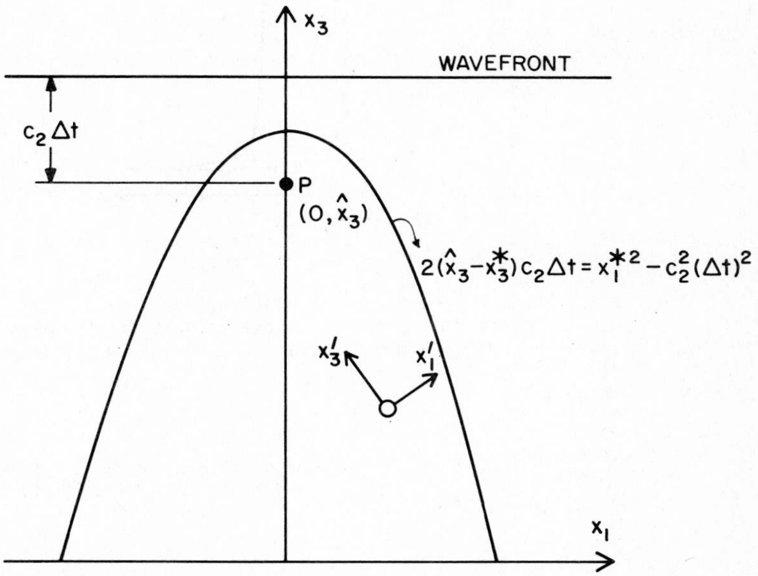

Fig. 4. Screw dislocations moving in the $-x_1'$ direction due to a wave propagating along x_3.

We assume that the dislocations are accelerated to a velocity v_d instantaneously. The velocity field of a screw dislocation, with axis parallel to the y direction, being at rest at the origin and starting moving at t=0 with velocity v_d along the -x direction is [10]

$$\dot{U}_y(x,z,t) = \frac{b_y}{2\pi} \frac{v_d z r^2 (t - x\frac{v_d}{c_2^2}) H(t - r/c_2)}{[(r^2 - v_d t x)^2 + v_d^2 z^2 (t^2 - r^2/c_2^2)]\sqrt{t^2 - r^2/c_2^2}} \tag{3}$$

where $r^2 = x^2+z^2$, H denotes the Heaviside step function, and b_y the Burgers vector for the screw dislocation.

Apparently from (3) the particle velocity has a square root singularity at the wavefront $t = r/c_2$. This singularity is the cause of finite precursor decay in the superposition of the effect of many dislocations, which we describe below.

At a time Δt after the incident wave has passed and at a point P at a distance $c_2\Delta t$ behind the wavefront, we compute the velocity field of the dislocations set in motion along the x_1' direction. The dislocations that contribute to $P(0, \hat{x}_3)$ at time $\hat{x}_3/c_2 + \Delta t$ lie inside the parabolic region of Fig. 4. The contribution of each dislocation is found from (3) with the appropriate coordinate transformation. This has to be integrated over the parabolic region to account for the superposition of the effects of the distributed dislocations. Subsequently the limit is taken as $\Delta t \to 0$. The limit is finite, although the area of the parabolic region shrinks to zero, since the integrand is singular. The end result is

$$\dot{U}_y\,(0,\, \hat{x}_3,\, \hat{x}_{3}/c_2) = \frac{b_y}{2} \int_0^{\hat{x}_3} \frac{NV_d \cos\psi}{1-\beta \sin\psi}\, dx_3 \tag{4}$$

and for the elastic precursor decay

$$\frac{d\dot{U}_y}{d\hat{x}_3} = \frac{b_y N V_d \cos\psi}{2(1-\beta\sin\psi)} \tag{5}$$

where $\beta = \frac{V_d}{c_2}$ and V_d and N are the dislocation velocity and line length density respectively at the wavefront.

If we compare (5) with the precursor decay relation resulting from (1) and (2), for $\psi = 0$ the two expressions coincide. For $\psi \neq 0$ there is a correction factor of $(1-\beta\sin\psi)$ in (4) which accounts for the fact that the dislocations moving in the direction of the incident wave contribute more than dislocations moving opposite to it.

Since most of the dislocations in real crystals are loops [1, 3], we next consider a model of the incident plane shear wave propagating in an elastic solid containing uniformly distributed circular loops ($\hat{N}$ per unit volume). The loops of initial radius ρ_0 lie on planes perpendicular to the direction $-x_3$ of propagation of the plane wave (Fig. 5). As the wave hits them, they start expanding and they emanate wavelets the shear fronts of which are tangent to the plane wave front.

The velocity field of a dislocation loop (Fig. 5) is given as an integral around the loop $L(\vec{x}',t')$ and over the history of the motion

$$\dot{U}_m(\vec{x},t) = \frac{\partial}{\partial x_\ell} \int_0^t dt' \oint_{L(t')} C_{ijk\ell}\, G_{km}\,(\vec{x}-\vec{x}',\, t-t') V_r(\vec{x}',t') b_i \varepsilon_{irs} dl_s'(t') \tag{6}$$

where $C_{ijk\ell}$ are the elastic coefficients, G_{km} the Green's function for a unit impulse force in a three-dimensional space, V_r the velocity vector at a point of the loop, and b_i the Burgers vector, which may be taken in the x_1-direction, i.e. $b_i = b_1$, without loss of generality. For isotropic material the only non-zero terms in (6) are $C_{1313}G_{11,3} + C_{1331}G_{31,1}$. For a circular loop suddenly expanding at $t' = 0$ with constant radial velocity $V_r = V$, $d\ell(t') = \rho'(t')d\theta$ with $\rho'(t') = \rho_0 + VH(t')$. The shear pulse that will produce a gliding motion of the loop must have a σ_{13} component. Analogously to the straight dislocations model we are interested in contributions to the particle velocity $\dot{U}_1$ at the point P (Fig. 5) and at a time Δt after the pulse has passed. It is found from (6) [9] that a loop in the plane x_3' contributes the velocity at $P(0, 0, \hat{x}_3)$ at time $t = \hat{x}_3/c_2 + \Delta t$:

$$\dot{U}_1(0, 0, \hat{x}_3, \hat{x}_{3/c_2} + \Delta t) = -\frac{C_{1313}}{4\pi\rho} b_1 \frac{2V}{c_2^3(t - \frac{x_3'}{c_2})} \cdot \frac{1}{\sqrt{\frac{2c_2\Delta t}{\hat{x}_3 - x_3'} - \omega^2}} \qquad (7)$$

where ρ is the density, and the angle ω is indicated in Fig. 5. Expression (7) is analogous to the one obtained for straight dislocations, that is both straight dislocations and loops have a square root singularity at the wavefront.

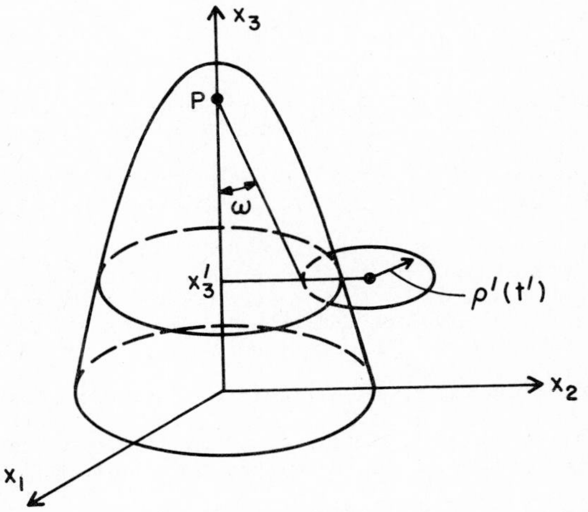

Fig. 5 Circular dislocation loop expanding in a plane normal to the incident wave.

After integrating over all the dislocation loops that intersect the domain of dependence (a paraboloid of revolution, Fig. 5), the end result for the precursor decay is

$$\frac{d\dot{U}_1}{d\hat{x}_3} = \pi b_1 V\hat{N}\rho_0 = b_1 \frac{VN}{2} \qquad (8)$$

where N denotes the dislocation line length per unit volume. Thus (8) coincides with (5) for $\psi = 0$, which could have been expected on physical grounds that the pulse does not "see" the curvature of the dislocation segments at the front.

An important non-intuitive result of this analysis is that only the dislocations at the wavefront contribute to the precursor decay at a given instant. The elastic precursor decays linearly with distance of propagation for uniform distribution of dislocations. In actual experiments [1, 3] where the dislocation density was higher near the impact surface, more decay must have occurred at the higher density region near the surface [8].

We can say, in conclusion, that these models provided further understanding into the dislocation mechanism that produces the decay of the elastic precursor and bridged in this particular phenomenon of dynamic plasticity continuum and dislocation approaches. While this model is adequate in describing the wavefront behavior of the plastic wave, description at the dislocation scale of the plastic wave at later times is increasingly complex since dislocation multiplication and interaction are involved.

ACKNOWLEDGMENT

This research was supported by the National Science Foundation.

REFERENCES

1. R.J. Clifton, Some results in plate impact experiments, in: Propagation of Shock Waves in Solids (edited by E. Varley), ASME AMD Vol. 17, p. 27 (1976).

2. W.G. Johnston and J.J. Gilman, Dislocation velocities, dislocation densities, and plastic flow in lithium fluoride crystals, J. Appl. Phys., 30, p. 129 (1959).

3. A. Kumar and R.J. Clifton, Dislocation motion and generation in LiF single crystals subjected to plate impact, J. Appl. Phys., 50, p. 4747 (1979).

4. J.E. Vortham, The dislocation nucleation hypothesis for shocked lithium fluoride, Ph.D. Thesis, Washington State University, Pullman, WA (1979).

5. W. Herrmann, Nonlinear stress waves in metals, in: Wave Propagation in Solids (edited by J. Miklowitz), p. 129, ASME (1969).

6. J.W. Taylor, Dislocation dynamics and dynamic yielding, J. Appl. Phys., 36, p. 3146 (1965).

7. E. Orowan, Problems of plastic gliding, Proc. Phys. Soc., A 52, p. 8 (1940).

8. R.J. Clifton and X. Markenscoff, Elastic precursor decay and radiation from nonuniformly moving dislocations, J. Mech. Phys. Sol. (to appear).

9. X Markenscoff and R.J. Clifton, Expanding circular dislocation loops and elastic precursor decay, (in preparation).

10. X. Markenscoff, The transient motion of a nonuniformly moving dislocation, J. Elast., 10, p. 193 (1980).

Continuum Models of Discrete Systems 4
eds. O. Brulin and R.K.T. Hsieh

CRITERIA OBTAINED FROM THE SMALL VOLUMES THEORY FOR CHOOSING ADEQUATE AVERAGE OPERATOR IN STATISTICAL CONTINUUM MECHANICS

P. Mazilu
Institut für Mechanik
Ruhr-Universität Bochum

1. INTRODUCTION

One of the fundamental problems in the statistical continuum mechanics is to associate to a given system of equations involving random parameters a deterministic mathematical model governing the average values. The succes in such an enterprise is fully conditioned by a proper choice of the average operator. This operator must satisfy the following conditions:

1. Commutes with the time or space differential operators which enter the mathematical model,
2. Allows the formulation of the boundary conditions in terms of averaged values,
3. Permits an expression of the averaged energy (or other thermodynamical functions) in terms of the averaged values.

It is obvious that one or other of the usual average operations (ensemble average, volume average, surface average, time average, etc.), cannot satisfy, in all cases, the above conditions. However, it is expected, that, for each particular phenomenon, one or a combination of the above mentioned average operator would be consistent with the above conditions.

Usually, in the steady problems of the statistical continuum mechanics (elastic equilibrium, steady heat transfer, dielectricity, etc.) the following averages are used: the *volume average*

$$\langle f \rangle = \frac{1}{V} \int_V f(x)\, dx, \tag{1.1}$$

where V denotes a representative volume element (R.V.E.), and *ensemble average*

$$\bar{f} = \lim_{N\to\infty} \frac{1}{N} \sum_{\nu=1}^{N} f \,, \tag{1.2}$$

where f_ν are realizations of the random function, f.

The volume average can, equally, be extended over the whole euclidian space E as follows:

$$<f> = \lim_{V\to\infty} \frac{1}{V} \int_V f(x)\, dx. \tag{1.3}$$

Each of these averagings recommend themselves through particular advantages .

The volume averaging applies to a single specimen. It helps via Hill's condition, in certain particular circumstances (see [1] and [4]) to express the average of the energy. The ensemble averaging commutes with all differential operators, it permits one to express the boundary conditions and also to describe non-uniform thermomechanical (dielectrical) states.

In order to take advantages of boths averagings, usually

$$<f> = \bar{f} \tag{1.4}$$

is assumed. In connection with this hypothesis (ergodic hypothesis) let us observe that, in any problem of the statistical continuum mechanics a difference between the "input" and "output" data can be marked. While for the input data, the ergodicity of experimental observations only need be valid, for the output data a special proof of ergodicity is required. In § 2 of the present work a counterexample in support of this statement is given. It implies that any direct extension of the ergodic hypothesis from input to output data is always valid. Consequently, many of the mathematical operations which usually are performed in statistical continuum mechanics are no longer valid. Such a difficulty could be avoided if one develops a complete theory by using, systematically, only one or other averaging. Unfortunately, this cannot be succesful since the volume averaging does not commute with the partial differentiation and, as it is pointed out in § 3, the ensemble average generally does not satisfy Hill's condition.

The last paragraph examines the possibility to define a new set of averaging operators for which the above mentioned difficulties are, avoided. Here some methods given by the small volumes theory are used.

By small volumes theory is meant such kind of theory of continua in which the mathematical models are derived by using the following axioms:

1. The balance equations are not applicable to volumes smaller than a certain limit volume (small volume);

2. The behaviour of the media is fully described by a discrete set of state-parameters associated with small volumes;
3. The discrete state parameters can be approximated by smooth fields with an error of the order of the small volume diameter.

Such theories originate from Lamé's and Navier's works (cf. Poincaré: "Lecons sur la théorie de l'élasticité") and has been recently [7] used for the derivation a new mathematical model of nonlaminar flow.

2. ERGODIC HYPOTHESIS

Let us denote by $\mathcal{M}(E)$ the set of bounded scalar or vector fields having continuous first order differential over entire euclidian space E

$$\mathcal{M}(E) = \{f \in C^1(E) \,|\, |f| \leq M\}.$$

We consider the following problem: For given vector field $\underset{\sim}{f} \in \mathcal{M}(E)$ obtained solution of the Poisson's equation

$$\Delta u = \operatorname{div} \underset{\sim}{f} \tag{2.1}$$

in E such that $u \in C^2(E)$ and grad $u \in \mathcal{M}(E)$. It is posible to prove that the solution of the above problem is unique (if one neglects linear terms) and its gradient is given by

$$\operatorname{grad} u(x_1) = \frac{1}{3} f(x_1) + \frac{1}{4\pi} \int_E \operatorname{grad}_1 (\operatorname{grad} \frac{1}{r_{12}}) f(x_2)\, dx_2 \tag{2.2}$$

where $r_{12} = |x_2 - x_1|$.

Let us denote by Γ the operator defined by

$$\Gamma f = -\frac{1}{3} f - \frac{1}{4\pi} \int_E \operatorname{grad}_1 (\operatorname{grad}_2 \frac{1}{r_{12}}) f(x_2)\, dx_2 \tag{2.3}$$

It is easy to prove that:

P_1. If grad $\varphi \in \mathcal{M}(E)$ then

$$\Gamma \operatorname{grad} \varphi = -\operatorname{grad} \varphi + \text{const}. \tag{2.4}$$

and

P_2. If rot $\underset{\sim}{\omega} \in \mathcal{M}(E)$ then

$$\Gamma \text{ rot } \underset{\sim}{\omega} = \text{const} . \qquad (2.5)$$

Let us consider now the vector field $\underset{\sim}{f}$ in the form

$$\underset{\sim}{f} = \text{grad}\,\varphi + \text{rot } \underset{\sim}{\omega} \qquad (2.6)$$

with gradφ, rot $\underset{\sim}{\omega} \in \mathcal{M}(E)$. According to the properties. P_1 and P_2 it follows that

$$\Gamma \underset{\sim}{f} = - \text{grad}\,\varphi + \text{const}.$$

Assuming, in addition, that φ and $\underset{\sim}{\omega} \in \mathcal{M}(E)$, then the following equality holds

$$<\Gamma \underset{\sim}{f}\ \Gamma \underset{\sim}{f}> = - <\underset{\sim}{f}\ \Gamma\ \underset{\sim}{f}> . \qquad (2.7)$$

In order to prove this we note that over any volume V

$$\int_V \text{rot } \underset{\sim}{\omega}\ \text{grad}\,\varphi\ dx = \int_{\partial V} \text{rot } \underset{\sim}{\omega}\ \varphi \underset{\sim}{n}\ ds$$

and, consequently,

$$\int_V \text{grad}\,\varphi\ \text{grad}\,\varphi\ dx = \int_V \underset{\sim}{f}\ \text{grad}\,\varphi\ dx - \int_{\partial V} \text{rot}\underset{\sim}{\omega}\ \varphi \underset{\sim}{n}\ ds . \qquad (2.8)$$

If we divide (2.8) by V and take (2.4) and (2.5) into account then (2.7) follows immediately in the limit when $V \to \infty$.

Let us now assume that $\underset{\sim}{f}$ defined by (2.6) satisfies the ergodic hypothesis (1.4). Since φ and $\underset{\sim}{\omega} \in \mathcal{M}(E)$ then $<\underset{\sim}{f}> = 0$ and (1.4) is valid only when

$$\overline{\text{grad}\,\varphi} + \overline{\text{rot } \underset{\sim}{\omega}} = 0 . \qquad (2.9)$$

If the ergodic hypothesis could be extended to the solution of (2.1) then

$$\overline{\Gamma \underset{\sim}{f}\ \Gamma \underset{\sim}{f}} = - \overline{\underset{\sim}{f}\ \Gamma\ \underset{\sim}{f}} \qquad (2.10)$$

must be true, or, equivalently

$$\Gamma_{12}\ \Gamma_{12'}\ \overline{\underset{\sim}{f}_2\ \underset{\sim}{f}_{2'}} + \Gamma_{12}\ \overline{\underset{\sim}{f}_1\ \underset{\sim}{f}_2} = 0 \qquad (2.11)$$

(the lower indices denote the variables x_1 or x_2).

It is easy to verify that (2.11) is satisfied only when

$$\overline{\operatorname{rot} \underset{\smile}{\omega} \operatorname{grad} \varphi} = 0$$

at every points of E. Obviously, such an equality is not a direct consequence of the ergodicity condition (2.9).

3. HILL's CONDITION FOR THE ENSEMBLE AVERAGING

Let us consider an elastic heterogeneous medium extending over space E and whose Hooke's tensor is a random tensor function L(x). Denote by $\bar{L}$ the enxemble average of L(x) and by $L^o(x)$ the difference $L^o(x) = L(x) - \bar{L}$. We shall assume that $\bar{L}$ defines an isotropic homogeneous material, i.e.

$$L_{ij}^{kl} = \lambda\, \delta_{ij}\, \delta_{kl} + \mu(\delta_{ik}\, \delta_{il} + \delta_{ie}\, \delta_{jk}),$$

where λ and μ are constants. For mathematical reasons we shall consider instead of the actually Hooke's tensor

$$L(x) = \langle L \rangle + L^o(x)$$

its generalization

$$L(x) = \langle L \rangle + \xi L^o(x),$$

where $\xi \in (0,1]$ is a real parameter. It is well known [2], [5], [6], that the stress and strain in such a material are

$$\begin{aligned} \varepsilon_1 &= \varepsilon^o + \xi\, \Gamma_{12}\, L_2^o \varepsilon^o + \xi^2\, \Gamma_{12}\, L_2\, \Gamma_{23}\, L_3^o\, \varepsilon^o + O(\xi^3), \\ \sigma_1 &= \bar{L}\ \varepsilon^o + \xi\, L^o\, \varepsilon^o + \xi\, \bar{L}\, \Gamma_{12'}\, L_{2'}^o\, \varepsilon^o + \\ &\quad + \xi^2\, L_1^o\, \Gamma_{12'}\, L_{2'}\, \varepsilon^o + \xi^2\, \bar{L}\ \Gamma_{12'}\, L_{2'}^o\, \Gamma_{23'}\, L_{3'}^o\, \varepsilon^o + O(\xi^3), \end{aligned} \tag{3.2}$$

where ε^o is a constant symmetric tensor and Γ denote the modified Green operator.

<u>REMARK</u>. The modified Green operator ist defined over the set of symmetric tensor fields σ by

$$\Gamma_{ijkl}\, \sigma_{kl} = \gamma_{(ij((kl)}\sigma_{kl} + \int_E \phi_{(ij)(kl)}(x_2 - x_1)\, \sigma_{kl}(x_2)\, dx_2 \tag{3.3}$$

where

$$\gamma_{(ij)(kl)} = \frac{1}{15\mu(\lambda+2\mu)} \; (\lambda+\mu)\delta_{ij}\delta_{kl} - \frac{3\lambda+8\mu}{2} \; (\delta_{ik}\delta_{jl} + \delta_{il}\delta_{jk}) \qquad (3.4)$$

and

$$\phi_{(ij)(kl)}(x) = \frac{1}{8\pi\mu} \; (\frac{1}{r})_{,lj} \; \delta_{ik} + (\frac{1}{r})_{,li} \; \delta_{jk} \; - \frac{\lambda+\mu}{8\pi\mu(\lambda+2\mu)} \; r_{,ijkl} \qquad (3.5)$$

In [2] the coefficients $\gamma_{(ij)(kl)}$ are taken to be zero and the functions $\phi_{(ij)(kl)}$ (denoted by Λ_{ijkl} in [2]) have a different signum. In [5] the coefficients $\gamma_{(ij)(kl)}$ (denoted by $- E_{ijkl}$ in [5]) are the same as (3.4), but the functions $\phi_{(ij)(kl)}$ (which are denoted by $- F_{ijkl}$) have a slightly different firm. The expressions (3.4) and (3.5) are calculated here by using the formula for differentiation of the integrals with weak singularity (see S. G. Michlin [8]) and according to the theory developed in [6].

The mean values of the strains and stresses are

$$\bar{\varepsilon} = \varepsilon^{o} + \xi^{2} \; \Gamma_{12} \; \Gamma_{13} \; L_2^{o} \; L_3^{o} \; \; \varepsilon^{o} + O(\xi^{3}),$$

and

$$\bar{\sigma} = \bar{L} \; \varepsilon^{o} + \xi^{2} \; \Gamma_{12'} \; L_1^{o} \; L_{2'}^{o} \; \varepsilon^{o} + \xi^{2} \; \bar{L} \; \Gamma_{12'} \; \Gamma_{2'3'} \; L_{2'}^{o} \; L_{3'}^{o} \; \; ^{o} + O(\xi^{3}).$$

It follows that stress-strain product $\bar{\sigma} \; \bar{\varepsilon}$ will be

$$\bar{\sigma} \; \bar{\varepsilon} = \bar{L} \; \varepsilon^{o} \; \varepsilon^{o} + \xi^{2} \; \bar{L} \; \Gamma_{12} \; \Gamma_{23} \; L_2^{o} \; L_3^{o} \; \varepsilon^{o} \; \varepsilon^{o} + \xi^{2} \; \Gamma_{12'} \; L_1^{o} \; L_{2'}^{o} +$$
$$+ \xi^{2} \; \bar{L} \; \Gamma_{12'} \; \Gamma_{2'3'} \; L_{2'}^{o} \; L_{3'}^{o} \; \varepsilon^{o} \; \varepsilon^{o} + O(\xi^{3}). \qquad (3.6)$$

If one computes first the product $\sigma \; \varepsilon$ and then the enxemble average $\overline{\sigma \; \varepsilon}$ one obtains

$$\overline{\sigma\varepsilon} = \bar{L} \; \varepsilon^{o} \; \varepsilon^{o} + \xi^{2} \; \Gamma_{12} \; L_1^{o} L_2^{o} \; \varepsilon^{o} \; \varepsilon^{o} + \xi^{2} \; \bar{L} \; \Gamma_{12'} \; \Gamma_{12'} \; L_2^{o} L_{2'}^{o} \; \varepsilon^{o} \; \varepsilon^{o} +$$
$$+ \xi^{2} \; \bar{L} \; \Gamma_{12} \; \Gamma_{23} \; L_2^{o} L_3^{o} \; \varepsilon^{o} \; \varepsilon^{o} + \xi^{2} \; \Gamma_{12'} \; L_1^{o} L_{2'}^{o} \; \varepsilon^{o} \varepsilon^{o} +$$
$$+ \xi^{2} \; \bar{L} \; \Gamma_{12'} \; \Gamma_{2'3'} \; L_{2'}^{o} L_{3'}^{o} \; \varepsilon^{o} \; \varepsilon^{o} + O(\xi^{3}). \qquad (3.7)$$

Comparing (3.6) with (3.7) and neglecting the terms $O(\xi^{3})$ it follows that Hill's condition $\overline{\sigma \boldsymbol{\varepsilon}} = \bar{\sigma} \; \bar{\varepsilon}$ is satisfied only when

$$\Gamma_{12} \; \overline{L_1^{o} L_2^{o}} \; \varepsilon^{o} \; \varepsilon^{o} + \bar{L} \; \Gamma_{12} \Gamma_{12'} \; \overline{L_2^{o} L_{2'}^{o}} \; \varepsilon^{o} \; \varepsilon^{o} = 0.$$

This relation has the same structure as (2.11) and seems not to be in any way implied by the ergodicity of L(x).

4. AVERAGING GIVEN BY SMALL VOLUME THEORY

Let us consider a disordered material for which it is possible to develop a macroscopical phenomenological theory having the representative volume element as small volume. The thermomechanical (or dielectrical) state of such a medium is characterized by

$\mathring{\Xi}$ - discrete set of macroscopic parameters,

$\hat{\Xi}$ - smooth fields approximating $\mathring{\Xi}$,

Ξ - microscopic parameters.

With the help of these parameters one can write various forms of the balance equations. These form are related as follows:

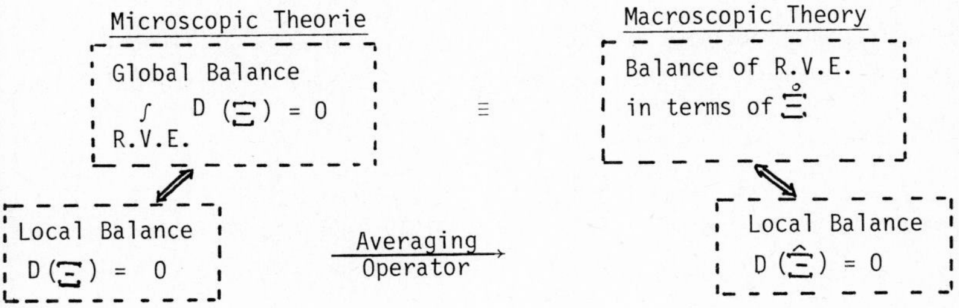

If we interpret now $\hat{\Xi}$ to be the averaged value of Ξ then, for each R.V.E.

$$< D(\Xi)> = D(\hat{\Xi}) \tag{4.1}$$

One can interpret this relation to be the commutation of averaging with the differential operators of the balance equations.

In the following we shall give two exemples where this procedure is applied.

1. Let us denote by σ the microscopic stress tensor, and by $\hat{\sigma}$ the smooth tensor field in a macroscopic theory. That means for elements ∂V of boundary surface of R.V.E.

$$\int_{\partial V} \sigma \underset{\sim}{n}\, ds = \int_{\partial V} \sigma \underset{\sim}{n}\, ds, \quad \int_{\partial V} \sigma \underset{\sim}{n} \times \underset{\sim}{r}\, ds = \int_{\partial V} \hat{\sigma} \underset{\sim}{n} \times \underset{\sim}{r}\, ds , \tag{4.2}$$

whence it follows that

$$<\text{div}\ \sigma> = \frac{1}{V} \int_{\partial V} \sigma \underset{\sim}{n}\, ds = \frac{1}{V} \int_{\partial V} \hat{\sigma} \underset{\sim}{n}\, ds = \text{div}\ \hat{\sigma} . \tag{4.3}$$

Therefore the volume averaging of the divergent operator applied to a microscopic stress tensor is equal to the divergent operator applied to the averaged stress tensor.

2. Let u be the microscopic displacement vector and $\hat{u}$ the displacement in the macroscopic theory, that means

$$\int_{\partial V} \hat{u}\, ds = \int_{\partial V} u\, ds \qquad (4.4)$$

over elements ∂V of R.V.E. Denoting by ε and $\hat{\varepsilon}$ respectively the corresponding strain tensors we have

$$\langle \text{def}\, u\rangle = \langle \varepsilon_{ij}\rangle = \frac{1}{2}\int_{\partial V}(u_i n_j + u_j n_i)ds = \frac{1}{2}\int_{\partial V}(\hat{u}_i u_j + \hat{u}_j n_i)ds = \hat{\varepsilon}_{ij} = \text{def}\, \hat{u}. \qquad (4.5)$$

Therefore, the mean values of the microscopic strains coincide with the deformation given by the mean displacements.

In the following we shall consider, as an application, the case of linear elasticity. The small volumes theory recommends, in this case, the following type of averaging

	Volume averaging	Surface averaging
Stress	———	$\hat{\sigma} = \frac{1}{s(\partial V)}\int_{\partial V} \sigma\, ds$
Displacement	———	$\hat{u} = \frac{1}{s(\partial V)}\int_{\partial V} u\, ds$
Strain	$\langle \varepsilon_{ij}\rangle = \frac{1}{V}\int_{\partial V}(u_i n_j + u_j n_i)ds$	———
Body Force	$\langle \text{div}\sigma\rangle = \frac{1}{V}\int_{\partial V}\hat{\sigma} n ds = -\langle f\rangle$	———
Energy	$\langle \sigma\varepsilon\rangle = \frac{1}{V}\int_V \sigma\ \varepsilon\, dx$	———

where $s(\partial v)$ denote the surface of ∂V.

In this case the balance equations, in terms of averaged values, are

$$\text{div}\, \hat{\sigma} = -\langle f\rangle, \qquad (4.6)$$

with the symmetric condition

$$\hat{\sigma}^T = \hat{\sigma}. \qquad (4.7)$$

The average strain is related in this case to gradient of average displacement via

$$< \varepsilon_{ij} > = \frac{1}{2}(\hat{u}_{i,j} + \hat{u}_{j,i}). \tag{4.8}$$

If we summarize the above results it follows that the small volumes theory recommends:

- surface averaging for the stresses and the displacements,
- volume averaging for the strains, body forces and energy.

REMARK. It is well known that some fundamental results of the statistical continuum mechanics (i.e. the derivation, via variational methods [3] of the upper and lower bounds for the effective elastic coefficients) requires for the stresses, the using of the volume average. This seems to be in contradiction with the above results . However it is possible to prove the following results which shows that, in special circumstances, this contradiction is only apparent: In a body having sufficiently great extent (in comparation with the R.V.E.) and which, subjected to the boundary conditions

$$u_i \big|_{\partial D} = \varepsilon^o_{ij}\, x_j, \qquad \varepsilon^o = \text{const.}$$

presents an uniform stress state $\hat{\sigma}$ = const, the volume and the surface averaging give the same result $<\sigma> = \hat{\sigma}$.

5. CONCLUSIONS

The paper proves that:

i) The ergodic hypothesis cannot be extended, without particular precautions, from input to output data of any equation involving random functions.

ii) For the ensemble averaging Hill's condition is not satisfied.

iii) A definition of the averaging which is in accordance with the small volumes theory recommend the surface averaging of the stresses. Such an averaging reduces only in particular circumstances to the volume averaging.

REFERENCES

1 Beran, M.J., Statistical continuum theories (Intersciences Publishers, New York 1968).

2 Eimer, Cz., Stresses in multi-phase media, Arch. Mech. Stos. 19 (1967), 521-538.

3 Hashin, Z., Shtrikman, S., A variational approach to the theory of elastic behaviour of polycristals. J. Mech. and Phys. Solids, 10 (1962), 335-342.

4 Kröner, E., Statistical Continuum Mechanics, International Centre for Mechanical Sciences, Udine 1971 (Springer Verlag, Wien-New York, 1971).

5 Kröner, E., Koch, H., Effective properties of disordered materials, SM Archiv, 1 (1976), 183-238.

6 Mazilu, P., On the theory of linear elasticity in statistically homogenuous media. Revue Roumaine des Mathēmatique Pures et Appliquées, 17 (1972) 261-273.

7 Mazilu, P., Small volumes in the derivation of the flow equations, in: Kröner, E. and Anthony, K.-H. (eds.) Continuum Models of Discrete Systems (University of Waterloo Press 1980).

8 Michlin, S.G., Multidimensional singular integrals and integral equations (Pergamon Press, 1965).

Part 2
HETEROGENEOUS MATERIALS, DEFECTS

Continuum Models of Discrete Systems 4
eds. O. Brulin and R.K.T. Hsieh

APPLICATIONS OF THE THEORY OF LINEAR ELASTIC MATERIALS WITH VOIDS

Stephen C. Cowin
Department of Biomedical Engineering
Tulane University
New Orleans, Louisiana

Jace W. Nunziato
Fluid Mechanics and Heat Transfer Division I
Sandia National Laboratories
Albuquerque, New Mexico

Pratap Puri
Department of Mathematics
University of New Orleans
New Orleans, Louisiana

The linear theory of elastic materials with voids is summarized and some applications are sketched. The theory differs from classical linear elasticity in that the volume fraction corresponding to the void volume is taken as an independent kinematic variable. The following applications of the theory are described briefly: response to homogeneous deformations, cylinder and sphere under internal pressure, pure bending of a beam, and small amplitude acoustic waves. In each of these applications, the change in the void volume induced by the deformation is determined.

INTRODUCTION

The linear theory of elastic materials with voids was developed by Cowin and Nunziato (1982) as a specialization of a non-linear theory presented earlier by Nunziato and Cowin (1979). The intended application of the theory is to linear elastic materials with small voids which are distributed throughout the material. When the void volume vanishes, the material is a linear elastic solid. The theory is distinguished from linear elasticity theory by the consideration of the void volume as an additional kinematic variable. The theory is distinguished from porous elastic materials by the fact that the void volume is a kinematic variable rather than a fixed parameter. In the next section the theory is summarized and, in four sections following, four specific applications are considered. These are the response of a linear elastic material with voids to quasi-static homogeneous deformations, the response of a cylinder and a sphere of this material to internal pressure and response of a beam of this material to pure bending, and the propagation of small amplitude acoustic waves in this material. In the final section of this paper the relationship between the theory presented and the effective moduli approach to composite materials is discussed.

SUMMARY OF THE THEORY

The basic concept underlying this theory is that of a material for which the bulk density ρ is written as the product of two fields, the density field of the matrix material γ and the volume fraction field ν,

$$\rho = \gamma \nu \tag{1}$$

The representation of the bulk density of the material introduces an additional degree of kinematic freedom in the theory. This idea was previously employed by Goodman and Cowin (1972, 1976) to develop a continuum theory for flowing granular materials. This linear theory of elastic materials with voids deals with small changes from a reference configuration of a porous body. In the reference configuration equation (1) is written as $\rho_R = \gamma_R \nu_R$ where the subscript indicates the reference configuration. We assume here that ν_R is spatially constant; the situation in which ν_R is not spatially constant is discussed by Cowin and Nunziatio (1982). As is customary in linear elasticity, the reference configuration is assumed to be stress free and strain free.

The independent kinematic variables in the linear theory are the displacement field $u_i(\underset{\sim}{x},t)$ from the reference configuration and the change in volume fraction from the reference configuration volume fraction $\phi(\underset{\sim}{x},t)$,

$$\phi(\underset{\sim}{x},t) = \nu(\underset{\sim}{x},t) - \nu_R \tag{2}$$

where $\underset{\sim}{x}$ is the spatial position vector in cartesian coordinates and t is time. The infinitesimal strain tensor $E_{ij}(\underset{\sim}{x},t)$ is determined from the displacement field u_i by the relation

$$E_{ij} = \tfrac{1}{2}(u_{i,j} + u_{j,i}) \tag{3}$$

where the comma followed by a lower case Latin letter indicates a partial deriv- derivative with respect to the indicated coordinate axes.

The equations of motion governing a linear elastic continuum with voids are the balance of linear momentum

$$\rho\ddot{u}_i = T_{ij,j} + \rho b_i \tag{4}$$

and the balance of equilibrated force

$$\rho k\ddot{\phi} = h_{i,i} + g + \rho\ell \tag{5}$$

where T_{ij} is the symmetric stress tensor, b_i is the body force vector, h_i is the equilibrated stress vector, k is the equilibrated inertia, g is the intrinsic equilibrated body force and ℓ is the extrinsic equilibrated body force. Each of these terms is discussed in detail by Nunziato and Cowin (1979) and Cowin and Nunziato (1982).

The constitutive equations for the linear isotropic theory of elastic materials with voids relate the stress tensor T_{ij}, the equilibrated stress vector h_i and the intrinsic equilibrated body force g to the strain E_{ij}, the change in volume fraction ϕ, the time rate of change of the volume fraction $\dot{\phi}$, and the gradient of the change in volume fraction $\phi_{,i}$; thus

$$T_{ij} = \lambda\delta_{ij}E_{kk} + 2\mu E_{ij} + \beta\phi\delta_{ij} , \tag{6}$$

$$h_i = \alpha\phi_{,i} , \tag{7}$$

$$g = -\omega\dot{\phi} - \xi\phi - \beta E_{kk} . \tag{8}$$

The coefficients λ, μ, α, β, ξ and ω, which all depend upon ν_R , must satisfy the following inequalities (cf. Cowin and Nunziato, 1982),

$$\begin{gathered} \mu \geq 0 , \alpha \geq 0 , \xi \geq 0 , \omega \geq 0 \\ (3\lambda + 2\mu) \geq 0 , (3\lambda + 2\mu)\xi \geq 12\beta^2 . \end{gathered} \tag{9}$$

The field equations governing the kinematic fields $u_i(\underset{\sim}{x},t)$ and $\phi(\underset{\sim}{x},t)$ are obtained by substituting the constitutive relations (6) through (8) into the equations of motion (4) and (5), thus

$$(\lambda + \mu)\underset{\sim}{\nabla}\underset{\sim}{\nabla}\cdot\underset{\sim}{u} + \mu\nabla^2\underset{\sim}{u} + \beta\phi + \rho\underset{\sim}{b} \quad \rho\ddot{\underset{\sim}{u}} \ , \tag{10}$$

and

$$\alpha\nabla^2\phi - \omega\dot{\phi} - \xi\phi - \beta\underset{\sim}{\nabla}\cdot\underset{\sim}{u} + \rho\ell = \rho k\ddot{\phi} \tag{11}$$

respectively. These two equations, one vector and one scalar, represent a system of four scalar equations in four unknowns, ϕ and the three components of u_i. The boundary and initial conditions on the displacement field $u_i(\underset{\sim}{x},t)$ are those of classical elasticity. The boundary condition on the field $\phi(\underset{\sim}{x},t)$ is that its derivative in a direction perpendicular to a boundary vanish,

$$\underset{\sim}{n}\cdot\underset{\sim}{\nabla}\phi = 0 \tag{12}$$

where $\underset{\sim}{n}$ is the unit normal to an external boundary. This boundary condition is discussed by Nunziato and Cowin (1979).

QUASI-STATIC HOMOGENEOUS DEFORMATIONS

Homogeneous deformations are deformations in which the displacement gradients, and hence the strain tensor E_{ij}, are independent of position $\underset{\sim}{x}$. Such deformations are important because they adequately model many problems of technological interest and, in many cases, experimental situations involving homogeneous deformations can be used to measure material coefficients. We consider here homogeneous deformations of a linear elastic material with voids for which ν_R is spatially constant. Specifically, we assume that a body of this material is subjected to a deformation process with the following properties: (i) the deformation process is homogeneous $E_{ij}(\underset{\sim}{x},t) = E^o_{ij}(t)$, (ii) the deformation process is quasi-static so the inertia terms $\rho\ddot{u}_i$ and $\rho k\ddot{\phi}$ can be neglected, and (iii) the deformation process occurs in the absence of body forces, i.e., both b_i and ℓ vanish. For this deformation process, the change in matrix volume fraction ϕ remains homogeneous and is given by

$$\phi = -\frac{\beta}{\omega}\int_o^t E^o_{kk}(\tau)\exp\left\{-\frac{\xi}{\omega}(t-\tau)\right\}d\tau \tag{13}$$

and the constitutive equation (6) for stress takes the form

$$T_{ij} = 2\mu E^o_{ij}(t) + \delta_{ij}\left\{\lambda E^o_{kk}(t) - \frac{\beta^2}{\omega}\int_o^t E^o_{kk}(\tau)\exp\left\{-\frac{\xi}{\omega}(t-\tau)\right\}d\tau\right\} \tag{14}$$

It is interesting to note the similarity of the stress relation (14) with those characterizing linear viscoelasticity. Specifically, an isotropic linear elastic material with voids has a linear volumetric viscoelasticity.

THE CLASSICAL PRESSURE VESSEL PROBLEMS

The traditional problems of the thick walled spherical and circular cylindrical shells under internal and external pressure can also be solved in the context of the theory of linear elastic materials with voids (c.f. Cowin and Puri, 1982). For both spherical and cylindrical shells and internal pressure $p_1(t)$ is applied at radius r_1, and an external pressure $p_2(t)$ is applied at radius r_2. The solution is quasi-static and the time domain is $0 \leq t < \infty$. In the case of the circular cylinder and axial strain of magnitude $\varepsilon(t)$ is also applied. The classical elasticity solutions to the spherical pressure vessel problem yields a radial displacement field u^{CE} of the form (c.f. Love, 1927)

$$u^{CE} = \frac{1}{3\lambda+2\mu}\left(\frac{p_1 r_1^3 - p_2 r_2^3}{r_2^3 - r_1^3}\right) r + \frac{1}{4\mu}\left(\frac{r_2^3 r_1^3 (p_1 - p_2)}{r_2^3 - r_1^3}\right)\frac{1}{r^2}, \tag{15}$$

and a radial stress T_{rr}^{CE} of the form

$$T_{rr}^{CE} = p_1 \frac{r_1^3}{r^3}\left(\frac{r_2^3 - r^3}{r_2^3 - r_1^3}\right) - p_2 \frac{r_2^3}{r^3}\left(\frac{r^3 - r_1^3}{r_2^3 - r_1^3}\right), \tag{16}$$

while the cylindrical pressure vessel problem has a radial displacement field u^{CE} of the form

$$u^{CE} = \left(\frac{p_1 r_1^2 - p_2 r_2^2}{r_2^2 - r_1^2} - \lambda\varepsilon\right)\frac{r}{2(\lambda+\mu)} + \frac{(p_1-p_2) r_1^2 r_2^2}{2\mu(r_2^2 - r_1^2)}\frac{1}{r} \tag{17}$$

and a radial stress T_{rr}^{CE} of the form

$$T_{rr}^{CE} = \frac{p_1 r_1^2 - p_2 r_2^2}{r_2^2 - r_1^2} - \frac{p_1 - p_2}{r_2^2 - r_1^2}\left(\frac{r_2^2 r_1^2}{r^2}\right). \tag{18}$$

The solution of these two classical problems in the context of the theory of linear isotropic elastic materials with voids yields the same stress distributions as predicted by classical elasticity. However, the volume fraction change $\phi(t)$ and the radial displacement $u(r,t)$ exhibit rate effects. If it is assumed that the initial volume fraction ϕ_o at time zero is spatially constant, then $\phi(t)$ is spatially constant for all time. For the spherical pressure vessel $\phi(t)$ is given by

$$\phi(t) = \phi_o e^{-\kappa_1 t} - \frac{3\beta}{\omega(3\lambda+2\mu)} \int_0^t \left(\frac{p_1(\tau) r_1^3 - p_2(\tau) r_2^3}{r_2^3 - r_1^3}\right) e^{-\kappa_1 (t-\tau)} d\tau \tag{19}$$

where

$$\kappa_1 = \frac{1}{\omega}\left(\xi - \frac{3\beta^2}{3\lambda+2\mu}\right), \tag{20}$$

and the radial displacement u(r,t) by

$$u(r,t) = u^{CE} - \frac{\beta}{3\lambda+2\mu}\phi(t)r \quad . \tag{21}$$

For the cylindrical pressure vessel problem $\phi(t)$ is given by

$$\phi(t) = \phi_o e^{-\kappa_1 t} - \frac{\beta}{\omega(\lambda+\mu)} \int_o^t \left(\frac{p_1(\tau)r_1^2 - p_2(\tau)r_2^2}{r_2^2 - r_1^2} + \mu\varepsilon(\tau)\right) e^{-\kappa_1(t-\tau)} d\tau \tag{22}$$

where

$$\kappa_1 = \frac{1}{\omega}\left(\xi - \frac{\beta^2}{\lambda+\mu}\right) \tag{23}$$

and the radial displacement u(r,t) by

$$u(r,t) = u^{CE} - \frac{\beta}{2(\lambda+\mu)}\phi(t)r \tag{24}$$

The salient feature of these solutions is the volumetric viscoelasticity introduced by the rate dependence in the volume fraction change.

PURE BENDING OF A BEAM

In the problems considered in this paper up to this point the change in void volume fraction has been spatially homogeneous. In order to illustrate the theory in a situation where the void volume fraction is not spatially homogeneous, we consider now the problem of pure bending of a beam composed of a linear isotropic elastic material with voids. In this problem a materially homogeneous body is subjected to an inhomogeneous deformation and the void fraction in the deformed state is inhomogeneous. The solution will be described because the associated equations are rather lengthy.

There are a number of interesting features of this solution. First, plane sections remain plane as in the classical elasticity solution. However, the stress distribution is not linear and the displacement has a rather more complicated dependence upon the transverse variable than that characterizing the classical theory. Second, the form of the transverse strain suggests that the depth of the beam may not change as much in the present theory as it would if it were an elastic material. Third, the stress normal to the lateral face does not vanish on that face, although it does not contribute a net force or moment to body. This implies that there will actually be distortions of the planar lateral faces. Fourth, the self equibrated stress vector $\underset{\sim}{h}$ has a negative maximum at the neutral axis of the beam and vanishes at the top and bottom of the beam.

SMALL AMPLITUDE ACOUSTIC WAVES

We consider briefly the main properties of small-amplitude acoustic waves in elastic materials with voids. Acoustic waves are solutions of (10) and (11), in the absence of body forces b_i and ℓ, for which the displacement u_i and the volume fraction change are of the form

$$u_i(\underset{\sim}{x},t) = \bar{u} d_i U(\underset{\sim}{x},t), \qquad \phi(\underset{\sim}{x},t) = \bar{\phi} U(\underset{\sim}{x},t) , \tag{25}$$

where

$$U(\underset{\sim}{x},t) = \mathrm{Re}\left\{ \exp[-(a + \frac{i\kappa}{c}) m_i x_i] \exp(i\kappa t) \right\} \tag{26}$$

Here d_i and m_i are unit vectors representing the <u>direction of displacement</u> and <u>direction of propagation</u>, respectively; a is the wave attenuation, c is the wave speed, κ is the frequency, and $\bar{u}$, $\bar{\phi}$ are wave amplitudes. The wave is said to be <u>longitudinal</u> if $d_i m_i = 1$; transverse if $d_i m_i = 0$.

Substitution of (25) and (26) into (10) and (11), setting the body forces b_i and ℓ equal to zero, yields

$$\{\rho \kappa^2 d_i + \mu(a + \frac{i\kappa}{c})^2 d_i + (\lambda+\mu)(d_i m_i)(a+\frac{i\kappa}{c})^2 m_i\} \bar{u} - \beta m_i (a+\frac{i\kappa}{c})\bar{\phi} = 0 \tag{27}$$

$$\{\rho k \kappa^2 - i\omega\kappa + \alpha(a + \frac{i\kappa}{c})^2 - \xi\}\bar{\phi} + \beta(d_i m_i)(a + \frac{i\kappa}{c})\bar{u} = 0 . \tag{28}$$

For transverse waves the volume fraction amplitude $\bar{\phi}$ must vanish; that is, <u>transverse waves propagate without affecting the porosity of the material</u>. further analysis reveals <u>that transverse waves propagate at a constant speed $c^2 = \mu/\rho$ and without attenuation</u>.

For longitudinal waves, the equations (27) and (28) result in

$$\{\rho \kappa^2 + (\lambda + 2\mu)(a + \frac{i\kappa}{c})^2\} \bar{u} - \beta(a + \frac{i\kappa}{c})\bar{\phi} = 0 , \qquad \beta(a + \frac{i\kappa}{c}) \bar{u} + \{\rho k \kappa^2 - i\omega\kappa + \alpha(a + \frac{i\kappa}{c})^2 - \xi\} \bar{\phi} = 0 . \tag{29}$$

A necessary and sufficient condition for this system to have a nontrivial solution for the amplitudes $\bar{u}$, $\bar{\phi}$ is that the determinant of the coefficients vanish:

$$\{\rho \kappa^2 + (\lambda+2\mu)(a+\frac{i\kappa}{c})^2\}\{\rho k\kappa^2 + \alpha(a+\frac{i\kappa}{c})^2 - i\omega\kappa - \xi\} + \beta^2(a+\frac{i\kappa}{c})^2 = 0 . \tag{30}$$

Excluding the damping term $i\omega\kappa$, this equation has been solved previously by Nunziato and Walsh (1977) in a study of one-dimensional acoustic waves in granular materials. Thus, the results of this study can be extended and we can conclude that both the attenuation a and the wave speed c are functions of frequency κ ; in other words, longitudinal acoustic waves are both attenuated and dispersed as a result of the changes in material porosity which accompany the wave. Furthermore, it is clear from (30) that it results in a fourth-order equation for the speed c and hence, two distinct types of waves are possible in the material; one associated with the elastic properties of the material, $\lambda + 2\mu$, and one associated with the properties governing changes in porosity, α and k. The possible existence of two waves in porous materials has been noted previously by Nunziato and Cowin (1979) and Nunziato and Walsh (1978).

THE EFFECTIVE MODULI APPROACH TO ELASTIC MATERIALS WITH VOIDS

In this section we address the question of the relationship between the theory whose applications are considered here and the effective moduli calculations of linear elastic materials with pores. The general philosophy of the effective moduli approach is to extend the use of homogeneous classical elasticity to inhomogeneous materials by replacing the elastic constants of the classical homogeneous theory by the effective elastic constants or moduli. The rationale employed in calculating the effective moduli is described by Christensen (1979) as "equivalent homogeneity." The idea of equivalent homogeneity for an inhomogeneous material such as a porous elastic solid is that the scale of the inhomogeneity is assumed to be several orders of magnitude smaller than the characteristic dimension of the problem of interest and, therefore, there exists an intermediate dimension over which the inhomogeneous properties can be averaged. On the length scale of the intermediate dimension, the elastic constants for the real inhomogeneous material can be replaced by the effective elastic moduli of the "equivalently homogeneous" model material.

While it is clear that the effective moduli calculations can be used in the present theory to determine the elastic moduli in reference configurations of known, uniform porosity from the elastic moduli of the matrix material, there are some limitations of the effective moduli approach. In the effective moduli approach to the problem of pure bending of a beam, the normal stress distribution is linear across the beam because the deformation induced void volume change is neglected. Our result shows that, while plane sections remain plane in both approaches, the void volume can change in a non-linear fashion across the beam depth and this non-linearity is reflected in the stress distribution. In the propagation of small-amplitude acoustic waves, we show that there are two longitudinal waves, whereas the effective moduli approach to elastic porous materials would suggest that there is only one wave speed. Both of these predictions of the theory of elastic materials with voids are consequences of the fact that the void volume was taken as an independent kinematic parameter. In the effective moduli calculations, the void volume is a physical property of the medium whose value is fixed independent of the deformation, stress and temperature.

In summary, the effective moduli approach is powerful because it can be used to predict effective material properties from the properties of the constituent materials. The theory of linear elastic materials with voids cannot do this and, in fact, certain of its material properties occuring in the theory must be determined experimentally for the specific porous material of interest. This is a limitation of the present theory that cannot be removed. However, the present theory can make predictions in situations where the effective moduli approach cannot. These situations will generally be those in which the deformation of the void volume is significant. It would appear that the two approaches may complement each other in certain situations.

ACKNOWLEDGMENT

This research was supported by National Science Foundation Grants to Tulane University and by the U.S. Department of Energy under contract DE-AC04-DP00789 to

Sandia National Laboratories, as a U.S. Department of Energy facility.

REFERENCES

[1] Cowin, S.C. and Nunziato, J.W., Linear elastic materials with pores, to appear in 1982.

[2] Nunziato, J.W. and Cowin, S.C., A non-linear theory of elastic materials with voids, Arch. Rational Mech. Anal., 72 (1979) 175-201.

[3] Cowin, S.C. and Puri, P., The classical pressure vessel problems for linear elastic materials with voids, to appear in 1982.

[4] Goodman, M.A., and Cowin, S.C., A continuum theory for granular materials, Arch. Rational Mech. Anal., 44 (1972) 249-266.

[5] Cowin, S.C. and Goodman, M.A., A variational principle for granular materials, ZAMP 56 (1976) 281-286.

[6] Love, A.E.H., A Treatise on the Mathematical Theory of Elasticity, (Cambridge, 1927).

[7] Nunziato, J.W. and Walsh, E.K., Small-amplitude wave behavior in one-dimensional granular materials, J. Appl. Mech., 44 (1977) 559-564.

[8] Nunziato, J.W. and Walsh, E.K., On the influence of void compaction and material non-uniformity on the propagation of one-dimensional acceleration waves in granular materials, Arch. Rational Mech. Anal., 64 (1977) 299-316; Addendum, ibid, 67 (1978) 395-397.

[9] Christensen, R.M., Mechanics of Composite Materials (Wiley, 1979).

Continuum Models of Discrete Systems 4
eds. O. Brulin and R.K.T. Hsieh

GAUGE INVARIANT FORMULATION OF CONTINUUM THEORY OF DEFECTS

B.K.D. Gairola

Institut für Theoretische und Angewandte Physik,
Universität Stuttgart, Stuttgart, W-Germany

INTRODUCTION

Physics is a field that thrives on analogies. Although the analogies are seldom perfect they often lead to generalizations and new insights. More than two decades ago the analogy between the space-time of general relativity and a crystalline medium was used by Kondo /1/, Bilby et al. /2/ and Kröner /3/ to formulate and solve many problems of defects. Nowadays the concept of vacuum is as complicated as a material medium and resembles it in many ways.

So it is tempting to consider the material such as a crystal as an analogue of vacuum. In fact Kröner /4/ has recently suggested this idea. The purpose of this paper is to exploit this analogy by using one of the most successful ideas in the field of elementary particle theory that has emerged in the last fifteen years. This is the principle of gauge invariance. It can be traced to the work of Weyl /5/. He formulated the gravitation of Dirac electron as a gauge field theory. This idea was generalized by Yang and Mills /6/ and Utiyama /7/. An excellent review of general relativity as a Poincaré gauge theory has been given by Hehl et al. /8/.

GAUGE GROUPS

Gauge theories deal with elementary particles which have intrinsic properties such as electric charge and isotopic spin. These internal properties are associated with internal degrees of freedom. One can then imagine that an elementary particle has its own internal space.

This idea can be generalized to any type of entity with internal or external freedom. If there is a continuous distribution of such entities in some continuum we can attach an internal space to each point of the continuum which may be called the base space. The internal space may have any dimensions not necessarily of the base space. We identify the base space with our three-dimensional material space. For the theory dealing with topological defects one can assume that the internal space is also three-dimensional. Each internal space can be subjected to transformations or mapping on itself which maintain its essential properties. These transformations form a group called internal group.

In many important applications the internal group is a Lie group. An element g of an r-parameter Lie group G can be expressed as

$$g = \exp(\theta^a f_a) \tag{1}$$

where Θ^a are parameters, f_a are group generators and the summation convention is used. The generator f_a satisfy the commutation relations

$$[f_a, f_b] = C_{ab}{}^c f_c. \qquad (2)$$

The coefficients $C_{ab}{}^c$ are called structure constants. The infinitesimal transformation is

$$g = 1 + \Theta^a f_a. \qquad (3)$$

If the transformation parameters are independent of the position $\underline{x}$ in the base space it is called global or gauge transformation of the first kind and if its parameters are functions of $\underline{x}$ it is called local or gauge transformation of the second kind.

Kröner /4/ has recently introduced the concept of an internal observer in the crystal vacuum who cannot perceive any compatible deformations. Therefore, the most general global group that one needs as an internal group is the general affine group GA which is the semi-direct product of the general linear group GL and the translation group T.

Its element g is given by

$$g = 1 + \Theta^a \partial_a + \Theta^a{}_b (x^b \partial_a + f_a{}^b) \qquad (4)$$

where

$$\begin{aligned} &[\partial_a, \partial_b] = 0 \\ &[f_a{}^b, f_c{}^d] = \delta_a^d f_c^b - \delta_c^b f_a^d \\ &[\partial_a, f_b^c] = \delta_a^c \partial_b - \delta_b^c \partial_a \end{aligned} \qquad (5)$$

MINIMAL COUPLING

Let us now introduce a set of orthonormal vectors $\underline{e}_a$ or triads at each point $\underline{x}$ of the base space which defines a coordinate system x^a in the internal space. These vectors in a way represent the internal observer who is able to measure the distances in a crystal for instance by counting atomic steps. In the base space we introduce a cartesian system of coordinates x^i with base vectors $\underline{e}_i$. Since the base vectors $\underline{e}_a$ can be arbitrarily chosen we can assume that they coincide with $\underline{e}_i$. Therefore, the two sets of vectors and their reciprocals satisfy the trivial relations

$$\begin{aligned} &\underline{e}_a = \delta_a^i \underline{e}_i, \qquad &\underline{e}^i = \delta_a^i \underline{e}^a \\ &\underline{e}^a = \delta_i^a \underline{e}^i \qquad &\underline{e}_i = \delta_i^a \underline{e}_a \\ &\underline{e}_a \cdot \underline{e}_b = \delta_{ab} \qquad &\underline{e}_i \cdot \underline{e}_j = \delta_{ij} \end{aligned} \qquad (6)$$

These relations define the holonomic gauge condition. It is preserved by the simultaneous action of a coordinate transformation and a triad deformation.
The vectors $\underline{e}_a$ and their tensor product provide a representation R(g) for the gauge groups.

Now let the defect distribution be described by a field $\Phi(\underline{x})$ which transforms tensorially with respect to the internal transformation. Accordingly its components will be referred to $\underline{e}_a$ only. For convenience we will suppress the indices. If the internal group is global we have

$$\Phi' = R(g)\Phi \tag{7}$$

and

$$\partial_i \Phi' = R(g)\,\partial_i \Phi \tag{8}$$

On the other hand under the action of local gauge group $\partial_i \Phi$ does not transform tensorially because now

$$\partial_i \Phi' = R(g)\partial_i \Phi + (\partial_i R(g))\Phi \tag{9}$$

Therefore an energy density $L_D(\Phi, \partial_i \Phi)$ associated with defects which is invariant under the action of global group will not be invariant under the action of local gauge group.

However, we can recover the invariance of the energy density by replacing the ordinary derivative by a modified derivative which behaves covariantly i.e. as a tensor. For this purpose we introduce a gauge potential Γ_i and define the covariant derivative as

$$D_i = (\partial_i + \Gamma_i) \tag{10}$$

The requirement that this should be a tensor i.e.

$$(\partial_i + \Gamma_i')\Phi' = (\partial_i + \Gamma_i)\Phi \tag{11}$$

implies

$$\Gamma_i' = R(g(\underline{x}))\Gamma_i R^{-1}(g(\underline{x})) - R(g(\underline{x}))\,\partial_i \overset{-1}{R}(g(\underline{x})) \tag{12}$$

The inhomogeneous term here cancels the second term in (9). The appearance of the gauge potential Γ_i is a consequence of the fact that under the action of local gauge transformation the triads $\underline{e}_a$ have been transformed to new triads which are no longer the same everywhere.

The gauge potential reflects the change in the geometrical structure of the base space because it defines the law of parallel transport via the equation

$$D_i \Phi = 0 \tag{13}$$

or equivalently

$$d\Phi = \Phi(\underline{x}+d\underline{x}) - \Phi(\underline{x}) = -dx^i\,\Gamma_i \Phi \tag{14}$$

It can be integrated along a curve C joining two points $\underline{x}_1$ and $\underline{x}_2$ and so may be said to transport ϕ from $\underline{x}_1$ to $\underline{x}_2$.

Let us assume that there is a particular gauge transformation which transforms Γ_i to zero. Then equation (12) reduces to

$$0 = R(g(\underline{x}))\,\Gamma_i R^{-1}(g(\underline{x})) - R(g(\underline{x}))\,\partial_i R^{-1}(g(\underline{x})) \tag{15}$$

and so Γ_i should have the special form

$$\Gamma_i = [\partial_i R^{-1}(g(\underline{x}))]\, R(g(\underline{x})) \tag{16}$$

If we regard this equation as a differential equation for $R^{-1}(g(x))$ then the question of existence of such a R reduces to the integrability condition

$$\begin{aligned}[\partial_i,\partial_j]R^{-1}(g(\underline{x})) &= \partial_i\left[\Gamma_j R^{-1}(g(\underline{x}))\right] - \partial_j\left[\Gamma_i R^{-1}(g(\underline{x}))\right] \\ &= F_{ij}R^{-1}(g(\underline{x}))\end{aligned} \tag{17}$$

where

$$F_{ij} = \partial_i\Gamma_j - \partial_j\Gamma_i + [\Gamma_i,\Gamma_j] \tag{18}$$

is called the Yang-Mills field or the curvature tensor. We conclude that Γ_i can be transformed to zero if and only if F_{ij} vanishes. Hence the curvature tensor describes an essential geometrical property of the base space.

Strictly speaking the integrability condition guarantees the existence of the solution only in local patches. It can not be extended to all space. It has to satisfy another type of integrability condition which we will derive shortly.

Therefore, to retain the gauge invariance of the energy density we must consider the triads $\underline{e}_a$ as transformed according to the appropriate local gauge group associated with the defect and thus consider $\underline{e}_a$ and $\underline{e}_i$ variables. Consequently we must replace the relations (6) by

$$\begin{aligned}\underline{e}_a &= e_a^i\underline{e}_i, & \underline{e}_i &= e_i^a\underline{e}_a \\ \underline{e}^a &= e_i^a\underline{e}^i, & \underline{e}^i &= e_a^i\underline{e}^a \\ \underline{e}_a\cdot\underline{e}_b &= g_{ab}, & \underline{e}_i\cdot\underline{e}_j &= g_{ij}\end{aligned} \tag{19}$$

As we move from x to x+ dx the change in the triads $\underline{e}_a$ is given by

$$d\underline{e}_a = \underline{e}_a(\underline{x}+d\underline{x}) - \underline{e}_a(\underline{x}) = dx^i\,\Gamma_{ia}^b\underline{e}_b \tag{20}$$

Accordingly the representation of the local gauge group has to be modified. For instance in the formulas (4) and (5) generator ∂_a will have to be replaced by $e_a^i D_i$. Thus

$$[\partial_i,\partial_j] \to [D_i,D_j] = F_{ij}^{ab}\,f_{ab} - F_{ij}{}^a D_a \tag{21}$$

shows how the integrability condition (18) becomes modified. The second term above supplies the remaining condition we mentioned earlier which is necessary for the existence of a global cartesian coordinate system. It is easy to verify that

$$F_{ij}{}^{a} = \partial_i e_j^a - \partial_j e_i^a + \Gamma_i{}^{ab} e_{bj} - \Gamma_j{}^{ab} e_{bi} \qquad (22)$$
$$= D_i e_j^a - D_j e_i^a$$

It may be seen that translation group when gauged in this way is no longer Abelian.

The relative change between two neighbouring base vectors $\underline{e}_i$ is

$$d\underline{e}_i = dx^i \Gamma_{ij}^k \underline{e}_k \qquad (23)$$

that means

$$\Gamma_{ij}^k = e_a^k (\partial_i e_j^a + \Gamma_i{}^a{}_b e_j^b) \qquad (24)$$
$$= e_a^k D_i e_j^a$$

defines the parallel transport via the covariant derivative

$$\nabla_i = \partial_i + \Gamma_i{}^{jk} f_{jk} \qquad (25)$$

where f_{jk} are the generators of the group of general coordinate transformation. The two kinds of derivatives are related by

$$\nabla_i \Phi_j{}^k = e_j^a e_a^k D_i \Phi_a{}^b \qquad (26)$$

This rule can be generalized to tensors of any rank.

If the gauge invariance is with respect to local GA we find from

$$e_i^a e_j^b g_{ab} = g_{ij} \quad \text{and} \quad D_i g_{ab} \neq 0 \qquad (27)$$

that

$$\nabla_i g_{jk} \neq 0 \qquad (28)$$

Putting

$$\nabla_i g_{jk} = Q_{ijk} \qquad (29)$$

and

$$\Gamma_{ij}^k - \Gamma_{ji}^k = 2T_{ij}{}^k \qquad (30)$$

we find

$$\Gamma_{ij}^k = g^{kl} P_{jil}^{ABC} \left[\frac{1}{2}\partial_A g_{BC} - g_{CD} T_{AB}{}^D + \frac{1}{2} Q_{ABC}\right] \qquad (31)$$

where

$$P_{jil}^{ABC} = \delta_j^A \delta_i^B \delta_l^C + \delta_i^A \delta_l^B \delta_j^C - \delta_l^A \delta_j^B \delta_i^C \qquad (32)$$

The tensor $T_{ij}{}^k$ is called torsion. The trace part of $Q_{ij}{}^k$ describes dilatation and the traceless part is the volume preserving shear deformation. The geometry of the base space is, therefore, characterized by the metric tensor, the torsion tensor and the curvature.

Now it is clear that the gauge invariant energy density of defects is obtained by the substitution

$$L_D(\phi, \partial_i\phi) = L_D(\phi, \delta^i_a \partial_i\phi) \rightarrow \mathcal{L}_D(\phi, e_a D_i\phi) \tag{33}$$

where

$$\mathcal{L}_D = eL_D \tag{34}$$

and

$$e = \det(e^a_i) = (\det(g_{ij}))^{1/2} \tag{35}$$

The factor e comes in because $\mathcal{L}_D$ should be a scalar density. This substituion is known as minimal coupling.

The only drawback to (33) is that we have not included the energy density due to the change in the geometrical properties of the base space. This defect is remedied by simply adding to (33) the energy density function which contains only the Yang-Mills tensors. This can be done in various ways. For instance we can take the variables e, ∂e, Γ and $\partial\Gamma$ and make the substitution

$$L_D(\phi, \partial_i\phi) \rightarrow \mathcal{L}_D(\phi, e^i_a D_i\phi) + \mathcal{L}_f(e^i_a, \Gamma^a_{ib}, \partial e^i_a, \partial\Gamma^a_{ib})$$

or in the conventional manner

$$L_D(\phi, \partial_i\phi) \rightarrow \mathcal{L}_D(\phi, e^i_a D_i\phi) + \mathcal{L}_f(e^i_a, F_{ij}{}^a, F_{ija}{}^b)$$

CONSERVATION LAWS

We now investigate the consequence of gauge invariance of energy density functions $\mathcal{L}_D$ and $\mathcal{L}_f$ by using Noether's theorem. To this purpose let us denote the variables collectively by U. Invariance under the action of gauge group requires

$$\int_{V'} dV\, \mathcal{L}_D(U', \partial_i U') - \int_V dV \mathcal{L}_D(U, \partial_i U) = 0 \tag{36}$$

Using the partial integration and divergence theorem we obtain the equivalent condition

$$\frac{\delta \mathcal{L}_D}{\delta U} + D_i(\theta^i \mathcal{L}_D + \frac{\partial \mathcal{L}_D}{\partial \partial_i U}\delta U) = 0 \tag{37}$$

Applying this result to

$$\mathcal{L}_D(\phi, e^i_a D_i\phi) = \mathcal{L}_D(\phi, \partial_i\phi, e^i_a, \Gamma^b_{ia})$$

we obtain

$$\frac{\delta \mathcal{L}_D}{\delta e^a_i}\delta e^a_i + \frac{\delta \mathcal{L}_D}{\delta \Gamma^{ab}_i}\delta \Gamma_i^{ab} + D_i\left(e^i_a \theta^a \mathcal{L}_D + \frac{\partial \mathcal{L}_D}{\partial \partial_i \phi}\delta\phi\right) = 0 \qquad (38)$$

Substituting

$$\delta e^a_i = \theta^a_b e^b_i + \theta^b F_{bi}{}^a$$

$$\delta \Gamma_i^{ab} = D_i\theta^{ab} + \theta^c F_{ci}{}^{ab} \qquad (39)$$

$$\delta\Phi = (\theta^{ab} fab + \theta^a Da)\Phi$$

in (38) we get

$$D_i \sigma_a^i = F_{ai}{}^b \sigma_b{}^i + F_{ai}{}^{bc} \mu_{bc}{}^i \qquad (40)$$

$$D_i \mu_{ab}{}^i - \sigma_{ab} = 0 \qquad (41)$$

where

$$\sigma_a^i = \frac{\delta \mathcal{L}_D}{\delta e^a_i} = e^i_a \mathcal{L}_D + \frac{\partial \mathcal{L}_D}{\partial \partial_i \phi} D_a \Phi \qquad (42)$$

may be called internal or eigenstress and

$$\mu_{ab}{}^i = \frac{\delta \mathcal{L}_D}{\delta \Gamma^{ab}_i} = \frac{\partial \mathcal{L}_D}{\partial \partial_i \Phi} f_{ab}\Phi \qquad (43)$$

is moment stress or couple stress. These quantities correspond to the canonical energy-impulse tensors in general relativity. They do not exist for the situation described by $L(\phi, \partial_i\phi)$.

The terms on the right hand side of (40) could be interpreted as some sort of force densities resembling Lorenz force or Matthisson force. In the case of dislocations it is called Peach-Koehler force.

With the rule given in the previous section of converting tensors from internal to base space we can write the conservation laws in the base space coordinates. Thus making the identification

$$\sigma_i{}^j = \mathcal{L}_D \delta_i{}^j + \frac{\partial \mathcal{L}_D}{\partial \partial_j \phi}\nabla_i \phi \qquad (44)$$

$$\mu_{ij}{}^k = \frac{\partial \mathcal{L}_D}{\partial \partial_k \Phi} f_{ij}\Phi \qquad (45)$$

we can write (40) and (41) as

$$\nabla_j \sigma_i{}^j = \sigma_k^j (T_{ij}{}^k + Q_{ij}{}^k) + \mu_{jk}{}^l R_{il}{}^{jk} \qquad (46)$$

and

$$\nabla_k \mu_{ij}{}^k - \sigma_{ij} = 0 \qquad (47)$$

where

$$R_{ijk}{}^{l} = e^a_k e^l_b F_{ija}{}^{b} \tag{48}$$

SOME SIMPLE EXAMPLES OF APPLICATIONS

Gauge theory of defect is a very general theory since only things needed are specifying the internal space or internal degrees of freedom and the gauge group appropriate to the defect. Therefore, it is applicable to a very wide class of materials and defects. We shall consider here a very simple material namely a simple Bravais crystal. Even in this simple material there can be various kinds of defects such as dislocations and point defects.

Dislocations are associated with glide along slip planes. This defect therefore, breaks the translational gauge symmetry of the first kind. In this case we start with an elastic energy density which is a function of the metric $\underline{e}^a \cdot \underline{e}^b$. The triads $\underline{e}^a$ may be taken along the lattice vectors. This energy density is, of course, quite arbitrary and can be put equal to zero at some later stage of the calculations. In this case triads themselves are fields describing the defects. Thus the energy density is $L_D(x^a, \partial_i x^a)$ or $L_D(x^a, \delta^a_i)$. If we keep to the homogeneous case the minimal coupling argument leads to the substitution

$$\delta^a_i \rightarrow e^a_i, \quad L_D(\delta^a_i) \rightarrow \mathcal{L}_D(e^a_i). \tag{49}$$

The gauge potential e^a_i defines only a Pfaffian form. The commutator of translational gauge group

$$[\partial_i, \partial_j] = - F_{ij}{}^{a} \partial_a \tag{50}$$

yields the Yang Mills tensor

$$F_{ij}{}^{a} = \partial_i e^a_j - \partial_j e^a_i \tag{51}$$

It can be easily verfied that the curvature tensor $F_{ija}{}^{b}$ now vanishes. Hence according to minimal argument we make the substitution

$$L_D(\delta^a_i) \rightarrow \mathcal{L}_D(e^a_i) + \mathcal{L}_f(F_{ij}{}^{a}) \tag{52}$$

Field equations are obtained by using the variation

$$\delta e^a_i = \partial_i \theta^a + \theta^b F_{bi}{}^{a} \tag{53}$$

we thus find

$$\sigma_a{}^{i} = \Sigma_a{}^{i} + \partial_j \chi_a{}^{ij} \tag{54}$$

$$D_i \sigma_a{}^{i} = F_{ai}{}^{b} \Sigma_b{}^{i} \tag{55}$$

where

$$\Sigma_a{}^{i} = e^i_a \mathcal{L}_f - F_{aj}{}^{b} \chi_b{}^{ji} \tag{56}$$

and

$$\chi_a^{\ ij} = \frac{\partial \mathcal{L}_f}{\partial F_{ij}^{\ a}} \tag{57}$$

converting these tensors to the base space coordinate system we can write

$$\sigma_{ij} = \Sigma_{ij} + \partial_k \chi_{ij}^{\ \ k} \tag{58}$$

$$\nabla_i \sigma_{ij} = T_{ji}^{\ \ k} \Sigma_k^{\ i} \tag{59}$$

where

$$\Sigma_{ij} = \mathcal{L}_f g_{ij} - T_{ik}^{\ \ l} \chi_l^{\ kj} \tag{60}$$

$$T_{ik}^{\ \ l} = \frac{1}{2} e_a^l (\partial_i e_k^a - \partial_k e_i^a) \tag{61}$$

is identified with the dislocation density-$\alpha_{ik}^{\ \ l}$. The equation (58) is the well known equation of balance of stress and couple stress.

Vanishing of $F_{ija}^{\ \ \ b}$ now corresponds to

$$R_{ijk}^{\ \ \ l} = 0 \tag{62}$$

Due to the antisymmetry in the indices R_{ijkl} this equation can be replaced by

$$G^{ij} = \frac{1}{4} \overset{(g)}{\epsilon}{}^{jnm} \overset{(g)}{\epsilon}{}^{ilk} R_{nmlk} = 0 \tag{63}$$

The antisymmetric part of this equation

$$G^{ij} - G^{ji} = 0 \tag{64}$$

implies that the divergence of dislocation density vanishes.

For further discussion we make the weak field approximation

$$e_i^a = \delta_i^a + \beta_i^a \tag{65}$$

where β is sufficiently small. It is usually called distortion. With this approximation we need not distinguish between the covariant and contravariant indices. The symmetric part

$$G^{ij} + G^{ji} = 0 \tag{66}$$

now reduces to

$$(\nabla \times \alpha)^s - \text{Inc}\,\varepsilon = 0 \tag{67}$$

where

$$\alpha_{ij} = -\epsilon_{kli} T_{klj} \tag{68}$$

is the dislocation density,

$$\text{Inc}\,\varepsilon = \nabla \times \varepsilon \times \nabla \tag{69}$$

and ε is the strain tensor $\varepsilon = \frac{1}{2}(\beta + \beta^T)$.
That means either $k = (\nabla x\alpha)^S$ or $\eta = \text{Inc}\,\varepsilon$ is a valid Yang Mills tensor. So instead of (52) we can write the total energy density according to minimal replacement as

$$L = L_D(\varepsilon) + L_f(\eta) \tag{70}$$

Using the variational principle we find

$$\sigma = \text{Inc}\,\chi \tag{71}$$

where

$$\sigma = \frac{\delta L_D}{\delta \varepsilon} \tag{72}$$

$$\chi = \frac{\delta L_f}{\delta \eta} \tag{73}$$

Equation (71) is the well known stress function representation. Furthermore, due to linearity conservation law (59) becomes

$$\partial_i \sigma_{ij} = 0 \tag{74}$$

It is interesting to note that (71) and (74) follow from the gauge theory in a rather independent manner. In continuum theory on the other hand one derives (71) from (74) via the Beltrami-Michell equations of compatibility.

The same equations are obtained if an extrinsic defect is described by a scalar field ϕ e.g. a dilatation field or a temperature field. The gauge invariant energy density associated with the defect is simply $\mathcal{L}_D(\phi, e^i_a \partial_i \phi)$ and gauge potential is, just like in the dislocation case, e^a_i. So the free part $\mathcal{L}_f$ is also the same and curvature $F_{ija}{}^b$ vanishes. Hence we end up again with the equation

$$(\nabla x \alpha^q)^S - \text{Inc} \quad = 0$$

The super script q on α is to indicate the fact that it is a quasi-dislocation density because in reality there are no dislocations.

Another interesting fact is that if the gauge theory is formulated in a hypothetical abstract space in which χ plays the role of distortion we would obtain the result

$$\eta = \text{Inc}\,\varepsilon, \quad \nabla \cdot \eta = o \tag{75}$$

The quantities occuring in (71) and (75) are not exactly the same because of a certain divergence term in the Noether's theorem. These equations in themselves reflect this fact because they do not have a unique solution in general. In fact this is a feature of all gauge theories. The reason is that we fixed only the translation and left the rotation arbitrary i.e. we gauged only a subgroup of the global gauge group. This results in a certain arbitrariness in the choice of gauge potential Γ_i or equivalently an arbitrariness in the absolute orientation of triads $\underline{e}_a$. Therefore, we must look for a way to constrain this arbitrariness of rotation. This is done by suitable gauge fixing condition which is also called symmetry breaking of the second kind. For instance if we make the assumption of physical linearity (Yang Mills approximation) and use the dual space formulation we can write

$L_D(\chi) = \chi \cdot \eta$

$L_f(\sigma) = \frac{1}{2} \sigma \cdot s \cdot \sigma, \qquad \sigma = \text{Inc}\,\chi$

The variational method yields the field equation

$\text{Inc } S \text{ Inc } \chi = \eta$

In this case the Kröner-Marguerre procedure is used for fixing the gauge. However, this procedure works because the translational gauge group is Abelian. It may not work for non-Abelian groups as is indicated by Gribov ambiguity in elementary particle physics /11/.

It also shows that whereas in the Yang Mills theories in elementary particle physics the coupling constant is simple, this is not the case here. For example we can write

$L_D = \frac{1}{2}\, \varepsilon \cdot C \cdot \varepsilon, \qquad L_f = \frac{1}{2}\, \eta \cdot H \cdot \eta$

Here C is the elastic constant and H is obtained by integrating the Green's function of the operator Inc C^{-1} Inc.

Finally it must be pointed out that the terms stress, stress function that we have used above are not quite legitimate. In reality the gauge invariant situation corresponds to the so-called natural state in which stresses vanish, and gauge non-invariant situation corresponds to ideal state. But if we apply a non-integrable transformation to the base space which makes it Euklidean, we can interpret the $\mathcal{L}_f$ term as due to the constraint that curvature vanishes. In other words σ etc. can be interpreted as Lagrange multipliers which represent constraint forces or stresses.

REFERENCES

/1/ K. Kondo, in RAAG Memoirs, Vols. 1 to 3, Gakujutsu Fukyu-kai, Tokyo, 1955-1962

/2/ B.A. Bilby, R. Bullough and E. Smith, Proc. Roy. Soc. London A231, 1955, p. 263

/3/ E. Kröner, Arch. Rat. Mech. 4, 1960, p. 273

/4/ E. Kröner, "Continuum Theory of Defects", Lectures held at the summer school on the Physics of Defects, Les Houches 1980

/5/ H. Weyl, Sitzungsber. Preuss. Akad. Wiss., 1918, p. 464, Z. Phys. 56, 1929, p. 330, Phys. Rev. 77, 1950, p. 699

/6/ C.N. Yang and R.L. Mills, Phys. Rev. 96, 1954, p. 191

/7/ R. Utiyama, Phys. Rev. 101, 1956, p. 1597

/8/ F.W. Hehl, P. von der Heyde, G.D. Kerlick and J.M. Nester, Rev. Mod. Phys. 48, 1976, p. 393

/9/ E. Kröner, Z. Phys. 139, 1954, p. 175

/10/ K. Marguerre, ZAMM 35, 1955, p. 242

/11/ V.N. Gribov, Nucl. Phys. B139, 1978, p. 1

Continuum Models of Discrete Systems 4
eds. O. Brulin and R.K.T. Hsieh
© North-Holland Publishing Company, 1981

YANG-MILLS TYPE MINIMAL COUPLING THEORY FOR MATERIALS WITH DEFECTS

Aida Kadić*

and

Dominic G. B. Edelen

Center for the Application of Mathematics
Lehigh University
Bethlehem, PA 18015

As is well known, classical elasticity theory starts with consideration of a current state vector $\underset{\sim}{x}(X^a)$ whose entries, $x^1(X^a)$, $x^2(X^a)$, $x^3(X^a)$, are the Cartesian coordinates of a particle at time $T = X^4$ that had coordinates (X^1, X^2, X^3) in some specified reference configuration. Here, (X^a) stands for (X^1, X^2, X^3, X^4) and lower case indices at the beginning of the alphabet have the range 1 through 4. The equations of motion can then be obtained by standard variational arguments based upon the Lagrangian density

(1) $$L_o = \frac{1}{2} \rho_o \, \partial_4 \underset{\sim}{x}^T \partial_4 \underset{\sim}{x} - U(e_{AB}) , \quad e_{AB} = \partial_A \underset{\sim}{x}^T \partial_B \underset{\sim}{x} - \delta_{AB} ,$$

where $U(e_{AB})$ is the elastic strain energy density. We use the notation $\partial_A := \partial/\partial X^A$, A=1,2,3; $\partial_a := \partial/\partial X^a$, a=1,2,3,4, and the superscript "T" denotes the transpose operation. The Lagrangian given by (1), and hence the whole theory, is manifestly invariant under the transformations $'\underset{\sim}{x}(X^a) = \underset{\sim}{R}\,\underset{\sim}{x}(X^a) + \underset{\sim}{b}$, $\underset{\sim}{R}^T\underset{\sim}{R} = \underset{\sim}{I}$ with $\underset{\sim}{R}$ and $\underset{\sim}{b}$ independent of (X^a). Noting that these transformations act on the state vector and are uncorrelated with transformations of the independent variables (X^a), they may be viewed as *gauge* transformations that leave the Lagrangian of elasticity theory invariant. Thus, although it is contrary to current practices in continuum mechanics, we may view the group $SO(3) \rhd T(3)$ as a natural, 6-parameter Lie group of gauge transformations of elasticity theory.

For technical purposes, it is necessary that we realize $G := SO(3) \rhd T(3)$ as a matrix Lie group. It is known that G can not be realized as a matrix Lie group on a 3-dimensional vector space. We therefore follow ROGULA [1] and consider the affine plane V in V_4 of all vectors of the form $\underset{\sim}{\Psi} = [x^1, x^2, x^3, 1]^T$. There is then a 1-to-1 correspondence between vectors $\underset{\sim}{x}$ in V_3 and vectors $\underset{\sim}{\Psi}$ in $V \subset V_4$. If M denotes the set of all 4-by-4 matrices $\underset{\sim}{A}$ of the form

$$\underset{\sim}{A} = \begin{pmatrix} \underset{\sim}{R} & \{\underset{\sim}{b}\} \\ [\underset{\sim}{0}] & 1 \end{pmatrix} \quad \text{then} \quad `\underset{\sim}{\Psi} = \underset{\sim}{A}\underset{\sim}{\Psi} \quad \text{gives} \quad `\begin{Bmatrix} \underset{\sim}{x} \\ 1 \end{Bmatrix} = \begin{Bmatrix} \underset{\sim}{R}\underset{\sim}{x}+\underset{\sim}{b} \\ 1 \end{Bmatrix},$$

and hence the set M can be placed in a 1-to-1 correspondence with the

*Permanent Address: Gradjevinski Fakultet, Sarajevo, Yugoslavia

transformations G. Thus, since the set M also forms a 6-parameter matrix Lie group with the same constants of structure as G

(2) $'\underset{\sim}{\Psi} = \underset{\sim}{A}\,\underset{\sim}{\Psi}$, $\quad \underset{\sim}{A} \in M$, $\quad \underset{\sim}{\Psi} \in V$

gives a faithful matrix representation of the gauge group G. For later purposes, we note that a basis for the Lie algebra of M is given by

$$\begin{pmatrix} \underset{\sim}{\gamma}_\alpha & \{\underset{\sim}{0}\} \\ [\underset{\sim}{0}] & 0 \end{pmatrix}, \quad \alpha = 1,2,3 \; ; \quad \begin{pmatrix} \underset{\sim}{0} & \underset{\sim}{t}_i \\ [\underset{\sim}{0}] & 0 \end{pmatrix}, \quad i = 1,2,3 \; ; \quad \underset{\sim}{t}_i = [\delta_{i1}, \delta_{i2}, \delta_{i3}]^T$$

where $(\underset{\sim}{\gamma}_1, \underset{\sim}{\gamma}_2, \underset{\sim}{\gamma}_3)$ is the standard basis for the 3-by-3 matrix Lie algebra of $SO(3)$. Since $\partial_a{}'\underset{\sim}{\Psi} = \underset{\sim}{A}\,\partial_a\underset{\sim}{\Psi}$, by (2), and $\partial_a\underset{\sim}{\Psi} = [\partial_a\underset{\sim}{x}, 0]^T$, the invariance of the Lagrangian L_o under action of the gauge group G is given by

(3) $L_o('\underset{\sim}{\Psi}, \partial_a{}'\underset{\sim}{\Psi}) = L_o(\underset{\sim}{A}\,\underset{\sim}{\Psi}, \underset{\sim}{A}\,\partial_a\underset{\sim}{\Psi}) = L_o(\underset{\sim}{\Psi}, \partial_a\underset{\sim}{\Psi})$.

The invariance of a Lagrangian under the action of a gauge group G is not mere curiosity; it is, in fact, the standard point of departure for the Yang-Mills minimal coupling universal gauge theory construction (see [2], [3], [4]). This construction consists of two parts. The first arises by allowing the parameters of the gauge group to be functions of position and time, in which case (2) is replaced by $'\underset{\sim}{\Psi}(x^a) = \underset{\sim}{A}(x^a)\,\underset{\sim}{\Psi}(x^a)$. Since $\partial_a{}'\underset{\sim}{\Psi} = \underset{\sim}{A}\,\partial_a\underset{\sim}{\Psi} + (\partial_a\underset{\sim}{A})\,\underset{\sim}{\Psi}$, $\underset{\sim}{A}$ no longer factors from the left, $L_o('\underset{\sim}{\Psi}, \partial_a{}'\underset{\sim}{\Psi}) \neq L_o(\underset{\sim}{\Psi}, \partial_a\underset{\sim}{\Psi})$, and the gauge invariance is lost. YANG and MILLS observed that a replacement of $\partial_a\underset{\sim}{\Psi}$ by $\partial_a\underset{\sim}{\Psi} + \underset{\sim}{\Gamma}_a\underset{\sim}{\Psi}$, where the $\underset{\sim}{\Gamma}_a$'s take their values in the Lie algebra of G and transform under the action of G by $'\underset{\sim}{\Gamma}_a = \underset{\sim}{A}\,\underset{\sim}{\Gamma}_a\,\underset{\sim}{A}^{-1} - (\partial_a\underset{\sim}{A})\,\underset{\sim}{A}^{-1}$, has the effect of restoring the factorization of $\underset{\sim}{A}$ from the left: $\partial_a{}'\underset{\sim}{\Psi} + {}'\underset{\sim}{\Gamma}_a{}'\underset{\sim}{\Psi} = \underset{\sim}{A}(\partial_a\underset{\sim}{\Psi} + \underset{\sim}{\Gamma}_a\underset{\sim}{\Psi})$. The matrix $\underset{\sim}{\Gamma} = \underset{\sim}{\Gamma}_a\,dx^a$ is then a matrix of connection 1-forms on the group G and the invariance of the Lagrangian under action by G is restored: $L_o('\underset{\sim}{\Psi}, \partial_a{}'\underset{\sim}{\Psi} + {}'\underset{\sim}{\Gamma}_a{}'\underset{\sim}{\Psi}) = L_o(\underset{\sim}{A}\underset{\sim}{\Psi}, \underset{\sim}{A}(\partial_a\underset{\sim}{\Psi} + \underset{\sim}{\Gamma}_a\underset{\sim}{\Psi})) = L_o(\underset{\sim}{\Psi}, \partial_a\underset{\sim}{\Psi} + \underset{\sim}{\Gamma}_a\underset{\sim}{\Psi})$. Thus, we obtain the famous *minimal replacement* $\partial_a\underset{\sim}{\Psi} \longmapsto \partial_a\underset{\sim}{\Psi} + \underset{\sim}{\Gamma}_a\underset{\sim}{\Psi}$.

We now apply these considerations to elasticity theory. Since the $\underset{\sim}{\Gamma}_a$ take their values in the Lie algebra of G, we have

$$(4) \qquad \underset{\sim}{\Gamma}_a = \begin{pmatrix} W^\alpha_a(x^b)\underset{\sim}{\gamma}_\alpha & \phi^i_a(x^b)\,\underset{\sim}{t}_i \\ [\underset{\sim}{0}] & 0 \end{pmatrix},$$

where the functions $W^\alpha_a(x^b)$, $\phi^i_a(x^b)$ are the components of the Yang-Mills compensating potentials. It is also clear from (4) that the compensating potentials $W^\alpha_a(x^b)$ arise through the breaking of the homogeneity of the action of the group $SO(3)$ while $\phi^i_a(x^b)$ arise through the breaking of the homogeneity of the action of the group $T(3)$. Since $\underset{\sim}{\Psi} = [x^1, x^2, x^3, 1]^T$ for elasticity theory, it only involves a simple computation to see that *the minimal replacement for elasticity* is

(5) $\partial_a x^i \longmapsto B^i_a = \partial_a x^i + \phi^i_a + W^\alpha_a \gamma_{\alpha j}^{\ i} x^j$.

The Lagrangian of elasticity theory for the inhomogeneous action of the gauge group G thus becomes

(6) $L_o = \frac{1}{2} \rho_o B^i_4 \delta_{ij} B^j_4 - U(E_{AB})$, $E_{AB} = B^i_A \delta_{ij} B^j_B - \delta_{AB}$.

The theory described by the Lagrangian given by (6) is clearly no longer elasticity theory. The question thus arises as just what phenomena are described by (6). If we introduce the 1-forms B^i by $B^i = B^i_a dX^a$, then $B^i = dx^i + \phi^i + W^\alpha \gamma_{\alpha j}^{\ i} x^j$ with $\phi^i = \phi^i_a dX^a$, $W^\alpha = W^\alpha_a dX^a$, and $dB^i = d\phi^i + (dW^\alpha) \gamma_{\alpha j}^{\ i} x^j - W^\alpha \wedge \gamma_{\alpha j}^{\ i} dx^j$. The 1-forms B^i are thus closed only if both ϕ^i and W^α are closed, in contrast with the displacement gradients, dx^i , of elasticity theory. However, the exterior derivatives of the B^i are not covariant with respect to the action of the gauge group G . If we introduce the connection 1-forms $\Gamma^i_j := W^\alpha \gamma_{\alpha j}^{\ i}$ of the subgroup $SO(3)$ of G , the quantities

(7) $\mathcal{D}^i := dB^i + \Gamma^i_j \wedge B^j = \Theta^i_j x^j + d\phi^i + \Gamma^i_j \wedge \phi^j$, $\Theta^i_j = d\Gamma^i_j + \Gamma^i_k \wedge \Gamma^k_j$, are covariant under the action of G . Now, the equations $\mathcal{D}^i = dB^i + K^i$, $d\mathcal{D}^i = \Omega^i$ have been shown (EDELEN [5]) to be the equations of defect dynamics in 4-dimensional form where B^i = distortion-velocity 1-forms, K^i = bend-twist-spin 2-forms, $\mathcal{D}^i$ = dislocation density and current 2-forms, and Ω^i = disclination current and density 3-forms. A direct comparison with (7) then gives $K^i = \Gamma^i_j \wedge B^j$ and we may view the Lagrangian given by (6) as characteristic of the "elastic" properties of materials with defects whose distortion-velocity 1-forms are B^i .

This view has some immediate and useful consequences. First, dislocations can be thought of as arising from the breaking of the homogeneity of the action of the translation group $T(3)$, as reflected in the occurrence of the 1-forms ϕ^i , while disclinations may be considered to arise as a consequence of the breaking of the homogeneity of the action of the rotation group $SO(3)$ and are reflected in the occurrence of the 1-forms W^α . In addition, since the components of the Piola-Kirchhoff stress tensor are defined in elasticity theory by $\sigma^A_i = \partial L_o/\partial(\partial_A x^i)$, while (5) shows that $\partial B^i_A/\partial(\partial_C x^j) = \delta^i_j \delta^C_A$, the minimal replacement argument yields

(8) $\sigma^A_i = \partial L_o/\partial B^i_A$.

The stress thus becomes a function of the distortion; a result now standard in defect dynamics but previously obtained on a somewhat *ad hoc* basis. A similar argument for momentum $(p_i = \partial L_o/\partial(\partial_4 x^i))$ gives $p_i = \rho_o \delta_{ij} B^j_4 = \rho_o \delta_{ij}(\partial_4 x^i + \phi^i_4 + W^\alpha_4 \gamma_{\alpha j}^{\ i} x^j)$ so that defect dynamics requires the replacement of the Newtonian velocity, $\partial_4 x^i$, by B^i_4 . This result is also standard in defect dynamics, although its previous justification appears based primarily on convenience. With $\phi^i = 0$, (7) shows that $\mathcal{D}^i = \Theta^i_j x^j$, and hence the dislocation currents and densities do not vanish. It is therefore impossible to have a state of pure disclination. Whenever disclinations are present ($W^{\alpha} \neq 0$), dislocations are there of necessity. These induced dislocations may be viewed as rotational dislocations.

The second half of the Yang-Mills construct consists of the *minimal coupling* argument. It is clear from the Lagrangian L_o given by (6) that rendering the associated action stationary with respect to the field variables ϕ^i_a and W^{α}_a will lead to the conditions $\phi^i_a = 0$, $W^{\alpha}_a = 0$ since L_o is only algebraic in these variables. By analogy with electrodynamics, YANG and MILLS found that they could construct a Lagrangian that depended on the first derivatives of the compensating potentials solely in terms of a single quadratic form in the components of the curvature associated with the connection 1-forms $\underset{\sim}{\Gamma}_a$ of the gauge group G. In their case, the group was the semisimple group $SU(2)$ so that there was only one nontrivial homogeneous quadratic form in the curvature that was also gauge invariant. In our case, a straightforward calculation shows that there are two: one characteristic of the dislocations and one characteristic of the disclinations. The minimal coupling construct then replaces the Lagrangian L_o by the final Lagrangian

$$
\begin{aligned}
& L = L_o - s_1 L_\phi - s_2 L_W , \\
& L_\phi = \frac{1}{2}\,\delta_{ij} D^i_{ab} k^{am} k^{bn} D^j_{mn} , \quad k^{ab} = \mathrm{diag}(-1,-1,-1,\tfrac{1}{y}) , \\
& L_W = \frac{1}{2}\, C_{\alpha\beta} F^{\alpha}_{ab} g^{am} g^{bn} F^{\beta}_{mn} , \quad g^{ab} = \mathrm{diag}(-1,-1,-1,\tfrac{1}{\zeta}) , \\
(9)\qquad & D^i_{ab} = \partial_a \phi^i_b - \partial_b \phi^i_a + \gamma_{\alpha j}{}^{i}(W^{\alpha}_a \phi^j_b - W^{\alpha}_b \phi^j_a) , \\
& F^{\alpha}_{ab} = \partial_a W^{\alpha}_b - \partial_b W^{\alpha}_a + C^{\alpha}_{\beta\gamma} W^{\beta}_a W^{\gamma}_b , \\
& \underset{\sim}{\gamma}_\alpha \underset{\sim}{\gamma}_\beta - \underset{\sim}{\gamma}_\beta \underset{\sim}{\gamma}_\alpha = C^{\rho}_{\alpha\beta}\, \underset{\sim}{\gamma}_\rho , \quad C_{\alpha\beta} = C^{\rho}_{\alpha\nu} C^{\nu}_{\beta\rho} ,
\end{aligned}
$$

where s_1 and s_2 are coupling constants. The theory thus involves only four free parameters (s_1, s_2, y, ζ) - it is truly a minimal coupling theory.

It is now a straightforward matter to compute the governing field equations; simply vary the associated action functional with respect to the field variables $(x^i, \phi^i_a, W^{\alpha}_a)$. Variation with respect to x^i gives the "balance of linear momentum" equations

$$(10)\qquad \partial_4 p_i - \partial_A \sigma^A_i = \gamma_{\alpha i}{}^{j}(W^{\alpha}_4 p_j - W^{\alpha}_A \sigma^A_j) .$$

Variation with respect to ϕ^i_a yields the "balance of dislocation" equations

$$(11)\qquad \partial_A R^{AB}_j + \partial_4 R^{4B}_j - \gamma_{\alpha j}{}^{i}(W^{\alpha}_A R^{AB}_i + W^{\alpha}_4 R^{4B}_i) = -\frac{1}{2}\,\sigma^B_j ,$$

(11) $$\partial_A R_j^{A4} - \gamma_{\alpha j}^{i} W_A^{\alpha} R_i^{A4} = \frac{1}{2} p_j \ , \quad R_i^{ab} = \partial L/\partial D_{ab}^{i} \ ,$$

while variation with respect to W_a^{α} gives the "balance of disclination" equations

(12)
$$\partial_A G_\eta^{AB} + \partial_4 G_\eta^{4B} + C_{\eta\alpha} C_{\varepsilon\gamma}^{\alpha} C^{\gamma\beta} (W_A^{\varepsilon} G_\beta^{AB} + W_4^{\varepsilon} G_\beta^{4B}) = \frac{1}{2} J_\eta^B \ ,$$
$$\partial_A G_\eta^{A4} + C_{\eta\varepsilon}^{\beta} W_A^{\varepsilon} G_\beta^{A4} = \frac{1}{2} J_\eta^4 \ , \quad G_\eta^{ab} = \partial L/\partial F_{ab}^{\eta} \ ,$$
$$J_\alpha^A = \gamma_{\alpha j}^{i} \{2(R_i^{AB} \phi_B^j + R_i^{A4} \phi_4^j) - \sigma_i^A x^j\} \ ,$$
$$J_\alpha^4 = \gamma_{\alpha j}^{i} \{2R_i^{4B} \phi_B^j + p_i x^j\} \ .$$

When the systems of equations (11) and (12) are rewritten in terms of the exterior differential calculus, they are easily seen to entail certain integrability conditions. The integrability conditions that obtain from the system (11) are not of any great significance. The integrability conditions that obtain from the system (12) are, however, another matter. When they are written out in detail, they may be shown to be equivalent to the conditions

(13) $$\sigma_i^A B_A^j = \sigma_j^A B_A^i \ .$$

Thus, the integrability conditions for the balance of disclination equations are just the statement of balance of moment of momentum under the minimal replacement $\partial_a x^i \longmapsto B_a^i$. It is of interest to note that previous theories of defects have ignored the question of balance of moment of momentum.

Because the theory derives from a variational principle, it has a well defined momentum-energy tensor, T_b^a. The important thing here is that the form of the Lagrangian, $L = L_o - s_1 L_\phi - s_2 L_w$, leads naturally to the additive decomposition

(14) $$T_b^a = T_{ob}^{\ a} - T_{\phi b}^{\ a} - T_{wb}^{\ a}$$

with
$$T_{ob}^{\ a} = \partial_b x^i \, \partial L_o/\partial(\partial_a x^i) - \delta_b^a L_o$$
$$= p_i \partial_4 x^i - \sigma_i^A \partial_A x^i - \delta_b^a L_o \ ,$$

(15) $$T_{\phi b}^{\ a} = s_1 \partial_b \phi_e^i \, \partial L_\phi/\partial(\partial_a \phi_e^i) - s_1 \delta_b^a L_\phi$$
$$= -2R_i^{ae} \, \partial_b \phi_e^i - s_1 \delta_b^a L_\phi \ ,$$
$$T_{wb}^{\ a} = s_2 \partial_b W_e^{\alpha} \, \partial L_w/\partial(\partial_a W_e^{\alpha}) - s_2 \delta_b^a L_w$$
$$= -2G_\alpha^{ae} \, \partial_b W_e^{\alpha} - s_2 \delta_b^a L_w \ .$$

Since any solution of the field equations (10)-(12) gives $\partial_a T_b^a = 0$, (14) leads to the "balance of forces"

(16) $$F_{ob} = F_{\phi b} + F_{wb} \ ,$$

$$F_{ob} = \partial_a T_{ob}^{\ a} , \quad F_{\phi b} = \partial_a T_{\phi b}^{\ a} , \quad F_{wb} = \partial_a T_{wb}^{\ a} .$$

We may thus view $\{F_{oa}\}$ as the "elastic excess" forces, $\{F_{\phi a}\}$ as the "dislocation forces" and $\{F_{wa}\}$ as the "disclination forces". When the field equations (10)-(12) are used to simplify the resulting expressions, we obtain

$$(17) \qquad F_{oa} = \sigma_i^B \partial_a \phi_B^i - p_i \partial_a \phi_4^i + \gamma_{\alpha j}^{\ i} x^j (\sigma_i^B \partial_a W_B^\alpha - p_i \partial_a W_4^\alpha) ,$$

$$(18) \qquad F_{\phi a} = \sigma_i^B \partial_a \phi_B^i - p_i \partial_a \phi_4^i + \gamma_{\alpha j}^{\ i} R_i^{bc} (\phi_c^j \partial_a W_b^\alpha - \phi_b^j \partial_a W_c^\alpha) ,$$

$$(19) \qquad F_{wa} = -J_\alpha^b \partial_a W_b^\alpha + C_{\beta\gamma}^\eta [2C_{\alpha\eta} C^{\gamma\sigma} G_\sigma^{bc} W_b^\beta \partial_a W_c^\alpha + G_\eta^{bc} \partial_a (W_b^\beta W_c^\gamma)] .$$

It is of interest to note that the first term, $\sigma_i^B \partial_a \phi_B^i$, in (18) reproduces the Peach-Koehler force [6], $\varepsilon_{ABC} b^i \sigma_i^C n^B$, for a disclination free distribution of dislocation with Burgers vector $\{b^i\}$ and unit tangent vector $\{n^A\}$ (i.e. $\partial_B \phi_C^i = b^i n^A \varepsilon_{ABC}$, $\phi_4^i = 0$). The various terms on the right-hand sides of equations (17)-(19) provide a full description of the various forces that act on the material, the dislocations, and the disclinations.

There are no approximations involved in the field equations (10) through (12) and they are indeed a formidable system of nonlinear partial differential equations. For purposes of simplicity, let us first eliminate the possibility of the material supporting disclinations. This is easily accomplished by simply putting $W_a^\alpha = 0$. Next, we make the standard linearization whereby linear elasticity theory obtains from nonlinear elasticity: $x^i = \delta_A^i X^A + u^i(X^a)$, $E_{AB} \doteq \delta_{jA}(\partial_B u^j + \phi_B^j) + \delta_{jB}(\partial_A u^j + \phi_A^j)$, and $\sigma_i^A \doteq \lambda \delta_i^A \delta_j^R(\partial_R u^j + \phi_R^j) + \mu(\delta_j^A \delta_i^R + \delta_{ij} \delta^{AR})(\partial_R u^j + \phi_R^j)$, $p_i = \rho_o \delta_{ij}(\partial_4 u^j + \phi_4^j)$. Finally, imposing the gauge conditions $\partial^A \phi_A^i = \frac{1}{y} \partial_4 \phi_4^i$ with $\partial^A := \delta^{AB} \partial_B$ and defining the parameters κ and c by $\kappa^2 = \mu/2s_1$, $c^2 = \mu/\rho_o$, the resulting field equations of the theory are

$$(20) \qquad (\nabla^2 - \frac{1}{y} \partial_4 \partial_4 - \kappa^2)\phi_E^m - \kappa^2(\delta_{Ej}\delta^{Rm} + \frac{\lambda}{\mu}\delta_E^m \delta_j^R)\phi_R^j$$

$$= \kappa^2(\frac{\lambda}{\mu}\delta_E^m \partial_j u^j + \delta_{Ej}\delta^{mk}\partial_k u^j + \delta_j^m \delta_E^R \partial_R u^j) ,$$

$$(21) \qquad (\nabla^2 - \frac{1}{y} \partial_4 \partial_4 - \kappa^2 \frac{y}{c^2})\phi_4^i = \kappa^2 \frac{y}{c^2} \partial_4 u^i ,$$

$$(22) \qquad (\nabla^2 - \frac{1}{c^2} \partial_4 \partial_4) u^m + (\frac{\lambda}{\mu} + 1)\partial^m \partial_j u^j$$

$$= \left[(\frac{y}{c^2} - 1)\delta_j^m \delta^{RA} - \frac{\lambda}{\mu}\delta^{mA}\delta_j^R - \delta_j^A \delta^{mR}\right]\partial_A \phi_R^j .$$

The system (20) is a coupled system of Klein-Gordon equations. If we write $\underset{\sim}{\Phi} := ((\phi_E^m))$, and let $\underset{\sim}{g}$ denote the matrix whose entries occur within the parentheses on the right-hand sides of (20), then (20) is satisfied by $\underset{\sim}{\Phi} = \underset{\sim}{\hat{\Phi}} + \frac{1}{3} \eta \underset{\sim}{I}$,

$\underset{\sim}{\Phi}^T = \underset{\sim}{\Phi}$, where

(23) $$(\nabla^2 - \frac{1}{y}\partial_4\partial_4 - 2\kappa^2)\hat{\underset{\sim}{\Phi}} = \kappa^2(\underset{\sim}{\sigma} - \frac{1}{3}(\mathrm{tr}\underset{\sim}{\sigma})\underset{\sim}{I}) ,$$

$$(\nabla^2 - \frac{1}{y}\partial_4\partial_4 - 2\kappa^2(1 + \frac{3\lambda}{2\mu}))\eta = \kappa^2(\mathrm{tr}\underset{\sim}{\sigma}) .$$

Hence, the "trace free" part of $\underset{\sim}{\Phi}$ is driven only by the "stress deviatoric" $\underset{\sim}{\sigma} - \frac{1}{3}(\mathrm{tr}\underset{\sim}{\sigma})\underset{\sim}{I}$. Finally, we observe that the scalar Klein-Gordon operators on the left-hand sides of (23) can be inverted by means of Green's functions with zero Cauchy data. When these results are substituted into the right-hand sides of equations (22), we obtain a system of integro-differential equations for the fields u^i . In the pure shear modes of response, the corresponding dispersion relation is given by

(24) $$(\frac{\omega^2}{c^2} - k^2)(\frac{\omega^2}{y} - k^2 - 2\kappa^2) + \kappa^2(\frac{y}{c^2} - 2)k^2 = 0$$

from which one easily identifies an acoustical branch and an "optical" branch. The dispersion relation for the dilation mode can also be obtained, but it is of 6th degree in ω and significantly more complicated than (24).

The four free parameters, y, ζ, s_1, s_2 of the theory play an important role in further understanding of defect phenomena. The constants y and ζ are dynamical in nature; they characterize the propagation of fronts of dislocation and disclination. On the other hand, s_1 and s_2 may be thought of as characteristic of the energy required in order to create "unit" dislocation and disclination, respectively, since $L = L_0 - s_1 L_\phi - s_2 L_W$.

The field equations (10)-(12) are a system of coupled nonlinear partial differential equations that appear very difficult to solve, yet they do reveal a lot of information. For instance, it follows from (11) that stresses drive dislocations (i.e., whenever there are nonzero stresses, the dislocation fields are nontrivial). Likewise (12) shows that both stress and dislocation drive disclinations. A fuller understanding of the significances of these driving processes demands some form of simplification. A natural scaling of the gauge group can be used to introduce an expansion parameter that results in approximation procedures that are, in a sense, uniform. If s_2 is taken as large compared to s_1, and s_1 large relative to the Lamé constant λ , definite and interesting results immerge. Classical elasticity theory is recovered in the lowest order approximation, while dislocations enter in the equations of next higher order. On the other hand, disclination fields are not picked up until third or higher order. Perhaps it is this fact that underlies the success in describing many defect phenomena solely in terms of dislocations. The fact that s_2 is large compared to s_1 and λ for this result to hold, may also be viewed as a statement that relatively large energies are necessary before disclinations become significant.

The results just discussed obtain under the assumption that the reference state is defect free, in which case all solutions of the homogeneous equations for the ϕ and W fields are put to zero. There are, however, known solutions of the homogeneous W equations [2] since (11) reduce to the Yang-Mills free field equations in this case. The nature of these solutions (magnostatic dipoles) in the context of the other field equations leads to structures that are highly suggestive of Frank-Reed sources. A definitive proof of such a conjecture is, of course, well in the future.

The results reported here constitute, in point of fact, an abstract of the full theory that is developed in [7].

References

[1] D. Rogula, "Large deformations of crystals, homotopy and defects" in *Trends in Applications of Pure Mathematics to Mechanics* (Pitman, London, 1976).

[2] C. N. Yang, "Gauge fields", in *Proc. Sixth Hawaii Topical Conf. Particle Physics* (Univ. of Hawaii, Honolulu, 1975).

[3] W. Dreschler and M. E. Mayer, *Fiber Bundle Techniques in Gauge Theories* (Lecture Notes in Physics No. 67, Springer, Berlin, 1977).

[4] A. Actor, *Rev. Mod. Phys.* *51* (1979), 461-525.

[5] D. G. B. Edelen, *Int. J. Engng. Sci.* *18* (1980), 1095-1116.

[6] M. Peach and J. S. Koehler, *Phys. Rev.* *80* (1950), 436-439.

[7] Aida Kadić, "A Complete Field Theory for Continua with Dislocations and Disclinations" (Thesis, Lehigh University, Bethlehem, PA, 1981).

Continuum Models of Discrete Systems 4
eds. O. Brulin and R.K.T. Hsieh

BOUNDARY VALUE PROBLEMS FOR HETEROGENEOUS MEDIA APPLICATION TO MECHANICAL AND ELECTRICAL ENGINEERING

H. LANCHON - A. MIRGAUX : L.E.M.T.A., 2, rue de la Citadelle, B.P. 850,
54011 NANCY CEDEX - FRANCE.
D. CIORANESCU - J. SAINT JEAN PAULIN : Analyse Numérique - Université Paris VI
4, Place Jussieu, 75230 PARIS CEDEX 05.
J.F. BOURGAT : I.N.R.I.A., Domaine de Voluceau - Rocquencourt - B.P. 105
78153 LE CHESNAY CEDEX. France.

This contribution has to be understood like an illustration of the general lectures given by J.L. LIONS about "Homogenization" and suppose a close collaboration between Physicits, Mathematiciens and Numéricians. Four engineering problems of different kinds are first presented ; the difficulties for computing the solutions because the big amount of heterogeneities are then emphasized and finally, the results obtained by the "homogenisation method" are explicity given. (J. Saint Jean Paulin is the only commun contributor for the 4 problems).

I. BRIEF DESCRIPTION OF THE PROBLEMS

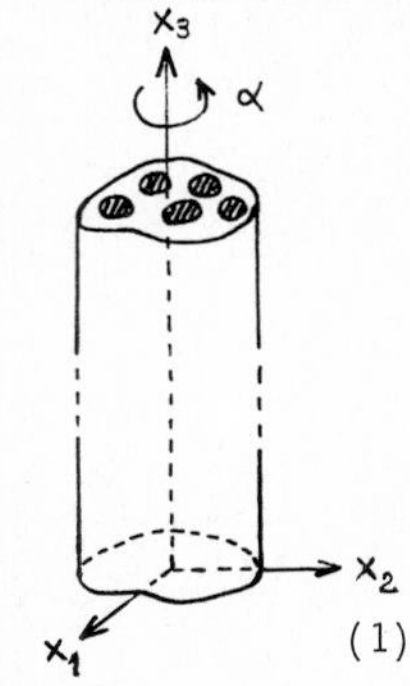

Figure 1

1. Elastic torsion of cylindrical "fibres reenforced bars" [1][2] [3]

The stress function θ is here the main unknown fonction and has to verified the following equation and boundary condition :

$$(P_1)\quad \begin{cases} \dfrac{\partial}{\partial x_i}\left[\dfrac{1}{\mu}\dfrac{\partial\theta}{\partial x_i}\right] + 2\alpha = 0 & \text{on } \Omega \\ \theta = 0 & \text{on } \Gamma \end{cases}$$

With summation on i = 1,2 and following notations (Fig. 1) :
Ω, *cross section of the cylindar* of boundary Γ
α, *twist angle* (given constant number)

(1) μ *élastic shear modulus* = $\begin{cases} \mu_f \text{ on the fibres (i.e. on the cross section } \Omega_l,\ l = 1,2,..,q) \\ \mu_m \text{ on the matrix (i.e. on the multiconnected cross section } \Omega_m) \end{cases}$
(Figure 2)

2. Elastic torsion of cylindrical bars with cavities [2][3][4]

With the same notations, we have now : $\mu f=0$ and then $1/\mu$ *is no longer a bounded function* ; for this reason, the formulation (P_1) has to be changed in :

$$(P_2)\quad \begin{cases} \Delta\theta + 2\mu_m\alpha = 0 & \text{on } \Omega_m \\ \theta = 0 \text{ on exterior part of } \Gamma_m \text{ (boundary of } \Omega_m) \\ \theta = C_l \text{ (arbitrary constant number) on } \Gamma_l & \text{for } l = 1,2,..,q \\ \displaystyle\int_{\Gamma_l} \frac{\partial\theta}{\partial\nu}\, ds = 2\,\mu_m\alpha\, A_l & \text{for } l = 1,2,..,q \end{cases}$$

(Figure 3)

Where Γ_l is the boundary of the cross section Ω_l of the lth cavity ; A_l is the area of Ω_l and $\vec{\nu}$ is the unit normal vector along Γ_l, pointed outward from Ω_m (the last two conditions express that the stress tensor, defined by the two derivatives $\frac{\partial\theta}{\partial x_i}$, is zero on the cavities, and univocally determined on Ω_m).

3. Transverse conductivity in superconducting multifilamentary composites [3][5][6]

We consider a C.M.S. (cylindrical multifinamentary superconductor) with q cylindrical superconducting fibers and a normal conducting matrix ; we submit it to a weak exterior parallel magnetic field :

$$\vec{H}_{ext}(t) = H(t)\,\vec{k}_3$$

The fact that H is weak, allows us to admit the approximation that the magnetic induction is not penetrating in the superconducting fibres (so $\vec{B} = 0$ on Γ_1 for $1 = 1,2,..,q$). Then, if the C.M.S. is very long the unknown field in the matrix

$$\vec{B} = \phi\,\vec{k}_3$$

has to verify the following boundary problem :

$$(P_3)\ \text{(Figure 3)} \begin{cases} \dfrac{\partial\phi}{\partial t} - \dfrac{1}{\sigma_m \mu_o}\Delta\phi = 0 & \text{on } \Omega_m \times\ ]0,T[\\ \phi = \mu_o H & \text{on the exterior part of } \Gamma_m \times]0,T[\\ \dfrac{\partial\phi}{\partial\nu} = 0 & \text{on } \Gamma_1 \times]0,T[\text{ for } 1 = 1,2,..\ n \\ \phi(.,0) = 0 & \text{on } \Omega_m \text{ (initial condition)} \end{cases}$$

σ_m is the conductivity of the matrix and μ_o the permitivity of the vaccuum.

4. <u>Heating behaviour of an electrical composite cable in a short-circuit</u> [3][7]

The cable is composed of a big number of electrical conducting fibres imbedded in a thermically insulating matrix ; however, in fact, the matrix is not a perfect insulating media and, during a short-circuit, the cable is heated by the effectif Joule dissipation in the matrix. If δ represente the "weak" thermal conductivity of the matrix, the temperature u_δ in the cable is given by the following undimensional transmission problem :

$$(P_4) \begin{cases} \dfrac{\partial u_\delta}{\partial t} - \dfrac{\partial}{x_i}[a(\delta)\dfrac{\partial u_\delta}{\partial x_i}] = f(u_\delta) & \text{in } \Omega \times\ ]0,T[\\ u_\delta = 0 & \text{on } \Gamma \times]0,T[\\ u_\delta(.,0) = 0 \text{ on } \Omega \text{ (initial condition)} \end{cases}$$

with (cf. Fig. 2)

$$(2) \qquad a(\delta) = \begin{cases} 1 & \text{on } \Omega_1 \\ \delta & \text{on } \Omega_m \end{cases} \qquad 1 = 1,2,..,q$$

$$(3) \qquad f(u) = \begin{cases} 1 + Ku & \text{on } \Omega_1 \\ 0 & \text{on } \Omega_m \end{cases} \qquad 1 = 1,2,..,q$$

K is here a given constant number characterizing the variation of the resistivity in the fibres with the temperature.

II. GENERAL AND PARTICULAR REMARKS ABOUT THE ABOVE PROBLEMS

Some theorems of existence and uniqueness have been given for the solutions of the four mentionned problems but the crucial question is of being able to compute actually the different solutions when A/q is very small [where A is the area of the cross section Ω and q the number of fibres or cavities].

More generally, it is important to reply the question : How to compute the solution of a boundary value problem for a material media in which the properties are varying very rapidely in the space ?

This question is relative to various physical situations like, for instance, composite materials, polycristals, porous media, suspensions in fluid [solid particules in liquids, smokes, clouds, media with bubles...]

<u>The "Homogenization" point of view is the following</u> :

If ε is a small parameter characterizing the microscale of the given media : [ε can be ratio A/q already mentionned or can represent the period in the case of a periodical distribution of heterogeneities, or also be the diameter of a representative volume element for a random-media],then all the preceeding problems are of the following global type.

(4) $$A_\varepsilon u_\varepsilon = f$$

where A_ε is a generalized operator depending on the scale ε and u_ε the corresponding solution.

In the exemples (P_1) *and* (P_4) : the operators A_ε depends on ε only by the coefficients and right hands of the equations.

In fact μ, $a(\delta)$ and $f(u)$ given respectively by (1)(2) and (3) depend on the distribution of fibers in Ω and then on ε ; these quantities are piecewise constant (or piecewise continuous in the case of $f(u)$) so the problems (P_1) and (P_4) has to be understood in the sense of distributions but on a fixed domain Ω (fig.2)

Moreover, the solution of (P_4) depends on a second small parameter δ, independant of the first one ε, and that,will lead us to an additional problem of singular perturbations.

In the exemples (P_2) *and* (P_3) the coefficients are fixed $[(\mu_m)^{-1}$ in (P_2) and $(\sigma_m\mu_o)^{-1}$ in (P_3)] ; but A_ε is depending on ε because the two problems have to be solved on the multiconnected domain Ω_m which is strongly depending on ε (fig.3)

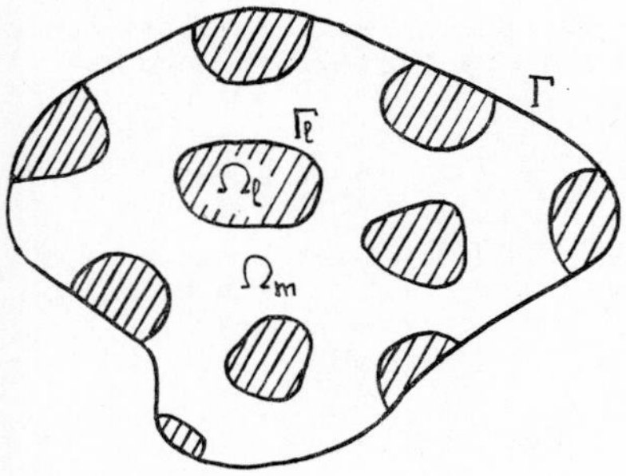

Figure 2

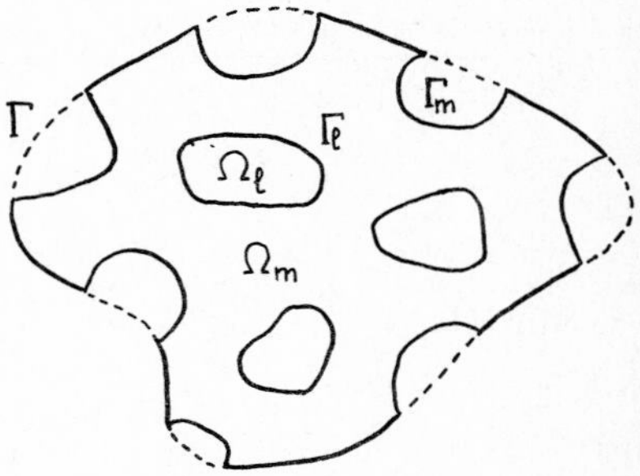

Figure 3

III. ROUGH IDEA ABOUT THE "HOMOGENIZATION METHODS"

Considering that the smaller is the scale ε, the larger are the difficulties to compute the corresponding solution u_ε of (4), a natural idea is to wander : *What is the limite behaviour of* u_ε *when* ε *goes to* 0? and then, in the case of an actual convergence of u_ε toward a definite element u_o : *Could* u_o *be interpreted as a good approximation of the solution* u_ε ?

Remark : There exist several ways for ε to go to 0. For exemples :

i) We way consider a family of composite or Hollowed media with finer and finer structure (like in a similitude in mathematics) keeping constant, in particular the relatives proportions of the different components.

ii) We may also consider suspensions with bigger and bigger concentrations of solids particles.

General results and comments : [cf., for instance [8]...[13] and again [3] for others exemples, proofs and general references].

In a relatively important number of such problems, it has been obtained that :

(5) "$\lim u_\varepsilon \longrightarrow u_o$ when $\varepsilon \to 0$ in some way",

Where u_o is determined as the solution of a boundary value problem, called generally "associated homogenized problem', defined on the fixed domain Ω (union either of matrix and fibres or solution and particles...).

In the most part of studied situations, the distribution of heterogeneities is periodic ;then :

i) the constant coefficients of the "homogenized problem" can be interpreted as the "effective" or "global" physical coefficients characterizing an "equivalent homogeneous media" ; their computation has to be made on the simple réference cell.

ii) once that these "homogenized coefficients" are known the numerical computation of the solution u_o does not implies any more problems.

iii) The maximum error $|u_\varepsilon - u_o|$ can be evaluated, giving an idea of the value of the approximation u_o ; but also correctors of different orders in ε can be computed in order to improve the approximation ; for instance u_1 for 1^{rst} order such that :

(6) $$u_\varepsilon(\underline{x}) = u_o(x) + \varepsilon\, u_1(x,\varepsilon) + o(\varepsilon^2).$$

IV. BRIEF DESCRIPTIONS OF RESULTS FOR THE ABOVE PARTICULAR PROBLEMS

1. Elastic torsion of cylindrical "fibres reenforced bars"

We have the following data :

Ω, square cross section with sides of unit length

Ω_1, square cross section of fibers with sides such that the total relative volume of fibers $\frac{V_f}{V}$ is either 0,11 or 0,7.

Matrix epoxy ($\mu_m = 2.2.10^5$ psi) ; Fibres of glass ($\mu_f = 4.10^6$ psi)

Taking the period ε successively equal to 1/2, 1/4 and 1/8. We have computed for each ε : the exact solution θ_ε, the homogenized one θ_o and the corrected one, up to the first order : $\tilde{\theta}_\varepsilon = \theta_o + \varepsilon\theta_1$.

The conclusions are :

i) the smaller is ε, the closer are θ_o, θ_ε and $\tilde{\theta}_\varepsilon$ (explicite curves given in [1] will be showed in the oral lecture).

ii) For $\varepsilon < \frac{1}{8}$, it is no longer possible to compute directly the exact solution θ_ε but then θ_o and $\tilde{\theta}_\varepsilon$ are already very good approximations.

iii) In the case $V_f/V = 0,7$ the homogenized shear modulus $\mu_{23} \# \mu_{13} = 0,965.10^6$ psi is reasonably closed of the experimental one given in [14] i.e. $0,95.10^6$

2. Elastic torsion of cylindrical bars with cavities

The technical difficulties have been solved in [4] and are developped in the above general lectures of J.L. Lions. Moreover, it has been proved in [2] that the solution of $[P_2]$ can be obtained from that of $[P_1]$ in passing to the limite when $\mu_f \to 0$.

3. Transverse conductivity in superconducting multifilamentary composites

The following curves (fig. 4) extracted from [5] show an explicite comparison between numerical computation given by the homogenization method, experimental result given in [6] and four classical pseudo empirical formula given by some physicists. The good agreement with Carr's Formula [15] and experiment results is quite encouraging.

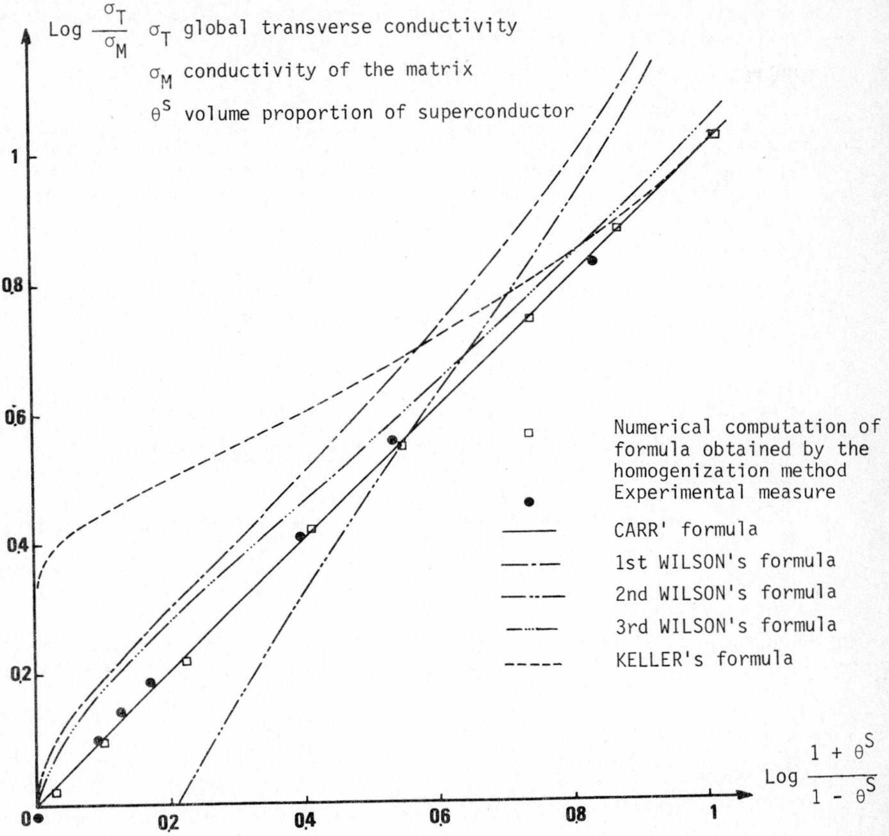

Figure 4 : Graphical comparison of the different results for transverse conductivity of a S.M.F..

4. Heating behavior of an electrical composite cable in a short circuit :
It has been proved in [3] that the homogenized solution u_o of (P_4) depends essentially of the relative smallness of the parameters ε and δ. If one introduce $s = \text{Log}\,\frac{\delta}{\varepsilon}$ and θ the relative volume of matrix then :

i) for $s > 2$ the homogenized solution is : $u_o(\underline{x},t) = \frac{1-\theta}{K}(e^{Kt} - 1)$

ii) for $0 < s < \frac{1}{3}$, the homogenized solution is : $u_o(\underline{x},t) = \frac{1}{K}[e^{(1-\theta)Kt} - 1]$

The following physical interpretation has been given for this surprizing result by A. Bossavit (Ingénieur who has submitted this problem before us [7]). The formulation (P_4) is an adimentional representation of the physical process ; the situations i) and ii) represent 2 differents scales of time ; i) gives the short term, when the matrix which is in principle a thermically insulator, had not yet the time to warm ; and ii) gives the long term, when the matrix which is in fact a bad insulator, had time to warm.

V. FINAL COMMENTS

The exemples given above show how the "Homogenization techniques" can be efficient in periodical situation. For random media, several good theories are now available about which a try of synthesis has been made in [16]. It is allowed to wonder if some basic ideas of the " Homogenization methods" could be a useful complement to these theories ; a first step on the way of such a unification has been done in [11].

VI. REFERENCES

[1] BOURGAT J.F. - LANCHON H. (1976) "*Application of the homogenization method to composite materials with periodic structure*". Rapport IRIA n° 208, Rocquencourt - France.

[2] CIORANESCU D. - SAINT JEAN PAULIN J. - LANCHON H. (1978) "*Elastic plastic torsion of heterogeneous cylindrical bars*", J. Inst. Maths. Applics, 24, pp. 353-378.

[3] SAINT JEAN PAULIN J. (1981) "*Etude de quelques problèmes de mécanique et d'électrotechnique liés aux méthodes d'homogénéisation*", Thèse de Doctorat d'Etat, Université Paris VI, France.

[4] CIORANESCU D. - SAINT JEAN PAULIN J. (1979) "*Homogenization in open sets with holes*", J. Math. Anal., Appl. 71, 2, pp. 530-607.

[5] MIRGAUX A. - SAINT JEAN PAULIN J. (1981) "*Asymptotic study of transverse conductivity in superconducting multifilamentary composites*", to appear in Int. J. Engng. Sci.

[6] DAVOUST M.E. - MAILFERT A. (1978) "*Conductivité électrique transverse dans les fils de matériaux composites multifilamentaires supraconducteurs*" Thèse de 3ème cycle, Paris VI.

[7] BOSSAVIT A. (1977, Bulletin de la direction des études et recherches, série C Mathématiques informatique n°1 - E.D.F. - France.

[8] BENSOUSSAN A. - LIONS J.L. - PAPANICOLAOU G. (1978), "*Asymptotic Analysis for periodic structures*", (North Holland).

[9] SANCHEZ PALENCIA E. (1980) "*Non homogeneous media and vibration theory*" Lecture notes in Physics 127, (Springer Verlag).

[10] DUVAUT G. (1976) "*Analyse fonctionnelle et mécanique des milieux continus. Application à l'étude des matériaux composites périodiques*", pp. 119-132, Congrès I.U.T.A.M., Delft - (North Holland Pub. Comp. Amsterdam).

[11] KRÖNER E. (1980) "*Effective elastic moduli of periodic and random media : A unification*", Mech. Res. Comm. 7, 5, pp. 323-327.

[12] CIORANESCU D. (1980) "*Calcul des variations sur des sous-espaces variables*", C.R. Acad. Sci. Paris t. 291, série A B n° 1 et 2.

[13] LEVY T. (1981) "*Lois de Darcy ou loi de Brinhman ?*" , C.R. Acad. Sc. Paris, (23 mars 1981).

[14] GARG S.K. - SVATBONAS V. - GURTMAN G.A. (1973) "*Analysis of structural Composite Materials*", p. 68, (Marcel Dekker Inc., New York).

[15] CARR W.J. (1975) "*Electromagnetic theory for filamentary superconductors*", Phys. Rev. B 11, (4), pp. 1547-1554.

[16] WILLIS J.R. (1980) "*Theoretical determination of the overall properties of composite Materials*", pp. 189-195, I.U.T.A.M. Congrès Toronto (North Holland)

Continuum Models of Discrete Systems 4
eds. O. Brulin and R.K.T. Hsieh

HOMOGENIZATION IN PERFORATED MATERIALS AND IN POROUS MEDIA

Jacques-Louis Lions

Collège de France
and
I.N.R.I.A.
Domaine de Voluceau
B.P. 105 - F-78150 Le Chesnay

We consider *asymptotic problems* for perforated materials or porous media, when there are a very large number of holes or obstacles arranged in a periodic manner. We consider both stationary and evolution cases. Using anstaz involving multi-scale variables, we obtain *homogenized equations* ; these equations can contain *non local terms* and can be easily expressed in terms of some "hidden" variables.

INTRODUCTION

We want to present here some asymptotic techniques which can be of some use in solving boundary value problems in perforated materials or in porous media when there are "many very small" holes (or obstacles) arranged in a periodic manner.

These techniques are based on some simple "ansatz" which are *multi-scale variables*- these multi-scale variables being here *space* variables.

Depending on the situations one is led to new problems for boundary layer terms, or to homogenized operators which can be of *non local* type ; this also leads to a new approach to the classical Darcy's law for flows in porous media.

The plan of the paper is as follows :

1. Perforated materials.
   - 1.1. Description of the domain Ω_ε.
   - 1.2. Model problem. Dirichlet boundary conditions.
   - 1.3. Model problem. Neumann boundary conditions.
2. Porous media (I). Stationary case.
   - 2.1. Setting of the problem.
   - 2.2. Homogenized system with two pressures.
   - 2.3. Homogenized system with one pressure. Darcy's law.
3. Porous media (II). Evolution case.
   - 3.1. Setting of the problem.
   - 3.2. Two pressures homogenized system.
4. Non local homogenized system in a perforated material.
   - 4.1. Setting of the problem.
   - 4.2. Homogenized system.
5. Various Remarks.

   Bibliography.

1. PERFORATED MATERIALS

1.1. Description of the domain Ω_ε.

Let Ω be a bounded[1] open set of $\mathbb{R}^n$ with smooth boundary Γ. We shall denote by Ω_ε the open set obtained from Ω by taking out a family of holes of "size" ε and arranged in a periodic manner. More precisely, we consider the unit cube $Y =]0,1[^n$ and let $\mathcal{O}$ be an open set in Y, with boundary S (cf. Fig.1).

We define $\chi(y) = \begin{cases} 1 & \text{in } Y\backslash\mathcal{O} \text{ and} \\ 0 & \text{in } \mathcal{O}. \end{cases}$

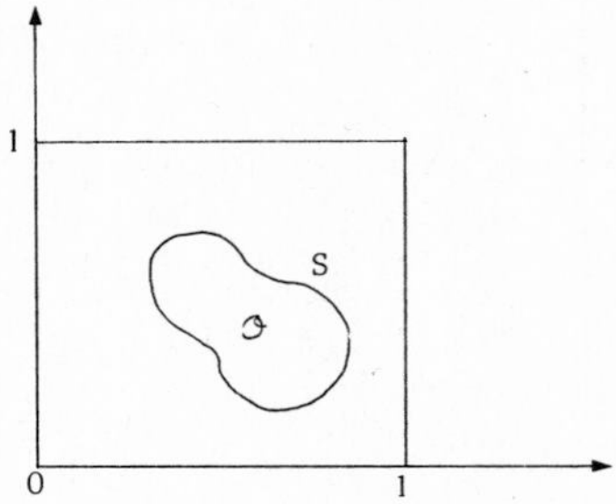

Figure 1.

We extend χ as a function on $\mathbb{R}^n$ with period 1 in all variables.

We then define :

$$\chi^2(x) = \chi\left(\frac{x}{\varepsilon}\right) \tag{1.1}$$

and

$$\Omega_\varepsilon = \{x \mid x \in \Omega \ , \ \chi^\varepsilon(x) = 1\}. \tag{1.2}$$

The boundary of Ω_ε consists now (in general) in two parts :

$$\delta\Omega_\varepsilon = \Gamma_\varepsilon \cup S_\varepsilon \ , \tag{1.3}$$

where Γ_ε denotes what remains of Γ when the holes are taken out (cf. Fig. 2) and where S_ε denotes the union of the boundaries of the holes (or the part of holes) which are taken out.

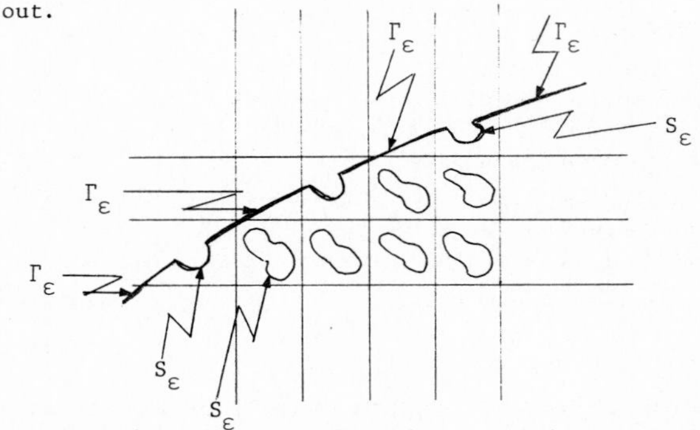

Figure 2.

We want now to study boundary value problems in Ω_ε and the asymptotic behaviour of their solution as $\varepsilon \to 0$.

In this section we study briefly two model problems chosen as simple as possible.

[1] This hypothesis is not indispensable.

1.2. Model problem. Dirichlet boundary conditions.

We consider the problem

(1.4) $$-\varepsilon^2 \Delta u_\varepsilon = f \quad \text{in} \quad \Omega_\varepsilon \,,$$

(1.5) $$u_\varepsilon = 0 \quad \text{on} \quad \partial\Omega_\varepsilon = \Gamma_\varepsilon \cup S_\varepsilon .$$

In (1.4) f denotes the *restriction* to Ω_ε of a function f defined in Ω. We shall suppose that f is smooth.

The *asymptotic expansion* is chosen by the following "Ansatz" :

(1.6) $$u_\varepsilon = u_o(x,y) + \varepsilon u_1(x,y) + \ldots$$

where y *is replaced by* x/ε and where

(1.7) $$\left| \begin{array}{ll} u_j(x,y) & \text{is defined for} \quad x \in \Omega \,, \quad y \in \mathcal{Y}, \\ u_j(x,y) & \text{is Y-periodic (i.e. admits period 1 in all y's variables)} \end{array} \right.$$

(1.8) $$u_j(x,y) = 0 \quad \text{if} \quad y \in S.$$

Remark 1.1.

So far no boundary condition is imposed for $x \in \Gamma = \partial\Omega$; we shall return to this (non trivial) point. If (1.6) is convergent, then

(1.9) $$u_\varepsilon = 0 \quad \text{for} \quad x \in S_\varepsilon$$

(since then $y = x/\varepsilon \in S$ and one has (1.8)). □

Remark 1.2.

The Ansatz (1.6) has been introduced in J.L. Lions [1]. It is a simple variant of the Ansatz systematically used in A. Bensoussan, J.L. Lions and G. Papanicolaou[1] □.

Formal computation.

We observe that $\Delta\phi(x,y)$ where $y = x/\varepsilon$ is computed by applying first the operator

(1.10) $$\varepsilon^{-2}\Delta_y + 2\varepsilon^{-1} \Delta_{xy} + \Delta_x \,,$$

where

(1.11) $$\Delta_{xy} = \frac{\partial^2}{\partial x_i \partial y_i} \quad \text{(we use the summation convention)}$$

and replacing in the result y by x/ε . Therefore $\varepsilon^2\Delta$ becomes

$$\Delta_y + 2\,\varepsilon\,\Delta_{xy} + \varepsilon^2\,\Delta_x$$

and by identification of the various powers of ε, (1.4) becomes

$$(1.12)\quad \left|\begin{array}{l} -\Delta_y u_0 = f \quad , \\ -\Delta_y u_1 - 2\Delta_{xy} u_o = 0 \ , \\ -\Delta_y u_2 - 2\Delta_{xy} u_1 - \Delta_x u_o = 0 \ , \\ \dots\dots\dots\dots\dots \end{array}\right.$$

with the u_j's subject to (1.7)(1.8). □

Solution of $(1.12)_1$.

In the equation $(1.12)_1$ x *plays the role of a parameter*. We define w_o by

$$(1.13)\quad \left|\begin{array}{l} -\Delta_y w_o(y) = 1 \quad \text{in} \quad \mathcal{Y} = Y\backslash \mathcal{O} \ , \\ w_o \ \text{is periodic (i.e. takes equal values together with its} \\ \quad \text{derivatives on opposite faces of } Y) \text{ and} \\ w_o = 0 \quad \text{on} \quad S. \end{array}\right.$$

Then

$$(1.14)\qquad u_o(x,y) = w_o(y)\ f(x).$$

Solution of $(1.12)_2$.

Equation $(1.12)_2$ becomes, using (1.14) :

$$(1.15)\qquad -\Delta_y u_1 = 2\,\frac{\partial w_o}{\partial y_i}(y)\,\frac{\partial f}{\partial x_i}(x) \ .$$

If we define w_i by

$$(1.16)\quad \left|\begin{array}{l} -\Delta_y w_i = 2\,\dfrac{\partial w_o}{\partial y_i} \quad \text{in } \mathcal{Y} \ , \\ w_i \ \ Y\text{-periodic}, \quad w_i = 0 \quad \text{on} \quad S \ , \end{array}\right.$$

then

$$(1.17)\qquad u_1(x,y) = w_i(y)\,\frac{\partial f}{\partial x_i}(x),$$

and we can proceed in this manner, assuming f to be smooth. □

Boundary conditions on Γ_ε.

Since x plays the role of a parameter in the above expansion, we have no flexibility left to impose boundary conditions on Γ_ε.

If we assume

(1.18) $$f \equiv 0 \quad \text{near } \Gamma$$

then all u_j's are identically zero near Γ (and on Γ_ε) *and* (1.5) *is satisfied.*

The asymptotic expansion is then convergent (cf. J.L. Lions [9] *and one has an error estimate in the Sobolev norm of* $H^1_o(\Omega_\varepsilon)$ *of order* ε^m *for the expansion* $u_o + \varepsilon u_1 + \ldots + \varepsilon^m u_m$.

If (1.18) *is not satisfied*, extra terms (*boundary layer terms*) have to be introduced.
This problem seems to be solved only in very particular cases. If one takes

(1.19) $$\Omega = \{x \mid x_n > 0\}$$

then $\Gamma_\varepsilon = \Gamma$ and one uses the following Ansatz (cf. J.L. Lions [14]) :

(1.20) $$u_\varepsilon = u_o + u_o^{BL} + \varepsilon(u_1 + u_1^{BL}) + \ldots$$

where the u_j's are defined as before and where the u_j^{BL}'s have the following properties :

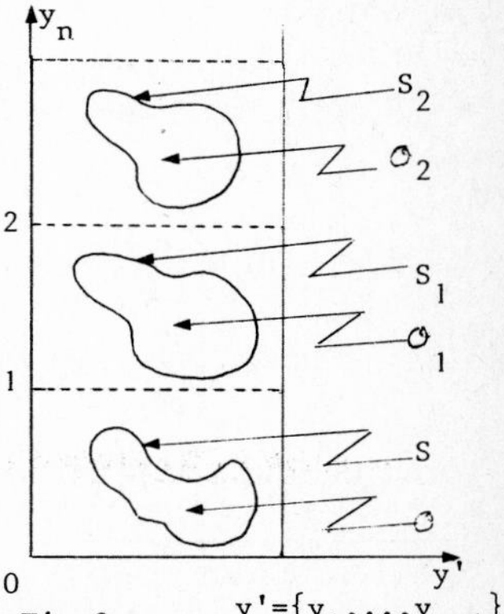

Fig.3.

we define $$\mathcal{G} = (]0,1[^{n-1} \times]0,+\infty[) \setminus (\mathcal{O} \cup \mathcal{O}_1 \cup \ldots)$$

(cf. Fig. 3) ; we suppose that

(1.21)
$u_j^{BL}(x,y)$ is defined for $x \in \Omega$, $y \in \mathcal{G}$,
$u_j^{BL}(x,y)$ is periodic in $y' = \{y_1, \ldots, y_{n-1}\}$,
$u_j^{BL}(x,y)$ decreases exponentially as $y_n \to +\infty$,

(1.22) $u_j^{BL}(x,y) = 0$ for $y \in \Sigma = S \cup S_1 \cup \ldots$ (cf. Fig. 3).

One chooses u_o^{BL} such that

(1.23) $\Delta_y u_o^{BL} = 0$,

u_o^{BL} subject to (1.21)(1.22) *and*

(1.24) $u_o^{BL} + u_o = 0$ if $x_n = 0$, $y_n = 0$;

and one can proceed in this manner. Cf. details in J.L. Lions [14].

1.3. Model problem. Neumann's boundary conditions.

We consider now the problem (compare to (1.4)(1.5))

(1.25) $-\Delta u_\varepsilon = f$ in Ω_ε,

(1.26) $u_\varepsilon = 0$ on Γ_ε (1) ,

(1.27) $\dfrac{\partial u_\varepsilon}{\partial \nu_\varepsilon} = 0$ on S_ε

where $\dfrac{\partial}{\partial \nu_\varepsilon}$ denotes the normal derivative to S_ε directed towards the exterior of Ω_ε.

Remark 1.3.

We have replaced $\varepsilon^2\Delta$ by Δ which of course amounts to changing u_ε into $\varepsilon^{-2}u_\varepsilon$. □

The Ansatz.

We try again the ansatz (1.6)(1.7), with conditions different than (1.8) on S , namely :

(1.28) $\dfrac{\partial u_o}{\partial \nu_{(y)}} = 0$, on S ,

(1.29) $\dfrac{\partial u_1}{\partial \nu_{(y)}} + \dfrac{\partial u_o}{\partial \nu_{(x)}} = 0$ on S , etc.. ,

where $\dfrac{\partial}{\partial \nu_{(y)}} = \nu_j(y)\dfrac{\partial}{\partial y_j}$, $\dfrac{\partial}{\partial \nu_{(x)}} = \nu_j(y)\dfrac{\partial}{\partial x_j}$, $\nu(y) = \{\nu_j(y)\} =$

(1) We could take as well Neumann's boundary condition on Γ_ε, replacing (1.25) by, say, $-\Delta u_\varepsilon + \lambda u_\varepsilon = f$, $\lambda>0$.

= unitary normal to S directed toward the exterior of $\mathcal{Y}$ (i.e. the interior of $\mathcal{O}$).

The identification of various powers of ε gives now :

(1.30)
$$\left|\begin{array}{l} -\Delta_y u_o = 0 \ , \\ -\Delta_y u_1 - 2\Delta_{xy} u_o = 0 \ , \\ -\Delta_y u_2 - 2\Delta_{xy} u_1 - \Delta_x u_o = f \ , \\ \cdots\cdots\cdots\cdots\cdots \ . \end{array}\right.$$
□

The equation $(1.30)_1$ with (1.7) and (1.28) gives

(1.31)
$$u_o(x,y) = u_o(x) \text{ does not depend on } y.$$

Equation $(1.30)_2$ becomes

(1.32)
$$-\Delta_y u_1 = 0 \ , \text{ subject to (1.7) and to (1.29) ;}$$

equation $(1.30)_3$ is

(1.33)
$$\left|\begin{array}{l} -\Delta_y u_2 - 2\Delta_{xy} u_1 - \Delta_x u_o = f \text{ subject to (1.7) and to} \\ \dfrac{\partial u_2}{\partial \nu_{(y)}} + \dfrac{\partial u_1}{\partial \nu_{(x)}} = 0 \text{ on } S. \end{array}\right.$$

Let us show that $\{u_o,u_1\}$ *is defined by the solution of an elliptic system.*

We introduce the Sobolev space $H^1_o(\Omega)$ of functions in $H^1(\Omega)$ which are zero on Γ, and

(1.34)
$$W(\mathcal{Y}) = \{v \mid v \in H^1(\mathcal{Y}) \ , \ v \text{ takes equal values on opposite faces of } Y\}.$$

We define $W^\circ(\mathcal{Y}) = W(\mathcal{Y})/\mathbb{R}$ and

(1.35)
$$V = H^1_o(\Omega) \times L^2(\Omega;W^\circ(\mathcal{Y})).$$

If g, h are two vector functions defined in $\Omega \times \mathcal{Y}$, we set

(1.36)
$$(g,h)_{\Omega\times\mathcal{Y}} = \iint_{\Omega\times\mathcal{Y}} g_i h_i \, dxdy.$$

Then applying Green's formula, we see *that* (1.32)(1.33) *is equivalent to find* $\{u_o, u_1\} \in V$ *such that*

(1.37)
$$\begin{cases} (\nabla_x u_o , \nabla_x v_o)_{\Omega\times\mathcal{Y}} + (\nabla_y u_1 , \nabla_x v_o)_{\Omega\times\mathcal{Y}} = (f, v_o)_{\Omega\times\mathcal{Y}} \ (^1) \\ (\nabla_y u_1 , \nabla_y v_1)_{\Omega\times\mathcal{Y}} + (\nabla_x u_o , \nabla_y v_1)_{\Omega\times\mathcal{Y}} = 0 \\ \forall \{v_o, v_1\} \in V. \end{cases}$$

This is an *elliptic system* on V. Indeed if we choose $v_o = u_o$, $v_1 = u_1$, and if we add up, (1.37) gives (^{2}) :

(1.38)
$$|\nabla_x u_o + \nabla_y u_1|^2_{\Omega\times\mathcal{Y}} = (f, u_o)_{\Omega\times\mathcal{Y}} ,$$

and everything rests on the following result :

(1.39) $|\nabla_x u_o + \nabla_y u_1|_{\Omega\times\mathcal{Y}}$ *defines a norm on* V *which is equivalent to the natural norm on* V.

To prove (1.39) let us introduce

(1.40) $\mathcal{Y}_k$ = strip in Y contained in $\mathcal{Y}$ (i.e. which does not intersect $\mathcal{O}$) and parallel to the y_k's axis

(cf. an example of $\mathcal{Y}_1$ on Fig. 4).

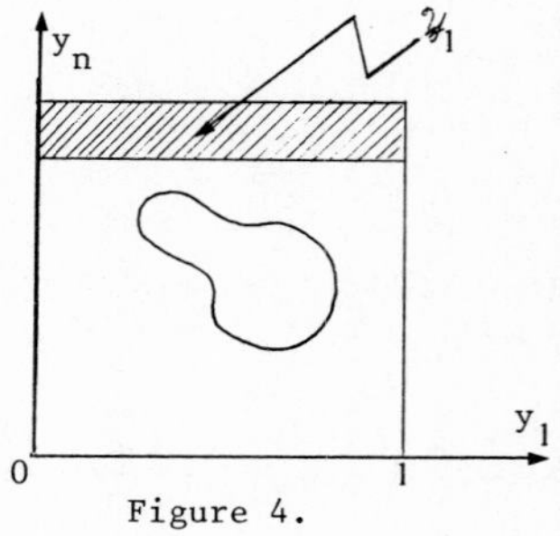

Figure 4.

If $\nabla_x u_o + \nabla_y u_1 = 0$ in $\Omega\times\mathcal{Y}$, then *in particular*

$$\frac{\partial u_o}{\partial x_k} + \frac{\partial u_1}{\partial y_k} = 0 \quad \text{in } \mathcal{Y}_k$$

hence

(1.41)
$$|\mathcal{Y}_k| \frac{\partial u_o}{\partial x_k} + \int_{\mathcal{Y}_k} \frac{\partial u_1}{\partial y_k} dy = 0 .$$

But $\int_{\mathcal{Y}_k} \frac{\partial u_1}{\partial y_k} dy = 0$ since u_1 is periodic, so that (1.41) implies $\frac{\partial u_o}{\partial x_k} = 0$

hence $\nabla_x u_o = 0$ and $\nabla_y u_1 = 0$.

(^{1}) $(f, v_o)_{\Omega\times\mathcal{Y}} = |\mathcal{Y}| \int_\Omega f\, v_o\, dx$, $|\mathcal{Y}|$ = measure of $\mathcal{Y}$.

(^{2}) $|g|_{\Omega\times\mathcal{Y}} = (g,g)^{\frac{1}{2}}_{\Omega\times\mathcal{Y}}$.

Therefore $u_o = 0$ and $u_1 = 0$ in the *quotient* space $L^2(\Omega; W^\circ(\mathcal{Y}))$.
In order to prove (1.39) it suffices to show that V is *complete* for the norm $|\nabla_x u_o + \nabla_y u_1|_{\Omega\times\mathcal{Y}}$. Let v_{om}, v_{1m} be a Cauchy sequence for this norm. Then

$$\frac{\partial v_{om}}{\partial x_k} + \frac{\partial v_{1m}}{\partial y_k} \quad \text{is a Cauchy sequence in } L^2(\Omega\times\mathcal{Y})$$

and therefore

$$\int_{\mathcal{Y}_k} \left(\frac{\partial v_{om}}{\partial x_k} + \frac{\partial v_{1m}}{\partial y_k}\right) dy \qquad \text{is a Cauchy sequence in } L^2(\Omega)$$

i.e. $\frac{\partial v_{om}}{\partial x_k}$ is a Cauchy sequence in $L^2(\Omega)$. Therefore $v_{om} \to v_o$ in $H_o^1(\Omega)$

and $\nabla_y v_{1m}$ is a Cauchy sequence in $(L^2(\Omega\times\mathcal{Y}))^n$ so that v_{1m} is a Cauchy sequence in $L^2(\Omega; W^\circ(\mathcal{Y}))$ hence the result follows. □

Remark 1.4.

System (1.37) is the *homogenized system* of the problem.
The function u_1 plays the role of the "hidden variable".
One can *eliminate* u_1 and one obtains then an *elliptic equation*

(1.42) $\qquad \mathcal{A} u_o = f \quad \text{in } \Omega ,$

(1.43) $\qquad u_o = 0 \quad \text{on } \Gamma ,$

where $\mathcal{A}$ is the *homogenized operator*. Cf. D. Cioranescu and J. Saint Jean Paulin [4] D. Cioranescu, H. Lanchon and J. Saint Jean Paulin [3] for other situations along similar lines. □

Remark 1.5.

One proves, by energy estimates (cf. D. Cioranescu and al. loc. cit.) that a suitable extension $P_\varepsilon u_\varepsilon$ of u_ε to Ω converges in $H_o^1(\Omega)$ to u_o as $\varepsilon \to 0$.

The question of *boundary layers* arises if one looks for higher order asymptotic expansions. This is an essentially open question. □

2. POROUS MEDIA (I). STATIONARY CASE

2.1. Setting of the Problem.

We consider Ω_ε defined as in Section 1 and we consider in Ω_ε the problem (*Stokes problem*)

(2.1) $$-\varepsilon^2 \Delta u_\varepsilon = f - \nabla p_\varepsilon ,$$

(2.2) $$\operatorname{div} u_\varepsilon = 0$$

where $u_\varepsilon = \{u_{\varepsilon 1}, \ldots, u_{\varepsilon n}\}$ is subject to

(2.3) $$u_\varepsilon = 0 \text{ on } \partial\Omega_\varepsilon = \Gamma_\varepsilon \cup S_\varepsilon .$$

In (2.1) f denotes the *restriction* to Ω_ε of a given vector function f defined in Ω.

We consider the "same" problem than in Section 1.2., namely the problem of finding an *asymptotic expansion for* u_ε (and also for p_ε) in terms of ε.

2.2. Homogenized system with two pressures.

We follow the analysis of J.L. Lions [10][13]. Similar conclusions with slightly different arguments are given in E. Sanchez-Palencia [21].

We look for u_ε in the form (1.6)(1.7)(1.8) where the u_j's are now *vectors*, and we look for p_ε in the form

(2.4) $$p_\varepsilon = p_o + \varepsilon p_1 + \ldots$$

where

(2.5) $$p_j = p_j(x,y) \text{ is defined in } \Omega \times \mathcal{Y} \text{ and is Y-periodic in } y.$$

We identify the various powers in the equations (2.1)(2.2) ; we obtain :

(2.6) $$\nabla_y p_o = 0$$

(2.7) $$\left| \begin{array}{l} -\Delta_y u_o = f - \nabla_y p_1 - \nabla_x p_o , \\ -\Delta_y u_1 - 2\Delta_{xy} u_o = - \nabla_y p_2 - \nabla_x p_1 \\ \cdots\cdots\cdots\cdots\cdots\cdots \end{array} \right.$$

$$(2.8)\quad \begin{cases} \operatorname{div}_y u_o = 0 , \\ \operatorname{div}_y u_1 + \operatorname{div}_x u_o = 0 , \\ \dots\dots\dots\dots \end{cases}$$

It follows from (2.6) that

$$(2.9)\quad p_o = p_o(x) \text{ does not depend on } y .$$

Equation $(2.8)_2$ admits a solution u_1 satisfying (1.7)(1.8)

iff $\int_{\mathcal{Y}} \operatorname{div}_x u_o \, dy = 0$ hence the system :

$$(2.10)\quad \begin{cases} -\Delta_y u_o = f - \nabla_x p_o - \nabla_y p_1 \\ \operatorname{div}_y u_o = 0 \quad \text{in } \Omega \times \mathcal{Y} , \\ \operatorname{div}_x \int_{\mathcal{Y}} u_o dy = 0 \quad \text{in } \Omega , \end{cases}$$

subject to

$$(2.11)\quad u_o \text{ is Y-periodic in } y \text{ and } u_o = 0 \text{ if } y \in S$$

and subject to

$$(2.12)\quad \begin{cases} \nu . \int_{\mathcal{Y}} u_o \, dy = 0 \quad \text{on} \quad \partial\Omega = \Gamma , \\ \text{where } \nu = \text{unitary normal to } \Gamma ; \end{cases}$$

only (2.12) deserves explanations : if u_o, p_o , p_1 satisfy (2.10) the only boundary condition on Γ which makes sense and is "close" to the condition $u_o = 0$ is precisely (2.12).

<u>*Remark 2.1.*</u>

To obtain $u_o = 0$ on Γ one has to introduce *boundary layer terms*. A solution, in a very particular case, is given in J.L. Lions [14] . □

One can show (cf. J.L. Lions [10][13]) that (2.10)(2.11)(2.12) *admits a unique solution* for u_o , and a solution defined modulo an additive constant for p_o and modulo an additive function of x for p_1. One can express the problem in a (simple looking) variational form as follows : we introduce

$$(2.13)\quad V(\mathcal{Y}) = \{\varphi \mid \varphi \in (H^1(\mathcal{Y}))^n , \ \varphi = 0 \text{ on } S , \ \varphi \text{ takes equal values on opposite faces of } Y, \ \operatorname{div}_y \varphi = 0 \text{ in } \mathcal{Y} \},$$

and

(2.14) $$\mathcal{V} = \{v \mid v \in L^2(\Omega;V(\mathcal{Y})),\ \operatorname{div}_x \int_{\mathcal{Y}} v(x,y)\,dy = 0 \text{ in } \Omega,\ \nu.\int_{\mathcal{Y}} v(x,y)\,dy = 0 \text{ on } \Gamma\}.$$

For u , $v \in \mathcal{V}$ we define

(2.15) $$a(u,v) = \iint_{\Omega\times\mathcal{Y}} \frac{\partial u_i}{\partial y_j}\frac{\partial v_i}{\partial y_j}\,dy\,dx.$$

Then the problem is *equivalent* to finding u_o such that

(2.16) $$a(u_o,v) = \iint_{\Omega\times\mathcal{Y}} f_i v_i\,dx\,dy \quad \forall v \in \mathcal{V}. \qquad \square$$

We have kept in (2.10) the "two pressures" p_o , p_1 and we have "eliminated" *both of them* in the variational formulation (2.16). We show now how one can also eliminate *only* p_1 , which leads to the Darcy's law.

2.3. Homogenized system with one pressure. Darcy's law.

Let us consider the first two equations in (2.10) where *we think of* x *as a parameter* :

(2.17) $$\begin{cases} -\Delta_y u_o = (f - \nabla_x p_o(x)) - \nabla_y p_1 , \\ \operatorname{div}_y u_o = 0 \quad \text{in } \mathcal{Y} \end{cases}$$

subject to (2.11).

Let us define $\phi_\lambda(y)$, $\pi_\lambda(y)$, where λ is given in $\mathbb{R}^n$, by

(2.18) $$\begin{cases} -\Delta_y \phi_\lambda = \lambda - \nabla_y \pi_\lambda , \\ \operatorname{div} \phi_\lambda = 0 \quad \text{in } \mathcal{Y} , \\ \phi_\lambda , \pi_\lambda \text{ are Y-periodic}, \quad \phi_\lambda = 0 \text{ on } S. \end{cases}$$

This system uniquely defines $\phi_\lambda(y) \in V(\mathcal{Y})$, and since $\phi_\lambda(y)$ depends linearly on λ, we have

(2.19) $$\begin{cases} \phi_\lambda(y) = \phi(y).\lambda \ , \ \phi(y) = \|\phi_{ij}(y)\| , \ i, j = 1, \dots, n , \\ \phi_{ij}(y) \in V(\mathcal{Y}). \end{cases}$$

With these notations (2.17) gives

(2.20) $$u_o(x,y) = \phi(y).(f(x) - \nabla p_o(x)).$$

If we take *averages in* y , we obtain *the Darcy's law* :

$$(2.21)\qquad \begin{cases} \int_{\mathcal{Y}} u_o(x,y)dy = \Psi(f - \nabla p_o) , \\ \Psi = \int_{\mathcal{Y}} \phi(y)dy = \text{"Darcy's matrix"} . \end{cases}$$

We now take into account $(2.10)_3$ and (2.12) which can be expressed *in Variational form* as follows

$$(2.22)\qquad (\Psi(f - \nabla p_o), \nabla q) = 0 \quad \forall\, q \in H^1(\Omega).$$

One can show that (2.22) uniquely *defines* (up to an additive constant) $p_o \in H^1(\Omega)$. Indeed one easily checks (cf. J.L. Lions, loc. cit.) *that* Ψ *is positive definite* so that (2.22) *is the Neumann problem for the elliptic operator* $-\operatorname{div}(\Psi\,\nabla)$. □

Remark 2.2.

One can prove (cf. L. Tartar [23]) that u_ε (extended by 0 outside Ω_ε) converges in $L^2(\Omega)$ towards $\int_{\mathcal{Y}} u_o(x,y)dy$. □

3. POROUS MEDIA (II). EVOLUTION CASE.

3.1. Setting of the problem.

We consider now the evolution case :

$$(3.1)\qquad \begin{cases} \dfrac{\partial u_\varepsilon}{\partial t} - \varepsilon^2 \Delta u_\varepsilon = f - \nabla p_\varepsilon , \\ \operatorname{div} u_\varepsilon = 0 \quad \text{in } \Omega_\varepsilon \times]0,T[, \end{cases}$$

subject to

$$(3.2)\qquad u_\varepsilon = 0 \quad \text{on} \quad \partial\Omega_\varepsilon \times]0,T[$$

and with the initial condition

$$(3.3)\qquad u_\varepsilon(x,o) = u^o(x) \quad \text{in } \Omega_\varepsilon .$$

As before f (resp. u^o) denotes the restriction to $\Omega_\varepsilon \times]0,T[$ (resp. Ω_ε) of a function f (resp. u^o) defined in $\Omega \times]0,T[$ (resp. in Ω).

We want to study the asymptotic expansion of u_ε *as* $\varepsilon \to 0$.

3.2. Two pressures homogenized system.

We use the ansatz :

(3.4) $$u_\varepsilon = u_o(x,y,t) + \varepsilon u_1(x,y,t) + \ldots$$

(3.5) $$p_\varepsilon = p_o(x,y,t) + \varepsilon p_1(x,y,t) + \ldots$$

where the u_j's and p_j's *are* Y-*periodic in* y and

(3.6) $$u_j = 0 \quad \text{if} \quad y \in S, \quad t \in]0,T[\;, \quad x \in \Omega.$$

We obtain *that* p_o *does not depend on* y and

(3.7) $$\left| \begin{aligned} & \frac{\partial u_o}{\partial t} - \Delta_y u_o = f - \nabla_x p_o - \nabla_y p_1 \;, \\ & \operatorname{div}_y u_o = 0 \quad \text{in} \quad \Omega\times\mathcal{Y}\times]0,T[\;, \\ & \operatorname{div}_x \int_{\mathcal{Y}} u_o dy = 0 \quad \text{in } \Omega\times]0,T[\;. \end{aligned} \right.$$

The *boundary conditions* are

(3.8) $$\left| \begin{aligned} & u_o \text{ is Y-periodic,} \quad u_o = 0 \text{ for } y \in S,\ x \in \Omega, \quad t \in]0,T[\;, \\ & \nu . \int_{\mathcal{Y}} u_o \, dy = 0 \quad \text{for} \quad x \in \Gamma \;, \quad t \times]0,T[\;, \\ & p_1 \quad \text{is} \quad \text{Y-periodic} \end{aligned} \right.$$

and the *initial condition* is

(3.9) $$u_o(x,y,o) = u^o(x) \quad \text{in} \quad \Omega \times \mathcal{Y}.$$

The system (3.7)(3.8)(3.9) *admits a unique solution*

(3.10) $$u_o \in L^2(0,T;\mathcal{V}), \quad \mathcal{V} \text{ defined by } (2.14) \;;$$

with the notation (2.15), the *variational form* of the problem is

(3.11) $$\left(\frac{\partial u_o}{\partial t}, v\right)_{\Omega\times\mathcal{Y}} + a(u_o,v) = (f,v)_{\Omega\times\mathcal{Y}} \qquad \forall\, v \in \mathcal{V}$$

with (3.9). □

One can eliminate only p_1 and one obtains for p_o an elliptic equation in x where t plays the role of a parameter (it is a *integro-differential* elliptic equation). Cf. J.L. Lions [14], E. Sanchez-Palencia [21] . □

4. NON LOCAL HOMOGENIZED SYSTEM IN A PERFORATED MATERIAL.

4.1. Setting of the problem.

Always with the same notations we consider the equation (which corresponds to a perforated elastic material with short range memory) :

$$\frac{\partial^2 u_\varepsilon}{\partial t^2} - \Delta \frac{\partial u_\varepsilon}{\partial t} - \Delta u_\varepsilon = 0 \quad \text{in} \quad \Omega_\varepsilon \times]0,T[\tag{4.1}$$

subject to *the boundary conditions* :

$$\left| \begin{array}{lll} u_\varepsilon = 0 & \text{on} & \Gamma_\varepsilon \times]0,T[\ , \\ \dfrac{\partial u_\varepsilon}{\partial \nu} = 0 & \text{on} & S_\varepsilon \times]0,T[\ , \end{array} \right. \tag{4.2}$$

and to *the initial conditions* :

$$u_\varepsilon(x,o) = u^o(x) \ , \qquad \frac{\partial u_\varepsilon}{\partial t}(x,o) = u^1(x) \quad \text{in} \ \Omega_\varepsilon \tag{4.3}$$

where u^o and u^1 actually denote the restrictions to Ω_ε of functions u^o and u^1 defined in Ω.

We look for an asymptotic expansion for u_ε . We are going to show that the homogenized equation (for the first term of the expansion) is an integro-differential equation of a *non local character*.

Remark 4.1.

The fact that materials with short range memory (and with long range memory as well) lead, *in the case of composite materials*, to *non local phenomena*, has been observed in J.L. Lions [11][12]. □

4.2. Homogenized system.

We try the ansatz :

$$u_\varepsilon = u_o(x,y,t) + \varepsilon u_1(x,y,t) + \ \ldots \tag{4.4}$$

where

(4.5) $u_j(x,y,t)$ is defined in $\Omega \times \mathcal{Y} \times]0,T[$, is Y-periodic in y ,

and where, with the notations of Section 1.3. :

$$(4.6)\quad \begin{cases} \dfrac{\partial u_o}{\partial \nu_{(y)}} = 0 & \text{if } y \in S, \quad x \in \Omega, \quad t \in]0,T[\ , \\ \dfrac{\partial u_j}{\partial \nu_{(y)}} + \dfrac{\partial u_{j-1}}{\partial \nu_{(x)}} = 0 & \text{if } y \in S, \quad x \in \Omega\ , \quad t \in]0,T[, \quad j=1,2, \ldots\ . \end{cases}$$

We obtain the equations :

$$(4.7)\quad \begin{cases} -\Delta_y \dfrac{\partial u_o}{\partial t} - \Delta_y u_o = 0\ , \\ -\Delta_y \dfrac{\partial u_1}{\partial t} - \Delta_y u_1 - 2\Delta_{xy} \dfrac{\partial u_o}{\partial t} - 2\Delta_{xy} u_o = 0\ , \\ -\Delta_y \dfrac{\partial u_2}{\partial t} - \Delta_y u_2 - 2\Delta_{xy} \dfrac{\partial u_1}{\partial t} - 2\Delta_{xy} u_1 - \Delta_x \dfrac{\partial u_o}{\partial t} - \Delta_x u_o + \dfrac{\partial^2 u_o}{\partial t^2} = 0. \end{cases}$$

The equations $(4.7)_1$ and $(4.6.)_1$ (and the Y-periodicity of u_o in y) lead to

$$(4.8)\qquad u_o = u_o(x,t) \quad \text{does not depend on } y.$$

Then $(4.7)_2$ reduces to

$$(4.9)\qquad -\Delta_y \frac{\partial u_1}{\partial t} - \Delta_y u_1 = 0.$$

We have now to solve $(4.9)(1.7)_3$ subject to $(4.6)_2$ for $j=1$ and $j=2$, with the initial conditions

$$(4.10)\qquad u_o(x,o) = u^o(x), \quad \frac{\partial u_o}{\partial t}(x,o) = u^1(x)\ ,$$

$$(4.11)\qquad u_1(x,y,o) = 0.$$

We use the notations (1.35)(1.36). We multiply $(4.7)_3$ by v_o and (4.9) by v_1 where $\{v_o,v_1\} \in V$ as defined in (1.35). After using Green's formula and $(4.6)_2$ we obtain :

$$(4.12)\qquad \begin{aligned} &\left(\frac{\partial^2 u_o}{\partial t^2}, v_o\right)_{\Omega\times\mathcal{Y}} + \left(\nabla_x \frac{\partial u_o}{\partial t} + \nabla_x u_o, \nabla_x v_o\right)_{\Omega\times\mathcal{Y}} + \\ &\quad + \left(\nabla_y \frac{\partial u_1}{\partial t} + \nabla_y u_1, \nabla_x v_o\right)_{\Omega\times\mathcal{Y}} = 0\ , \end{aligned}$$

and

$$(4.13)\qquad \left(\nabla_y \frac{\partial u_1}{\partial t} + \nabla_y u_1, \nabla_y v_1\right)_{\Omega\times\mathcal{Y}} + \left(\nabla_x \frac{\partial u_o}{\partial t} + \nabla_x u_o, \nabla_y v_1\right)_{\Omega\times\mathcal{Y}} = 0.$$

This evolution system admits a unique solution

$$(4.14)\quad \left|\begin{array}{l} \{u_o, u_1\}\in L^\infty(0,T;V)\ ,\quad \{\frac{\partial u_o}{\partial t}\ ,\ \frac{\partial u_1}{\partial t}\}\in L^2(0,T;V)\ , \\ \frac{\partial u_o}{\partial t}\in L^\infty(0,T;L^2(\Omega\times\mathcal{Y})). \end{array}\right.$$

Indeed if we take $v_o = \frac{\partial u_o}{\partial t}$ and $v_1 = \frac{\partial u_1}{\partial t}$ in (4.12) and (4.13), we obtain :

$$(4.15)\quad \frac{1}{2}\frac{d}{dt}\left[\left|\frac{\partial u_o}{\partial t}\right|^2_{\Omega\times\mathcal{Y}} + \left|\nabla_x u_o + \nabla_y u_1\right|^2_{\Omega\times\mathcal{Y}}\right] + \left|\nabla_x \frac{\partial u_o}{\partial t} + \nabla_y \frac{\partial u_1}{\partial t}\right|^2_{\Omega\times\mathcal{Y}} = 0.$$

hence the result follows by using (1.39). □

Remark 4.2.

One can eliminate u_1 as follows : if we introduce $\phi_j(y,t)$ by

$$(4.16)\quad \left|\begin{array}{l} -\Delta_y \frac{\partial \phi_j}{\partial t} - \Delta_y \phi_j = 0 \quad \text{in } \mathcal{Y}\times]0,T[\ , \\ \frac{\partial \phi_j}{\partial \nu(y)} = \nu_j(y) \quad \text{on } S\ \times]0,T[\ , \\ \phi_j \text{ is } Y\text{-periodic}\ , \\ \phi_j(y,o) = 0 \quad \text{in } \mathcal{Y}\ , \end{array}\right.$$

then

$$(4.17)\quad u_1(x,y,t) = \int_o^t \phi_j(y,t-\sigma)\,\frac{\partial u_o}{\partial x_j}(x,\sigma)\,d\sigma$$

and using (4.17) *into* (4.12) *gives the integro partial differential equation (in variational form) satisfied by* u_o. □

5. VARIOUS REMARKS.

5.1. *Non linear models* can be studied by similar techniques. For *non newtonian flows* in porous media, we refer to E. Sanchez-Palencia and the A. [17] , and for other non linear models to E. Sanchez-Palencia [22] and to the A. [15][16] .

5.2. The techniques of homogenization seem to give some light on some *turbulence problems*. Cf. P. Perrier and O. Pironneau [20] , G. Papanicolaou and O. Pironneau [18] .

5.3. For the case of *"random materials"* we refer to G. Papanicolaou and S.R. Varadhan [19].

5.4. For the cases of non periodic holes, we refer to the book of Hruslov and Marcenko [6] ant to D. Cioranescu and F. Murat [3] . (They also considered the homogenization of unilateral problems).

5.5. We also refer to the papers of E. de Giorgi [5] and of V.V.Jikov, S.M.Koslov, O.A. Oleinik and Ha Tsen Gnoan [7] and to the Bibliography therein.

REFERENCES.

[1] Bensoussan A., Lions J.L., Papanicolaou G., *Asymptotic Analysis for Periodic Structures*. North-Holland (1978).

[2] Cioranescu D., Lanchon H. and Saint Jean Paulin J.S, Elastic-plastic torsion of heterogeneous cylindrical bars. J. Inst. Math. Applications, 1979 (24)p.353-378.

[3] Cioranescu D., Murat F., Séminaire H. Brézis-J.L. Lions, 1980-1981 - Dec.1980.

[4] Cioranescu D., Saint Jean Paulin J.S., Homogenization in open sets with holes. J.M.M.A. 71 (02) October 1979, pp. 530-607.

[5] de Giorgi E., Convergence problems for functionals and operators. Proceedings International Meeting on Recent Methods in non linear Analysis. Ed. de Giorgi, E. Magenes and U. Mosco, ed. Pitagora, Bologna (1979), pp. 131-188.

[6] Hruslov E., Marcenko V.A., *Boundary value problems in domains with granulated boundary*. In Russian. Kiev, 1974.

[7] Jikov V.V., Koslov S.M., Oleinik O.A., and Ha Tsen-Gnoan, Homogenization and G-convergence of differential operators. Ouspechi Mat. Nauk, 1979 (34), pp. 65-133.

[8] Lions J.L., Introductory remarks to asymptotic analysis of periodic structures. Lecture Porabka, Kozubnik, September 1977.

[9] Lions J.L., Asymptotic expansions in perforated media with a periodic structure. Rocky Mountains Journal of Mathematics, 10, 1, 1979.

[10] Lions J.L., Some problems connected with Navier Stokes equations. Proc. IVth Escuela Latino-Americana de Matematica, 1978, Analisis y sus Aplicaciones Lima, 1979, pp. 222-286.

[11] Lions J.L., Homogenéisation non locale. Proc. Int. Meeting on Recent Methods in non linear analysis, ed. de Giorgi, E. Magenes, U. Mosco, ed. Pitagora, Bologna, 1979, pp. 189-203.

[12] Lions J.L., Remarks on some asymptotic problems in composite materials and in perforated materials. Proc. IUTAM Symposium Northwestern University, Ed. Nemmat-Nasser, North-Holland, 1979.

[13] Lions J.L., Quelques problèmes liés aux équations de Navier-Stokes. Lectures at GNAFA IV, Naples, March 1980.

[14] Lions J.L., *Some Methods in the Mathematical Analysis of Systems and their Control*. Sciences Press, Pekin, 1981.

[15] Lions J.L., Optimal control of non well posed distributed systems and related non linear partial differential equations. Lecture at Los Alamos, March 1981. Proceedings of the Conference.

[16] Lions J.L., Homogenization problems. Lecture at GAMM. Würzburg, March 81.

[17] Lions, J.L., Sanchez-Palencia, E., Ecoulement d'un fluide visco-plastique de Bingham dans un milieu poreux. J.M.P.A., 1981.

[18] Papanicolaou G., Pironneau O. To appear.

[19] Papanicolaou G., Varadhan S.R., Boundary value problems with rapidly oscillating random coefficients. C.P.A.M., 1981.

[20] Perrier P., Pironneau O., Couplage des grosses et petites structures turbulentes par l'homogénéisation. C.R.A.S. Paris, 1978.

[21] Sanchez-Palencia E., *Non homogeneous media and vibration theory*. Lecture Notes in Physics, 127 (1980).

[22] Sanchez-Palencia E., To appear.

[23] Tartar L., Incompressible fluid flow in a porous medium. Convergence of the homogenization process. Appendix of Lecture Notes in Physics, by E. Sanchez-Palencia, 127, 1980.

-=-=-

Part 3
FLUIDS, COUPLED FIELDS

Continuum Models of Discrete Systems 4
eds. O. Brulin and R.K.T. Hsieh

ASYMPTOTIC METHOD APPLIED TO A PROBLEM OF SOLAR FROST PRODUCTION
(Méthode asymptotique appliquée à un problème de production solaire de froid)

Jean Marie ABILLON *'
Georges DUVAUT **'

présenté par Georges DUVAUT

We apply an asymptotic expansion method to study the temperature field in a periodic structure composed of zeolite grains in a mixture of gaz and vapor. Adsorbsion-desorbsion cycles give to this apparatus the features of a solar refrigerator.

1) PRESENTATION DU PROBLEME PHYSIQUE

Les adsorbants solides possèdent de grands avantages en production solaire de froid, en particulier les couples zéolithe-eau ou zéolithe-méthanol {1} qui peuvent être directement utilisés dans des capteurs plans solaires conventionnels.

Ce travail présente un modèle de fonctionnement d'un tel capteur, dans la phase de désorbtion du cycle de production de froid, phase pouvant durer toute une journée {2}.
Un tel capteur est assimilé à une cuve parallélépipédique remplie de grains de zéolithe supposés répartis périodiquement. Une période de base, grossie, est représentée par le parallélépipède

$$Y =]o, Y_1[\times]o, Y_2[\times]o, Y_3[$$

où $Y^{(2)}$ représente le grain de zéolithe, $Y^{(1)}$ sa partie complémentaire et $\gamma=\partial Y^{(2)}$ son bord. Dans la cuve, désiquée par Ω, l'ensemble des grains de zéolithe est $\Omega_\varepsilon^{(2)}$, de bord $\gamma_\varepsilon=\partial\Omega_\varepsilon^{(2)}$ et de complémentaire $\Omega_\varepsilon^{(1)}$. La structure de Ω est périodique, de période εY, où, la période de base Y étant fixée, ε est un scalaire tendant vers zéro.
La région $\Omega_\varepsilon^{(1)}$ est occupée par un fluide gazeux (vapeur d'eau, méthanol, etc.) pouvant être adsorbée par la zéolithe. La masse de vapeur adsorbée par unité de surface de zéolithe dépend de la température et de la pression {3}. En phase de désorbsion, la cuve est reliée à un récipient (condenseur) à température constante et à la pression de vapeur saturante de la vapeur. La pression dans la cuve est par conséquent supposée évoluer peu et le phénomène est seulement régi par les variations de température.

---

*' Professeur au Centre Universitaire Antille Guyane, Pointe à Pitre, Guadeloupe

**' Professeur à l'Université P. et M. Curie, 4 Place Jussieu, 75005 Paris

On utilise les notations suivantes :

$\partial_1\Omega$: bords latéraux et fond de la cuve supposés thermiquement isolés,

$\partial_2\Omega$: partie supérieure de la cuve exposée au rayonnement solaire,

$n=(n_i)$: normale unitaire éxtérieure à $\partial\Omega_\varepsilon^{(1)}$.

$\phi(t)$: flux de chaleur solaire, dépendant du temps, sur $\partial_2\Omega$.

$\theta^\varepsilon(x,t)$: champ de température dans la cuve,

$C(y)$: chaleur volumique sur Y, prolongée par Y-périodicité. On pose

(1) $$C^\varepsilon(x)=C(x/\varepsilon).$$

$k_{ij}(y)$: matrice symétrique de conductivité thermique, donnée sur Y et prolongée par Y-périodicité. On pose

(1)' $$k_{ij}^\varepsilon(x)= k_{ij}(x/\varepsilon).$$

q_i^ε : flux de chaleur

(2) $$q_i^\varepsilon= -k_{ij}^\varepsilon(x)\,\frac{\partial\theta^\varepsilon}{\partial x_j},$$

m : masse de vapeur adsorbée à la surface de la zéolithe,

(3) $$dm = -\varepsilon K_1 d\theta^\varepsilon ,$$

K_1 est une constante positive dans un faible domaine de variation de la température. Cette hypothèse est justifiée dans la mesure où la période de fonctionnement dans la phase de désorbsion peut être découpée en intervalles de temps suffisamment courts.

Q : chaleur fournie au système par unité de temps et unité de surface de $\partial\Omega_\varepsilon^{(2)}$,

(4) $$Q = \vec{q}^{(1)}\cdot\vec{n}-\vec{q}^{(2)}\cdot\vec{n} \quad \text{sur } \gamma_\varepsilon ,$$

où les indices (1) et (2) désignent les limites sur γ_ε des valeurs prises sur $\Omega_\varepsilon^{(1)}$ et $\Omega_\varepsilon^{(2)}$ respectivement.

D'autre part {3} ,

(5) $$Q = -K_2\,\frac{dm}{dt}, \quad K_2 \text{ constante positive.}$$

Les équations et conditions aux limites du problème pour le champ de température $\theta^\varepsilon(x,t)$ sont alors,

(6) $$C^\varepsilon(x)\,\frac{\partial\theta^\varepsilon}{\partial t} - \frac{\partial}{\partial x_i}\left(k_{ij}^\varepsilon(x)\,\frac{\partial\theta^\varepsilon}{\partial x_j}\right) = 0 \quad \text{dans } \Omega$$

(7) $$q_i^\varepsilon n_i = 0 \quad \text{sur } \partial_1\Omega$$

(8) $$q_i^\varepsilon n_i = -\phi(t) \quad \text{sur } \partial_2\Omega$$

(9) $$\vec{q}^{(1)}\cdot\vec{n} - \vec{q}^{(2)}\cdot\vec{n} = \varepsilon L \frac{d\theta^\varepsilon}{dt} \quad \text{sur } \gamma_\varepsilon$$

(10) $$\theta^{(1)} = \theta^{(2)} \quad \text{sur } \gamma_\varepsilon$$

avec la conditio n initiale

(11) $$\theta^\varepsilon(x,o) = \Theta o(x) \; .$$

2) FORMULATION VARIATIONNELLE

En désignant par V l'espace de Sobolev $H^1(\Omega)$, on a

Propriété 1 Le champ de température $\theta^\varepsilon(x,t)$ solution de (6) - (11) satisfait

(12) $$\begin{cases} \theta^\varepsilon(x,t) \in V, \; \forall t \in [o,t_1] \; ; \; \theta^\varepsilon(x,o) = \Theta o(x) \; , \\ \displaystyle\int_\Omega C^\varepsilon(x)\frac{\partial\theta^\varepsilon}{\partial t} v \, d\, x + \varepsilon\int_{\gamma_\varepsilon} L\frac{\partial\theta^\varepsilon}{\partial t} v \, d\, s + a^\varepsilon(\theta^\varepsilon,v) = F(v), \; \forall v \in V, \end{cases}$$

où on a posé

(13) $$a^\varepsilon(\theta,v) = \int_\Omega k_{ij}^\varepsilon(x) \frac{\partial\theta}{\partial x_j}\frac{\partial v}{\partial x_i} \, dx, \quad F(v) = \int_{\partial_2\Omega} v\phi(t) \, d\, s \; .$$

Sur les données du problème on fait les hypothèses suivantes

(14) $$\begin{cases} C(y), \; k_{ij}(y) \in L^\infty(Y) \; ; \; k_{ij} = k_{ji} \; ; \; L = \text{Constante positive.} \\ \exists \alpha_1 > o, \; C(y) \geq \alpha_1, \; k_{ij}\, x_i\, x_j \geq \alpha_1 x_j\, x_j, \; \forall \, x_j. \\ \phi(t) \in L^\infty(\mathbb{R}) \; ; \; \Theta o \in L^\infty(\Omega) \cap V, \end{cases}$$

ce qui permet de montrer l'existence d'une unique θ^ε solution de (12) dans la classe

$$\theta^\varepsilon \in L^2(o,t_1;\, V) \cap L^\infty(o,t_1;\, L^\infty(\Omega)) \; .$$

3) DEVELOPPEMENT ASYMTOTIQUE

On cherche $\theta^\varepsilon(x,t)$ sous forme d'un développement asymptotique à double échelle, (Cf. {4}),

(15) $$\theta^\varepsilon(x,t) = \theta_o(x,y;t) + \varepsilon\theta_1(x,y;t) + \ldots \; ; \; y = \frac{x}{\varepsilon} \, ,$$

où θ_i est, pour tout (x,t) une fonction y-périodique en y. Notant que pour toute fonction $\chi(x,y = \frac{x}{\varepsilon})$ une dérivée partielle en x_i devient

(16) $$\frac{\partial \chi}{\partial x_i} + \frac{1}{\varepsilon}\frac{\partial \chi}{\partial y_i},$$

on peut réécrire l'opération figurant dans (6) sous la forme

(17) $$\varepsilon^{-2} A_o + \varepsilon^{-1} A_1 + \varepsilon^o A_2 + C(y)\frac{\partial}{\partial t},$$

(18) $$\begin{cases} A_o = -\dfrac{\partial}{\partial y_i}\left(k_{ij}(y)\dfrac{\partial}{\partial y_i}\right), \quad A_2 = -k_{ij}(y)\dfrac{\partial^2}{\partial x_i \partial x_j} \\ A_1 = -\dfrac{\partial}{\partial y_i}\left(k_{ij}(y)\dfrac{\partial}{\partial x_j}\right) - \dfrac{\partial}{\partial x_i}\left(k_{ij}(y)\dfrac{\partial}{\partial y_j}\right), \end{cases}$$

et l'opérateur frontière (9) sous la forme,

(19) $$\varepsilon^{-1} B_o + \varepsilon^o B_1 - L\frac{\partial}{\partial t}$$

$$B_o = n_i k_{ij}^{(2)}\frac{\partial}{\partial y_j} - n_i k_{ij}^{(1)}\frac{\partial}{\partial y_j}, \quad B_1 = n_i k_{ij}^{(2)}\frac{\partial}{\partial x_j} - n_i k_{ij}^{(1)}\frac{\partial}{\partial x_j}.$$

En identifiant les termes de même degré en ε dans les développements obtenus, on arrive au résultat suivant pour le développement de θ^ε.

Propriété 2

Le champ de températures solution θ^ε admet, lorsque ε tend vers zéro, un développement asymtotique de la forme (15) où θ_o ne dépend que de (x,t) et est solution de

(20) $$\begin{cases} \left(\overline{c} + \dfrac{\text{mes } \gamma}{\text{mes } y} L\right)\dfrac{\partial \theta_o}{\partial t} + \dfrac{\partial q_i^{(o)}}{\partial x_i} = o \\ q_i^{(o)} = -K_{ij}\dfrac{\partial \theta_o}{\partial x_j} \\ q_i^{(o)} n_i = o \quad \text{sur} \quad \partial_1\Omega \\ q_i^{o} n_i = -\phi(t) \quad \text{sur} \quad \partial_2\Omega \\ \theta_o(x,t) = {}_o(x). \end{cases}$$

On a posé

(21) $$\overline{c} = \frac{1}{\text{mes } Y}\int_Y C_{(y)}\,dy = \text{moyenne de } C(y) \text{ sur } Y.$$

$$K_{ij} = \frac{1}{\text{mes } Y}\left[\iint_Y k_{ij}(y)\,dy - \int_Y k_i\,(y)\frac{\partial \chi_j}{\partial y}\,dy\right].$$

Les fonctions $\chi_j(y)$ sont solution du problème elliptique sur Y,

(22) $$\begin{cases} \chi_j \text{ est } Y\text{ - périodique} \\ A_o\chi_j = -\dfrac{\partial}{\partial y_i} k_{ij}(y) \text{ dans } Y^{(1)} \text{ et } Y^{(2)} \\ B_o\chi_j = n_i\,(k_{ij}^{(2)} - k_{ij}^{(1)}) \text{ sur } \gamma \end{cases}$$

Le deuxième terme $\theta_1(x,y)$ est donné par

(23) $$\theta_1 = -\frac{\partial\theta_o(x,t)}{\partial x_y}\,\chi_j(y)\ .$$

4) -CALCUL DE LA MASSE DE VAPEUR DESORBEE

D'après (3) , la masse de vapeur désorbée, par unité de temps à l'instant t, est donnée, pour la totalité du massif Ω par

$$M_\varepsilon = -\int_{\gamma_\varepsilon}\frac{dm}{dt}\,ds = \int_{\gamma_\varepsilon}\varepsilon K_1\,\frac{d\theta^\varepsilon}{dt}\,ds\ .$$

La quantité de chaleur correspondant à cette désorbsion est

$$\mathcal{L}_\varepsilon = \int_{\gamma_\varepsilon} Q\,ds = \int_{\gamma_\varepsilon} K_2\,\frac{dm}{dt}\,ds = K_2\,M_\varepsilon\ .$$

La détermination de M_ε et $\mathcal{L}_\varepsilon$ revient donc à calculer χ_ε

(24) $$\chi_\varepsilon = \varepsilon\int_{\gamma_\varepsilon}\frac{d\theta^\varepsilon}{dt}\,ds$$

ou même sa limite lorsque ε tend vers zéro.On obtient le résultat suivant,

Propriété 3

Lorsque ε tend vers zéro, on a

(26) $$X = \lim_{\varepsilon\to o} X_\varepsilon = \frac{\text{mes }\gamma}{\text{mes }Y}\int_\Omega\frac{\partial\theta_o}{\partial t}\,dx\ .$$

Le membre de droite peut être évalué grâce à (20) pour donner

(25) $$X = \frac{\text{mes }\gamma}{c\text{ mes }Y + L\text{ mes }\gamma}\,\phi(t)\ ,$$

ce qui implique

$$\lim_{\varepsilon\to o} M_\varepsilon = K_1 X \ , \ \lim_{\varepsilon\to o} \mathcal{Q}_\varepsilon = K_1 K_2 X.$$

REFERENCES

{1} Brech, D.W., Zeolite molecular sieves, Wiley Interscience Publ. (1974)

{2} Guilleminot, J.J., Meunier, F., Mischler, B., Utilisation d'un cycle intermittent zéolithe 13X-H_2O pour la réfrigération solaire. XVe Congrès International du froid. Venise 23-29 September (1979)

{3} Guilleminot, J.J., Caractérisation de l'état stationnaire liquide gaz-adsorbant lors de l'adsorbsion de gaz facilement condensable sur les zéolithes. Thèse de Docteur Ingénieur. Université de Dijon (1978).

{4} Bensoussan, A., Lions, J.L., Papanicolaou, G., Asymptotic analysis for periodic structures, North-Holland, Amsterdam.

Continuum Models of Discrete Systems 4
eds. O. Brulin and R.K.T. Hsieh

VARIATIONAL PRINCIPLES IN NONLINEAR ELASTOSTATICS OF CLOSELY INTERACTING SOLID CONTINUA

GUO Zhong-heng

Department of Mathematics
Peking University
Peking, China

Institut für Mechanik
Ruhr-Universität
Bochum, FR Germany

1. INTRODUCTION

The different physical properties generally give rise to a difference in average displacement of each composite constituent. In order to describe this phenomenon in composite materials, Bedford and Stern [1] in 1970 developed a theory called closely interacting multi-continuum theory. According to this theory, composite materials are modeled by superimposed continua, each of them is allowed to undergo an individual motion, and all interactions occur only between congruent particles. The concept of congruent particles has been introduced to indicate the material particles of different constituents which have the same material coordinates. The close coupling in the purely mechanical non-polar case may be imagined as a connection between congruent particles by a string. In 1977, Tiersten and Jahanmir [2] proposed a similar theory under an additional assumption that the relative motion of the separate constituents is required to be infinitesimal in order that the composite not rupture. However, this restriction is not necessary for our purpose. The aim of the present paper is to generalize the existing variational principles in nonlinear elasticity [3] to closely interacting solid continua. The starting point is some integral relation. The procedure of derivation will be illustrated by a two-constituent medium. An extension to N-constituent media is straightforward.

2. PRELIMINARIES

We consider a two-constituent medium. Each of the constituents "*" undergoes an individual motion:

$$\underset{\sim}{x}^* = \underset{\sim}{x}^*(\underset{\sim}{X}, t) \qquad (* = 1,2) \tag{2.1}$$

with displacement vector $\underset{\sim}{u}^* = \underset{\sim}{x}^* - \underset{\sim}{X}$.
In (2.1) the position $\underset{\sim}{X}$ in the reference configuration $\mathcal{R}$ identifies the congruent particles and t denotes the time. In what follows the constituent-superscript "*" always takes values 1, 2; Σ implies summation over "*" from 1 to 2. In the case the motions (2.1) are independent of each other, they would be determined separately in the usual manner by Cauchy's equations of motion:

$$\underset{\approx}{t}^*\underset{\sim}{\nabla}^* + \underset{\sim}{f}^*/J^* = \rho^*\ddot{\underset{\sim}{x}}^*, \tag{2.2}$$

the boundary conditions (boundary $A = A_u + A_t$ in $\mathcal{R}$):

$$\underset{\sim}{u}^*|_{A_u} = \mathring{\underset{\sim}{u}}, \qquad \underset{\sim}{t}^*_n|_{A_t} = \underset{\approx}{t}^*\underset{\sim}{n}|_{A_t} = \mathring{\underset{\sim}{t}}_n, \tag{2.3}$$

the constitutive relation:

$$\underset{\approx}{T}^* = \frac{dW^*}{d\underset{\approx}{E}^*} \qquad (\underset{\approx}{T}^* = J^* \overset{-1}{\underset{\approx}{F}}{}^* \underset{\approx}{t}^* \overset{-1}{\underset{\approx}{F}}{}^{*T}) \tag{2.4}$$

and the appropriate initial conditions, where $\underset{\approx}{t}^*$, $\underset{\approx}{T}^*$, $\underset{\sim}{t}_n^*$, $\underset{\sim}{f}^*$, J^*, $\underset{\sim}{\nabla}^*$, ρ^*, W^*, $\underset{\approx}{F}^* = \underset{\sim}{x}^* \otimes \underset{\sim}{\nabla}$ and $\underset{\approx}{E}^*$ denote, respectively, Cauchy stress tensor, Kirchhoff stress tensor, stress vector, volume force per unit volume (in $\mathcal{R}$), Jacobian, absolute differentiation, current mass density, elastic potential per unit volume (in $\mathcal{R}$), deformation gradient and Lagrangean strain tensor

$$\underset{\approx}{E}^* = \frac{1}{2}(\underset{\approx}{F}^{*T}\underset{\approx}{F}^* - \underset{\approx}{I}) = \frac{1}{2}\,[\underset{\sim}{u}^* \otimes \underset{\sim}{\nabla} + \underset{\sim}{\nabla} \otimes \underset{\sim}{u}^* + (\underset{\sim}{\nabla} \otimes \underset{\sim}{u}^*)\,(\underset{\sim}{u}^* \otimes \underset{\sim}{\nabla})] \tag{2.5}$$

for *th constituent. Here the scalar product is denoted by juxtaposition of the concerned quantities instead of the usual dot, "$\underset{\sim}{\nabla}$" indicates absolute differentiation in $\mathcal{R}$, "T" denotes transpose, $\underset{\sim}{n}$ is the unit normal, and $\underset{\approx}{I}$ is the unit tensor. We shall also use the Piola and Jaumann stress tensors:

$$\underset{\approx}{\tau}^* = J^*\,\underset{\approx}{t}^*\,\underset{\approx}{F}^{-1\,*T} = \underset{\approx}{F}^*\underset{\approx}{T}^*, \tag{2.6}$$

$$\underset{\approx}{S}^* = \frac{1}{2}(\underset{\approx}{T}^*\underset{\approx}{U}^* + \underset{\approx}{U}^*\underset{\approx}{T}^*) = \frac{1}{2}(\underset{\approx}{\tau}^{*T}\underset{\approx}{R}^* + \underset{\approx}{R}^{*T}\underset{\approx}{\tau}^*), \tag{2.7}$$

where $\underset{\approx}{U}^*$ is the right stretch tensor appearing in the polar decomposition: $\underset{\approx}{F}^* = \underset{\approx}{R}^*\underset{\approx}{U}^*$ ($\underset{\approx}{R}^*$- rotation tensor)..

As has been assumed in our case, $\underset{\sim}{f}^*$ in (2.2) are the unknown coupling forces ($\underset{\sim}{f}^1 \equiv \underset{\sim}{f}$ is the force with which particle $\underset{\sim}{x}^2$ acts on the particle $\underset{\sim}{x}^1$ and $\underset{\sim}{f}^2 \equiv -\underset{\sim}{f}$). Therefore, the problem can not be solved for the individual constituents separately and we are faced with a coupled problem. For the present we write the conservation law of energy for the combined medium:

$$\frac{d}{dt}[\sum \int_{v^*} (K^* + \rho^*\,\varepsilon^*)dv] = \sum \int_{\partial v^*} \dot{\underset{\sim}{x}}^*\underset{\sim}{t}_n^* da, \tag{2.8}$$

where v^* is an arbitrary current volume of the *th constituent corresponding to the same material volume V in $\mathcal{R}$, ε^* is the internal energy per unit mass and $K^* = (1/2)\,\rho^*\dot{\underset{\sim}{x}}^*\dot{\underset{\sim}{x}}^*$ is the kinetic energy per unit volume. Introducing the internal energy per unit volume in $\mathcal{R}$ for the combined medium: $W \overset{df}{=} \sum J^*\rho^*\varepsilon^*$ and the relative displacement of the congruent particles

$$\underset{\sim}{w} \overset{df}{=} \underset{\sim}{x}^2 - \underset{\sim}{x}^1 = \underset{\sim}{u}^2 - \underset{\sim}{u}^1, \tag{2.9}$$

making use of the divergence theorem, and performing the material differentiation, Eq. (2.8) reduces to

$$\begin{aligned}\int_V \dot{W}dV &= \sum \int_{v^*} [\underset{\approx}{t}^* : (\dot{\underset{\sim}{x}}^* \otimes \underset{\sim}{\nabla}^*) - \underset{\sim}{f}^*\dot{\underset{\sim}{x}}^*/J^*]dv \\ &= \int_V [\sum \underset{\approx}{\tau}^* : (\dot{\underset{\sim}{x}}^* \otimes \underset{\sim}{\nabla}) + \underset{\sim}{f}\dot{\underset{\sim}{w}}]dV\end{aligned} \tag{2.10}$$

and, consequently,

$$\dot{W} = \sum \underset{\approx}{\tau}^* : \overline{\underset{\sim}{x}^* \otimes \underset{\sim}{\nabla}}^{\bullet} + \underset{\sim}{f}\dot{\underset{\sim}{w}} = \sum \underset{\approx}{\tau}^* : \dot{\underset{\approx}{F}}^* + \underset{\sim}{f}\dot{\underset{\sim}{w}}. \tag{2.11}$$

Here ":" denotes the double scalar product of two tensors. It is also easy to show [3] that

$$\dot{W} = \sum \underset{\approx}{T}^* : \dot{\underset{\approx}{E}}^* + \underset{\sim}{f}\dot{\underset{\sim}{w}} = \sum \underset{\approx}{S}^* : \dot{\underset{\approx}{U}}^* + \underset{\sim}{f}\dot{\underset{\sim}{w}}. \tag{2.12}$$

Clearly, W must satisfy the principle of material objectivity. In this paper W is assumed, as in [4], to have either of the following forms:

$$W = W(\underset{\approx}{E}^*, \underset{\sim}{w}) \equiv W_1(\underset{\approx}{E}^1) + W_2(\underset{\approx}{E}^2) + W_{12}(\underset{\sim}{w}), \tag{2.13}$$

$$W = W(\underset{\approx}{U}^*, \underset{\sim}{w}) \equiv W_1(\underset{\approx}{U}^1) + W_2(\underset{\approx}{U}^2) + W_{12}(\underset{\sim}{w}). \tag{2.14}$$

Substituting the material derivative of (2.13) or (2.14) into (2.12), we obtain

$$\underset{\approx}{T}^*(\underset{\approx}{E}^*) = \frac{\partial W}{\partial \underset{\approx}{E}^*}, \qquad \underset{\sim}{f}(\underset{\sim}{w}) = \frac{\partial W}{\partial \underset{\sim}{w}}, \tag{2.15}$$

$$\underset{\approx}{S}^*(\underset{\approx}{U}^*) = \frac{\partial W}{\partial \underset{\approx}{U}^*}, \qquad \underset{\sim}{f}(\underset{\sim}{w}) = \frac{\partial W}{\partial \underset{\sim}{w}}. \tag{2.16}$$

We shall deal only with the elastostatics. In this case, W has a meaning of elastic potential, and treating the material derivative dot "." as the variational delta "δ", (2.11, 12) may be interpreted as the increment of W caused by an increment of $\delta \underset{\sim}{u}^*$. For brevity, the symbol "δ" will be frequently replaced by a superposed dot. In virtue of one-to-one stress-strain correspondence in elasticity, $\underset{\approx}{T}^*(\underset{\approx}{E}^*)$ or $\underset{\approx}{S}^*(\underset{\approx}{U}^*)$ and $\underset{\sim}{f}(\underset{\sim}{w})$ are invertible. For the complementary formulation we introduce, through Legendre transformation, the first and the second complementary energy:

$$W^c(\underset{\approx}{T}^*, \underset{\sim}{f}) = \Sigma \underset{\approx}{T}^* : \underset{\approx}{E}^* + \underset{\sim\sim}{fw} - W[\underset{\approx}{E}^*(\underset{\approx}{T}^*), \underset{\sim}{w}(\underset{\sim}{f})], \tag{2.17}$$

$$\widetilde{W}^c(\underset{\approx}{S}^*, \underset{\sim}{f}) = \Sigma \underset{\approx}{S}^* : \underset{\approx}{U}^* + \underset{\sim\sim}{fw} - W[\underset{\approx}{U}^*(\underset{\approx}{S}^*), \underset{\sim}{w}(\underset{\sim}{f})], \tag{2.18}$$

and, consequently,

$$\underset{\approx}{E}^*(\underset{\approx}{T}^*) = \frac{\partial W^c}{\partial \underset{\approx}{T}^*}, \qquad \underset{\sim}{w}(\underset{\sim}{f}) = \frac{\partial W^c}{\partial \underset{\sim}{f}}, \tag{2.19}$$

$$\underset{\approx}{U}^*(\underset{\approx}{S}^*) = \frac{\partial \widetilde{W}^c}{\partial \underset{\approx}{S}^*}, \qquad \underset{\sim}{w}(\underset{\sim}{f}) = \frac{\partial \widetilde{W}^c}{\partial \underset{\sim}{f}}. \tag{2.20}$$

Regarding W_* in (2.13) as a composed function of $\underset{\approx}{F}^*$, Eq. (2.11) yields

$$\underset{\approx}{\tau}^*(\underset{\approx}{F}^*) = \frac{\partial W}{\partial \underset{\approx}{F}^*} = \underset{\approx}{F}^* \frac{\partial W}{\partial \underset{\approx}{E}^*}, \qquad \underset{\sim}{f}(\underset{\sim}{w}) = \frac{\partial W}{\partial \underset{\sim}{w}} \tag{2.21}$$

When $\underset{\approx}{\tau}^*(\underset{\approx}{F}^*)$ are invertible, in view of $\underset{\approx}{S}^* : \underset{\approx}{U}^* = \underset{\approx}{\tau}^* : \underset{\approx}{F}^*$, the transformation (2.18) can be written as

$$\widetilde{W}^c(\underset{\approx}{\tau}^*, \underset{\approx}{f}) = \Sigma \underset{\approx}{\tau}^* : \underset{\approx}{F}^* + \underset{\sim\sim}{fw} - W[\underset{\approx}{F}^*(\underset{\approx}{\tau}^*), \underset{\sim}{w}(\underset{\sim}{f})] \tag{2.22}$$

and then we have

$$\underset{\approx}{F}^*(\underset{\approx}{\tau}^*) = \frac{\partial \widetilde{W}^c}{\partial \underset{\approx}{\tau}^*}, \qquad \underset{\sim}{w}(\underset{\sim}{f}) = \frac{\partial \widetilde{W}^c}{\partial \underset{\sim}{f}} \tag{2.23}$$

In the case $\underset{\approx}{\tau}^*(\underset{\approx}{F}^*)$ is not invertible, the right-hand side of (2.18), in comformity with (2.7) and the invertibility of $\underset{\approx}{S}^*(\underset{\approx}{U}^*)$, depends on $\underset{\approx}{\tau}^*$, $\underset{\approx}{R}^*$, $\underset{\sim}{f}$, regarded simultaneously as independent variables. Then the second complementary energy may be written as

$$\widetilde{W}^c[\underset{\approx}{S}^*(\underset{\approx}{\tau}^*, \underset{\approx}{R}^*), \underset{\sim}{f}] = \Sigma \underset{\approx}{\tau}^* : [\underset{\approx}{R}^* \underset{\approx}{U}^*(\underset{\approx}{S}^*)] + \underset{\sim\sim}{fw} - W[\underset{\approx}{U}^*(\underset{\approx}{S}^*), \underset{\sim}{w}(\underset{\sim}{f})]. \tag{2.24}$$

In terms of Piola stress tensor, the stress boundary condition has the form: $\underset{\approx}{\tau}^* \underset{\sim}{N}_{|A_t} = \overset{\circ}{\underset{\sim}{T}}{}^*_N$, where $\underset{\sim}{N}$ is the corresponding unit normal in $\mathcal{R}$.

3. PRINCIPLE OF VIRTUAL WORK

Let the combined medium occupy in $\mathcal{R}$ a domain V with boundary $A = A_u + A_t$, and the solution to the coupled problem be called the real displacement $\underset{\sim}{u}^*$, the real Piola stress $\underset{\approx}{\tau}^*$ and the real coupling forces $\underset{\sim}{f}^*$. The nomenclature "real" is used in order to distinguish it from the following definitions:

(1) Kinematically admissible displacement fields $\hat{\underset{\sim}{u}}^*$: which are those that satisfy the geometric boundary condition

$$\hat{\underset{\sim}{u}}^*|_{A_u} = \mathring{\underset{\sim}{u}}. \tag{3.1}$$

(2) Statically admissible stress and force fields $\check{\underset{\approx}{\tau}}^*$, $\check{\underset{\sim}{f}}^*$: which are those that satisfy the force equilibrium equation

$$\check{\underset{\approx}{\tau}}^*\underset{\sim}{\nabla} + \check{\underset{\sim}{f}}^* = 0 \tag{3.2}$$

and the traction boundary condition (dead loading)

$$\check{\underset{\approx}{\tau}}^*\underset{\sim}{N}|_{A_t} = \mathring{\underset{\sim}{T}}{}^*_N. \tag{3.3}$$

For any independent $\hat{\underset{\sim}{u}}^*$, $\check{\underset{\approx}{\tau}}^*$, $\check{\underset{\sim}{f}}^*$, the following integral relation is valid

$$\Sigma\left(\int_{A_u} \mathring{\underset{\sim}{u}}\, \check{\underset{\approx}{\tau}}^*\underset{\sim}{N}dA + \int_{A_t} \hat{\underset{\sim}{u}}^*\, \mathring{\underset{\sim}{T}}{}^*_N dA\right) = \Sigma\int_V [\check{\underset{\approx}{\tau}}^*: (\hat{\underset{\sim}{u}}^* \otimes \underset{\sim}{\nabla}) - \check{\underset{\sim}{f}}^*\hat{\underset{\sim}{u}}^*]dV. \tag{3.4}$$

Consecutive substitution of $\underset{\sim}{u}^* + \dot{\underset{\sim}{u}}^*$ and $\underset{\sim}{u}^*$ for $\hat{\underset{\sim}{u}}^*$ in Eq. (3.4) and subtraction of the resulting equations from each other lead, for the real stress and coupling force, to the virtual displacement principle:

$$\Sigma\int_{A_t} \dot{\underset{\sim}{u}}^*\mathring{\underset{\sim}{T}}{}^*_N dA = \Sigma\int_V [\underset{\approx}{\tau}^*: (\dot{\underset{\sim}{u}}^*\otimes \underset{\sim}{\nabla}) - \underset{\sim}{f}^*\dot{\underset{\sim}{u}}^*]dV. \tag{3.5}$$

In a similar manner, from (3.4) we obtain the virtual stress principle:

$$\Sigma\int_{A_u} \mathring{\underset{\sim}{u}}\, \dot{\underset{\approx}{\tau}}^*\underset{\sim}{N}dA = \Sigma\int_V [\dot{\underset{\approx}{\tau}}^*: (\underset{\sim}{u}^*\otimes \underset{\sim}{\nabla}) - \dot{\underset{\sim}{f}}^*\underset{\sim}{u}^*]dV. \tag{3.6}$$

Both principles, originating from (3.4), may be conventionally called the principles of virtual work.

4. CLASSICAL VARIATIONAL PRINCIPLES

Inserting (2.11) into (3.5), we obtain

$$0 = \int_V \dot{W}dV - \Sigma\int_{A_t} \dot{\underset{\sim}{u}}^*\mathring{\underset{\sim}{T}}{}^*_N dA = \delta\left(\int_V WdV - \Sigma\int_{A_t} \underset{\sim}{u}^*\mathring{\underset{\sim}{T}}{}^*_N dA\right), \tag{4.1}$$

since $\mathring{\underset{\sim}{T}}{}^*_N$ are dead loadings. The usage of $\underset{\sim}{f}\underset{\sim}{w}$ or $\underset{\sim}{f}^*\underset{\sim}{u}^*$ is a mere matter of convenience, because $\underset{\sim}{f}\underset{\sim}{w} = -\Sigma\,\underset{\sim}{f}^*\underset{\sim}{u}^*$. The functional

$$I(\underset{\sim}{u}^*) = \int_V W(\underset{\approx}{F}^*, \underset{\sim}{w})dV - \Sigma\int_{A_t} \underset{\sim}{u}^*\mathring{\underset{\sim}{T}}{}^*_N dA \tag{4.2}$$

defined for the class of admissible displacements is called the total potential energy. The formula $(4.1)_2$ states that the real displacements make I stationary. The converse statement is also true: the stresses $\underset{\approx}{\tau}^*$ and forces $\underset{\sim}{f}^*$ - corresponding, through the constitutive relations (2.21), to the admissible displacements $\underset{\sim}{u}^*$ which make I stationary - are admissible and satisfy the moment equilibrium condition. In order to show this, using

$$\dot{W} = \Sigma\frac{\partial W}{\partial \underset{\approx}{F}^*}: \dot{\underset{\approx}{F}}{}^* + \frac{\partial W}{\partial \underset{\sim}{w}} : \dot{\underset{\sim}{w}} = \Sigma[(\dot{\underset{\sim}{u}}^*\underset{\approx}{\tau}^*)\, \underset{\sim}{\nabla} - \dot{\underset{\sim}{u}}^*(\underset{\approx}{\tau}^*\, \underset{\sim}{\nabla} + \underset{\sim}{f}^*)], \tag{4.3}$$

we compute the variation of I $(\underset{\sim}{u}^*)$:

$$\dot{I} = \Sigma\,[\int_{A_t} \dot{\underset{\sim}{u}}^*(\underset{\approx}{\tau}^*\underset{\sim}{N} - \overset{\circ}{\underset{\sim}{T}}{}^*_N)dA - \int_V \dot{\underset{\sim}{u}}^*(\underset{\approx}{\tau}^*\underset{\sim}{\nabla} + \underset{\sim}{f}^*)dV]. \tag{4.4}$$

From $\dot{I} = 0$ and the arbitrariness of $\dot{\underset{\sim}{u}}^*$ it follows that

$$\underset{\approx}{\tau}^*\underset{\sim}{\nabla} + \underset{\sim}{f}^* = 0 \quad \text{in } V \quad \text{and} \quad \underset{\approx}{\tau}^*\underset{\sim}{N}\big|_{A_t} = \overset{\circ}{\underset{\sim}{T}}{}^*_N. \tag{4.5}$$

From (2.21) and the symmetry of $\partial W/\partial \underset{\approx}{E}^*$ follows the moment equilibrium condition:

$$\underset{\approx}{\tau}^*\underset{\approx}{F}^{*T} = \underset{\approx}{F}^*\underset{\approx}{\tau}^{*T} \quad \text{in } V. \tag{4.6}$$

These stress fields $\underset{\approx}{\tau}^*$ and coupling force fields $\underset{\sim}{f}^*$ are real because they not only correspond to admissible displacements $\underset{\sim}{u}^*$, but also satisfy the conditions (4.5,6). The variational principle associated with the functional I is called the stationary principle of the total potential energy.

We proceed now to derive the stationary principle of the complementary energy, i. e. Reissner's principle, from the virtual stress principle. Substituting

$$\dot{\underset{\approx}{\tau}}^*:(\underset{\sim}{u}^*\otimes\underset{\sim}{\nabla}) = \underset{\approx}{E}^*:\dot{\underset{\approx}{T}}^* + \frac{1}{2}\,\delta\,\{[(\underset{\sim}{\nabla}\otimes\underset{\sim}{u}^*)(\underset{\sim}{u}^*\otimes\underset{\sim}{\nabla})]:\underset{\approx}{T}^*\}, \tag{4.7}$$

$$\dot{W}^C = \Sigma\,\underset{\approx}{E}^*:\dot{\underset{\approx}{T}}^* + \underset{\sim}{w}\dot{\underset{\sim}{f}} \tag{4.8}$$

into (3.6), we obtain

$$\delta\left\{\int_V [W^C + \frac{1}{2}\,\Sigma((\underset{\sim}{\nabla}\otimes\underset{\sim}{u}^*)(\underset{\sim}{u}^*\otimes\underset{\sim}{\nabla})):\underset{\approx}{T}^*]dV - \Sigma\int_{A_u}\overset{\circ}{\underset{\sim}{u}}\underset{\approx}{F}^*\underset{\approx}{T}^*\underset{\sim}{N}dA\right\} = 0. \tag{4.9}$$

The functional

$$I^C(\underset{\approx}{T}^*, \underset{\sim}{f}, \underset{\sim}{u}^*) = \int_V\{W^C(\underset{\approx}{T}^*, \underset{\sim}{f}) + \frac{1}{2}\,\Sigma[(\underset{\sim}{\nabla}\otimes\underset{\sim}{u}^*)(\underset{\sim}{u}^*\otimes\underset{\sim}{\nabla})]:\underset{\approx}{T}^*\}dV - \Sigma\int_{A_u}\overset{\circ}{\underset{\sim}{u}}\underset{\approx}{F}^*\underset{\approx}{T}^*\underset{\sim}{N}dA \tag{4.10}$$

defined for admissible displacements, stresses and coupling force is called the total complementary energy. The formula (4.9) states that the real $\underset{\approx}{T}^*$, $\underset{\sim}{f}$ and $\underset{\sim}{u}^*$ fields make I^C stationary. In order to prove the converse, we introduce the Lagrangean multipliers $\underset{\sim}{y}^*$ and $\underset{\sim}{z}^*$ to release the variables $\underset{\approx}{T}^*$ and $\underset{\sim}{f}$ from constraints. This leads to a functional without subsidiary conditions:

$$\tilde{I}^C = \int_V\{W^C(\underset{\approx}{T}^*, \underset{\sim}{f}) + \Sigma[\tfrac{1}{2}((\underset{\sim}{\nabla}\otimes\underset{\sim}{u}^*)(\underset{\sim}{u}^*\otimes\underset{\sim}{\nabla})):\underset{\approx}{T}^* + \underset{\sim}{y}^*((\underset{\approx}{F}^*\underset{\approx}{T}^*)\underset{\sim}{\nabla} + \underset{\sim}{f}^*)]\}dV - \Sigma[\int_{A_u}\overset{\circ}{\underset{\sim}{u}}\underset{\approx}{F}^*\underset{\approx}{T}^*\underset{\sim}{N}dA + \int_{A_t}\underset{\sim}{z}^*(\underset{\approx}{F}^*\underset{\approx}{T}^*\underset{\sim}{N} - \overset{\circ}{\underset{\sim}{T}}{}^*_N)dA], \tag{4.11}$$

in which $\underset{\approx}{T}^*$, $\underset{\sim}{f}$, $\underset{\sim}{u}^*$, $\underset{\sim}{y}^*$ and $\underset{\sim}{z}^*$ are independent free variables. Varying only $\underset{\sim}{u}^*$, we calculate the variation of $\tilde{I}^C$:

$$\delta\tilde{I}^C = \Sigma\left\{\int_V[(\underset{\sim}{u}^* - \underset{\sim}{y}^*)\otimes\underset{\sim}{\nabla}]:\dot{\underset{\approx}{\tau}}^*dV + \int_{A_u}(\underset{\sim}{y}^* - \overset{\circ}{\underset{\sim}{u}})\dot{\underset{\approx}{\tau}}^*\underset{\sim}{N}dA + \int_{A_t}(\underset{\sim}{y}^* - \underset{\sim}{z}^*)\dot{\underset{\approx}{\tau}}^*\underset{\sim}{N}dA\right\}, \tag{4.12}$$

where $\dot{\underset{\approx}{\tau}}^* = \dot{\underset{\approx}{F}}^*\underset{\approx}{T}^*$ are the increments of $\underset{\approx}{\tau}^*$ caused by $\dot{\underset{\sim}{u}}^*$ and may be regarded as entirely arbitrary. From $(\underset{\sim}{u}^* - \underset{\sim}{y}^*)\otimes\underset{\sim}{\nabla} = 0$ in V, $\underset{\sim}{y}^* = \overset{\circ}{\underset{\sim}{u}}$ on A_u and $\underset{\sim}{z}^* = \underset{\sim}{y}^*$ on A_t, as a result of $\delta\tilde{I}^C = 0$, follows the physical interpretation of the Lagrangean multipliers: $\underset{\sim}{y}^* = \underset{\sim}{u}^*$ in V and $\underset{\sim}{z}^* = \underset{\sim}{u}^*$ on A_t. Thus, we have the functional

$$\tilde{I}^C(\underset{\approx}{T}^*, \underset{\sim}{f}, \underset{\sim}{u}^*) = \int_V\{W^C(\underset{\approx}{T}^*, \underset{\sim}{f}) + \Sigma[\tfrac{1}{2}((\underset{\sim}{\nabla}\otimes\underset{\sim}{u}^*)(\underset{\sim}{u}^*\otimes\underset{\sim}{\nabla})):\underset{\approx}{T}^* + \underset{\sim}{u}^*((\underset{\approx}{F}^*\underset{\approx}{T}^*)\underset{\sim}{\nabla} + \underset{\sim}{f}^*)]\}dV - \Sigma[\int_{A_u}\overset{\circ}{\underset{\sim}{u}}\underset{\approx}{F}^*\underset{\approx}{T}^*\underset{\sim}{N}dA + \int_{A_t}\underset{\sim}{u}^*(\underset{\approx}{F}^*\underset{\approx}{T}^*\underset{\sim}{N} - \overset{\circ}{\underset{\sim}{T}}{}^*_N)dA] \tag{4.13}$$

$$= \int_V\{W^C(\underset{\approx}{T}^*, \underset{\sim}{f}) + \Sigma[\tfrac{1}{2}((\underset{\sim}{\nabla}\otimes\underset{\sim}{u}^*)(\underset{\sim}{u}^*\otimes\underset{\sim}{\nabla})):\underset{\approx}{T}^* - (\underset{\sim}{u}^*\otimes\underset{\sim}{\nabla}):(\underset{\approx}{F}^*\underset{\approx}{T}^*) + \underset{\sim}{u}^*\underset{\sim}{f}^*]\}dV + \Sigma[\int_{A_u}(\underset{\sim}{u}^* - \overset{\circ}{\underset{\sim}{u}})\,\underset{\approx}{F}^*\underset{\approx}{T}^*\underset{\sim}{N}dA + \int_{A_t}\underset{\sim}{u}^*\overset{\circ}{\underset{\sim}{T}}{}^*_N\,dA]. \tag{4.14}$$

Calculating the variation of $\tilde{I}^c$:

$$\delta\tilde{I}^c = \sum\int_V\left\{\left[\frac{\partial W^c}{\partial T^*} - \frac{1}{2}(\nabla\otimes u^* + u^*\otimes\nabla + (\nabla\otimes u^*)(u^*\otimes\nabla))\right]:\dot{T}^* + [(F^*T^*)\nabla + f^*]\dot{u}^*\right\}dV$$

$$+\int_V\left(\frac{\partial W^c}{\partial f} - w\right)\dot{f}dV + \sum\left[\int_{A_u}(u^* - \mathring{u})\delta(F^*T^*)NdA - \int_{A_t}\dot{u}^*(F^*T^*N - \mathring{T}_N^*)dA\right], \quad (4.15)$$

we finally arrive at

$$(F^*T^*)\nabla + f^* = 0 \qquad \text{in } V, \quad (4.16)$$

$$F^*T^*N|_{A_t} = \mathring{T}_N^*, \qquad u^*|_{A_u} = \mathring{u}, \quad (4.17)$$

$$\frac{\partial W^c}{\partial f} = u^2 - u^1, \frac{\partial W^c}{\partial T^*} = \frac{1}{2}[\nabla\otimes u^* + u^*\otimes\nabla + (\nabla\otimes u^*)(u^*\otimes\nabla)] \quad \text{in V}. \quad (4.18)$$

Eqs. (4.18) confirm that T^*, f and u^* which make $\tilde{I}^c$ (and then I^c) stationary correspond to each other through the constitutive relation (2.19). Thus T^*, f and u^* solve the problem.

5. LEVINSON'S PRINCIPLE

Assume now that $\tau^*(F^*)$ are invertible. Using (2.23), we have

$$\delta\tilde{W}^c = \sum[\dot{\tau}^*:(u^*\otimes\nabla) + \text{tr}\,\dot{\tau}^*] + w\dot{f}. \quad (5.1)$$

A substitution of (5.1) into (3.6) yields

$$\delta\left\{\int_V[\tilde{W}^c(\tau^*, f) - \sum\text{tr}\tau^*]dV - \sum\int_{A_u}\mathring{u}\tau^*NdA\right\} = 0. \quad (5.2)$$

For the functional

$$I_1^c(\tau^*, f) = \int_V[\tilde{W}^c(\tau^*, f) - \sum\text{tr}\tau^*]dV - \sum\int_{A_u}\mathring{u}\tau^*NdA \quad (5.3)$$

defined for the class of admissible stresses and coupling force, Eq. (5.2) states that the real τ^* and f make I_1^c stationary. In the following the converse will be proved: the admissible τ^* and f which make I_1^c stationary satisfy the moment equilibrium condition and have associated displacement fields satisfying the geometric boundary condition. Supposing $\tilde{W}_1^c(\tau^1)$ and $\tilde{W}_c^c(\tau^2)$ are composed functions of τ^* through $K^* = \frac{1}{2}\tau^{*T}\tau^*$, we have

$$F^* = \frac{\partial W^c}{\partial \tau^*} = \tau^*\frac{\partial W^c}{\partial K^*}. \quad (5.4)$$

From the symmetry of $\partial\tilde{W}^c/\partial K^*$, the moment equilibrium condition follows immediately:

$$\tau^*F^{*T} = F^*\tau^{*T} \qquad \text{in V}. \quad (5.5)$$

In order to be able to calculate the displacements from

$$u^*(X) = \int_{X_0}^{X}(F^* - I)dX + \mathring{u}(X_0), \quad (5.6)$$

F^* obtained from (5.4) must satisfy the integrability condition:

$$F^*\times\nabla = 0 \qquad \text{in V}. \quad (5.7)$$

For the purpose of release from constraints, we introduce stress functions P^* and force functions $Q^*(Q^1 = Q, Q^2 = -Q)$:

$$\tau^* = P^*\times\nabla - Q^*\otimes\nabla, \qquad f^* = Q^*\otimes\nabla\nabla, \quad (5.8)$$

which satisfy (3.2) identically. Thus the variation of I_1^c is

$$\dot{I}_1^c = \Sigma\{\int_V [(\underset{\approx}{F}^* - \underset{\approx}{I}):(\dot{\underset{\approx}{P}}^* \times \underset{\sim}{\nabla} - \dot{\underset{\sim}{Q}}^* \otimes \underset{\sim}{\nabla}) - \underset{\sim}{u}^*(\dot{\underset{\sim}{Q}}^* \otimes \underset{\sim\sim}{\nabla\nabla})]dV - \int_{A_u} \mathring{\underset{\sim}{u}}\underset{\approx}{\tau}^* NdA\}$$

$$= \Sigma\{\int_V (\underset{\approx}{F}^* \times \underset{\sim}{\nabla}):\dot{\underset{\approx}{P}}^* dV + \oint_A [(\underset{\approx}{F}^* - \underset{\approx}{I}):(\dot{\underset{\approx}{P}}^* \times \underset{\sim}{N}) - \underset{\sim}{u}^*(\dot{\underset{\sim}{Q}}^* \otimes \underset{\sim}{\nabla})\underset{\sim}{N}]dA - \int_{A_u} \mathring{\underset{\sim}{u}}\dot{\underset{\approx}{\tau}}^* \underset{\sim}{N}dA\}. \tag{5.9}$$

From the volume integrals the integrability conditions (5.7) follow. Thus two vector fields $\underset{\sim}{u}^*(\underset{\sim}{X})$ are determined from (5.6). These fields become the displacement fields provided they satisfy the geometric boundary condition. In order to show this, substituting these fields into the first surface integrals of (5.9), the first integrand functions assume the form:

$$(\underset{\approx}{F}^* - \underset{\approx}{I}):(\dot{\underset{\approx}{P}}^* \times \underset{\sim}{N}) = \underset{\sim}{u}^*(\dot{\underset{\approx}{P}}^* \times \underset{\sim}{\nabla})\underset{\sim}{N} - [(\underset{\sim}{u}^*\dot{\underset{\approx}{P}}^*) \times \underset{\sim}{\nabla}]\underset{\sim}{N}. \tag{5.10}$$

Since "the flux of the curl of any vector field through a closed surface equals zero" (Stokes theorem), the surface integrals reduce to

$$\oint \underset{\sim}{u}^*(\dot{\underset{\approx}{P}}^* \times \underset{\sim}{\nabla} - \dot{\underset{\sim}{Q}}^* \otimes \underset{\sim}{\nabla})\underset{\sim}{N}dA - \int_{A_u} \mathring{\underset{\sim}{u}}\dot{\underset{\approx}{\tau}}^* \underset{\sim}{N}dA = \int_{A_u} (\underset{\sim}{u}^* - \mathring{\underset{\sim}{u}})\dot{\underset{\approx}{\tau}}^* \underset{\sim}{N}dA \tag{5.11}$$

and, consequently, $\underset{\sim}{u}^*|_{A_u} = \mathring{\underset{\sim}{u}}$, where $\dot{\underset{\approx}{\tau}}^*\underset{\sim}{N}|_{A_t} = 0$ has been used. Thus the proof of Levinson's principle is completed.

6. FRAEIJS DE VEUBEKE'S PRINCIPLE

In the case $\underset{\approx}{\tau}^*(\underset{\approx}{F}^*)$ are not invertible, Levinson's principle is invalid. Using (2.24) in (5.2), we get

$$\delta\{\int_V [\tilde{W}^c(\underset{\approx}{S}^*(\underset{\approx}{\tau}^*, \underset{\approx}{R}^*), \underset{\sim}{f}) - \Sigma\, \mathrm{tr}\underset{\approx}{\tau}^*]dV - \Sigma\int_{A_u} \mathring{\underset{\sim}{u}}\underset{\approx}{\tau}^* \underset{\sim}{N}dA\} = 0 \tag{6.1}$$

and functional

$$I_2^c(\underset{\approx}{\tau}^*, \underset{\approx}{R}^*, \underset{\sim}{f}) = \int_V \{\tilde{W}^c[\underset{\approx}{S}^*(\underset{\approx}{\tau}^*, \underset{\approx}{R}^*), \underset{\sim}{f}] - \Sigma\mathrm{tr}\underset{\approx}{\tau}^*\}dV - \Sigma\int_{A_u} \mathring{\underset{\sim}{u}}\underset{\approx}{\tau}^* \underset{\sim}{N}dA \tag{6.2}$$

with $\underset{\approx}{\tau}^*$, $\underset{\approx}{R}^*$ and $\underset{\sim}{f}$ as independent variables. Eq. (6.1) states that the real stresses $\underset{\approx}{\tau}^*$, rotations $\underset{\approx}{R}^*$ and coupling force $\underset{\sim}{f}$ make I_2^c stationary. The converse is true: the admissible $\underset{\approx}{\tau}^*$, $\underset{\approx}{R}^*$ and $\underset{\sim}{f}$ which make I_2^c stationary satisfy the moment equilibrium condition and possess unique associated displacement fields $\underset{\sim}{u}^*$ consistent with geometric boundary conditions. Thus these $\underset{\approx}{\tau}^*$, $\underset{\approx}{R}^*$, $\underset{\sim}{f}$ are the solution to the problem. The associated displacements mean the displacements calculated in the following way: starting from $\underset{\approx}{\tau}^*$ and $\underset{\approx}{R}^*$, by virtue of $(2.7)_2$ and the invertibility of $\underset{\approx}{S}^*(\underset{\approx}{U}^*)$ compute $\underset{\approx}{R}^*\underset{\approx}{U}^*$, and then perform the integration

$$\underset{\sim}{u}^*(\underset{\sim}{X}) = \int_{\underset{\sim}{X}_0}^{\underset{\sim}{X}} (\underset{\approx}{R}^*\underset{\approx}{U}^* - \underset{\approx}{I})d\underset{\sim}{X} + \mathring{\underset{\sim}{u}}(\underset{\sim}{X}_0). \tag{6.3}$$

Now the integrability condition is

$$(\underset{\approx}{R}^*\underset{\approx}{U}^*) \times \underset{\sim}{\nabla} = 0 \qquad \text{in } V. \tag{6.4}$$

Introducing $\underset{\approx}{P}^*$ and $\underset{\sim}{Q}^*$, as in (5.8), and making use of

$$\Sigma(\frac{\partial \tilde{W}^c}{\partial \underset{\approx}{S}^*}:\dot{\underset{\approx}{S}}^* - \mathrm{tr}\dot{\underset{\approx}{\tau}}^*) + \frac{\partial \tilde{W}^c}{\partial \underset{\sim}{f}}\dot{\underset{\sim}{f}} = \Sigma\{(\underset{\approx}{R}^*\underset{\approx}{U}^* - \underset{\approx}{I}):(\dot{\underset{\approx}{P}}^* \times \underset{\sim}{\nabla})$$

$$+ \frac{1}{2}(\underset{\approx}{\tau}^*\underset{\approx}{U}^*\underset{\approx}{R}^{*T} - \underset{\approx}{R}^*\underset{\approx}{U}^*\underset{\approx}{\tau}^{*T}):(\dot{\underset{\approx}{R}}^* \underset{\approx}{R}^{*T}) - [\underset{\sim}{u}^*(\dot{\underset{\sim}{Q}}^* \otimes \underset{\sim}{\nabla})]\underset{\sim}{\nabla}\}, \tag{6.5}$$

we calculate the variation of I_2^c:

$$\dot{I}_2^c = \sum\left\{\int_V [((\underset{\approx}{R}^*\underset{\approx}{U}^*) \times \underset{\sim}{\nabla}):\dot{\underset{\approx}{P}}^* + \frac{1}{2}(\underset{\approx}{\tau}^*\underset{\approx}{U}^*\underset{\approx}{R}^{*T} - \underset{\approx}{R}^*\underset{\approx}{U}^*\underset{\approx}{\tau}^{*T}):(\dot{\underset{\approx}{R}}^*\underset{\approx}{R}^{*T})]dV \right.$$
$$\left. + \oint_A [(\underset{\approx}{R}^*\underset{\approx}{U}^* - \underset{\approx}{I}):(\dot{\underset{\approx}{P}}^* \times \underset{\sim}{N}) - \underset{\sim}{u}^*(\dot{\underset{\sim}{Q}}^* \otimes \underset{\sim}{\nabla})\underset{\sim}{N}]dA - \int_{A_u} \mathring{\underset{\sim}{u}}\dot{\underset{\approx}{\tau}}^*\underset{\sim}{N}dA\right\}. \qquad (6.6)$$

From the volume integral we arrive at the integrability condition (6.4) and the moment equilibrium condition:

$$\underset{\approx}{\tau}^*\underset{\approx}{U}^*\underset{\approx}{R}^{*T} = \underset{\approx}{R}^*\underset{\approx}{U}^*\underset{\approx}{\tau}^{*T} \qquad \text{in } V. \qquad (6.7)$$

If we replace $\underset{\approx}{F}^*$ by $\underset{\approx}{R}^*\underset{\approx}{U}^*$, the further procedure to obtain the displacement fields $\underset{\sim}{u}^*(\underset{\sim}{X})$ satisfying the geometric boundary condition is precisely the same as for the case of Levinson's principle. Thus Fraeijs de Veubeke's principle is entirely proved.

It can be seen that we have derived all the variational principles for closely interacting solid continua, as has been done for nonlinear elasticity in [3], from the virtual work principle in a unifying manner.

ACKNOWLEDGEMENTS

The author is indebted to Professor Th. Lehmann for his permanent interest and hearty hospitality. Substantial support from the Deutsche Forschungsgemeinschaft and Alexander von Humboldt-Stiftung is gratefully acknowledged.

REFERENCES

[1] Bedford, A. and Stern, M., Acta Mech. 14 (1972) 85-102.

[2] Tiersten, H.F. and Jahanmir, M., Arch. Rat. Mech. Anal. 65 (1977) 153-192.

[3] Guo, Zhong-heng, Arch. Mech. 32 (1980) 577-596.

[4] McCarthy, M.F. and Tiersten, H.F., in Proc. Conf. "CMDS3", SM Study No. 15 (Univ. of Waterloo Press, Canada, 1980) 531-549.

Continuum Models of Discrete Systems 4
eds. O. Brulin and R.K.T. Hsieh

NONLOCAL ELECTRODYNAMICS OF VISCOUS FLUIDS

Richard K. T. Hsieh

Department of Mechanics
Royal Institute of Technology
S-10044 Stockholm
Sweden

A nonlocal theory of electrodynamics of viscous fluids is developed on the basis of the Chu formulation of the Maxwell equations and the general principles of continuum mechanics. Nonlocal balance laws and jump conditions are obtained. Boundary conditions and constitutive equations satisfying the principle of objectivity are derived.

INTRODUCTION

It is well known that the classical theory of fluids ignores long range intermolecular forces and therefore is valid only for very long wavelengths. Similarly classical continuum mechanics needs modification if it is to account for the inner structure of fluids, e.g. suspensions.

It is also well known that the classical Maxwell theory of electromagnetism predicts a constant wave velocity and consequently a constant refractive index in an isotropic nondissipative medium while experiments indicate that this index is frequency dependent. For magnetic materials the existence of magnetic domains, instabilities and spin waves cannot be explained through the classical formalism.

To remedy the mechanical inadequacies both on the molecular and granular levels, several continuum theories were proposed. Concerned with the molecular aspects we have the nonlocal theories developed by among others, Kunin [1], Kröner [2], Edelen and Laws [3], Eringen [4].

Dealing with the inner structure of materials we have an extensive literature on polar theories, see e.g. Nowacki and Olszak [5], Hsieh [6].

As to incorporate the electromagnetic effects, either quantum mechanical approach is used or the molecular theories are modified.

The purpose of the present paper is the construction of the nonlocal theory for viscous fluids subject to electromagnetic interactions.

The method is based on the works of Eringen [7] and Edelen [8]. The features that distinguish nonlocal theories from the classical ones are the statements of postulates for the whole body, the occurence of localization residuals in the local equations that are obtained from the global statements and the formulation of constitutive relations that allow the dependence on the properties of the whole body. In addition this theory is adequate to describe and analyze classical effects from the continuum region all the way to the molecular scale. The Maxwell equations used are based on the Chu formulation, cf. Pao and Hutter [9] and Hsieh [10].

We assume that rigid body electromagnetism can be applied to deformable bodies in motion and for the sake of simplicity we avoid relativity here. The resulting descriptions are thus accurate up to terms of the order v^2/c^2 with v as material velocities and c being the speed of light in vacuum.

In the second section starting with the global electromechanical balance laws for the entire body we obtain the equivalent local laws. These equations contain nonlocal residuals accounting for the distant molecular interactions. The integrals of the nonlocal residuals over the entire body vanish. In the last section we develop a set of constitutive equations for nonlocal electromagnetic viscous fluids and determine the nonlocal residuals.

BALANCE LAWS

The integral form of the balance laws in continuum theory of electromagnetism for moving media can be written as

Gauss-Faraday's law:

$$\oint_{S-\Gamma} \underline{B}.d\underline{S} = 0 \qquad (1)$$

Faraday's law:

$$\oint_{C-\gamma} \underline{E}^e.\, d\underline{x} = -\frac{d}{dt}\int_{S-\Gamma} \underline{B}.d\underline{S} \qquad (2)$$

Gauss' law:

$$\oint_{S-\Gamma} \underline{D}.d\underline{S} = \int_{V-\Gamma} \rho_f\, dV \qquad (3)$$

Ampére's law:

$$\oint_{C-\gamma} \underline{H}^e.d\underline{x} = \frac{d}{dt}\int_{S-\Gamma} \underline{D}.d\underline{S} + \int_{S-\Gamma} \underline{J}_f^e.d\underline{S} \qquad (4)$$

Conservation of charge:

$$\frac{d}{dt}\int_{V-\Gamma} \rho_f dV + \int_{S-\Gamma} \underline{J}^e.d\underline{S} = 0 \qquad (5)$$

In the preceding, $\underline{B}$ is the magnetic induction, $\underline{D}$ the electrical displacement, ρ_f the free electrical charge, $\underline{E}^e$ and $\underline{H}^e$ are respectively the effective electric field and magnetic field while $\underline{J}^e$ is the free effective current. The electromagnetic laws are expressed in MKSA units.

The motion and deformation of a continuum are governed by four balance laws.

Conservation of mass:

$$\frac{d}{dt}\int_{V-\Gamma} \rho dV = 0 \qquad (6)$$

Balance of linear momentum:

$$\frac{d}{dt}\int_{V-\Gamma} \rho \underline{v}\, dV = \oint_{S-\Gamma} \underline{t}_k\, dS_k + \int_{V-\Gamma} \rho \underline{f}\, dV \tag{7}$$

Balance of angular momentum:

$$\frac{d}{dt}\int_{V-\Gamma} \underline{x} \times \rho \underline{v}\, dV = \oint_{S-\Gamma} \underline{x} \times \underline{t}_k\, dS_k + \int_{V-\Gamma} (\underline{x} \times \rho \underline{f} + \rho \underline{l})\, dV \tag{8}$$

Conservation of energy:

$$\frac{d}{dt}\int_{V-\Gamma} \rho(\tfrac{1}{2}\,\underline{v}.\underline{v} + \epsilon)dV = \oint_{S-\Gamma} (\underline{t}_k.\underline{v} + q_k)dS_k + \int_{V-\Gamma} \rho(\underline{f}.\underline{v} + r_h + r_e)\, dV \tag{9}$$

In the preceding ρ is the mass density, $\rho\underline{v}$ the linear momentum density, $\underline{t}_k$ the stress vector, $\underline{f}$ the body force per unit mass, $\underline{l}$ the body couple per unit mass, ϵ the internal energy density, $\underline{q}$ the heat vector, r_h the heat energy supply per unit mass and r_e is the energy supply generated by the electromagnetic fields.

The stress vector $\underline{t}_k$ includes the action in the form of a surface force due to electromagnetic fields. The rate of work done by the couple $\underline{l}$ is contained in the source term r_e. All energy flux flowing across the surface other than the heat flux $\underline{q}$ is represented by the term $\underline{t}_k.\underline{v}$

For a general thermodynamic process, a fifth balance law in the form of an inequality (Clausius-Duhem) must be added

$$\frac{d}{dt}\int_{V-\Gamma} \rho\eta dV - \oint_{S-\Gamma} \frac{1}{T}\,\underline{q}.\underline{dS} \geq \int_{V-\Gamma} \frac{\rho r_h}{T}\, dV \tag{10}$$

where η is the entropy density and T the absolute temperature.

In (9) and (10) the heat vector $\underline{q}$ includes heating of purely electromagnetic nature.

In the Chu formulation the effective fields are defined as

$$\underline{E}^e = \underline{E} + \underline{v} \times \mu_o \underline{H};\ \underline{H}^e = \underline{H} - \underline{v} \times \epsilon_o \underline{E};\ \underline{J}^e = \underline{J}_f - \rho_f \underline{v} \tag{11}$$

where $\underline{E}$, $\underline{H}$ and $\underline{J}_f$ are fields measured in laboratory frame and μ_o and ϵ_o are two universal constants. For future convenience we also introduce the polarization vector $\underline{P}$ and the magnetization vector $\underline{M}$ by

$$\underline{D} = \epsilon_o\, \underline{E} + \underline{P};\ \underline{B} = \mu_o(\underline{H} + \underline{M}) \tag{12}$$

All integrations are taken over material volume V, material surface S and material circuit C with the exception of points contained on the discontinuity surface Γ and discontinuity curve γ. V, S and C are thus moving with the materials contained in them.

The time differentiations should then be carried out by applying the transport theorems [11]

$$\frac{d}{dt}\int_{V - \Gamma} \Psi \, dV = \int_{V - \Gamma} \{\frac{\partial \Psi}{\partial t} + \nabla . (\underline{v}\Psi)\} dV$$

$$\frac{d}{dt}\int \underline{q}.\underline{dS} = \int \overset{*}{\underline{q}}.\underline{dS}$$

$$\overset{*}{\underline{q}} \equiv \frac{\partial \underline{q}}{\partial t} + (\nabla.\underline{q})\underline{v} + \nabla \times (\underline{q} \times \underline{v}) \tag{13}$$

In (13) $\overset{*}{\underline{q}}$ is known as the convective derivative of a vector $\underline{q}$ and Ψ can be any smooth scalar, vector or tensor function.

Applying Green Gauss theorem and Stoke's theorem the integral balance laws (1) to (9) can be transformed to either of the following two forms

$$\int_{V - \Gamma} \{\frac{\partial \Psi}{\partial t} + \nabla . (\underline{v}\Psi) - \tau_{k,k} - g\} dV + \int_{\Gamma} [\Psi(v_k - u_k) - \tau_k] \, n_k dS = 0$$

$$\int_{S - \Gamma} (\overset{*}{\underline{q}} - \text{curl } \underline{h} - \underline{r}).\underline{dS} + \int_{\Gamma} [\underline{q} \times (\underline{v} - \underline{u}) - \underline{h}] . \underline{k} \, dx = 0 \tag{14}$$

Where $\underline{k}$ is the unit tangent vector on γ, $\underline{u}$ is the velocity of the discontinuity surface Γ. The brackets indicate the jump across Γ in (14a) and across γ in (14b).

The local balance laws of classical theory are obtained by positing that the balance laws (1)-(10) with the transformations (14) are valid for every part of the body. When the long range intermolecular interactions are strong this is not permissible. However these expressions can be localized by introducing the localization residuals which account for the effects of nonlocal fields. The integrals of the nonlocal residuals over the entire body vanish.

The determination of the nonlocal residuals is an integral part of the nonlocal theory. Thus (1)-(9) are equivalent to

Gauss-Faraday:

$$\nabla.\underline{B} = \hat{\underline{m}} \quad \text{in } V - \Gamma$$

$$[\underline{B} + \hat{\underline{B}}] . \underline{n} = 0 \quad \text{on } \Gamma \tag{15}$$

Faraday:

$$\nabla \times \underline{E}^e + \overset{*}{\underline{B}} = \hat{\underline{b}} \quad \text{in } V - \Gamma$$

$$[\underline{E} + \underline{u} \times \mu_o\underline{H} + \hat{\underline{E}}] \times \underline{n} = 0 \quad \text{on } \Gamma \tag{16}$$

Gauss:

$$\nabla.\underline{D} - \rho_f = \hat{\rho}_f \quad \text{in } V - \Gamma$$

$$[\underline{D} + \hat{\underline{D}}] . \underline{n} = 0 \quad \text{on } \Gamma \tag{17}$$

Ampére:

$$\nabla \times \underline{H}^e - \overset{*}{\underline{D}} - \underline{J}^e = \hat{\underline{J}}^e \quad \text{in } V - \Gamma$$

$$[\underline{H} - \underline{u} \times \varepsilon_o \underline{E} + \hat{\underline{H}}] \times \underline{n} = 0 \quad \text{on } \Gamma \tag{18}$$

Charge:

$$\nabla . \underline{J}^e + \frac{\partial \rho_f}{\partial t} + \nabla . (\underline{v}\, \rho_f) = \hat{\sigma} \quad \text{in } V - \Gamma$$

$$[\underline{J}^e + \hat{\underline{\Sigma}}] . \underline{n} = 0 \quad \text{on } \Gamma \tag{19}$$

Mass:

$$\frac{\partial \rho}{\partial t} + \nabla . (\rho \underline{v}) = \hat{\rho} \quad \text{in } V - \Gamma$$

$$[\rho(v_k - u_k) - \hat{\rho}_k] n_k = 0 \quad \text{on } \Gamma \tag{20}$$

Linear Momentum:

$$\underline{t}_{k,k} + \rho(\underline{f} - \dot{\underline{v}}) = \hat{\rho}\underline{v} - \rho\hat{\underline{f}} \quad \text{in } V - \Gamma$$

$$[\underline{t}_k - \rho\underline{v}(v_k - u_k) + \rho\hat{\underline{f}}_k]\, n_k = 0 \quad \text{on } \Gamma \tag{21}$$

Angular momentum:

$$\underline{x}_{,k} \times \underline{t}_k - \rho \underline{l} = \rho \underline{x} \times \hat{\underline{f}} - \rho \hat{\underline{l}} \quad \text{in } V - \Gamma$$

$$[\rho \underline{x} \times \{\underline{t}_k - \underline{v}\,(v_k - u_k)\} + \hat{\underline{l}}_k]\, n_k = 0 \quad \text{on } \Gamma \tag{22}$$

Energy:

$$\rho\dot{\epsilon} - \underline{t}_k . \underline{v}_{,k} - q_{k,k} - \rho r_h - \rho r_e =$$

$$= \rho\hat{r}_h - \rho\hat{r}_e - \rho\underline{v}.\hat{\underline{f}} - \hat{\rho}\,\{\epsilon - \frac{v^2}{2}\} \quad \text{in } V - \Gamma$$

$$[\underline{t}_k . \underline{v} + q_k - \rho\,(\epsilon + \frac{1}{2} v^2)\,(v_k - u_k) + \hat{r}_{k(h)} + \hat{r}_{k(e)}]\, n_k = 0 \quad \text{on } \Gamma \tag{23}$$

2nd law of thermodynamics:

$$\rho\dot{\eta} - \nabla . (\frac{1}{T}\underline{q}) - \frac{1}{T}\rho r_h + \hat{\rho}\eta - \frac{1}{T}\rho\hat{s} \geq 0 \quad \text{in } V - \Gamma$$

$$[\rho\eta(v_k - u_k) - \frac{1}{T}\underline{q} - \hat{s}_k]\, n_k \geq 0 \quad \text{on } \Gamma \tag{24}$$

In nonlocal theory, nonlocal residuals of various fields are introduced to localize the global laws. These are denoted by letters carrying a carat $\hat{}$. Thus, $\hat{\rho}$ represents the mass production at a point of the body by all other points of the body. Similarly $\hat{f}$ is the nonlocal body force at a point arising from the long-range forces. The nonlocal residuals for the couple, energy and entropy are respectively denoted by $\hat{l}$, $\hat{r}_h + \hat{r}_e$ and $\hat{s}$. Corresponding to these terms on any discontinuity surface Γ sweeping the body with a velocity $\underline{u}$ we have the nonlocal residuals denoted by $\hat{\rho}$, $\hat{\underline{f}}_k$, $\hat{\underline{l}}_k$, $\hat{r}_h + \hat{r}_e$ and $\hat{s}$. The electromagnetic volume residuals are $\hat{m}$, $\hat{\rho}_f$, $\hat{\sigma}$. Their surface residuals are given by $\hat{\rho}_f$, $\hat{r}_{k(h)}$, $\hat{r}_{k(e)}$, $\hat{\underline{B}}_k$, $\hat{\underline{D}}_k$ and line residuals, by $\hat{\underline{E}}$ and $\hat{\underline{H}}$. $\hat{\rho}_f$ is the induced nonlocal charge at a point x of the body due to charges at all other points. Similarly $\hat{m}$ is viewed as the magnetic pole strength at x induced by the rest of the body. $\hat{\underline{J}}_f$ is the contributions from all other points of the body to the currents at x and $\hat{\underline{b}}$ to the magnetic flux, cf. [7]

The nonlocal residuals of fields are subject to the conditions that over the manifolds of their definitions they vanish, i.e.

$$\int_{V-\Gamma} (\hat{\rho}, \rho\hat{\underline{f}}_k, \rho\hat{\underline{l}}_k, \rho\ (\hat{r}_h + \hat{r}_e), \rho\hat{s}, \hat{\underline{m}}, \hat{\rho}_f, \hat{\sigma})\, dV = 0$$

$$\int_{S-\Gamma} (\hat{\underline{b}}, \hat{\underline{J}}_f, \hat{\underline{\Sigma}}) .\, d\underline{S} = 0$$

$$\int_{\Gamma} (\hat{\rho}_k, \hat{\underline{f}}_k, \hat{\underline{l}}_k, \hat{r}_{k(h)}, \hat{r}_{k(e)}, \hat{\underline{B}}_k, \hat{\underline{D}}_k)\, n_k\, dS = 0$$

$$\int_{\gamma} (\hat{\underline{E}}, \hat{\underline{H}}) .\, \underline{k}\, dx = 0 \tag{25}$$

The stress tensor t_{kl} is introduced as usual by

$$\underline{t}_k = t_{kl}\, \underline{i}_l$$

where $\underline{i}_l$ are the unit base vectors in the spatial coordinates x_k which are taken to be rectangular.

Eliminating ρr_h between (23) and (24) we obtain

$$-\frac{\rho}{T}(\dot{\epsilon} - T\dot{\eta}) + \frac{1}{T}\underline{t}_k.\underline{v}_{,k} + \frac{q_k}{T^2}T_{,k} + \frac{1}{T}\rho r_e + \hat{\rho}\eta - \frac{1}{T}\rho\hat{s} +$$

$$+\frac{\rho}{T}(\hat{r}_h + \hat{r}_e) - \frac{\rho}{T}\underline{v}.\hat{\underline{f}} - \frac{1}{T}\hat{\rho}\{\epsilon - \frac{v^2}{2}\} \geq 0 \tag{26}$$

Let us now derive an explicit expression for r_e. In view of eqs (15a) and (17a) we define the polarization density $\underline{P}$ as

$$\underline{P} = \lim_{|\underline{d}|\to 0} N\,(q + \hat{q})\,\underline{d} \tag{27a}$$

where the two nonlocal electric charges $\pm(q + \hat{q})$ are separated at a distance $\underline{d}$, N is the number of dipoles per unit volume and $(q + \hat{q})|\underline{d}|$ is assumed to remain finite in the limit. Similarly the magnetization density may be defined in terms of two magnetic poles (charges) $\pm(p + \hat{p})\ \underline{d}$

$$\mu_o\underline{M} = \lim_{|\underline{d}|\to 0} N^*(p + \hat{p})\,\underline{d} \tag{27b}$$

where N^* is the number of the magnetic dipoles per unit volume. Equations (28) are the nonlocal definitions of the Chu formalism of magnetization and polarization . Since the electric and magnetic dipoles are charges bound to the particle, we have two more conservation laws

$$\frac{d}{dt}\int_V N\, dV = 0 \quad \frac{d}{dt}\int_V N^* dV = 0 \tag{28}$$

Equations (28a) can be written equivalently as

$$\frac{\partial}{\partial t} N + (Nv_k)_{,k} = \hat{N} \quad \frac{\partial}{\partial t} N^* + (N^* v_k)_{,k} = \hat{N}^* \tag{29}$$

where $\hat{N}$ and $\hat{N}^*$ are respectively the nonlocal production of electric and magnetic dipoles and verify relations of type (25a).

With account of eqs.(11) and (27), the forces acting in the charges of the electric and magnetic dipoles i.e. inside the body, are

$$\underline{f}_{q+\hat{q}} = (q + \hat{q})\ [\underline{E} + \underline{v} \times \mu_o \underline{H}];\ \underline{f}^*_{p+\hat{p}} = (p + \hat{p})\ [\underline{H} - \underline{v} \times \epsilon_o \underline{E}] \tag{30}$$

where $\underline{E}$ and $\underline{H}$ are the electric and magnetic fields inside a moving body satisfying (15) - (18) and $\underline{v}$ is the velocity of the material particle.

We are now in a position to calculate the body force due to the electric dipoles. Let the negative charge $(q + \hat{q})^-$ be at $\underline{x}$, while the positive charge $(q + \hat{q})^+$ is at $\underline{x} + \underline{d}$. The respective velocities of the moving charges are $\underline{v}^- = \underline{v}$ and $\underline{v}^+ = \underline{v} + \dot{\underline{d}}$. Using (30), we obtain by means of a Taylor series expansion the force on each charge as

$$\underline{f}^-_{(q+\hat{q})} = (q + \hat{q})\quad [\underline{E}(\underline{x}) - \underline{v} \times \mu_o \underline{H}(\underline{x})]$$

$$\underline{f}^+_{(q+\hat{q})} = (q + \hat{q})\ \underline{E}(\underline{x}) + (q + \hat{q})\underline{d} \cdot \nabla\underline{E}(\underline{x}) + (q + \hat{q})\ \underline{v} \times \mu_o \underline{H}(\underline{x}) +$$

$$+ (q + \hat{q})\ \underline{v} \times (\underline{d}.\nabla)\ \mu_o \underline{H}(\underline{x}) + (q + \hat{q})\dot{\underline{d}} \times \mu_o \underline{H}\ (\underline{x}) + \ldots$$

The force on each dipole being $\underline{f}^+_{(q+\hat{q})} + \underline{f}^-_{(q+\hat{q})}$, the total electric dipole force on the volume element is $\rho \underline{f}_p = \lim_{|\underline{d}|\to 0} N(\underline{f}^+_{(q+\hat{q})} + \underline{f}^-_{(q+\hat{q})})$ we then find

$$\rho \underline{f}_p = \underline{P}.\ \nabla\underline{E} + \underline{v} \times (\underline{P}.\nabla\ )\ \mu_o \underline{H} + [\rho \frac{d}{dt}\ (\frac{\underline{P}}{\rho}) + \underline{P}\ \frac{\hat{\rho}}{\rho} + \underline{P}^o]\quad \times\quad \mu_o \underline{H} \tag{31}$$

where $\underline{P}^o$ is defined as $\underline{P}^o = \lim_{|\underline{d}|\to 0}(q + \hat{q})\ \underline{d}\ \hat{N}$ and we used the relations (20a) and (29a) and the definition (27a). Similarly the total magnetic dipole force on the volume element is $\rho \underline{f}_M = \lim_{|\underline{d}|\to 0} N^*(\underline{f}^+_{(p+\hat{p})} + \underline{f}^-_{(p+\hat{p})})$ it can then be found that

$$\rho \underline{f}_M = \mu_o \underline{M}.\ \nabla\underline{H} - \epsilon_o\ \underline{v} \times (\mu_o \underline{M}.\nabla)\underline{E} - [\rho \frac{d}{dt}\ (\frac{\mu_o \underline{M}}{\rho}) + \underline{M}\ \frac{\hat{\rho}}{\rho} + \underline{M}^o]\ \times\ \epsilon_o \underline{E} \tag{32}$$

where $\underline{M}^o$ is defined as $\underline{M}^o = \lim_{|\underline{d}|\to 0} (p + \hat{p})\underline{d}\ \hat{N}^*$ and we used the relations (30), (20a) (29b) and the definition (27b).

In addition to the forces due to polarization and magnetization there is the force due to the nonlocal free charge $(\rho_f + \hat{\rho}_f)\,\underline{E}$ and that due to the nonlocal free current $(\underline{J}_f + \hat{\underline{J}}_f) \times \mu_o\underline{H}$. The total force of electromagnetic origin on an element of unit volume is thus

$$\rho\underline{f} = \rho\underline{f}_p + \rho\underline{f}_M + (\rho_f + \hat{\rho}_f)\underline{E} + (\underline{J}_f + \hat{\underline{J}}_f) \times \mu_o\underline{H} \tag{33}$$

The rate of work done per unit volume due to polarization is given by

$$\rho W_p = \lim_{|\underline{d}|\to 0} N\,[\underline{f}^-(q + \hat{q})\cdot\underline{v}^- + \underline{f}^+(q + \hat{q})\cdot\underline{v}^+]$$

$$= [\rho\frac{d}{dt}(\frac{\underline{P}}{\rho}) + \underline{P}\,\frac{\hat{\rho}}{\rho} + \underline{P}^o].\underline{E} + (\underline{P}.\ \nabla\underline{E}).\ \underline{v} \tag{34}$$

The rate of work done per unit volume due to magnetization is given by

$$\rho W_M = \lim_{|\underline{d}|\to 0} N^*\,[\underline{f}^-(p + \hat{p})\cdot\underline{v}^- + \underline{f}^+(p + \hat{p})\cdot\underline{v}^+]$$

$$= [\rho\frac{d}{dt}(\frac{\mu_o\underline{M}}{\rho}) + \underline{M}\,\frac{\hat{\rho}}{\rho} + \underline{M}^o].\underline{H} + (\mu_o\underline{M}.\ \nabla\underline{H}).\underline{v} \tag{35}$$

The relations (34) and (35) are derived in a similar way to (31) and (32).

In addition there is the power supplied by the free current $(\underline{J}_f + \hat{\underline{J}}_f).\underline{E}$. Thus the total power input due to the polarization, magnetization and free current is

$$\rho W = \rho W_p + \rho W_M + (\underline{J}_f + \hat{\underline{J}}_f).\underline{E} \tag{36a}$$

We are now in a position to determine the source r_e. With the account of

$$\rho W = \rho W_p + \rho W_M + (\underline{J}_f + \hat{\underline{J}}_f).\underline{E} =$$

$$= \rho\underline{f}.\underline{v} + \rho r_e \tag{36b}$$

and eliminating r_e from (36b) we have

$$\rho r_e = [\rho\frac{d}{dt}(\frac{\underline{P}}{\rho}) + \underline{P}\,\frac{\hat{\rho}}{\rho} + \underline{P}^o].\ \underline{E}^e + [\rho\frac{d}{dt}(\frac{\mu_o\underline{M}}{\rho}) + \underline{M}\,\frac{\hat{\rho}}{\rho} + \underline{M}^o].\ \underline{H}^e$$

$$+ [\underline{J}^e + (\hat{\underline{J}}_f - \hat{\rho}_f\underline{v})].\underline{E}^e \tag{37}$$

where the effective fields $\underline{H}^e$, $\underline{E}^e$ and $\underline{J}^e$ are given by (11)

We now introduce the free energy Ψ as

$$\Psi = \epsilon - T\eta - \underline{E}^e.\ (\frac{\underline{P}}{\rho}) - \underline{H}^e.\ (\frac{\mu_o\underline{M}}{\rho}) \tag{38}$$

Substituting (38) into (26) and (37), the Clausius Duhem inequality can be written as

$$-\rho(\dot{\Psi} + \eta\dot{T}) \pm \underline{t}_k.\underline{v}_{,k} - \underline{P}.\dot{\underline{E}}^e - \mu_o\ \underline{M}.\dot{\underline{H}}^e + (\underline{P}\,\frac{\hat{\rho}}{\rho} + \underline{P}^o).\ \underline{E}^e + (\underline{M}\,\frac{\hat{\rho}}{\rho} + \underline{M}^o).\ \underline{H}^e$$

$$+ \frac{q_k}{T} T_{,k} + \hat{\rho}\eta T - \rho\hat{s} + \rho(\hat{r}_h + \hat{r}_e) - \rho\underline{v}.\, \underline{\hat{f}} - \frac{1}{T}\, \hat{\rho}\{\epsilon - \frac{v^2}{2}\} +$$

$$+ [\underline{J}^e + (\underline{\hat{J}}_f - \hat{\rho}_f\underline{v})].\, \underline{E}^e \geq 0 \qquad (39)$$

The inequality (39) will be used to restrict the constitutive equations. The constitutive equations that satisfy (39) for all independent changes in the body will be called thermodynamically admissible.

For simplicity here we only study the case when heat conduction, nonlocal mass, magnetic and electric dipole creations are excluded i.e. $\underline{q} = 0$, $\hat{\rho} = 0$, $\hat{N} = 0$ and $\hat{N}^* = 0$.

In this case the energy equation and the entropy inequality (39) write

$$\rho\dot{\epsilon} + \rho\underline{v}.\underline{\hat{f}} - \underline{t}_k.\underline{v}_{,k} - \rho\,(r_h + r_e) - \rho(\hat{r}_h + \hat{r}_e) = 0 \qquad (40)$$

$$- \rho(\dot{\Psi} + \eta\dot{T}) + \underline{t}_k.\underline{v}_{,k} - \underline{P}.\underline{\dot{E}}^e - \mu_o\underline{M}.\underline{\dot{H}}^e - \rho\hat{s} +$$

$$+ \rho(\hat{r}_h + \hat{r}_e) - \rho\underline{v}.\underline{\hat{f}} + [\underline{J}^e + (\underline{\hat{J}}_f - \hat{\rho}_f\underline{v})].\, \underline{E}^e \geq 0 \qquad (41)$$

CONSTITUTIVE EQUATIONS

The constitutive theory for viscous fluids is characterized by the velocity gradients. Based on the entropy inequality (41), the constitutive variables for the fluid might be

$$\{\rho^{-1},\, d_{ij},\, T,\, E^e_i,\, H^e_i,\, r^{-1}(\underline{x}'),\, \beta_k\,(\underline{x}'),\, \gamma_{kl}\,(\underline{x}'),\, E^o_k\,(\underline{x}'),\, H^o_k\,(\underline{x}')\}$$

where

$$r^{-1}\,(\underline{x}') = \rho^{-1}\,(\underline{x}') - \rho^{-1}$$

$$\beta_k\,(\underline{x}') = (\underline{x}' - \underline{x}).\,\underline{v}_{,k} - v_k(\underline{x}') - v_k$$

$$\gamma_{kl}\,(\underline{x}') = v_{k,l}\,(\underline{x}') - v_{k,l}$$

$$2\, d_{ij} = v_{i,j} + v_{j,i}$$

$$E^o_k(\underline{x}') = E^e_k\,(\underline{x}',t) - E^e_k$$

$$H^o_k(\underline{x}') = H^e_k\,(\underline{x}',t) - H^e_k$$

and $\rho^{-1}\,(x'_m)$, $v_k(x'_m)$, $\beta_k(x'_m)$, $\gamma_{kl}(x'_m)$, $E^e_i(x'_m)$, $H^e_i(x'_m)$ are respectively the density, the velocity,... at x'_m and ρ^{-1}, v_k, β_k, γ_{kl}, E^e_i and H^e_i are respectively the density, the velocity,... at x_j. The three first arguments of (42) are those of a classical Stokesian fluid, the next two are due to classical electromagnetic sources and the remaining arguments are due to nonlocal sources.

We thus have

$$\Psi = \Psi(\rho^{-1},\, d_{ij},\, T,\, E^e_i,\, H^e_i,\, r^{-1}(x'_m),\, \beta_i(x'_m),\, \gamma_{ij}(x'_m),$$

$$E^o_i(x'_m),\, H^o_i(x'_m)\,) \qquad (42)$$

with similar forms for the other constitutive statements. Here Ψ is a functional.

We now calculate

$$\dot{\Psi} = \frac{\partial\Psi}{\partial T}\dot{T} + \frac{\partial\Psi}{\partial_{\rho}-1}\,\dot{\overline{\rho^{-1}}} + \frac{\partial\Psi}{\partial d_{kl}}\,\dot{d}_{kl} + \frac{\partial\Psi}{\partial E_i^e}\,\dot{E}_i^e + \frac{\partial\Psi}{\partial H_i^e}\,\dot{H}_i^e +$$

$$+ \int_{V-\Gamma} \frac{\delta\Psi}{\delta r^{-1}}\,[\rho^{-1}, d_{ij}, T, \ldots, r^{-1}(\underline{x}), \ldots, H_k^{\rho}(\underline{x}), \lambda_k]\,\dot{\overline{r^{-1}}}(\underline{\lambda})\,dV(\underline{\lambda}) +$$

$$+ \int_{V-\Gamma} \frac{\delta\Psi}{\delta\beta_k}\,\dot{\beta}_k(\underline{\lambda})dV(\underline{\lambda}) + \int_{V-\Gamma} \frac{\delta\Psi}{\delta\gamma_{kl}}\,\dot{\gamma}_{kl}(\underline{\lambda})dV(\underline{\lambda}) +$$

$$+ \int_{V-\Gamma} \frac{\delta\Psi}{\delta E_k^{\rho}}\,\dot{E}_k^{\rho}(\underline{\lambda})dV(\underline{\lambda}) + \int_{V-\Gamma} \frac{\delta\Psi}{\delta H_k^{\rho}}\,\dot{H}_k^{\rho}(\underline{\lambda})dV(\underline{\lambda}) \qquad (43)$$

where δ/δ indicates functional (Fréchet) partial derivative. The functional gradients appearing in these integrals are also functions of a vector λ_i. Using the equation of continuity (20a) with $\hat{\rho} = 0$ and substituting eq. (43) into (41) we obtain

$$- \frac{\rho}{T}(\eta + \frac{\partial\Psi}{\partial T})\dot{T} - \frac{\rho}{T}\,[\hat{f}_l - v_{l,k}\int_{V-\Gamma}\frac{\delta\Psi}{\delta\beta_k}dV(\underline{\lambda})]v_l$$

$$+ \frac{1}{T}\,[t_{kl} - \frac{\partial\Psi}{\partial_{\rho}-1}\delta_{kl} + \delta_{kl}\int_{V-\Gamma}\frac{\delta\Psi}{\delta r^{-1}}\,dV(\underline{\lambda}) - \rho\int_{V-\Gamma}\frac{\delta\Psi}{\delta\beta_k}\,v_l(\underline{\lambda})dV(\underline{\lambda})]\,v_{l,k}$$

$$- \frac{\rho}{T}\,\dot{v}_k\int_{V-\Gamma}\frac{\delta\Psi}{\delta\beta_k}dV(\underline{\lambda}) - \frac{\rho}{T}\{\frac{1}{2}\frac{\partial\Psi}{\partial d_{kl}} + \frac{\partial\Psi}{\partial d_{lk}}\} + \{\int_{V-\Gamma}[(\lambda_k - x_k)\,\frac{\delta\Psi}{\delta\beta_l} -$$

$$- \frac{\delta\Psi}{\delta\gamma_{kl}}]\,dV(\underline{\lambda})\}\,\dot{\overline{v_{k,l}}} - \frac{1}{T}\,[P_l + \rho\,\frac{\partial\Psi}{\partial E_l^e} + \rho\int_{V-\Gamma}\frac{\delta\Psi}{\delta E_l^{\rho}}\,dV(\underline{\lambda})]\,.\,\dot{E}_l^e$$

$$- \frac{1}{T}\;\mu_o M_l + \rho\frac{\partial\Psi}{\partial H_l^e} + \rho\int_{V-\Gamma}\frac{\delta\Psi}{\delta H_l^{\rho}}dV(\underline{\lambda})]\,.\,\dot{H}_l^e$$

$$- \frac{\rho}{T}\int_{V-\Gamma}[\frac{\delta\Psi}{\delta r^{-1}}\,\rho^{-1}(\underline{\lambda})v_k(\underline{\lambda})_{,k} + \frac{\delta\Psi}{\delta\beta_k}\,\dot{v}_k(\underline{\lambda}) + \frac{\delta\Psi}{\delta\gamma_{k,l}}\,\dot{\overline{v_{k,l}}}(\underline{\lambda}) - \frac{\delta\Psi}{\delta E_l^{\rho}}\,\dot{E}_l^e(\underline{\lambda}) -$$

$$- \frac{\delta\Psi}{\delta H_l^{\rho}}\,\dot{H}_l^e(\underline{\lambda})]\,dV(\underline{\lambda}) + \frac{\rho}{T}\,r_e - \frac{\rho}{T}\,\hat{s} + \frac{1}{T}\,(\rho\hat{r}_h + \rho\hat{r}_e) \geq 0 \qquad (44)$$

This inequality is linear in $\dot{T}$, v_k, $\dot{\overline{v_{k,l}}}$, $\dot{E}_l^e$ and $\dot{H}_l^e$. If r_e, $\hat{r}_h$, $\hat{r}_e$ and $\hat{s}$ are independent of these quantities, then (44) cannot be maintained for all possible variations of these quantities unless

$$\eta = - \frac{\partial\Psi}{\partial T}$$

$$\hat{f}_k = v_{l,k}\int_{V-\Gamma}\frac{\delta\Psi}{\delta\beta_k}\,dV(\underline{\lambda}) = 0$$

$$\frac{1}{2}\,(\frac{\partial\Psi}{\partial d_{kl}} + \frac{\partial\Psi}{\partial d_{lk}}) + \int_{V-\Gamma}[(\lambda_k - x_k)\,\frac{\delta\Psi}{\delta\beta_l} - \frac{\delta\Psi}{\delta\gamma_{kl}}]\,dV(\underline{\lambda}) = 0$$

$$P_l = - \rho\frac{\partial\Psi}{\partial E_l^e} - \rho\int_{V-\Gamma}\frac{\delta\Psi}{\delta E_l^{\rho}}dV(\underline{\lambda})$$

$$\mu_o M_l = - \rho\frac{\partial\Psi}{\partial H_l^e} - \rho\int_{V-\Gamma}\frac{\delta\Psi}{\delta H_l^{\rho}}\,dV(\underline{\lambda})$$

$$ {}_Dt_{kl}v_{l,k} + J_i^e E_i^e + \rho[(\hat{r}_h + \hat{r}_{e'}) - {}_D\hat{r}] - \rho\hat{s} \geq 0 \tag{45}$$

where

$$ {}_Dt_{kl} = t_{kl} - \frac{\partial\Psi}{\partial\rho}\text{-}1\ \delta_{kl} + \delta_{kl}\int_{V-\Gamma} \frac{\delta\Psi}{\delta r}\text{-}1 dV(\underline{\lambda})$$

$$ {}_D\hat{r} = \int_{V-\Gamma} [\frac{\delta\Psi}{\delta r}\text{-}1\ \rho^{-1}(\underline{\lambda})\ v_k(\underline{\lambda})_{,k} + \frac{\delta\Psi}{\delta\beta_k}\dot{v}_k(\underline{\lambda}) + \frac{\delta\Psi}{\delta\gamma_{kl}}\dot{\overline{v_{kl}}}(\underline{\lambda})] +$$

$$ + v_{l,k}\int_{V-\Gamma}\frac{\delta\Psi}{\delta\beta_k} v_l(\underline{\lambda})dV(\underline{\lambda}) + \int_{V-\Gamma} [\frac{\delta\Psi}{\delta \overset{o}{E}_l}\dot{E}_l^e(\underline{\lambda}) + \frac{\delta\Psi}{\delta \overset{o}{H}_l}\dot{H}_l^e(\underline{\lambda})]\ dV(\underline{\lambda}) \tag{46}$$

Thus, the constitutive equations of nonlocal electromagnetic fluids are thermodynamically admissible if and only if (45) are satisfied.

The Ψ in (42), (43) should be invariant under an orthogonal transformation. It is well known that the deformation rate measures d_{kl} is objective, the constitutive variables E_i^e and H_i^e have to be replaced by the objective quantities denoted by a tilt $\sim$, $\tilde{E} = E_i^e E_i^e$, $\tilde{H} = H_i^e H_i^e$ and $\tilde{K} = H_i^e E_i^e$. A simple calculation shows that the deformation rate measures $\beta_k(x_i')$, $\gamma_{kl}(x_i')$ as well as $\overset{o}{E}_k$ and $\overset{o}{H}_k$ are objective quantities. Furthermore it shows that with this functional form for the constitutive equations of nonlocal electromagnetic fluids, the dependence on the spin tensor defined by $2\ W_{kl}(x_i') = v_{k,l}(x_i') - v_{l,k}(x_i')$ on the fluid is not eliminated and that in opposition to the classical case [9] the nonlocal electromagnetic body couple is not vanishing and thus $2t_{[kl]} = t_{kl} - t_{lk}$ is not vanishing too. In addition, we remark that the body force residual $\hat{f}$ for fluid vanishes, from (45c) that in a nonlocal theory the free energy can depend on the rate variables such as d_{kl}, β_k and γ_{kl} and that eqs. (45a, d and e) imply that η, P_i and M_i for a fluid can be determined from a free energy function just like for an elastic solid.

With the use of (45) and (46), the energy equation (26) reduces to

$$\rho T\dot{\eta} - {}_Dt_{kl}d_{kl} - \rho r_h - J_i^e E_i^e - \rho(\hat{r}_h + \hat{r}_{e'} - {}_D\hat{r}) = 0 \tag{47}$$

We now posit that this energy equation should be invariant under rigid motions. Because all other quantities are invariant this leads to

$$\hat{r}_h + \hat{r}_{e'} = \hat{r}_{h_o} + \hat{r}_{e'_o} + {}_D\hat{r} \tag{48}$$

where $\hat{r}_{h_o}$ and $\hat{r}_{e'_o}$ are scalar-valued constitutive functionals of the same type as Ψ and satisfy the invariance requirements. The nonlocal residuals $\hat{f}$ and $\hat{r}_h + \hat{r}_e$ are thus determined by (45b) and (48). Because $\hat{r}_h + \hat{r}_e$ are determined within a class of scalar-valued constitutive functionals, the nonlocal residuals constitute an equivalent class. Using (48), the entropy inequality (45f) reduces to

$$ {}_Dt_{kl}v_{l,k} + J_i^e E_i^e + \rho(\hat{r}_{h_o} + \hat{r}_{e'_o} - \hat{s}) \geq 0 \tag{49}$$

and the energy equation (47) reduces to

$$\rho T\dot{\eta} - {}_Dt_{kl}d_{kl} - \rho r_h - J_i^e E_i^e - \rho(\hat{r}_h + \hat{r}_{e'}) = 0 \tag{50}$$

The stresses ${}_Dt_{kl}$ and the effective current are general functionals of ρ, d_{ij}, E_i^e, H_i^e, T, $r^{-1}(x_i')$, $\beta_k(x_i')$, $\gamma_{kl}(x_i')$, $\overset{o}{H}_k(x_i')$ and $\overset{o}{E}_k(x_i')$ satisfying the inequality (50)

As an example (49) is satisfied if

$$ {}_{D}t_{ij} = {}_{D}\tilde{t}_{ij}(d_{mn},T,H^e_m,\ldots) \qquad {}_{D}\tilde{t}_{ij}d_{ij} > 0 $$

$$ J^e_p = \tilde{J}_p\ (E^e_m,T,H^e_m) \qquad \tilde{J}^e_p E^e_p \geq 0 \qquad (51) $$

where (51b) and (51d) will place conditions on the nonlocal moduli of the fluid. We remark that such moduli must also obey the axiom of attenuating neighborhood, i.e. the continuity requirement that nonlocal effects diminish rapidly with the distance.

REFERENCES

[1] Kunin, I. A., *The Theory of Elastic Media with Microstructure* in: Kröner, E. (ed.), Mechanics of Generalized Continua (Springer-Verlag, Berlin, 1968) 321 - 329

[2] Kröner, E., *Elasticity Theory of Materials with Long Range Cohesive Forces*, Int. J. Solids Struct. 3 (1967) 731 - 742

[3] Edelen, D. G. B. and Laws, N., *On the Thermodynamics of Systems with Non-locality*, Arch. Rat. Mech. Anal. 43 (1971) 24 - 35

[4] Eringen, A. C., *On Nonlocal Fluid Mechanics*, Int. J. Engng. Sci. 10 (1972) 561 - 575

[5] Nowacki, W. and Olszak, W. (eds.) *Mechanics of Micropolar Elasticity* (Int'l Center Mech. Sci. Courses, Udine, 1974)

[6] Hsieh, R. K. T., *Continuum Mechanics of a Magnetically Saturated Fluid*, IEEE Trans. Magn. 16 (1980) 207 - 210

[7] Eringen, A. C., *Theory of Nonlocal Electromagnetic Elastic Solids*, J. Math. Phys. 14 (1973) 733 - 740

[8] Edelen, D. G. B., *Solutions of the Clausius-Duhem Inequality for Electro-mechanical Systems*, L. Appl. Engng. Sci. 1 (1973) 497 - 503

[9] Pao, Y. H. and Hutter, K., *Electrodynamics for Moving Elastic Solids and Viscous Fluids*, Proc. IEEE. 63, 7 (1975) 1011 - 1021

[10] Hsieh, R. K. T., *Micropolarized and Magnetized Media* in: Brulin, O. and Hsieh, R. K. T. (eds.), Mechanics of Micropolar Media (Int'l Center Mech. Sci. Courses, Udine, 1979)

[11] Eringen, A. C., *Mechanics of Continua* (Wiley, New York, 1967) 77

Continuum Models of Discrete Systems 4
eds. O. Brulin and R.K.T. Hsieh

SCATTERING OF MAGNETOELASTIC WAVES ON DISLOCATIONS

G.A.Maugin and A.Fomethe+

Université Pierre-et-Marie-Curie,
Laboratoire de Mécanique Théorique associé au CNRS,
Tour 66, 4 Place Jussieu , 75230 Paris Cedex 05, France.

The construction of a phenomenological theory of the elasto-viscoplasticity of ferromagnetic crystals is achieved by combining ideas from the modern theory of finitely deformable elastoviscoplastic materials and from the magnetoelasticity of ferromagnetic perfect crystals with a view to modelling a defectuous ferromagnetic lattice.An externally applied intense magnetic field breaks the material symmetry and favors the existence of induced piezomagnetism which,in turn, in a study of coupled magnetoacoustic modes, allows one to place in evidence the attenuation of spin waves by dislocations.

INTRODUCTION

Two discrete ordered structures co-exist in a perfect ferromagnetic crystal, the lattice and the magnetic ordering pertaining to a ferromagnetic state. The purpose of this paper is an attempt at the modelling of the effects of defects (e.g., dislocations) in the first structure on the dynamical properties related to the second structure. More precisely, a continuum theory is constructed from first principles and is applied to the description of the damping of spin waves (the coherent oscillations associated with the magnetic ordering [1] which is caused by the scattering of the said waves on dislocations. To achieve such a purpose one can basically follow two largely differing lines of thought. On the one hand, one may quantify the magnetostriction energy for a given dislocation strain field and thereafter apply the so-called Holstein-Primakoff transformation to study spin waves on a quantum-mechanical basis (see [2] and [3]). The present paper considers an entirely different point of view in that the problem of formulating a theory of crystally defectuous ferromagnets is approached phenomenologically by using the most recent developments in the theory of finitely deformable elastoviscoplastic materials (see,e.g.,[4]-[6]) and combining them with the now well-established magnetoelasticity of ferromagnetic perfect crystals (see, e.g.,[7]-[9]). The reason for developing a finite-strain theory for treating coupled-wave propagation problems has been made clear in those works since a linearization process about a finitely magnetized and stressed state is required. The main ingredients from nonlinear elastoviscoplasticity are those concerning the use of thermodynamics with internal state variables (e.g., dislocation densities, plastic rate of deformation), the multiplicative decomposition of the total deformation gradient when viscoplastic effects are taken into account, the application of these concepts to the description of the elastoviscoplasticity of monocrystals and its relationship with microscopic theory and, finally, the introduction of a multiple plastic potential in plasticity. The ingredients needed in the theory of micromagnetism essentially are those pertaining to the macroscopic representation of spin-lattice interactions and ferromagnetic exchange forces of quantum origin. Once the astute combination of these ingre-

+Presently at the Institut Polytechnique, Yaoundé, Cameroun.

dients is achieved in the form of an elastoviscoplastic theory of ferromagnetic
monocrystals, the study of coupled magnetoacoustic modes that may propagate in the
form of free, small-amplitude, oscillations about an initial spatially uniform
state of magnetization (a single ferromagnetic domain obtained by the application
of an intense magnetic field), will result in the entanglement of acoustic and
spin branches with the allied magnetoacoustic resonance phenomenon (compare [10]).
In addition, the viscoplastic nature of the body will cause a damping of coupled
modes and, for long wavelengths, a damping of those waves which are essentially
spin waves, hence the result looked for. This damping naturally occurs by the
intermediate of piezomagnetism which is induced during the symmetry breaking by
the intense initial magnetic field. Only essential ideas and results are reported
in the present paper. The reader is referred to Refs.[11] and [12] for an exhaus-
tive treatment.

THE ELASTOVISCOPLASTICITY OF FERROMAGNETIC CRYSTALS

The general continuum equations which govern deformable ferromagnets in the quasi-
magnetostatic approximation are as follows at all regular points in the bulk of the
material (Cartesian tensor or dyadic notation, Lorentz-Heaviside units):

. continuity:

$$\dot{\rho} + \rho \, \underline{\nabla}.\underline{U} = 0 \qquad\qquad ; (1)$$

. Euler-Cauchy equations of motion:

$$t_{ij,j} + f_i = \rho \, \dot{U}_i \qquad (i,j=1,2,3) \;, \qquad \underline{f} = \rho(\, \underline{\mu}.\underline{\nabla})\underline{B} \qquad\qquad ; (2)$$

. Maxwell's equations:

$$\underline{\nabla} \times \underline{B} = \underline{\nabla} \times (\, \rho \, \underline{\mu} \,) \;, \qquad \underline{\nabla}.\underline{B} = 0 \;, \quad \underline{H} = \underline{B} - \underline{M} \;, \quad \underline{M} = \rho \, \underline{\mu} \qquad ; (3)$$

. Spin-precession equation:

$$\dot{\underline{\mu}} = \underline{\omega} \times \underline{\mu} \;, \quad \underline{\omega} = -\gamma \, \underline{H}^{\text{eff}} \qquad\qquad , (4)$$

where the nonsymmetric Cauchy stress tensor t_{ij} and the effective magnetic field
$\underline{H}^{\text{eff}}$ admit the following general decompositions:

$$t_{ij} = \sigma_{ij} + \rho \, {}^{L}B_{[i} \, \mu_{j]} - |B_{[i|k|} \, \mu_{j],k} \quad \neq t_{ji} \qquad (5)$$

and

$$\underline{H}^{\text{eff}} = \underline{H}^{(i)} + {}^{L}\underline{B} + \rho^{-1} \, \text{div} \, |\underline{B} \qquad , \quad \underline{H}^{(i)} = \underline{H}^{e} - \rho \, \underline{N}.\underline{\mu} \qquad . \quad (6)$$

Here $\underline{\sigma}$ is the intrinsic symmetric stress tensor , ${}^{L}\underline{B}$ is the so-called local magnetic field (representative of spin-lattice interactions) , γ is the gyromagnetic
ratio , $\underline{H}^{e}$ is the externally applied magnetic field, $\underline{N}$ is the symmetric demagneti-
zation tensor (which accounts for the shape of the specimen) and $|\underline{B}$ is the spin-
interaction tensor representative of exchange ferromagnetic forces. $\underline{B}$ is the magnetic induction and $\underline{\mu}$ is the magnetization per unit mass. The energy equation and
the corresponding Clausius-Duhem inequality (local statement of the second principle
of thermodynamics) are given by [3]

$$\rho \, \dot{e} = \sigma_{ij} \, D_{ij} - \rho \, {}^{L}B_i \, \hat{m}_i + |B_{ij} \, \hat{M}_{ij} + \rho \, h - q_{k,k} \qquad (7)$$

and

$$- \rho(\dot{\Psi} + \eta \, \dot{\theta}) + \sigma_{ij} D_{ij} - \rho \, {}^{L}B_i \, \hat{m}_i + |B_{ij} \, \hat{M}_{ij} - \theta^{-1} \, q_k \, \theta_{,k} \geq 0 \qquad , \quad (8)$$

in which e , Ψ , η and $0 < \theta \ll \theta_c$ are the internal energy , free energy and
entropy densities and the absolute temperature (θ_c = Curie ferromagnetic temperature). We have set

$$D_{ij} = U_{(i,j)} \quad , \quad \Omega_{ij} = U_{[i,j]} \quad , \quad U_i = \left.\frac{\partial \mathfrak{X}_i(\underline{X},t)}{\partial t}\right|_{\underline{X}} \quad ,$$
$$\hat{m}_i = \dot{\mu}_i - \Omega_{ij}\mu_j \quad , \quad \dot{(\cdot)} = \frac{\partial}{\partial t}(\cdot) + \underline{U}.\underline{\nabla}(\cdot) \quad , \tag{9}$$
$$\hat{M}_{ij} = (\dot{\mu}_i)_{,j} - \Omega_{ik}\mu_{k,j} \quad .$$

The general motion between a free-field reference configuration K_R and the current configuration K_t is described by $\underline{x} = \mathfrak{X}(\underline{X},t)$ with deformation gradient $\mathbb{F}$. It is possible to prove the existence , on account of boundary conditions, of a rigid-body, uniformly magnetized ,stable (by magnetic perturbations) solution S_o of the above system , that corresponds to a so-called ferromagnetic fundamental phase (reference configuration without magnetic domains) K_o if a sufficiently strong spatially uniform field $\underline{H}^e_o$ is applied. Then the finite elastoviscoplasticity of ferromagnetic monocrystals is constructed as follows under certain hypotheses: (H1) the crystal satisfies the solution S_o but contains structural defects such as dislocations;(H2) the influence of work-hardening (Barkhausen effect,cf.[13]) on the saturation value of the magnetization $|\underline{\mu}| = \mu_s$ is neglected;(H3) the internal or proper stresses and strains due to structural defects in K_o are discarded; (H4) relaxation effects related to spin-lattice interactions are discarded. The current configuration K_t results from the dynamical-kinematical process when, from K_o , the crystal is subjected to an elastoviscoplastic deformation under the influence of external loadings, of a nonuniform temperature field and a nonuniform magnetic field which superimposes itself on $\underline{H}^e_o$. Like in pure viscoplasticity [4]-[5], a "natural"(intermediate) local configuration of the ferromagnetic crystal , K^*_t , is obtained by relaxation of magnetoelastic processes(simultaneous relaxation of the fields $\underline{\sigma}$, ${}^L\underline{B}$ and $\underline{\mathbb{B}}$ related to elastic strains). The orientation of K^*_t is fixed by assuming that crystalline directions remain fixed in the course of the viscoplastic deformation. The kinematics then is entirely described by the elements of the multiplicative decomposition

$$\underline{\mathbb{F}} = \underline{\mathbb{F}}^e \underline{\mathbb{F}}^p \quad , \tag{10}$$

where $\underline{\mathbb{F}}^e$ (elastic) and $\underline{\mathbb{F}}^p$ (viscoplastic) are no longer gradients. The viscoplastic strain rate is given by

$$\dot{\underline{\chi}}^p = \underline{\mathbb{F}}^e \dot{\underline{\mathbb{F}}}^p \underline{\mathbb{F}}^{p-1} \underline{\mathbb{F}}^{e-1} \quad . \tag{11}$$

After a long algebra the following can be proven[11]: The magnetothermoelastic constitutive equations of a deformable monocrystalline ferromagnetic body which is subjected to elastoviscoplastic strains are given by

$$t_{ij} = \sigma_{ij} - \rho\,\mu_{[i}\,{}^L B_{j]} \quad , \tag{12}$$

$$\sigma_{ij} = \rho\left[\frac{\partial\hat{\psi}}{\partial \mathbb{E}^e_{\alpha\beta}}\mathbb{F}^e_{i\alpha}\mathbb{F}^e_{j\beta} + \frac{\partial\hat{\psi}}{\partial m^*_\alpha}\mu_{(i}\mathbb{F}^e_{j)\alpha}\right] \; , \quad {}^L B_i = -\frac{\partial\hat{\psi}}{\partial m^*_\alpha}\mathbb{F}^e_{i\alpha} \; , \tag{13}$$

$$\mathbb{B}_{ij} = 2\rho\,\frac{\partial\hat{\psi}}{\partial M^*_{\alpha\beta}}\mu_{i,k}\mathbb{F}^e_{j\alpha}\mathbb{F}^e_{k\beta} \quad , \qquad \eta = -\left(\frac{\partial\hat{\psi}}{\partial\theta} + \frac{\partial\varphi}{\partial\theta}\right) \tag{14}$$

with

$$\psi = \hat{\psi}(\mathbb{E}^e_{\alpha\beta} \; , \; m^*_\alpha \; , M^*_{\alpha\beta} \; , \theta) + \varphi(\theta,\underline{\alpha}) \quad , \tag{15}$$

where

$$\mathbb{E}^e_{\alpha\beta} = \frac{1}{2}(\mathbb{F}^e_{i\alpha}\mathbb{F}^e_{i\beta} - \delta_{\alpha\beta}) \; , \quad m^*_\alpha = \mathbb{F}^e_{i\alpha}\mu_i \; , \quad M^*_{\alpha\beta} = \mu_{i,j}\mathbb{F}^e_{j\alpha}\mathbb{F}^e_{k\beta}\mu_{i,k} \quad . \tag{16}$$

Assuming that heat conduction can be discarded, we have the following remaining dissipation inequality

$$(t_{ij} + \mu_{k,i} \mathbb{B}_{k,j}) \dot{\chi}^{p}_{ij} - \rho [\partial \psi / \partial \alpha^{(j)}] A^{(j)} \geqq 0 \tag{17}$$

along with the evolution equation and activation criterion

$$\dot{\alpha}^{(j)} = \begin{cases} A^{(j)}(T^{(j)}_{g}, \theta , \underline{\alpha}) & \text{for } T^{(j)}_{g} > T^{(j)}_{\mu} \\ 0 & \text{otherwise} \end{cases} , \tag{18}$$

where $\underline{\alpha}$ denotes the n-vector of dislocation densities ($\alpha^{(j)}$;j=1,2,..,n), $\underline{n}^{(j)}$ and $\underline{g}^{(j)}$ denote unit directions normally to a gliding plane and in the direction of gliding for the (j)-th gliding system in the viscoplastic crystal; $T^{(j)}_{\mu}$ is the resolved shear stress associated with nonlocalized obstacles and the generalized resolved shear stress is defined by (T = transpose)

$$T^{(j)}_{g} = \underline{g}^{(j)}.[(\rho_o/\rho)\{\underline{\mathbb{F}}^{eT} [\underline{t} + (\nabla \underline{\mu})^T \underline{\mathbb{B}}](\underline{\mathbb{F}}^{-1e})^T \}. \underline{n}^{(j)}] \quad . \tag{19}$$

The viscoplastic strain rate and the motion of dislocations are related by

$$\dot{\underline{\mathbb{F}}}^{p} \underline{\mathbb{F}}^{-1p} = b \sum_j \dot{\alpha}^{(j)} L^{(j)} \underline{g}^{(j)} \otimes \underline{n}^{(j)} , \tag{20}$$

where b is Burgers' vector and $L^{(j)}$ is the total length of moving dislocation lines in the (j)-th system at time t. In establishing equations (12)-(20) we have assumed that : (i) the viscoplastic deformation is incompressible; (ii) the thermodynamical state of the medium depends on the total strain $\mathbb{E}$ and the viscoplastic "gradient" $\mathbb{F}^{p}$ only through the finite elastic strain $\mathbb{E}^{e}$; (iii) elastic and magnetic material coefficients do not depend on the viscoplastic strain (compare [14]);(iv) viscoplastic strains occur only through simple gliding of dislocations in a finite number ,n, of gliding systems (the uprise motion of screw dislocations is excluded); (v) the overall motion of dislocations is quasi-stationary; (vi) the localized obstacles are overtaken thanks to thermal agitation. The theory thus obtained is rewarding in spite of its apparent complexity.The second contribution in the left-hand side of eq.(17) is the time-rate of change of the "stocked" free energy (cf. [14]) in reason of the splitting (15) of the free energy.

COUPLED MAGNETOACOUSTIC WAVES

In studying harmonic waves of complex circular frequency $\Omega = \omega + i\Gamma$ and wave vector $\underline{k}$ (wave number k) corresponding to perturbations ($\underline{u}$, $\underline{\bar{\mu}}$, $\underline{h}, \underline{\bar{\alpha}}$) superimposed on ($\underline{I}, \underline{\mu}_{o}$, $\underline{H}_{o}$, $\underline{\alpha}_{o}$) , where $\underline{u}$ is the infinitesimal displacement and $\underline{I}$ is the identity, we assume that $\underline{k}$ lies along the same unit direction $\underline{d}$ as $\underline{\mu}_{o}$ and $\underline{H}_{o}$ and obtain first linearized equations about S_{o} . These linearized equations govern perturbations which propagate through an hexagonal system, the hexagonal symmetry being induced by symmetry breaking of the ideal material symmetry (here selected as the isotropy with respect to the field-free reference configuration for the nonlinear theory). The gliding of dislocations then occur in planes orthogonal to $\underline{d}$, with three potential gliding systems, and we set $\underline{n}^{(j)} = \underline{d}$. We further assume that only one system (j) is active so that the index (j) can be omitted. We finally define the viscosity coefficient η by (compare [14])

$$\eta^{-1} = L_o b (\partial A^{(j)}/ \partial T^{(j)}_{g})_o > 0 \quad (\text{j chosen}) \quad . \tag{21}$$

Elasticity is recovered by letting η to go to infinity. In the linearization we have

$$u_{i,j} = f^{e}_{ij} + f^{p}_{ij} , \tag{22}$$

and after a long algebra [12] we obtain the following linearized system:

$$\rho_o(\ddot{f}^e_{im} + \ddot{f}^p_{im}) = (\hat{C}_{ijkl}+C^*_{ijkl} - M_o^2\,\delta_{ij}\,\delta_{kl})f^e_{kl,jm} + M_o\,d_j\,h_{j,im}$$
$$- M_o^2\,f^p_{kk,im} + \rho_o M_o[\epsilon\,d_j\,\bar{\mu}_{i,jm} + (\epsilon+\bar{b})d_i\,\bar{\mu}_{j,jm}] \quad , (23)$$

$$\rho_o\,\dot{\bar{\mu}}_p = (\gamma M_o)\,\varepsilon_{pmi}\,d_m[h_i - M_o((\epsilon+\bar{b})\,\delta_{ki}d_j + \epsilon\,\delta_{ij}d_k)f^e_{jk}$$
$$+ \rho_o\,(\xi\,\delta_{jk} + \bar{\xi}\,d_jd_k)\,\bar{\mu}_{i,jk} - \rho_o\,\hat{b}\,\bar{\mu}_i\,] \quad , (24)$$

$$h_{i,i} + \rho_o\,\bar{\mu}_{i,i} - M_o(f^e_{kk,i} + f^p_{kk,i})d_i = 0\ ,\ \underline{\nabla}\times\underline{h} = \underline{0} \quad , (25)$$

$$\dot{f}^p_{ij} = b\,L_o\,\dot{\bar{\alpha}}\,g_id_j\ ,\ \dot{\bar{\alpha}} = (L_ob\eta)^{-1}[(\hat{C}_{ijkl}+C^*_{ijkl})f^e_{kl}\,g_id_j + \rho_oM_o\epsilon\,\bar{\mu}_ig_i]\ ,(26)$$

where ε_{pmi} is the permutation symbol, $\hat{C}_{ijkl}$ and C^*_{ijkl} are tensors of <u>effective</u> elasticities , ϵ is a piezomagnetic coefficient, $\hat{b} = \bar{b} + (H_o / M_o)$ is an effective magnetic anisotropy constant and ξ and $\bar{\xi}$ are exchange constants. In deriving eq.(23) the gradient of the motion equation has been taken (compare [15]) . In carrying harmonic wave-like solutions in the above system and applying a projection technique for the decomposition of the modes, we arrive at the following dispersion relations [12]:

(i) <u>for longitudinal elastic modes</u>:

$$\omega(k) = \pm\,c_L\,k\ ,\quad \Gamma(k) = 0 \quad ;(27)$$

(ii) <u>for transverse magnetoacoustic modes polarized in the</u> ($\underline{d}\times\underline{g}$)-<u>direction</u>:

$$[\Omega^2 - \omega_T^2(k)][\Omega \mp \omega_S(k)] \mp \varepsilon_p\,\omega_M\,\omega_T^2(k) = 0 \quad ,\ \omega_M \equiv \gamma M_o \quad ,(28)$$

where

$$\varepsilon_p = (\epsilon^2\,M_o^2/\rho_o\,c_T^2) > 0\ ,\ \omega_T^2 = c_T^2k^2\ ,\ \omega_S(k) = \omega_M[(\xi+\bar{\xi})k^2 + \hat{b}] \quad ;(29)$$

This is similar to the result obtained in the purely elastic case(cf.[9]) so that those transverse modes do not allow for the distinction between the elastically perfect case and the elastoviscoplastic case. We refer the reader to Ref.[9] for a qualitative discussion of the dispersion relation (28) and the accompanying <u>resonance</u> and <u>repulsion</u> phenomena between acoustic and spin branches at the critical wave number $k_o^* \simeq \omega_S(0)/c_T$ for left-circularly polarized modes.

(iii) <u>for transverse magnetoacoustic modes polarized in the g-direction:</u>

On defining the <u>primary relaxation time</u> τ_o by (compare Mandel[14],pp.113-114)

$$\tau_o = (\,\eta\,/\rho_o\,c_T^2) > 0 \quad , (30)$$

we arrive at the complex dispersion relation

$$[\Omega \mp \omega_S(k)][\Omega^2 - i(\Omega/\tau_o) - \omega_T^2(k)] \mp \varepsilon_p\,\omega_M[\omega_T^2 + i(\Omega/\tau_o)] = 0 \quad . \quad (31)$$

The study of right-circularly polarized modes (lower signs) shows that only a slight added dispersion of the pure elastoviscoplastic mode occurs (in particular, there is no resonance) . For left-circularly polarized modes, a resonance occurs between spin waves and the elastoviscoplastic branch at a critical wave number k^* which is slightly larger than that corresponding to the purely elastic case (k_o^*) . There is repulsion of the order of $\sqrt{\varepsilon_p}$ between the coupled branches at k^* (see Figure I) and for small wave numbers (long wavelengths) , it is found that the damping of the mode (I) — in Figure I — which then is practically a spin-wave mode, is approximately given by

$$\Gamma(k) \simeq \varepsilon_p \frac{\omega_M \; \omega_S(0)}{\tau_o[\omega_S^2(0)+\tau_o^{-2}]} > 0 \qquad (32)$$

We have therefore placed in evidence the damping of spin waves by dislocations for long wavelengths (the relaxation time τ_o is related to the viscosity η which in turn is related to the viscosplastic flow evaluated at the initial state —cf. eq.(21). This damping is proportional to the nondimensional piezomagnetic coupling coefficient ε_p , which results from the symmetry breaking by the intense initial magnetic field ${}^P\underline{H}_o$. We have obtained here the first result of its kind in the magnetoelasticity of bodies with defects. In the above analysis we have assumed that restoring effects were absent and that the linearization was performed about a fundamental ferromagnetic phase with a practically vanishing viscoplastic threshold (the latter hypothesis is feasible if we are sure to stay within the viscoplastic domain;cf. the works of M.Piau).

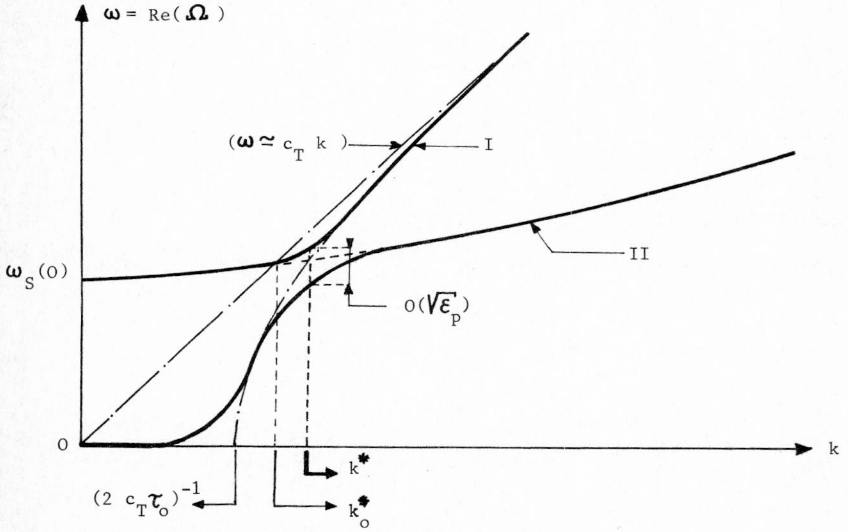

Figure I.- Real dispersion relation and resonance for coupled elastovisco-plastic and spin waves in a ferromagnet.

REFERENCES

[1] Akhiezer A.I.,Bar'yakhtar V.G. and Peletminskii S.V., Spin Waves (Translation from the Russian, North-Holland , Amsterdam ,1968).

[2] Morkowski J., Broadening of Ferromagnetic Resonance Line by Dislocations, Acta Phys.Polonica,35 (1969),565-583.

[3] Maugin G.A.,Classical Magnetoelasticity in Ferromagnets with Defects, in: Parkus H.(ed.), Electromagnetic Interactions in Elastic Solids,pp.243-324, (Springer-Verlag, Wien,1979).

[4] Teodosiu C.,Physical Theory of the Finite Elastic-Viscoplastic Behavior of Single Crystals,Rozprawy inż(Engin.Transactions,Poland),23(1975),151-184.

[5] Teodosiu C. and Sidoroff F.,A Theory of Finite Elastoviscoplasticity of Single Crystals, Int.J.Engng.Sci.,14 (1976),165-176.

[6] Rice J.R.,Inelastic Constitutive Relations for Solids: An Internal-Variable Theory and its Application to Metal Plasticity, J.Mech.Phys.Solids, 19(1971), 433-455.

[7] Maugin G.A., Micromagnetism, in : Eringen A.C.(ed.), Continuum Physics,Vol.III, pp. 213-312 (Academic Press, New York, 1976).

[8] Maugin G.A., A Continuum Approach to Magnon-Phonon Couplings-I, Int.J.Engng. Sci., 17 (1979),1073-1091.

[9] Maugin G.A., A Continuum Approach to Magnon-Phonon Couplings-II,Int.J.Engng. Sci., 17 (1979),1093-1108.

[10] Maugin G.A.,Elastic-Electromagnetic Resonance Couplings in Electromagnetically Ordered Media, in: Rimrott F.P.J. and Tabarrok B.(eds.),Theoretical and Applied Mechanics (XV th ICTAM),(North-Holland, Amsterdam,1980).

[11] Maugin G.A. and Fomethe A.,On the Elastoviscoplasticity of Ferromagnetic Crystals, J.Phys.Sol.State Phys;(submitted for publication in ,1981).

[12] Fomethe A., and Maugin G.A.,Influence of Dislocations on Magnon-Phonon Couplings- A Phenomenological Approach, J.Phys.Sol.State Phys.,(submiited for publication in,1981).

[13] Friedel J.,Dislocations (Translation from the French, Pergamon Press Ltd, Oxford,1964).

[14] Mandel J., Plasticité classique et Viscoplasticité (1971 CISM Courses and Lectures, Udine,Italy),(Springer-Verlag, Wien ,1972).

[15] Nozdriev V.F., and Fedorichtchienko N.V., Molecular Acoustics (in Russian, Moscow, 1974).

Continuum Models of Discrete Systems 4
eds. O. Brulin and R.K.T. Hsieh

STRUCTURE OF MAGNETICALLY STABILIZED FLUIDIZED SOLIDS

Ronald E. Rosensweig
Exxon Corporate Research - Science Laboratories
Linden, New Jersey, USA

Gary R. Jerauld
Department of Chemical Engineering
University of Minnesota
Minneapolis, Minnesota, USA

Markus Zahn
Department of Electrical Engineering
Massachusetts Institute of Technology
Cambridge, Massachusetts, USA

Magnetically stabilized fluidized solids are absent of fluid or solids backmixing which normally occur in bubbling fluidized beds, yet the beds are flowable. The presence or absence of structuring and degree of uniformity in these beds is important in determining their applicability as chemical reactors or contactors for physical processes. This work derives structure from permeability measurements and direct observations of sectioned castings. The results are compared with a computer simulation model.

INTRODUCTION

The magnetic stabilization of the state of uniform fluidization has been treated as a problem in continuum mechanics (Reference 1). Using linear stability theory the averaged equations of motion successfully predict the existence of operating regimes in which minimum fluidization speed U_M is unaffected by magnetization in uniform field while a magnetically stabilized regime appears having upper boundary U_T representing the transition speed for onset of bubbling. Limitations of the continuum description concern the structure of the bed and the resultant influence on bed stability and performance of the bed as a contactor (References 2 and 3). To obtain insight into these features we begin by exploring a computational model of magnetized stabilized bed (MSB) structure.

SIMULATION OF BED STRUCTURE

A computer simulation of structure in beds is developed for two dimensional packings by placing mono-size magnetically susceptible spheres in a box. This work is considered a preliminary to modeling the full dynamics of an MSB including fluid-particle interaction forces and three dimensional effects.

This model introduces a particle at a randomly chosen position above the bed, then permits the particle to seek a local potential minimum. Once situated at the local minimum a particle is assumed to remain in the same position. The potential is calculated by summing magnetic and gravitational potentials. The particles are considered magnetically soft, so with magnetization specified as much less than the applied field the dipoles all point vertically. Periodic sidewall boundary conditions are incorporated to simulate semi-infinite packings.

The total of gravitational and magnetic potential may be expressed nondimensionally as

$$\tilde{E}_T = Y\tilde{E}_G + \tilde{E}_M \tag{1}$$

$$\tilde{E}_T = \frac{144\,E_T}{\pi\mu_o M^2 D_p^{\,3}}$$

$$\tilde{E}_G = \frac{24\rho g D_p}{\mu_o M^2} \tag{2a,b,c}$$

$$\tilde{E}_M = \sum_{i=1}^{n} \frac{(1-3\cos^2\theta_i)}{R_i^{\,3}}$$

and $$Y = \frac{y}{D_p} \qquad R_i = \frac{r_i}{D_p} \tag{3a,b}$$

The number of dipoles within a radius of influence of a given particle is n; practically the radius need not be greater than 5 particle diameters. θ is the angle between the vertical and the line of centers of two particles, ρ density, M magnetization, D_p particle diameter, y vertical distance, and r_i particle separation (S.I. units). $\tilde{E}_G$ represents the ratio of gravitational energy of elevating a particle the distance of its own diameter to the magnetic repulsion energy of side by side, tangent, magnetized particles. From experiments with MSBs we know roughly that magnetizations $\mu_o M$ of less than 0.01 Tesla (100 gauss) yield insufficient stabilization while magnetizations in excess of 0.1T (1000 gauss) tend to produce unflowable beds. With particle size in the range 100-1000 μm the range of values of $\tilde{E}_G$ is from 0.003 to 30.

As shown in Fig. 1, a particle in the model is permitted to roll along the interface in the direction of lower potential until coming to rest at a position of local potential minimum. Unlike in models of nonmagnetic beds (Reference 4) the particle need not find a two point support at equilibrium. The model calculates the potential at small intervals along the interface and uses a cubic spline to interpolate the potential between fixed points. The bed is seeded by placing particles at the bottom with random spacings ranging from zero to one particle diameter.

Structures generated by this model are reproduced in Figure 2. At values of $\tilde{E}_G$=10, corresponding to a large ratio of gravity to magnetic energy, the structure appears random, compact, and free of orientational effects. As magnetization increases and hence $\tilde{E}_G$ decreases the packing becomes more open. At sufficiently large values of magnetization the structure polymerizes into vertical chains or open columns.

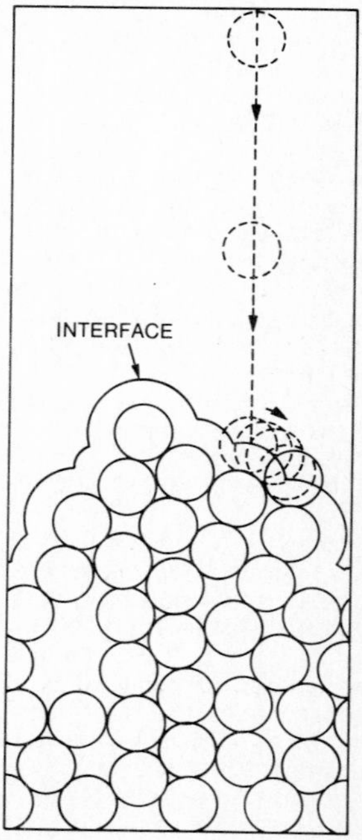

Fig. 1 - Simulation Model Illustrating Addition of a Particle to the Bed

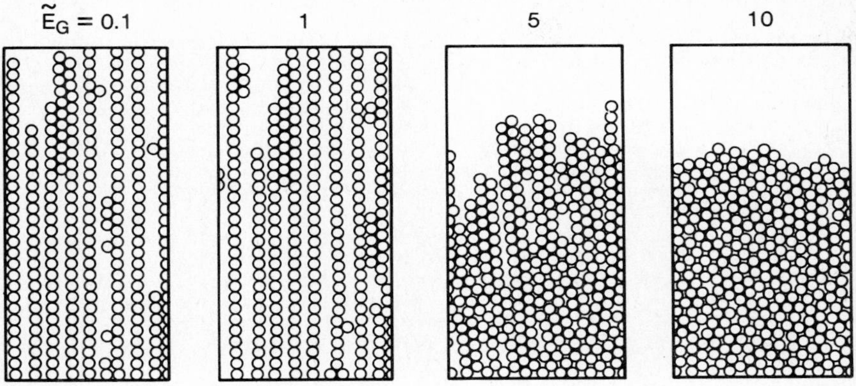

Fig. 2 - Structures Generated by the Simulation Model

Figure 3 plots void fraction of the structures as a function of the modulus $\tilde{E}_G$. The plot reveals a transition between column structure and random packing corresponding to the following values of $\tilde{E}_G$.

$\tilde{E}_G < 1$	Open columns	
$1 < \tilde{E}_G < 10$	Partially structured	(4)
$\tilde{E}_G > 10$	Dense, isotropic packing	

The numerical value of asymptotic void fraction in dense random packing is consistent with results reported in the literature (Reference 5) for equal spheres in two dimensions ($\varepsilon_0 \simeq 0.18$). This voidage is much lower than values typical of equal spheres in random three dimensional packing. Accordingly, it is concluded that two dimensional simulation cannot yield quantitative agreement with three dimensional experiments but may show qualitative similarities.

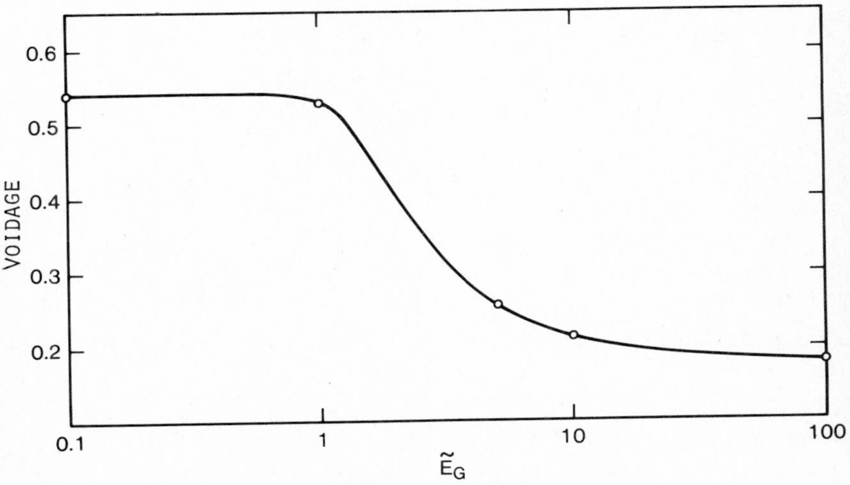

Fig. 3 - Voidage Fraction of Simulated Structures

STRUCTURE RELATED TO BED MAGNETIC PERMEABILITY

Measurements of bed magnetic permeability may be used to characterize the bed structure experimentally. The experiments measure the terminal voltage and current through a coil surrounding the bed as a function of an alternating source of voltage connected in series with the coil, a standard resistor, and a capacitor to lower the circuit impedance. The coil serves as both the source of magnetic field and the generator of the sensed signal. The value of bed inductance determined in this manner may then be used to compute μ_b, the permeability of the bed, with a correction made for presence of the nonmagnetic container wall.

It is assumed that the particle material has a constant magnetic permeability μ and the bed a uniform voidage ε. The bed magnetic permeability μ_b will depend on some weighted average of the particle and free space permeabilities and on the particle structure within the vessel. Various limiting cases of particle structure then lead to closed form expressions for bed permeability. These are: particles magnetically in parallel, particles magnetically in series, and a cavity model.

$$\frac{\mu_b}{\mu_o} = \varepsilon + \frac{\mu}{\mu_o}(1-\varepsilon) \qquad \text{(parallel)} \qquad (5)$$

$$\frac{\mu_b}{\mu_o} = \frac{\mu}{\mu\varepsilon + \mu_o(1-\varepsilon)} \qquad \text{(series)} \qquad (6)$$

For the cavity model consider a small magnetizable particle or chain of particles with demagnetizing coefficient d_p placed within a hollow free space cavity with demagnetizing coefficient d_c. A linear chain of m particles approximates a prolate ellipsoid.

The demagnetizing factor for a prolate ellipsoid having major axis m times the two equal minor axes, and magnetized parallel to its major axis is (Reference 6),

$$d = \frac{1}{m^2-1}\left[\frac{m}{\sqrt{m^2-1}} \ln\left(m + \sqrt{m^2-1}\right) - 1\right] \qquad (7)$$

The particle magnetic field H_p is related to the applied bed magnetic field H_o as

$$H_p = H_o + d_c M - d_p M_p = H_o + M_p\left[d_c(1-\varepsilon) - d_p\right] \qquad (8)$$

where M the bulk averaged magnetization of the bed is related to the particle magnetization as $M = (1-\varepsilon)M_p$. If the particle has magnetic permeability μ then

$$M_p = \left(\frac{\mu-\mu_o}{\mu_o}\right) H_p \qquad (9)$$

Thus, the relative bed magnetic permeability $1 + M/H_o$ in the cavity model is

$$\frac{\mu_b}{\mu_o} = \left[1 + \frac{(1-\varepsilon)}{\left(\frac{\mu_o}{\mu-\mu_o}\right) - d_c(1-\varepsilon) + d_p}\right] \qquad \text{(cavity model)} \qquad (10)$$

Because the steel particles are highly magnetizable with effective permeability on the order of 100 or greater we examine the theoretical expressions of eqs. (5), (6), and (10) in the limit of large magnetic permeability.

$$\lim_{\mu \to \infty} \frac{\mu_b}{\mu_o} = \begin{cases} \infty & \text{(parallel)} \\ \dfrac{1}{\varepsilon} & \text{(series)} \\ 1 + \dfrac{(1-\varepsilon)}{d_p - d_c(1-\varepsilon)} & \text{(cavity model)} \end{cases} \qquad (11)$$

For each particle size the experimental procedure was as follows. Measure air inductance of the empty vessel (7.6 cm I.D. by 34.3 cm length). Fill vessel with steel particles of the particular size level to the top. Measure inductance at constant current. Remove some particles and weigh, then fluidize and expand the bed to the original level. The experimental results of repeating this procedure to a maximum expansion of ε=0.59 are plotted in Fig. 4. The applied field intensity is 0.01T rms at 60 Hz. Considering the experimental scatter it may be seen that the bed magnetic permeability is essentially independent of particle size. The numerical value of $\tilde{E}_G$ ranged from 0.057 to 1.54 in these tests.

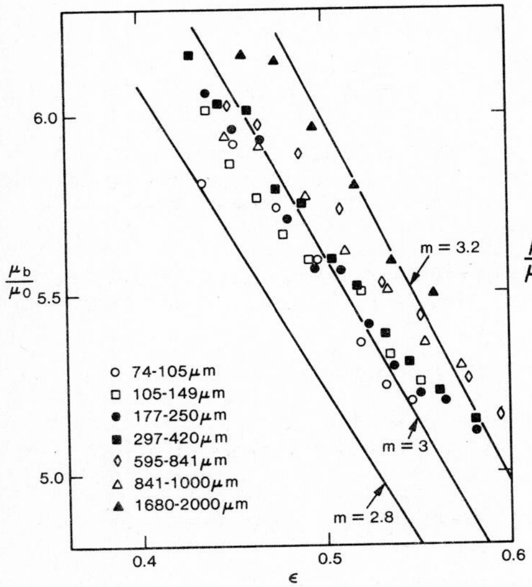

Fig. 4 - Measured Relative Bed Permeability for Steel Particles. Solid Lines Represent Theory for d_c = 0.

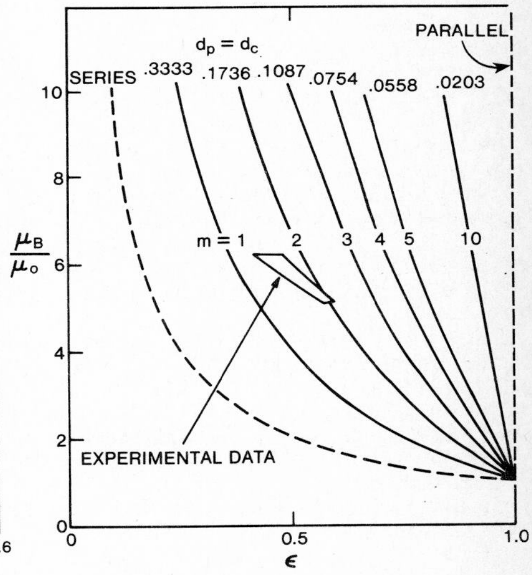

Fig. 5 - Model Values of Relative Bed Permeability for Steel Spheres. Solid Curves Represent Theory for $d = d_p = d_c$.

The data in Fig. 4 display decrease of permeability with increased expansion of the stabilized bed. Measurements in a fluidized magnetized bed of ferrite excited with a 50 Hz alternating field yielded a small increase in permeability (Reference 7). We attribute this variance in behavior to a difference in the state of the beds; magnetite beds are agitated in alternating fields due to the particle coercivity and could be expected to structure with attendant change in the bed permeability.

All data must fall within the limiting series and parallel curves as plotted with dashed lines in Fig. 5. The data enveloped in Fig. 5 shows that measured permeability is about 1/3 to 2/3 greater at a given value of void function than predicted by the cavity model based on a spherical particle in a spherical cavity. If it is assumed that $d_p = d_c$ then the data in Fig. 5 indicate a gradual increase in m or effective chain length up to a value of 2 at the highest bed expansion. The cavity surrounding a cluster of particles need not have the same shape as the cluster although it is plausible that the cavity must be at least as elongated as the cluster or chain of particles. In the limit of a highly elongated cavity $d_c = 0$ the final expression in Eq. 11 for relative permeability reduces to $1 + (1-\varepsilon)/d_p$. The solid lines in Fig. 4 plot this relationship for values of d_p corresponding to m = 2.8, 3, and 3.2. It may be seen from the figure that the data exhibit a trend of increasing m as the bed expands with a maximum value of m of about 3.2 near $\varepsilon = 0.6$.

The permeability measurements thus are consistent regardless of the detailed assumption of cavity shape with the picture of a gradual structuring in the bed as the bed is expanded with apparent maximum chain length in particle diameters of 2 to 3 plus.

STRUCTURE DETERMINED FROM MATRIX SECTIONS

The structure of MSBs in certain cases may be examined rather directly upon casting an entire bed in a matrix of a polymerizing fluid. Bed expansion is first established in a modest applied field intensity, then the bed is magnetically frozen with high field ($\sim$0.1 T) so gas flow can be ceased; the bed length stays constant with the structure apparently unchanged. Desirable properties of the bed solids for this purpose are ease of mechanical sectioning, high optical contrast, particle homogeneity, and high saturation magnetization. Beds of steel spheres are well suited by all these criteria and were used in these studies.

Figure 6 compares the random appearance of a section of an unexpanded bed to the somewhat structured appearance of the most expanded bed. Beds having expansion of 16%, 30% and 37% appeared isotropic when inspected with a magnifying loupe. Chaining structure, however, was noticeable in the bed having expansion of 44.7%, with a typical chain containing about 7 particles.

It is noted that the observed structure is relatively small in scale; a chain of seven particles represents only 0.5% of the expanded bed length of 11.6 cm.

Additional castings of beds were progressively turned on a lathe to circular cylindrical shape. These samples were measured and weighed and values of voidage computed. It was found that void fraction is constant ($\pm$.02) over the cross section for any individual sample tested regardless of the degree of bed expansion which ranged up to 40%. The data establishing the homogeneity of the particle beds are shown in Table I.

Table I - Homogeneity of MSB Particle Beds Determined Gravimetrically from Castings Machined to Cylindrical Diameter D.

Sample No.	D=1.9 cm	1.7	1.4	1.2	0.9	Expansion
1	ε=0.40	0.40	0.40	0.41	0.42	0%
2	0.46	0.46	0.46	0.46	0.45	11%
3	0.49	0.49	0.49	0.50	0.51	18%
4	0.57	0.56	0.56	0.57	0.60	40%

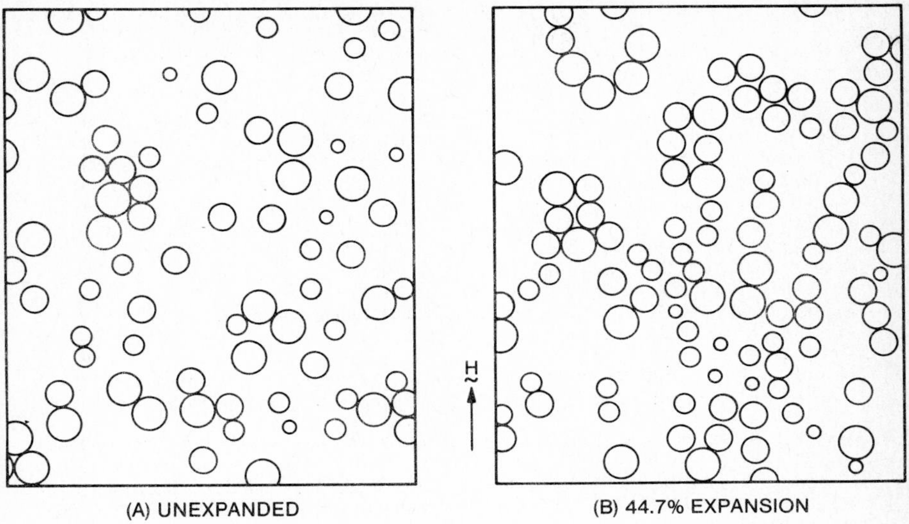

Fig. 6 - Particle Structure in MSB's Traced from Photographs of Plane Sectioned Castings. D_p = 149-177 microns. H = 0.0056 Tesla. U_M = 18 cm/s. U_T = 49 cm/s. The Unexpanded Bed Voidage is 0.43. Superficial Velocity in (B) is 44 cm/s.

We may attempt to compare the observations of bed structure obtained from the sectioned samples with the results of the computer simulation study, as well as with the measurements of bed magnetic permeability. Experimentally, from Fig. 4 the bed permeability ranged from about 5.2 to 6.4. Accordingly, in the applied induction field of 0.0056 Tesla (56 gauss) the bed magnetization is 0.0291 to 0.0358 Tesla. Since bed void fraction ranges from 0.40 to 0.60 in the experiments and permeability is greatest when voidage is least it may be calculated that particle magnetization ranged from 0.0628 to 0.0739 Tesla. Substituting into the expression for $\tilde{E}_G$ of eq. 2b with average particle size D_p of 163 micron and density ρ of 7870 kg/m^3 yields values of $\tilde{E}_G$ ranging from 0.069 to 0.096. These values are near that of the left most simulation in Fig. 2. In comparison the bed displays much less chaining. It is concluded that the two dimensional model overpredicts the degree of structuring that actually occurs.

CONCLUSION

Simulation produces a transition from isotropic structure to ordering and yields a quantitative criterion for the onset of transition. However, the model assumptions of two dimensionality and sequential addition of particles overpredict the tendency of chaining compared to measurements.

Permeability models are in good agreement with permeability measurements and indicate that a small degree of chaining develops, on the order of 2 to 3 particles per chain. In common with the simulation model, the degree of chaining increases as the bed void fraction increases.

Direct observation of sectioned castings reveals great structural homogeneity and a degree of chaining that is intermediate to the chaining deduced from permeability measurements and the chaining resulting from the simulation model.

REFERENCES

[1] Rosensweig, R. E., Magnetic Stabilization of the State of Uniform Fluidization, I&EC Fund., Vol. 18, No. 3 (1979) 260-269.

[2] Lucchesi, P. J., W. H. Hatch, F. X. Mayer, and R. E. Rosensweig, Magnetically Stabilized Beds-New Gas Solids Contacting Technology, Proc. of the 10th World Petroleum Congress, Bucharest, 1979, Vol. 4, Heyden and Sons, Philadelphia, Pa. (1979).

[3] Rosensweig, R. E., J. H. Siegell, W. K. Lee, and T. Mikus, Magnetically Stabilized Fluidized Solids, A.I.Ch.E. Symposium Series, Vol. 77, No. 205, Recent Advances in Fluidization and Fluid-Particle Systems (1981).

[4] Jodery, W. S. and E. M. Tory, Simulation of Random Packing of Spheres. Simulation, 1 (1979) 1-12.

[5] Visscher, W. M. and M. Bolsterli, Random Packing of Equal and Unequal Spheres in Two and Three Dimensions. Nature, 239 (1972) 504-507.

[6] Bozorth, R. M., Ferromagnetism (D. Van Nostrand, New York, 1953), p. 849.

[7] Filippov, M. V., The Effective Magnetic Permeability of a Fluidized Bed of Ferromagnetic Material. Izvestiya Akademii Nauk. Latviiskoi SSR (Fizika), 12 (1961) 52-54. Translated from Russian.

Continuum Models of Discrete Systems 4
eds. O. Brulin and R.K.T. Hsieh

DYNAMICS OF LIQUIDS

A. Sjölander

Institute of Theoretical Physics
Chalmers University of Technology
S-412 96 Göteborg
Sweden

When one discusses the macroscopic flow of fluids - gases or liquids - this is usually based on the hydrodynamic equations. Besides being capable of describing the flow pattern around various obstacles, these equations also account for the propagation of ordinary sound waves and the diffusion of heat. Since the general equations are non-linear they include such complicated physical phenomenon as turbulent flow and other non-linear effects. However, they do contain some shortcomings. So for instance, the way the pressure and entropy depends on density and temperature and the values of the transport coefficients must all be obtained from other sources. It goes beyond hydrodynamics to determine these and we have here to take recourse either to general statistical physics or to purely experimental information. We know that hydrodynamics can only account for flows which vary slowly both in space and time on a microscopic scale and there is no way to find the criteria for its validity from the hydrodynamic equations themselves.

Any fluid consists of individual atoms or molecules and any motion, irrespective of whether we are talking about microscopic or macroscopic flows, originates from the motions of these atoms and from their collisions. A large number of atoms may move together in a cooperative fashion so that we can even observe the flow of the fluid by our own eyes. However most of the energy content lies in the very erratic motion of the individual atoms and we would need very particular experimental tools to observe this directly. In our daily experience it appears in the form of the temperature of the fluid.

I would here like to discuss some of these aspects and review in a rather general fashion our present microscopic understanding of atomic motions in liquids and gases, and then connect this to the macroscopic behaviour. It will be done in a descriptive way in order to avoid any detailed mathematics, which would probably obscure the physics involved rather than clarify it. In this connection it is very appropriate to mention the name of Boltzmann, since he pioneered this field and since he died exactly 75 years ago. He took seriously - and he had to suffer for that - the conjecture that all matter consists of atoms and that the thermal properties are nothing else than the manifestation of the motions of these atoms. Introducing the statistical quantity $f(\vec{r}\vec{p}t)$, the density of atoms at position $\vec{r}$ at time t and having the momentum $\vec{p}$, Boltzmann formulated a general equation of motion for this density function, the famous Boltzmann equation (1872):

$$\frac{\partial f}{\partial t} + \vec{v}\cdot\vec{\nabla}_r f + \vec{F}\cdot\vec{\nabla}_p f = C(f,f). \qquad (1)$$

Here, $\vec{v} = \vec{p}/m$ in the velocity of the atom and $\vec{F}$ is an external force acting on the atom. With the right hand side being zero, the transport equation contains nothing more than the information that each individual atom moves under the external force $\vec{F}$ without affecting each other. The effect of the collisions between the atoms is contained in C(f,f) and this is a bilinear integral operator on $f(\vec{r}\vec{p}t)$. It contains explicitly the interaction between the atoms.

Boltzmann made several very explicit assumptions, when deriving equation (1) from

Newton's equations, and this made it valid only for a low density gas. Let me summarize the assumptions:

i) Only two atoms interact simultaneously and the probability that three or more atoms come together and interact at the same time was considered insignificantly small.

ii) The velocities of the two atoms, which approach each other in a collision, were assumed to be completely uncorrelated, the so called "Stosszahlansatz". After the collision there is a certain correlation between the velocities but this was assumed to be lost in the subsequent collisions with other atoms. In this way the evolution of the system was expressed in terms of a sequence of independent binary collisions between the atoms and the basic microscopic information was connected with the scattering cross section of the binary collision.

iii) The finite extension both in space and time of a binary collision event was ignored. This implies that one avoided describing the details of this event, which in ordinary gases means ignoring what happens on a length scale of the order 5 Å and on a time scale of the order 10^{-13} sec.

The Boltzmann equation clarified in a transparent way how the very chaotic behaviour of the atoms on an atomistic scale could give rise to a macroscopic flow. Even though there is a rapid exchange of energy and momenta between the particles in each binary collision, the total energy and total momentum is not changed. This is built into the collision term C(f,f). As a consequence of this the long wavelength motions, involving the collective flow of many atoms, change on a much longer time scale. The hydrodynamic equations, which refer to long wavelength variation in density, momentum, and energy, follow directly from the Boltzmann equation and is a consequence of the conservation of particle number, total momentum, and total energy in each binary collision. All other dynamical variables are damped out on a microscopic time scale and this is the reason why only the hydrodynamic ones - density, current, temperature - become relevant on macroscopic length and time scales. However, the values of the transport coefficients, like the shear viscosity, depends on what happens within a collision time, i.e. within a time of the order 10^{-13} sec. Similarly, the microscopic equation is needed for deciding when the hydrodynamic variables are sufficient to describe the system.

Even though Boltzmann's equation has given very accurate predictions for the transport coefficients of dilute gases, there are some shortcomings of great conceptual interest. Let me through Figure 1 illustrate a sequence of collisions and focus particular attention on the two atoms which are marked in black. Let us further assume that just before they collide the first time, one of the two atoms has a considerably larger velocity than the other. Transfer of energy and momentum to the slower atom will then occur and the latter atom will in its subsequent collisions with other atoms share some of this energy and momentum to these. When the rapid atom collides later on, it may do that with one of those atoms which got part of the shared energy and momentum. If this happens, it violates the Boltzmann "Stosszahlansatz" and it is an effect which is not included in Boltzmann's equation; at least not in the linearized version of the equation. We would on intuitive basis expect that all this is quite insignificant at low densities and this is essentially true. However, it has the interesting consequence that the transport coefficients become non-analytic functions of the density and also that the so called Burnett coefficients are infinite. These coefficients enter when one goes beyond ordinary hydrodynamics and include next order terms in the velocity gradient. Opposite to what one would believe on intuitive ground the ordinary transport coefficients diverge for a strictly two-dimensional gas. This implies that the linearized Navier-Stokes equation is not appropriate for describing "sound" wave propagation in two dimensions.

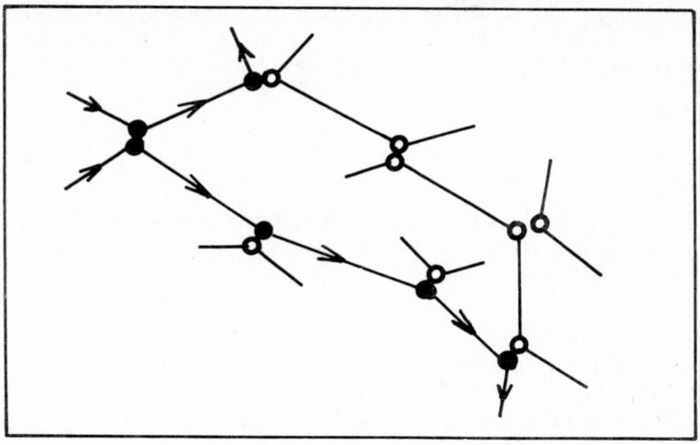

Figure 1
Illustrates correlated binary collisions, which cause violation of the "Stosszahlansatz".

If we now turn to dense gases and liquids, we cannot expect the Boltzmann equation to be valid. However, except for a few modifications, the conceptual aspects which Boltzmann introduced still hold. The interaction between the atoms consists of a strong, very short range repulsive part, which is realized when two atoms with large velocities bump into each other, and a weaker somewhat longer range attractive part. It is the latter part which is responsible for the appearance of separate gas and liquid phases, whereas the strong repulsive part is of main importance for the crystalline phase. We should expect that binary collisions are still very significant but that we can no more ignore the fact that around any single atom we can find simultaneously a number of other atoms within the interaction range. One finds from experiments that a certain local ordering exists among the atoms, somewhat analogous to what we have in solids, and this is normally expressed in terms of the pair correlation function or the so called static structure factor. The latter is directly obtained through scattering of X-rays or neutrons in liquids. It can also be calculated from statistical physics, if the interaction potential between the atoms is known. The local ordering extends over a microscopic distance, 1-15 Å and has a significant effect on the microscopic motion. It introduces one important modification of the Boltzmann equation. Each atom experiences besides an external force an effective force from the surrounding atoms similar to what happens in charged plasmas. It introduces in equation (1) an extra term

$$\{- \int d\vec{r}'\vec{\nabla}v_{eff}(\vec{r}-\vec{r}')\ n(\vec{r}'t)\}\cdot\vec{\nabla}_p f(\vec{r}\vec{p}t)\ , \tag{2}$$

where $n(\vec{r}t)$ is the particle number density and $v_{eff}(r)$ is in a simple way related to the pair correlation function.

Let me digress for a moment and ask the question how one can test the above point. The most detailed information is obtained by scattering of slow neutrons against a liquid sample and measuring the number of neutrons for various scattering directions and energy transfers. The slow neutrons are obtained in large numbers from a reactor and the experimental data are normally given in terms of the so called dynamical structure factor $S(q,\omega)$. This quantity describes how a thermal fluctuation in the liquid density and of a given wavelength $\lambda=2\pi/q$ propagates and dissipates. For long wavelength it manifests the propagation and damping of ordinary sound waves as well as diffusion of temperature fluctuations, which through the local thermal expansion also affects the density variations. It gives rise to a three-peak structure in $S(q,\omega)$ and is most easily observed through inelastic scattering of light (see Figure 2)

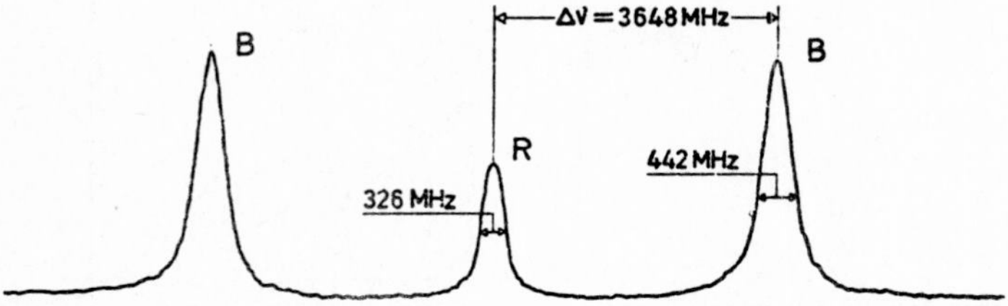

Figure 2
Shows $S(q,\omega)$ from scattering of light on silvernitrate (from G. Knape, J. Acoust. Soc. Am. 60, 83 (1976)).

In neutron scattering one is able to study the wavelength region of 1-20 Å and Figure 3 shows experimental results on liquid rubidium. The fairly broad resonance peak at non-zero frequences reveals that very short wavelength "sound" waves can propagate through the liquid. The damping is however quite significant but it is much less that what ordinary hydrodynamics would predict. Actually, hydrodynamics would only give a broad structureless peak around $\omega=0$ in complete disagreement with the experiments. The resonance disappears as we come to wavelengths which are less than the interparticle distance. The basic cause for these short wavelength "sound" waves is the existence of the effective force in equation (2) and the "sound" wave dispersion depends strongly on the strength and form of $v_{eff}(r)$.

These "sound" waves are more closely related to the short wavelength lattice vibrations, phonons, in crystalline solids and which contain the main part of the thermal energy of the solid. This brings in the concept of "collision-dominated" and "collisionless" sound. The former is ordinary sound and requires that collisions, represented by $C(f,f)$ in the Boltzmann equation, are sufficiently effective to keep the system in thermal equilibrium locally. The damping of the sound waves is then small only for long wavelengths. The collisionless sound arises from the effective force in equation (2) and is important only for higher densities and the damping decreases as the wavelength becomes smaller. This kind of sound shows up more clearly in a quantum fluid as helium and experiments have beautifully revealed the transition from ordinary sound to collisionless sound as one goes to shorter and shorter wavelengths.

The "binary" collisions play an essential role for explaining the damping of the short wavelength "sound" waves. Since we here are concerned with fluctuations on the time scale of the order 10^{-12} - 10^{-13} sec., we cannot ignore the extension in space and time of these binary collisions. This modifies the form of the collision term in equation (1) and, in particular, it introduces a time integration over the binary collision time. More interesting from a conceptual point of view is that those events, which lead to a violation of the "Stosszahlansatz", now become quite significant. The following microscopic picture of the atomic motion has emerged from extensive theoretical and experimental work over the last decade or so. As is illustrated in Figure 4 a single atom collides with another atom, transferring

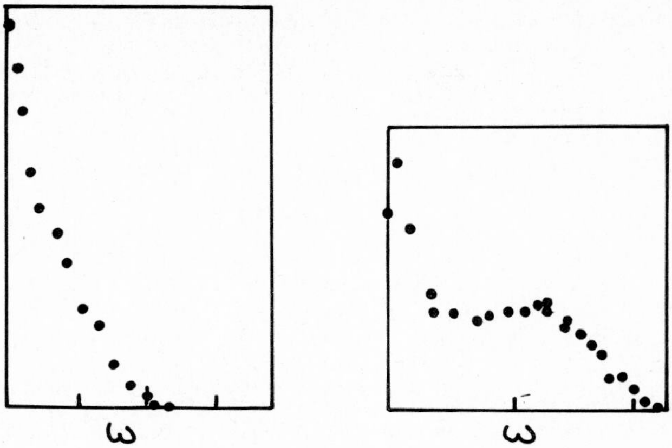

Figure 3
Shows $S(q,\omega)$ from scattering of neutrons on rubidium (from J.R.D. Copley and J.M. Rowe, Phys. Rev. Letters 32, 49 (1974); Phys. Rev. A9, 1656 (1974)). The left figure is for $q=1.25$ Å^{-1}($\lambda=4.1$ Å) and the right figure is for $q=1.00$ Å^{-1}($\lambda=6.3$ Å). The frequency unit, illustrated by the vertical lines on the ω-axis, is $5\cdot 10^{13}$ rad./sec.

energy and momenta to its surrounding, and a collective flow pattern builds up around the first atom. It takes several collision times to create this collective motion in the surrounding, but once created it dies out more slowly. It gives rise to a significant memory effect and it causes us to modify the collision term in

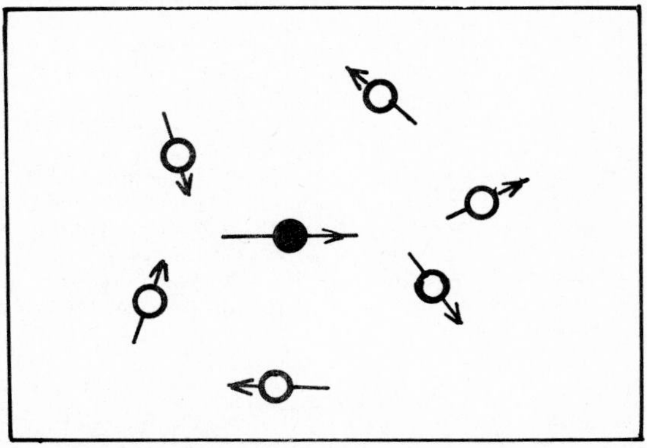

Figure 4
Illustrates the collective flow pattern of atoms surrounding a particular atom, the black one. The arrows indicate the direction of velocity.

the Boltzmann equation. The physics involved has some similarity to what happens around a sphere which moves in an ordinary Navier-Stokes fluid. Density, current, and temperature disturbances are created in the fluid by the sphere and these propagate through the fluid and affect the motion of the sphere at a later time. The macroscopic problem can be solved and leads to the following drag force on the sphere

$$\vec{F}(t) = -\ 6\pi\eta R\vec{v}(t) - \frac{2\pi}{3}\rho R^3 \frac{d\vec{v}(t)}{dt} - \int_0^t dt'\gamma(t-t')\ \vec{v}(t')\ , \tag{3}$$

where

$$\gamma(t) = \text{const}\ t^{-3/2}\ . \tag{4}$$

Here, R is the radius of the sphere and $\vec{v}(t)$ is its velocity at time t. The fluid is characterized by its mass density ρ and shear viscosity η. The constant before $t^{-3/2}$ contains besides the shear viscosity the self diffusion constant of the sphere. In particular, we notice that the force depends on the velocity at earlier times and that this memory effect dies out fairly slowly. This is due to the slowly decaying backflow effect around the sphere. Microscopically, we have to replace the sphere by the true atom and treat the surrounding from an atomistic point of view. However conceptually the physics is essentially the same. Of course, collisions between individual atoms enter now and the local structure of the atoms enters as well. All atoms behave on the average in the same way and we have the picture where each atom transfers energy and momenta to neighbouring atoms and builds up a surrounding collective backflow around itself. From this emerges ordinary hydrodynamics when we consider long wavelength motions.

Since the compressibility becomes very large near the gas-liquid critical point, the long wavelength motions in the backflow become strongly magnified and it leads to an exceptionally long memory effect. This is precisely the reason why ordinary hydrodynamics becomes invalid. It has, however, significant effects on the microscopic motions also far from the critical point and the values of the ordinary

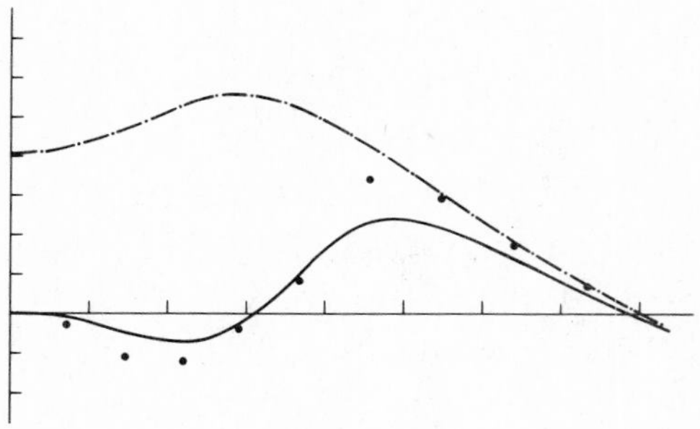

Figure 5
Shows the width $\Delta(q)$ of the neutron incoherent structure factor $S_s(q,\omega)$ versus q. The points are "experimental" data from computer simulations on liquid argon (from D. Levesque and L. Verlet, Phys. Rev. A2, 2514 (1970)). The full line is from theory which includes backflow effects besides uncorrelated binary collisions. The dashed-dot curve is from theory which ignores backflow (from G. Wahnström and L. Sjögren, preprint). The units for q, indicated by vertical lines on the q-axis, is 1 Å^{-1}.

hydrodynamic transport coefficients are affected significantly. All this has been clearly revealed in various computer simulations of the atomic motions in liquids and dense gases. So for instance, the self diffusion constant shows a characteristic density dependence when compared with what only binary collisions would give, the so called Enskog value. For high densities as in liquids the true self diffusion constant is smaller than the Enskog value and it reflects the fact that each atom is to some extent trapped by the surrounding atoms like in solids. In dense gases, on the other hand, the opposite happens. A backflow current is built up around each atom and it hits the atom from behind and increases the diffusion constant above the Enskog value. Incoherent scattering of neutrons measures how single atoms diffuse in a liquid and it is usually expressed in terms of a corresponding structure factor $S_s(q,\omega)$. The latter is found to be quite structureless with a peak around $\omega=0$ and a width $\Delta(q)$, which depends on the wavelength considered. The width has, however, a characteristic q-dependence (see Figure 5), which arises from the backflow effect. If one assumes that only uncorrelated binary collisions occur - the dashed curve in the figure - the theoretical result does not even qualitatively resemble the experimental one. The most accurate "experimental" information on this point is obtained from computer simulations.

The various aspects of liquid dynamics, which have been discussed above, are included in a generalized Boltzmann equation:

$$\left(\frac{\partial}{\partial t} + \frac{1}{m}\vec{p}\cdot\vec{\nabla}_r\right)f(\vec{r}\vec{p}t) - \int d\vec{r}\,\vec{\nabla}v_{eff}(\vec{r}-\vec{r}')\; n(\vec{r}'t)\cdot\vec{\nabla}_p f(\vec{r}\vec{p}t) - \int_0^t dt' \int d\vec{r}' d\vec{p}'\; \vec{\nabla}_p\cdot L(\vec{r}\vec{p}t;\vec{r}'\vec{p}'t')\cdot\left\{\frac{1}{mk_BT}\vec{p}' + \vec{\nabla}_{p'}\right\}f(\vec{r}'\vec{p}'t') = 0 \tag{5}$$

The last term replaces the Boltzmann collision term and the tensor function L contains both binary collisions as well as backflow effects. There is no possibility to write down the exact form for the latter quantity, but one has come quite far in clarifying its characteristic form and in finding approximate values for simple liquids. It contains information on how a single atom and its surrounding move in an average sense as time passes from t' to t .

In a certain sense the loop is closed. The developement which all started with Boltzmann has been concluded with a modified Boltzmann equation. The basic original ideas still hold and certain additional features have been added in order to describe effects which become significant at higher densities. This also unifies the treatment of ordinary fluids and charged fluids, so called plasmas. Only minor conceptual points have to be added to make it applicable to quantum fluids as well.

References:

Instead of referring to the original articles, which necessarily would mean writing a long list, I will here only refer to a few basic books or review articles.

1. March, N.M. and Tosi, M.P., Atomic Dynamics in Liquids (The Mac Millan Press LTD, 1976).
2. Hansen, J.P. and McDonald, I.R., The Theory of Simple Liquids (Academic Press, 1976).
3. Berne, B.J., ed., Statistical Mechanics PartB: Time-dependent Processes (Plenum Press, 1977).
4. Luscher, E. and Coufal, H., ed., Liquid and Amorphous Metals (NATO advanced study institute series, Series E: Applied sciences no. 36, 1980).
5. Boon, J.P. and Yip, S., Molecular Hydrodynamics (McGraw Hill Inc., 1980).

Continuum Models of Discrete Systems 4
eds. O. Brulin and R.K.T. Hsieh

INFLUENCE OF MAGNETIC FIELD ON DIFUSSION IN COSSERAT MEDIUM

Jarosław Stefaniak

Institute of Technical Mechanics
Technical University of Poznań
Poznań, Poland

In this paper the diffusion of gas- or fluid-components into the Cosserat medium is considered. The body is also influenced by the action of an initial constant magnetic field of intensity $\underset{\sim}{H}_0$. The starting point for the considerations are the equations and relations of electrodynamics. The relations are drawn out on the basis of methods developed by W.Nowacki /thermodiffusion/ and H.Parkus /magneto-thermoelasticity/.

INTRODUCTION

Let us consider an isotropic, elastic Cosserat medium exposed to the action of a constant magnetic field H_0 . Inside of the body the mass forces $\underset{\sim}{X}$, mass couples $\underset{\sim}{Y}$ and the heat sources r act. Each of these causes may change the strain γ_{ij} , the „twist" $\varkappa_{ij}$, the temperature T , the concentration c and the initial magnetic field $\underset{\sim}{H}_0$ as well. Our considerations are limited to a conductor of electricity with one component /i.e. gas or fluid/ diffusing in it. The fundamental literature consists of papers of W.Nowacki [1,2,3,4,5], H.Parkus [6,7], S.R. de Groot and P.Mazur [8].

EQUATIONS AND RELATIONS OF ELECTRODYNAMICS

The properties of electromagnetic field are described by:
Maxwell equations

$$(1)\qquad \nabla\times\underset{\sim}{H} = \underset{\sim}{j} + \frac{\partial \underset{\sim}{D}}{\partial t}, \qquad \nabla\times\underset{\sim}{E} = -\frac{\partial \underset{\sim}{B}}{\partial t}, \qquad \nabla\cdot\underset{\sim}{D} = \varrho_e, \qquad \nabla\cdot\underset{\sim}{B} = 0,$$

and relations

$$(2)\qquad \underset{\sim}{D} = \varepsilon_0\underset{\sim}{E} + \underset{\sim}{P}, \qquad \underset{\sim}{B} = \mu_0(\underset{\sim}{H} + \underset{\sim}{M}), \qquad \underset{\sim}{j} = \lambda_0\underset{\sim}{E},$$

where
$\underset{\sim}{H}$, $\underset{\sim}{E}$ - magnetic and electric field strength respectively
$\underset{\sim}{B}$ - magnetic induction
$\underset{\sim}{D}$ - electric displacement
ϱ_e - charge density
$\underset{\sim}{j}$ - electric current density
$\underset{\sim}{P}$ - polarization
$\underset{\sim}{M}$ - magnetization
ε_0 - dielectric constant of vacuum
μ_0 - permeability of vacuum
λ_0 - electrical conductivity.

For the isotropic body of linear electrical and magnetical properties we also have

$$(3)\qquad \underset{\sim}{P} = \chi_e \underset{\sim}{E}, \qquad \underset{\sim}{M} = \chi_m \underset{\sim}{H}.$$

Then $(2)_1$ and $(2)_2$ may be expressed as follows

$$(4)\qquad \underset{\sim}{D} = \varepsilon \underset{\sim}{E}, \qquad \underset{\sim}{B} = \bar{\mu} \underset{\sim}{H},$$

where
$\varepsilon = \varepsilon_0 + \chi_0$ - dielectric constant
$\bar{\mu} = \mu_0(1+\chi_m)$ - permeability
χ_e, χ_m - electric and magnetic susceptibilities.

If the velocity v satisfies the inequality $\frac{v}{c_0} \ll 1$, where c_0 is the speed of light in vacuum, and the influence of temperature is taken into account, then the relations (3), (4) and $(2)_3$ take the form [6] :

$$\underset{\sim}{P} = \chi_e(\underset{\sim}{E} + \underset{\sim}{v} \times \underset{\sim}{B}), \qquad \underset{\sim}{M} = \chi_m(\underset{\sim}{H} - \underset{\sim}{v} \times \underset{\sim}{D}),$$

$$(5)\qquad \underset{\sim}{D} = \varepsilon \underset{\sim}{E} + \chi_e(\underset{\sim}{v} \times \underset{\sim}{B}), \qquad \underset{\sim}{B} = \bar{\mu}\underset{\sim}{H} - (\varepsilon\bar{\mu} - \frac{1}{c_0^2})\underset{\sim}{v} \times \underset{\sim}{E},$$

$$\underset{\sim}{j} = \lambda_0(\underset{\sim}{E} + \underset{\sim}{v} \times \underset{\sim}{B} - \varkappa \nabla T).$$

In our considerations $\underset{\sim}{v} = \frac{d\underset{\sim}{u}}{dt} = \dot{u}$ denotes the speed of displacement, then the condition $\frac{v}{c_0} \ll 1$ is allways satisfied.

BALANCE OF ENERGY

The first law of thermodynamics /balance of energy/, generalized on the Cosserat medium and with regard to diffusion, may be expressed as follows [5,6]:

$$(6)\qquad \begin{aligned} &\frac{d}{dt}\int_V \left(\varrho\frac{v^2}{2} + J\omega^2 + \varrho U + U_e\right)dV = \int_V (\varrho r + \varrho X_i v_i + \varrho Y_i \omega_i) n_i \, dV + \\ &+ \int_{\partial V} \left(\sigma_{ij} v_j + \mu_{ij}\omega_j + \varepsilon_{ijk} E_j H_k - q_i - \sum_{k=1}^{n} m^k \eta_i^k + U_e v_i\right) n_i \, d\partial V, \end{aligned}$$

where
ϱ - mass density
J - rotational inertia
U - internal energy per unit of mass
U_e - electromagnetic energy per unit of volume
σ_{ij} - stress tensor
μ_{ij} - couple stress tensor
$\omega_i = \dot{\varphi}_i$ - time derivative of rotation
q_i - heat flow
m^k - chemical potential of component k
η_i^k - diffusion flow of component k.

The relation holds if the following assumptions are satisfied:
i/ There are not chemical reactions between the components
ii/ the terms connected with the „exchange energy" are neglected.

Because considerations are limited to one gas- or fluid-component diffusing into the solid, the sume in relation (6) is reduced to two terms only. Making use in relation (6) of the Gauss theorem and of the formula

$$\frac{d}{dt}\int_V U_e\, dV = \int_V \frac{\partial U_e}{\partial t}\, dV + \int_{\partial V} U_e v_i n_i\, d\partial V$$

one obtains the local form of balance of energy:

$$\varrho\frac{dU}{dt} + \frac{\partial U_e}{\partial t} + (\varrho\dot{v}_i - \varrho X_i - \sigma_{ji,j})v_i + (J\dot{\omega}_i - \varrho Y_i -$$

$$(7)\qquad -\varepsilon_{ijk}\sigma_{jk} - \mu_{ji,j})\omega_i - \sigma_{ji}\dot{\gamma}_{ji} - \mu_{ji}\dot{\varkappa}_{ji} - \varepsilon_{ijk}E_j H_k +$$

$$+ q_{i,i} + \sum_{k=1}^{2} m^k_{,i} + \sum_{k=1}^{2} m^k \eta^k_i - \varrho r = 0,$$

where

$$\gamma_{ji} = u_{i,j} - \varepsilon_{kji}\varphi_k, \qquad \varkappa_{ji} = \varphi_{i,j}.$$

In order to transform the term $\frac{\partial U_e}{\partial t}$ there are to determine the electromagnetic properties of the body. Assuming, that the conductor is considered one has $\underset{\sim}{P} = \underset{\sim}{0}$ and consequently

$$\frac{\partial U_e}{\partial t} = \varepsilon_0 \underset{\sim}{E}\cdot\frac{\partial \underset{\sim}{E}}{\partial t} + \mu_0 \underset{\sim}{H}\cdot\frac{\partial \underset{\sim}{H}}{\partial t} =$$

$$= -\nabla\cdot(\underset{\sim}{E}\times\underset{\sim}{H}) - \underset{\sim}{E}\cdot\underset{\sim}{j} - \mu_0\underset{\sim}{H}\cdot\frac{\partial \underset{\sim}{M}}{\partial t}.$$

After elementary transformations the relation (7) may be written as follows:

$$\varrho\frac{dU}{dt} + (\varrho v_i - \varrho X_i - \sigma_{ji,j} - \varepsilon_{ijk} j_j B_k)v_i +$$

$$+ (J\dot{\omega}_i - \varrho Y_i - \varepsilon_{ijk}\sigma_{jk} - \mu_{ji,j})\omega_i - \sigma_{ji}\dot{\gamma}_{ji} - \mu_{ji}\dot{\varkappa}_{ji} -$$

$$-\mu_0 H_i\frac{\partial M_i}{\partial t} + \sum_{k=1}^{2} m^k_{,i}\eta^k_i + \sum_{k=1}^{2} m^k \eta^k_{i,i} + q_{i,i} - \varrho r - \frac{j^2}{\lambda_0} - \varkappa \underset{\sim}{j}\cdot\nabla T = 0.$$

The essential effect on diffusion do have the mobility of gas- or fluid-component, but not the motion of atoms of the elastic body. Them we can assume, that the flow η_2 is equal to zero and $c_2 = \frac{\varrho_2}{\varrho}$ does vary in time. Therefore two equations of conservation of mass [8]

$$\varrho\frac{dc}{dt} = -\nabla\cdot\underset{\sim}{\eta}, \qquad \varrho\frac{dc_2}{dt} = -\nabla\cdot\underset{\sim}{\eta}_2$$

reduce to one equation

$$(8)\qquad \varrho\frac{dc}{dt} = -\nabla\cdot\underset{\sim}{\eta},$$

where $c = \frac{\varrho_1}{\varrho}$ denotes concentration of the diffusing component, $\underset{\sim}{\eta}$ - the flow of it, and $\varrho = \varrho_1 + \varrho_2$

Introducing now the free energy per unit of mass

$$F = U - TS,$$

where S is the entropy per unit of mass, and assuming it as a function of variables $\gamma_{ij}, \varkappa_{ij}, T, c, M_i$, i.e.:

$$F = F(\gamma_{ij}, \varkappa_{ij}, T, c, M_i)$$

one obtains

$$\begin{aligned}(9)\quad & \left(\varrho \frac{\partial F}{\partial \gamma_{ji}} - \sigma_{ji}\right)\dot{\gamma}_{ji} + \left(\varrho \frac{\partial F}{\partial \varkappa_{ji}} - \mu_{ji}\right)\dot{\varkappa}_{ji} + \varrho\left(\frac{\partial F}{\partial T} + S\right)\dot{T} + \\ & + \varrho\left(\frac{\partial F}{\partial c} - m\right)\dot{c} + \left(\varrho \frac{\partial F}{\partial M_i} - \mu_0 H_i\right)\dot{M}_i + (\varrho \dot{v}_i - \varrho X_i - \sigma_{ji,j} - \\ & - \varepsilon_{ijk} j_j B_k + \mu_0 H_j M_{j,i}) v_i + (J\dot{\omega}_i - \varrho Y_i - \varepsilon_{ijk}\sigma_{jk} - \mu_{ji,j})\omega_i + \\ & + \varrho T \dot{S} + q_{i,i} + m_{,i}\eta_i - \varrho r - \frac{j^2}{\lambda_0} - \varkappa j_i T_{,i} = 0 .\end{aligned}$$

EQUATIONS OF MOTION, CONSTITUTIVE EQUATIONS, HEAT CONDUCTION EQUATION

The quantities $\gamma_{ij}, \varkappa_{ij}, T, c, M_i$ are independent variables of free energy F . Therefore the coefficients at $\dot{\gamma}_{ji}, \dot{\varkappa}_{ji}, \dot{T}, \dot{c}, \dot{M}_i$ in equation (9) have to be equal to zero. The coefficients at v_i and ω_i also vanish /the invariance conditions/. Certainly, the „free term" is equal to zero as well. As a consequence we have: equations of motion

$$(10)\quad \sigma_{ji,j} + \varepsilon_{ijk} j_j B_k - \mu_0 H_j M_{j,i} + \varrho X_i = \varrho \dot{v}_i ,$$

$$(11)\quad \varepsilon_{ijk}\sigma_{jk} + \mu_{ji,j} + \varrho Y_i = J\dot{\omega}_i ,$$

constitutive equations

$$(12)\quad \begin{aligned} & \sigma_{ji} = \varrho \frac{\partial F}{\partial \gamma_{ji}} , \quad \mu_{ji} = \varrho \frac{\partial F}{\partial \varkappa_{ji}} , \quad m = \frac{\partial F}{\partial c} , \\ & S = -\frac{\partial F}{\partial T} , \quad H_i = \frac{\varrho}{\mu_0}\frac{\partial F}{\partial M_i} , \end{aligned}$$

heat conduction equation

$$(13)\quad \varrho T \dot{S} = -q_{i,i} - m_{,i}\eta_i + \varrho r + \varkappa j_i T_{,i} + \frac{j^2}{\lambda_0} .$$

Let us notice that the assumed micropolar properties of the body do not occur in the heat conduction equation (13). Now we proceed to the transformation of equation (13). It may be expressed in the form of

$$\varrho \dot{S} = -\left(\frac{q_i}{T}\right)_{,i} + \frac{\lambda_0 \varrho r + j^2}{\lambda_0 T} + \Omega ,$$

where

$$\Omega = -\frac{1}{T}\left(q_i \frac{T_{,i}}{T} + \eta_i m_{,i} - \mathcal{H} j_i T_{,i}\right)$$

is the „entropy production". According to its general form in linear theory [8]:

$$\Omega = \sum_{i=1}^{n} \underset{\sim}{J}_i \cdot \underset{\sim}{X}_i \,,$$

where $\underset{\sim}{J}_i$, $\underset{\sim}{X}_i$ denote flows and forces respectively, one has

$$\underset{\sim}{J}_1 = \frac{\underset{\sim}{q}}{T} - \mathcal{H}\underset{\sim}{j} \,, \qquad \underset{\sim}{J}_2 = \underset{\sim}{\eta} \,,$$

$$\underset{\sim}{X}_1 = -\nabla T \,, \qquad \underset{\sim}{X}_2 = -\nabla m \,.$$

Moreover, assuming the linear dependance between flows and forces:

$$\underset{\sim}{J}_i = \sum_{j=1}^{n} L_{ij} \underset{\sim}{X}_j \,,$$

one obtains

$$\underset{\sim}{J}_1 = \frac{\underset{\sim}{q}}{T} - \mathcal{H}\underset{\sim}{j} = L_{11}\nabla T - L_{12}\nabla m \,,$$

(14)
$$\underset{\sim}{J}_2 = \underset{\sim}{\eta} = -L_{21}\nabla T - L_{22}\nabla m \,,$$

where $L_{ij} = L_{ji}$.

The replacement of $\underset{\sim}{J}_1$ and $\underset{\sim}{J}_2$ in equations (13) and (8) by relations (14) yields

(15)
$$T\varrho\dot{S} = L_{11} T \nabla^2 T + L_{22} T \nabla^2 m - \mathcal{H}\,\nabla\cdot\underset{\sim}{j}\,T + L_{11}\nabla T\cdot\nabla T + L_{12}\nabla T\cdot\nabla m + L_{21}\nabla T\cdot\nabla m + L_{22}\nabla m\cdot\nabla m + \frac{j^2 + \lambda_0 \varrho r}{\lambda_0} \,,$$

(16)
$$\varrho\dot{c} = L_{21}\nabla^2 T + L_{22}\nabla^2 m \,.$$

The sets of equations (10), (11), (12), (15), (16) together with Maxwell equations (1) describe the considered fields. The particular form of constitutive equations (12) is determined by the form of free energy F . If it is assumed as a quadratic form. the equations reduce to linear form. We have then [5]:

(17)
$$\varrho F = \frac{\mu+\alpha}{2}\gamma_{ji}\gamma_{ji} + \frac{\mu-\alpha}{2}\gamma_{ji}\gamma_{ij} + \frac{\lambda}{2}\gamma_{kk}\gamma_{mm} + \frac{\gamma+\varepsilon}{2}\mathcal{H}_{ji}\mathcal{H}_{ji} + \frac{\gamma-\varepsilon}{2}\mathcal{H}_{ji}\mathcal{H}_{ij} + \frac{\beta}{2}\mathcal{H}_{kk}\mathcal{H}_{mm} - \nu_\Theta\gamma_{kk}\Theta - \chi_\Theta\mathcal{H}_{kk}\Theta - \nu_c\gamma_{kk}c - \chi_c\mathcal{H}_{kk}c + \frac{a'}{2}\varrho c^2 - \frac{n'}{2}\varrho\Theta^2 - d'\varrho c\Theta + \frac{\alpha_1}{2}M_i M_i \,,$$

where $\Theta = T - T_0 \ll T_0$, T_0 is the temperature of „natural" state.

Making use of the expression (17) for (12) one obtains:

$$
\begin{aligned}
\sigma_{ji} &= (\mu+\alpha)\gamma_{ji} + (\mu-\alpha)\gamma_{ij} + (\lambda\gamma_{kk} - \nu_\Theta\Theta - \nu_c c)\delta_{ij}, \\
\mu_{ji} &= (\gamma+\varepsilon)\varkappa_{ji} + (\gamma-\varepsilon)\varkappa_{ij} + (\beta\varkappa_{kk} - \chi_\Theta\Theta - \chi_c c)\delta_{ij}, \\
m &= -\frac{\nu_c}{\varrho}\gamma_{kk} - \frac{\chi_c}{\varrho}\varkappa_{kk} - d'\Theta + a'c, \\
S &= \frac{\nu_\Theta}{\varrho}\gamma_{kk} + \frac{\chi_\Theta}{\varrho}\varkappa_{kk} + n'\Theta + d'c, \\
H_i &= \frac{\alpha_1}{\mu_0}M_i .
\end{aligned} \tag{18}
$$

Substituting the relations (18) into equations (15) and (16) one obtains the heat conduction equation and diffusion equation in which the chemical potential m and entropy S do not appear. These equations, reduced to a linear form, are

$$
\begin{aligned}
&(k_1\nabla^2\Theta - n_1\dot{\Theta}) + (k_2\nabla^2 c - n_2\dot{c}) - (k_3\nabla^2\gamma_{kk} + n_3\dot{\gamma}_{kk}) - \\
&-(k_4\nabla^2\varkappa_{kk} + n_4\dot{\varkappa}_{kk}) - \frac{j^2 + \lambda_0\varrho r}{\lambda_0} = 0,
\end{aligned} \tag{19}
$$

$$
(l_1\nabla^2 c - m_1\dot{c}) + l_2\nabla^2\Theta - l_3\nabla^2\gamma_{kk} - l_4\nabla^2\varkappa_{kk} = 0 . \tag{20}
$$

The presented approach is a very formal one. From the physical point of view the effect of the terms γ_{kk} and $\varkappa_{kk}$ in equations (19) and (20) needs individual discussion.

REFERENCES

[1] Nowacki W., Certain problem of thermodiffusion in solids, Arch.Mech. 23, 6 /1971/.

[2] Nowacki W., Two- dimentional problem of micropolar magneto-elasticity, Bull.Acad.Polon.Sci., Sér. Sci Techn. 19, 4 /1971/ 161-168.

[3] Nowacki W., Dynamic problem of thermodiffusion in elastic solids, Proc. Vibr. Probl. 15, 2 /1974/, 105-128.

[4] Nowacki W., Thermodiffusion in solids /in polish/, Mech.Teoret. Stos. 13, 2 /1975/, 143-158.

[5] Nowacki W., Theory of micropolar elasticity, in polish /PWN, Warszawa, 1971/.

[6] Parkus H., Magneto-thermoelasticity, /Springer-Verlag, Wien, New York - Udine 1972/.

[7] Parkus H., Thermoelastic equations for ferromagnetic bodies, Arch.Mech. 24, 5-6 /1972/ 819-825.

[8] de Groot S.R., Mazur P., Non- equilibrium thermodynamics /North - Holland, Amsterdam 1962/.

Continuum Models of Discrete Systems 4
eds. O. Brulin and R.K.T. Hsieh

SINGULAR SURFACES IN MAGNETIC FLUIDS

P.D.S. Verma and M. Singh

Department of Mathematics
Simon Fraser University
Burnaby, B.C., Canada

Using the method developed by Thomas, the general conditions are obtained for the existence of singular surfaces in magnetic fluids. It is shown that singular surfaces exist for finite values of magnetization.

INTRODUCTION

Magnetic fluids are prepared by suspending the ferromagnetic grains in a non-magnetic electrically non-conducting liquid (Rosensweig, (1966)). The concentration of particles is of the order of 10^{18} cm^{-3}. The particles which are magnetic and approximately of domain size, remain homogeneously dispersed in the fluid carrier even in the presence of a strong magnetic field or strong magnetic field gradients. The magnetic force originates within the particles which attempt to slip relative to the fluid thereby transmitting a drag to the fluid and thus causing the dispersion to move as a whole. The macroscopic behaviour of these fluids can be adequately described by treating the dispersion as a true continuum.

A continuum theory of magnetic fluids illustrated with a few simple flows of paramagnetic fluids was developed and presented by Jenkins (1971, 72). Later, Verma and Vedan (1978, 79) and Verma (1980) applied this theory to examine several additional flow problems. In this presentation, we investigate the existence of singular surfaces in magnetic fluids following a procedure given by Thomas (1961). The general conditions governing the existence of such surfaces are derived. Explicit examples of such types of singular surfaces occur in the analysis of nonlinear waves in conductive magnetizable fluid (Tarapov and Patsegon (1980)). We conclude that singular surfaces exist in magnetic fluids when the magnetization is finite. The singular surfaces become material surfaces when the magnetization tends to be infinite.

BASIC EQUATIONS

We recapitulate here the basic equations formulated by Rosensweig (1966).

Equation of continuity:

$$\dot{\rho}+\rho\,\dot{x}_{k,k} = 0 \quad . \tag{1}$$

Equation of motion:

$$\rho\,\ddot{x}_i = t_{ik,k} + \rho f_i + \rho m_k H^{(1)}_{i,k} \tag{2}$$

Equation governing magnetization:

$$\rho\,\ddot{m}_i = -4\pi M_s m_i/(\chi_o(m_s-m)) - 2\alpha^2 m_i^* /m + H_i^{(0)} + H_i^{(1)} \tag{3}$$

Constitutive equations:

$$t_{ik} = -p\,\delta_{ik} + 2\mu_f d_{ik} - (2\rho\alpha^2/m)\,m_{[k}m^*_{i]} \qquad (4)$$

Here, cartesian tensor notation is used. ρ is the fluid density, f_i the external body force per unit mass, m_i and M_i the magnetization per unit mass and per unit volume, respectively, t_{ik} the stress tensor, m_s and M_s the saturation magnetization per unit mass and per unit volume, respectively, p the hydrostatic pressure, μ_f the viscosity, δ_{ik} the kronecker delta, α, β the material constants, and χ_o the initial susceptibility. The comma denotes differentiation, superposed dot the material time derivative, and the bracketed indices are to be anti-symmetrized. The stretching tensor d_{ij} and m_i^* representing the corotational derivative of m_i are defined by

$$m_i^* = \dot{m}_i - w_{ik}\,m_k, \qquad (5)$$

$$2d_{ij} = \dot{x}_{i,j} + \dot{x}_{j,i}\,, \qquad (6)$$

where the spin tensor w_{ik} is given by

$$2w_{ik} = \dot{x}_{i,k} - \dot{x}_{k,i}\,, \qquad (7)$$

In equations (2) and (4), $H_i^{(0)}$ stands for the external field that would be present even in the absence of the fluid and $H_i^{(1)}$ represents the self-field due to the magnetic fluid.

The total magnetic field H_i is

$$H_i = H_i^{(0)} + H_i^{(1)} \qquad (8)$$

We consider the paramagnetic fluid confined in some region V with surface $\Sigma(t)$. The external magnetic field is determined from the known distribution of the electric current in the region exterior to V. The self-field $H_i^{(1)}$ satisfies the following conditions:

$$\varepsilon_{ijk}\,H^{(1)}_{k,j} = 0 \quad \text{everywhere}, \qquad (9)$$

with

$$H^{(1)}_{k,k} = -4\pi M_{k,k} \quad \text{in V}, \qquad (10)$$

and

$$H^{(1)}_{k,k} = 0 \quad \text{outside V}. \qquad (11)$$

WEAK DISCONTINUITIES

We assume that the quantities p_i, m_i, $H_i^{(1)}$, v_j $(= \dot{x}_j)$, $v_{k,m}$ $(=\dot{x}_{k,m})$, $\frac{\partial v_r}{\partial t}$, and all the partial derivatives of order M of these functions are continuous in the neighbourhood of a moving smooth two-sided surface $\Sigma(t)$. Here $M = 0,1,\ldots,N-1$. It is stipulated that at each point of $\Sigma(t)$, one or more of the N th order partial derivatives have a finite discontinuity across the surface, Though otherwise continuous in the neighbourhood of $\Sigma(t)$. Such a $\Sigma(t)$ is called a surface of weak discontinuity of order N. The discontinuities in ρ, f_i, m_s, m, $H_i^{(0)}$, $H_i^{(1)}$, u_f, β, χ_o, and α are not considered. The symbols ν and v denote, respectively, the unit outward normal and the rate of displacement of the surface $\Sigma(t)$.

We examine the case when N = 1. There shall exist quantities $\underline{A}$, $\underline{B}$, $\underline{C}$, and D defined on $\Sigma(t)$ and not all zero such that

$$[m_{i,k}] = A_i \nu_k, \quad [\partial m_i/\partial t] = -A_i V, \tag{12}$$

$$[v_{k,rs}] = B_k \nu_r \nu_s, \quad [\partial v_{k,r}/\partial t] = -B_k \nu_r V, \quad [\partial^2 v_k/\partial t^2] = B_k V^2, \tag{13}$$

$$[H^{(1)}_{i,k}] = C_i \nu_k, \quad [\partial H^{(1)}_i/\partial t] = -C_i V, \tag{14}$$

$$[p,_k] = D \nu_k, \quad [\partial p/\partial t] = -DV. \tag{15}$$

In equations (12)-(15), the square bracket denotes the jump in the quantity enclosed. From these equations and equations (1) to (11), we obtain for incompressible paramagnetic fluids:

$$[v_{k,kj}] = B_k \nu_k \nu_j = 0 , \tag{16}$$

$$[t_{ik,k}] = -\rho m_k C_i \nu_k , \tag{17}$$

$$[\partial m_i/\partial t] + v_p [m_{i,p}] = A_i (v_p \nu_p - V) , \tag{18}$$

$$[\ddot{m}_i] = [D^2 m_i/Dt^2] = A_i (v_p \nu_p - V)^2 , \tag{19}$$

$$\begin{aligned}[t_{ik,k}] = {} & (-D - B_k \nu_k - \rho^2 B_p m_p m_k \nu_k/2m) \nu_i + B_i\{\mu_f + \\ & + (\rho\alpha^2/2m)(m_p \nu_p)^2\} + A_i \{-(\rho\alpha^2/m) m_k \nu_k (v_p \nu_p - V)\} + \\ & + m_i \{(\rho\alpha^2/2m)(B_p m_p - m_p \nu_p B_k \nu_k) + (\rho\alpha^2/m) A_k \nu_k \cdot \\ & \cdot (v_p \nu_p - V)\} .\end{aligned} \tag{20}$$

From these, we obtain the first order conditions of compatibility:

$$B_k \nu_k = 0 , \tag{21}$$

$$\begin{aligned}& A_i \{-(\rho\alpha^2/m) m_k \nu_k (v_p \nu_p - V)\} + B_i \{\mu_f + (\rho\alpha^2/2m)(m_p \nu_p)^2\} + \\ & + C_i (\rho m_k \nu_k) + \nu_i (-D - (\rho\alpha^2/2m) B_p m_p m_k \nu_k) + m_i \{(\rho\alpha^2/2m) B_p m_p + \\ & + (\rho\alpha^2/m) A_k \nu_k (v_p \nu_p - V)\} = 0 ,\end{aligned} \tag{22}$$

$$\{\beta(v_p \nu_p - V) + (2\alpha^2/m)\} A_i (v_p \nu_p - V) = 0 , \tag{23}$$

$$\varepsilon_{ijk} C_k \nu_j = 0 , \tag{24}$$

$$C_k \nu_k + 4\pi\rho A_k \nu_k = 0 , \tag{25}$$

The higher order compatibility conditions such as discussed by Truesdell and

Toupin (1960) are not investigated here.

The three different ways of satisfying the equation (23) are either $v_p \nu_p - V = 0$ or $\beta(v_p \nu_p - V) + 2\alpha^2/m = 0$, or $A_i = 0$. Taking $v_p \nu_p = V$, which implies that $\Sigma(t)$ is a material surface, equation (22) yields

$$B_i \{\mu_f + (\rho\alpha^2/2m)(m_p \nu_p)^2\} + C_i(\rho m_k \nu_k) + \nu_i(-D-(\rho\alpha^2/2m)B_p m_p m_k \nu_k) +$$

$$m_i(\rho\alpha^2/2m)B_p m_p = 0. \tag{26}$$

Contracting equation (26) with ν_i, and applying equation (21), we get

$$D = \rho m_k \nu_k C_i \nu_i . \tag{27}$$

Substitution for D from (27) into (22) gives

$$B_i \{\mu_f + (\rho\alpha^2/2m)(m_p \nu_p)^2\} + (\rho\alpha^2/2m)B_p m_p (m_i - m_k \nu_k \nu_i) = 0 , \tag{28}$$

Taking

$$\beta_i = \gamma(m_i - m_k \nu_k \nu_i), \tag{29}$$

where γ is the non-zero scalar, equation (28) furnishes.

$$\mu_f + \frac{1}{2} m \alpha^2 = 0. \tag{30}$$

However, if (29) does not hold, then equation (28) gives

$$\mu_f + (\rho\alpha^2/2m)(m_p \nu_p)^2 = 0, \tag{31}$$

together with either

$$B_p m_p = 0 \tag{32}$$

or

$$m_i = m_k \nu_k \nu_i = m \nu_i \quad \text{Since } \rho\alpha^2/2m \neq 0. \tag{33}$$

For some ideal fluids, it is impossible to satisfy equation (30) or (31). We, therefore, disregard these possibilities. Thus, $\Sigma(t)$ is a material surface.

Another way to satisfy equation (23) is to take

$$V = V_p \nu_p + 2\alpha^2/m\beta. \tag{34}$$

From equation (34), it is clear that in case of finite magnetization, the velocity of the surface of weak discontinuity, Σ , is more than that of the material surface, thus indicating the possibility of existence of singular surface in magnetic fluids. When magnetization tends to become large, the surface of weak discontinuity becomes the material surface. Before we confirm the existence, let us look to the third possibility of satisfying the equation (23) viz.,

$$A_i = 0. \tag{35}$$

From equations (22) and (35), it follows that

$$B_i \{\mu_f + \frac{\rho\alpha^2}{2m} (m_p \nu_p)^2\} + C_i\{\rho m_k \nu_k\} + \{\nu_i \quad -D -\frac{\rho\alpha^2}{2m} B_p m_p m_k \nu_k\}$$

$$+ m_i \{\frac{\rho\alpha^2}{2m}\} B_p m_p = 0. \tag{36}$$

Equation (25) then gives

$$C_k \nu_k = 0 , \tag{37}$$

while equation (24) implies that

$$C_k \nu_k = Q \quad \text{(Some non zero scalar)}. \tag{38}$$

The two equations (37) and (38) contradict each other, thus confirming our earlier conclusions.

REFERENCES

[1] Jenkins, J. T., "A theory of magnetic fluids", Arch Ration. Mech. Analysis, 46, (1972), 42.

[2] Jenkins, J. T., "Some simple flows of a paramagnetic fluids", Le J. de Phys., 32, (1971), 931.

[3] Rosensweig, R.E., "Magnetic fluids", Int. Sci. Technol., 55, (1966), 48.

[4] Thomas, T. Y., "Plastic flow and fracture of solids", Acad. Press, New York-London, (1961), 37.

[5] Tarapov, I. Ye and Patsegon N.F., "Nonlinear waves in conducting magnetizable fluids", IEEE Transactions on Magnetics, Mag-16 (1980), 309.

[6] Truesdell, C. and Toupin, R.A., "The Classical field theories", Hand-buch der Physik, edited by S. Flugge Springer-Ver-lag (1960) Band III/1, 491.

[7] Verma, P.D.S., and Vedan M.J., "Paramagnetic fluid flow through an annulus", Proc. Ind. National Sci. Acad. Pt. A, 44, (1978), 239.

[8] Verma, P.D.S., "Response of magnetic fluids to mechanical, magnetic and thermal forces", IEEE Transactions on Magnetics, Mag-16, (1980), 317.

Part 4
BIOLOGICAL SYSTEMS, COTINUUM MODELS OF DISCRETE SYSTEMS

Continuum Models of Discrete Systems 4
eds. O. Brulin and R.K.T. Hsieh

A CONTINUUM MODEL OF THE HUMAN SPINE AS AN ANISOTROPIC BEAM

Stig Hjalmars

Department of Mechanics
Royal Institute of Technology
Stockholm
Sweden

A beam model for the lateral bending of the human spine is developed, where the microstructure is assumed to be a row of solid bodies, the vertebrae, connected by stiff bars, rigidly fixed in the solids but connected by joints with small angle-proportional moments, originating from the discs. The continuum limit is deduced, giving the equation for an anisotropic beam. Reference is made to an application by Lindbeck of this model to functional scoliosis, where the model is shown to reproduce the characteristic scoliotic equilibrium form, which, as also for the normal spine, is likely to be unstable.

The most flexible part of the human spine i.e. the 11 vertebrae Th_7 - Th_{12} and L_1 - L_5 between the pelvis and the rib cage, seems from the macroscopic point of view to behave as a sort of column or vertical beam. However, the properties of this beam show a pronounced anisotropy in the sense that its subjectively manifest low bending stiffness is much smaller as compared to its high longitudinal incompressibility than it would be in a beam, made of isotropic material. Such extreme anisotropy will of course give the spine rather different mechanical properties than those of an isotropic beam, e.g. as regards the geometrical form, internal stress and stability in equilibrium under different geometrical conditions and loads, without and with muscle action. All these properties are of course of paramount orthopedic interest, but seem at present not sufficiently well understood.

The mechanical properties of the human spine have been extensively studied e.g. by Schultz et al [1,2] by means of a mathematical model of the microstructure of the spine, where the vertebrae are idealized as small rigid bodies and the different connections as small deformable elements. Of course such direct computations from the microlevel should be utmost valuable in estimating the influence of local abnormal or correcting changes on the microlevel. However, their comparative computational complexity makes them less useful for the purpose of drawing general and perspicuous conclusions about the mechanical properties of the normal spine, mentioned above. As an alternative tool for such a general analysis we will here propose an anisotropic beam model for the lateral bending of the spine, starting from a fairly realistic, although idealized model of its microstructure.

Of course the microstructure of the spine is a rather subtle piece of mechanical engineering, but its general features may be schematically described as follows. The relative lateral motion of two adjacent vertebrae is constrained as a plane, one-parametric motion by the intervertebral joints and rather inextensible ligaments. Considering the vulnerability of the discs we may confidently assume that the constraints of the motion are so constructed that the relative instant centre is located almost in the centre of the intermediate disc, which means that the constraints take up the big longitudinal compressive force, but give almost no resistance against rotation, i.e. the bending. The small bending stiffness, i.e. the small resistance against rotation around the instant centre, is then mainly given by the symmetrical prismatic deformation of the disc, without average compression, i.e. without compression of its central part. It is evident that in the case of disturbances in the constraints of the motion, causing e.g. an asymmetric location of the instant centre and thus an asymmetric deformation of the disc, the disc will have to take up a part of the longitudinal load. This will considerably increase its internal pressure. Disturbances in the constraints of the relative vertebral motion should then be an important cause of the usual disc damages.

The schematic description of the kinematics of the vertebrae, given above, makes it reasonable to adopt the micromodel, shown in Figure 1.

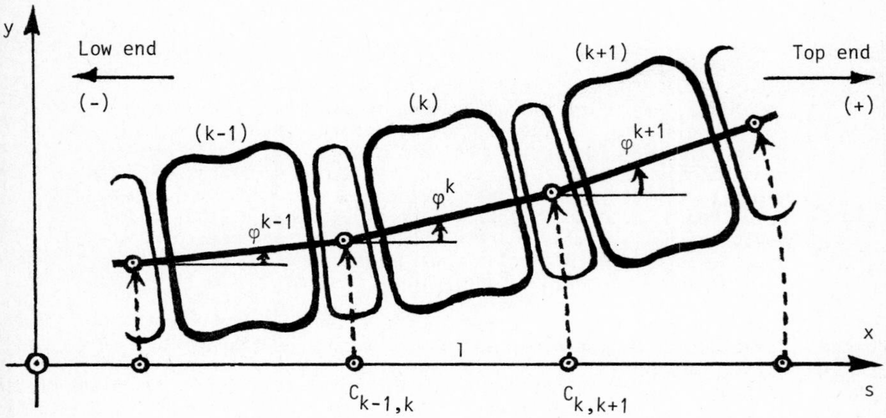

Figure 1
Micromodel for the lateral motion of the spine
$C_{k,l}$ is the instant centre for the relative motion of the vertebrae (k) and (l)

Here the vertebrae are modelled as rigid bars of invariable length l, connected by joints, representing the instant centres of the relative motion of the vertebrae. The discs in their function of counteracting the bending are modelled as moment couples in the joints, negatively proportional to the relative angle deflection.

We assume that the moment couples, acting on the vertebra (k) from (k - 1) and (k + 1) respectively, have the form:

$$m^{(k,k-1)} = -k\left(\varphi^k - \varphi^{k-1}\right), \qquad m^{(k,k+1)} = k\left(\varphi^{k+1} - \varphi^k\right). \tag{1}$$

Furthermore, assuming that the external forces, acting on the top end (+) and low end (-) of the spine, have the components P_x^+, P_y^+ and P_x^-, P_y^-, we have for the forces, acting on (k) from (k + 1) and (k - 1):

$$p_x^{(k,k+1)} = -p_x^{(k,k-1)} = P_x^+ = -P_x^-, \qquad p_y^{(k,k+1)} = -p_y^{(k,k-1)} = P_y^+ = -P_y^-. \tag{2}$$

The equilibrium moment condition for the vertebra (k) around $C_{k-1,k}$ is:

$$-k\left(\varphi^k - \varphi^{k-1}\right) + k\left(\varphi^{k+1} - \varphi^k\right) - lP_x^+ \sin\varphi^k + lP_y^+ \cos\varphi^k = 0. \tag{3}$$

Under muscle action of course corresponding external forces should be added.

Turning the difference equation (3) into a differential continuum equation by means of the expansion,

$$\varphi^{k\pm1} = \left(1 \pm 1 \frac{\partial}{\partial x} + \frac{1^2}{2} \frac{\partial^2}{\partial x^2} \pm \ldots\right) \varphi^k \ , \tag{4}$$

and noting that x is equal to the invariable arc length, s, we get with the <u>bending stiffness</u> $b = kl$:

$$b \frac{d^2\varphi}{ds^2} - P_x^+ \frac{dy}{ds} + P_y^+ \frac{dx}{ds} = 0 \ . \tag{5}$$

Integration gives:

$$b \frac{d\varphi}{ds} - P_x^+ (y + a) + P_y^+ x = 0 \ . \tag{6}$$

Now, substitute y instead of φ by means of

$$\frac{d\varphi}{ds} = \frac{d^2y}{ds^2} \left[1 - \left(\frac{dy}{ds}\right)^2\right]^{-1/2} \ , \tag{7}$$

and approximate up to the order $(dy/ds)^2$ as compared to 1. It can be readily seen that at the actual bendings even a reasonable contribution of third degree in the angle differences in (1) would not add significantly to this approximation. We thus obtain the following differential equation for a muscle-relaxed position:

$$b \frac{d^2y}{ds^2} - \left[P_x^+ (y + a) - P_y^+ x\right]\left[1 - \frac{1}{2} (dy/ds)^2\right] = 0 \ . \tag{8}$$

The boundary conditions can be given as a proper set of values on positions (y) directions (dy/ds) and loads (P_x, P_y, M) at the $(+)$ and $(-)$ ends. As can be seen from the continuum limits of (1) and from (6) they satisfy

$$M^\pm = \pm b(d\varphi/ds)^\pm \ , \quad M^\pm = P_x^\pm (y^\pm + a) - P_y^\pm x^\pm \ , \tag{9}$$

by means of which the M-conditions can be reduced to y- and P-conditions.

A first application of this model has been made by Lindbeck [3], who has used it for the mechanical buckling analysis of a spine with functional scoliosis, caused by different length of the legs. The equilibrium form of the spine shows in this case a very characteristic feature with an inflexion in the middle, and a deflection at the top towards the side of the higher leg, as is seen from Figure 2, which is extracted from a typical X-ray picture of a muscle-relaxed position.

With forces $P_y^+ = 0$, P_x^+, and moment M^+ given by the upper part of the trunk, considered as a rigid body with the centre of gravity at a distance $q \sim 0.1$ m from the top end, Lindbeck solves the equation (8) for different values of the parameter $n^2 = -P_x^+/b$. He finds a bifurcative behaviour with one or three equilibrium states in different intervals of n^2, as shown in Figure 3. For the value $n^2 \sim 176\ m^{-2}$ he finds in an interval with three solutions a very good reproduction of the observed scoliotic form of Figure 2, i.e. by the intermediate equilibrium form, which of course is very likely to be unstable. The value $n^2 \sim 176\ m^{-2}$ corresponds to a bending stiffness, which is about 10 times less than that given by rubber, which seems reasonable. It is interesting to note that Nature in the engineering of the spine column chooses a construction, which is not mechanically stable, but achieves the stability by means of a control system, commanded by the nervous system, and working with small corrective muscle actions.

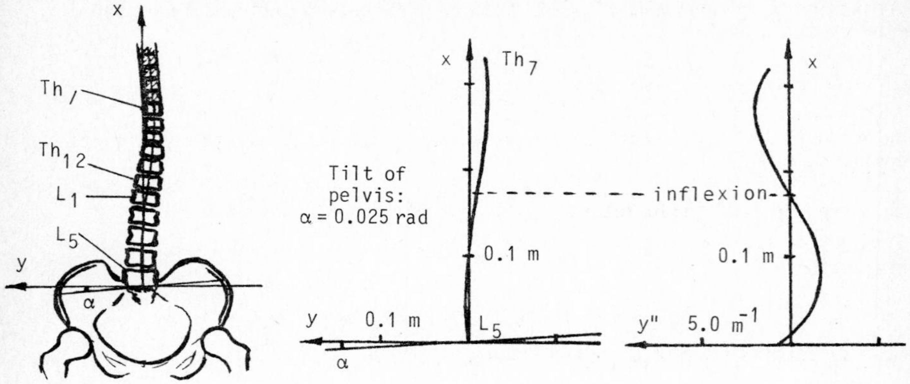

Figure 2
An extract from X-ray picture of spine suffering from scoliosis, caused by different length of legs

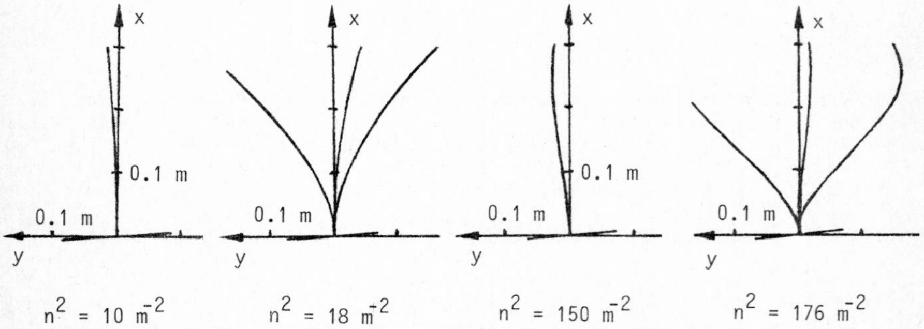

Figure 3
Calculated equilibrium forms for different $n^2 = -P_x^+/b$
Bifurcations at $n^2 \approx 17$, 50 and 170 m^{-2}

REFERENCES

[1] Belytschko, T.B., Andriacchi, T.P., Schultz, A.B., and Galante, J.O., Analog studies of forces in the human spine: Computational techniques, J. Biomechanics 6 (1973) 361-371.

[2] Schultz, A.B., Belytschko, T.B., and Andriacchi, T.P., Analog studies of forces in the human spine: Mechanical properties and motion segment behavior, J. Biomechanics 6 (1973) 373-383.

[3] Lindbeck, L., A theoretical model of a human spine suffering from a functional scoliosis, TRITA-MEK-80-04. Report to be obtained freely from Department of Mechanics, Royal Institute of Technology, S-100 44 Stockholm, Sweden.

Continuum Models of Discrete Systems 4
eds. O. Brulin and R.K.T. Hsieh

INTRINSIC CONSTITUTIONAL MEANING OF CONTINUUM MODELS OF DISCRETE SYSTEMS

Kazuo Kondo
C. P. N. P.
1570 Yotsukaido, Yotsukaido City, Chiba-ken,
284 Japan

I. HIGHER ORDER SPACE ASPECTS WHERE SPATIAL COORDINATES WORK AS PULTIPLE PARAMETERS

1. Compromise Required between Continuum Picture and Discrete Construction

The handling of discrete systems in terms of continuous models needs to start with a compromise in which each discrete constituent is not a mere point but an object equipped with physical properties while the object as a whole is associated with a point in space for specifying it. The points so used for specification labels are shown by black spots in Figure 1, each with an atmosphere of dots indicating physical properties. But the labels being represented by real coordinates accorded to parts of continuous space, all mathematically conventional techniques of continuity creep into the analysis.

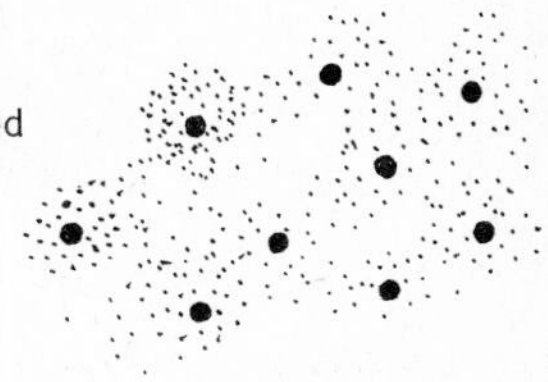

Figure 1

Any information so obtained is given by an integral

$$S = \int \Omega \, dX \qquad (1.1)$$

where dX is the element of 3-dimensional spatial volume and the integrand Ω may depend on some parameters ξ^i $(i = 1, \ldots, N)$ and their spatial derivatives

$$\xi^i_{\alpha(r)} = \partial \xi^i / \partial u^{\alpha_1} \ldots \partial u^{\alpha_r}$$

where u^α runs over the three spartial coordinates. The upper bound of the order r need not a priori be restricted. Similarly for the upper bound of N. They can be some finite numbers according to the microscopic penetration that is carried into effect. At any event, we can proceed with an L-dimensional integration so that

$$dX = du^1 du^2 \ldots du^L$$

where L = 3 unless time dependency is also considered.

An important feature should be remarked that a continuum is a set of an immense number of one and the same species of element or particle although they are labelled with different spatial and/or temporal coordinates u^α (α = 1,2,3 or 4). Hence the fundamental function Ω should not depend on u^α:

$$\partial \Omega / \partial u^\alpha = 0 \qquad (1.2)$$

in spite of its dependency on $\xi^i_{\alpha(r)}$. The restriction to (1.2) may be referred to as a discrete uniformity condition. Invariant information needs to be secured under all conceivable transformations of parameters $\xi^i_{\alpha(r)}$ and of u^α.

2. Extended Expoint and Extended Expoint Transformation

Under an arbitrary analytic transformation of ξ^i(which henceforth we call coordinates):

$$\xi^i \to \xi^a = \xi^a(\xi^i),$$

they are transformed by the naturally derived relations among them which are called an extended expoint transformations, whose construction has been studied by the mathematician. The generalized concept represented by $\xi^i_{\alpha(r)}$ is an extended expoint. Any quantity $p^i_{\alpha(r)}$ that is transformed in the same manner as this is also considered. It is an extended extensor.

3. Zermelo-Géhéniau Conditions and Multiparametric Higher Order Spaces

The invariance condition under transformation of u^α has also been studied and is given by the set of equations

$$\Delta_\alpha^{\beta(H)}\Omega = \delta^1_H\delta^{\beta(H)}_\alpha\Omega, \tag{3.1}$$

where α, β and H run from 1 to M, the Kronecker delta are self-evident, and

$$\Delta_\alpha^{\beta(H)} = \sum_{r=H}^{H} \binom{r}{H}\xi^i_{\alpha\gamma(r-H)}\bar{\partial}/\partial\xi^i_{\beta(H)\gamma(r-H)}, \tag{3.2}$$

where $\xi^i_{\alpha\gamma(r-H)}$ is equivalent to $\xi^i_{\alpha(r-H+1)}$ and $\bar{\partial}/\partial\xi^i_{\beta(H)\gamma(r-H)}$ to $\partial/\partial\xi^i_{\alpha(r)}$ with a relevant modification of the numerical factors.

The equations of (3.1) may be referred to as the Zermelo-Géhéniau conditions[1] [1]. Under (1.2) and (3.1) our model assumes the structure of the higher order space of order M and N dimensions with multiple parameters u^α. It is denoted by $K^{(M)}_{L,N}$ by A. Kawaguchi.[2]

By explorations of the restrictions by these equations in regard to their epistemological implications, we are led to criteria recognized in fundamental physics including, inter alia, the dynamical principle. They are subject to a constitutional restriction imposed by the Zermelo-Géhéniau conditions.[3]

II. MECHANICS OF CONTINUA BY OSCULATION

4. More Macroscopic Observation and Osculation

Owing to the foregoing construction, any stringent values of Ω, $\xi^i_{\alpha\beta(r-H)}$ etc. at a fixed field point are hardly required but an average over an arbitrary small L-dimensional domain, ΔX ≠ 0, suffices. The size of ΔX can be indefinite so that (3.1) can be made up to an integral equation:

$$\underbrace{\int\cdots\int}_{\leftarrow L \rightarrow} \delta(\Delta_\alpha^{\beta(H)}\Omega - \delta_H\delta^{\beta(H)}_\alpha\Omega)dX = 0. \tag{4.1}$$

The macroscopic observer has another lack of ability that he cannot penetrate more microscopic constructions of the disturbances such as are represented by more higher order phases. He may trace them as far as revealed through the lower order terminology but fails to recognize higher orders directly. They are revealed only through their expressions in terms of lower order quantities, i.e., by their osculation to lower orders. Such a principle of osculation is indeed latent also in classical dynamical theory which consists in osculating accelerations and higher derivatives of coordinates in regard to the time parameters. In compliance with this situation, all the higher partial derivatives

$$\xi^i_{\alpha\beta(r-1)} \quad \text{for} \quad r \geq 2$$

should be osculated to the lower orders to be expressed in terms of ξ^i and ξ^i_α so that, if the osculated quantities are indicated by ~ over the kernel letters, such as ξ, LΩ etc., then $d\tilde{\xi}^i_{\alpha\beta(r-1)}$ and $d\tilde{\Omega}$ are all linear in $d\xi^i$ and $d\xi^i_\alpha$ with coefficients depending solely on ξ^i and ξ^i_α.

[1,2] We would refrain from enumerating the extensive literature on higher order space theory. For an expository material, see [1].

[3] Notice the implication of the term *constitutional* that stands here for a meaning more primary than accorded to the term *constitutive* in the conventional usage in most of the contributions to micropolar mechanics.

Especially for $H = 1$ and $\alpha = \beta$, $\delta^{\alpha}_{\alpha} = L$, the equation (4.1) is reduced to:

$$\underset{\leftarrow L\rightarrow}{\int\ldots\int}\{(f_i - L\partial\tilde{\Omega}/\partial\xi^i)\delta\xi^i + (T^{\alpha}_i - \partial\tilde{\Omega}/\partial\xi^i_{\alpha})\delta\xi^i_{\alpha}\}dX = 0 \qquad (4.2)$$

(where f_i and T^{α}_i are all osculated quatities). By integration by parts, from (4.2) we have

$$\underset{(L-1)}{\oint} \delta\xi \{T^{\alpha}_i - \partial(L\tilde{\Omega})/\partial\xi^i_{\alpha}\}dS_{\alpha} + \underset{\leftarrow L\rightarrow}{\int\ldots\int}\delta\xi^i\{\partial[\partial(L\tilde{\Omega})/\partial\xi^i_{\alpha}]/\partial u^{\alpha}$$

$$- \partial T^{\alpha}_i/\partial u^{\alpha} - \partial(L\tilde{\Omega})/\partial\xi^i + f_i\}dX = 0. \qquad (4.3)$$

The equation (4.3) needs to hold whatever the virtual disturbance $\delta\xi^i$ in the field as well as on the (L-1)-dimensional boundary denoted by dS_{α}. There follow then the definition of the <u>stress</u>:

$$T_i = \partial(L\tilde{\Omega})/\partial\xi^i \qquad (4.4)$$

and the field equation

$$\partial\{\partial(L\tilde{\Omega})/\partial\xi^i_{\alpha}\}/\partial u^{\alpha} - \partial(L\tilde{\Omega})/\partial\xi^i = - f_i + \partial T^{\alpha}_i/\partial u^{\alpha} \qquad (4.5)$$

which has evidently the structure of the Lagrangian dynamical equation modified with the term $Q_i = -f_i + \partial T^{\alpha}_i/\partial u^{\alpha}$. If $Q_i = 0$, then (4.5) is derived from a simple variational criterion

$$\delta\underset{\leftarrow L\rightarrow}{\int\ldots\int}(L\tilde{\Omega})dX = 0 \qquad (4.10)$$

(where the boundaries are not varied) so that $L\tilde{\Omega}$ happens to be the same as the <u>Lagrangian density</u> in the conventional sense.

5. Higher Order Construction of Field Force, Stress and Strain

The quantity Q_i consists of two parts having higher order constructions. Being coupled with $\delta\xi^i$ to contribute $\delta(L\tilde{\Omega}) = f_i\delta\xi^i$ to the density, f_i is the resistance to a <u>virtual displacement</u> $\delta\xi^i$. Hence it can be regarded as a field force per unit L-dimensional volume. Its 3-dimensional part is an ordinary <u>body force</u>.

Returning to the original expression even before the osculation, we have

$$\underset{r}{\delta}(L\Omega) = \delta\xi_{\alpha\gamma(r-1)}T^{\alpha\gamma(r-1)}_i + \xi^i_{\alpha\gamma(r-1)}\delta T^{\alpha\gamma(r-1)}$$

where

$$T^{\alpha\gamma(r-1)} = r\partial\Omega/\partial\xi^i_{\alpha\gamma(r-1)}. \qquad (5.1)$$

With r+1 indices i and $\alpha\gamma(r-1)$, the 3-dimensional parts of the quantities $\xi_{\alpha\gamma(r-1)}$ and $T^{\alpha\gamma(r-1)}_i$ can be responsible for higher order strains and higher order stresses mutually coupled to contribute $\delta(L\Omega)$ to the Lagrangian density.[4]

The apparent stress in the ordinary sense (generalized to cover an antisymmetric stress) is linearly contributed by all of these higher order stresses according to the formula:

$$T^{\lambda}_i = \sum_{(r-1)}\sum_{r}(\partial\tilde{\xi}^i_{\alpha\gamma(r-1)}/\partial\xi^i_{\lambda})T^{\alpha\gamma(r-1)}_i, \qquad (5.2)$$

viz., the apparent stress is a projection of all these higher order constructions.

6. Reinterpretation of Higher Order Features as Fields around Discrete Dynamical Systems

The apparent external force f_i and stress T^{α}_i have originated from Zermelo-Géhéniau construction $\Delta^{\beta(H)}_{\alpha}\Omega$ (especially for $H = 1$) under osculation of higher order features. Tracing more detailly their variational aspects after some calculations we have a volume and a surface integral which should vanish in spite of arbitrary variation

[4] In particular, $\xi_{[i\alpha]\gamma}$ implies a *dislocation density* and $T^{[i\alpha]\gamma}$ coupled with it a *couple stress*.

$\delta\xi^i$ so that we have the field equation

$$\sum_{r=1}^{M} A^{\alpha(r)}\xi_{\alpha(r)} + \sum_{q=1}\sum_{r=1} B^{\alpha(r)\rho(q)}\xi_{\alpha(r)\rho(q)} = -k \qquad (6.1)$$

to be satisfied by the disturbance ξ (where the index i is suppressed) and a set of boundary condition equations, where $A^{\alpha(r)}$, $B^{\alpha(r)\rho(q)}$ and k are appropriately derived from $L\Omega$.

The lowest possible case, i.e., M = 1, gives the field equation of second order. The Laplace-Poisson and the classical wave field are particular realizations of this constitution under isotropy. A more penetrating fourth order description includes a non-homogeneous Klein-Gordon equation:

$$\Box\psi - \alpha^2\psi = -k, \qquad \psi = \Box\xi.$$

III. DISINTEGRATION INTO ONE-PARAMETRIC LINE ELEMENTS

7. Detaching a $K_N^{(M)}$ from $K_{L,N}^{(M)}$

Since transformations are allowed for u^α to go from one set to another, the parametric line need not be Cartesian. If a consecution by one of them, say u, is described without reference to other parameters, it is to detach one-parametric submanifold $K_{1,N}^{(M)}$ which is also denoted by $K_N^{(M)}$. Of the informational features by (1.1) only

$$s = \int F\, du \qquad (7.1)$$

remains then as significant, where F is the specific feature of Ω restricted to a single parameter u so that, in place of the entire set of the extended expoint parameters $\xi^i_{(r)}$, it depends only on those which are given by

$$\xi^i_{(r)} = d^r\xi^i/du^r \quad (r = 0, 1, \ldots, M)$$

and which may be called simply <u>expoint coordinates</u>. The concept so defined is an <u>expoint</u> treated sometimes more geometrically as a <u>line element of order</u> M.

Upon this detachment of the line element, we need to consider simply the (M+1)N expoint transformation relations between expoint coordinates

$$\xi^i_{(r)} \quad \text{and} \quad \xi^a_{(s)} \quad (r, s = 0, 1, \ldots, M;\ i, a = 1, \ldots, N).$$

The discrete uniformity condition (1.2) is reduced to only a single relation

$$\partial F/\partial u = 0. \qquad (7.2)$$

Also, we require not the entire system of Zermelo-Géhéniau conditions but only those which are called simply Zermelo conditions consisting of only M equations. However, a modification we shall next consider is more intrinsically related to our problem.

8. Disintegration accompanied by a Neighbourhood

The physics of the discrete element represented by ξ^i includes at least the feasibility of arbitrary small variation of its location which to represent three (i.e. L = 3) of the N variables ξ^i need to be applied. They afford the lower bound that should be preserved in spite of the disintegration which is the result of abolition of more physical constraints. Hence for the detached line element

$$N = L = 3$$

can be assumed. On the other hand, owing to the compromise from the fundamental discrete structure to a continuum approximation, the disturbances ξ^i for any definit u are not uniquely restricted but a slight deviation $\delta\xi^i$ at least needs to be involved, the fundamental function F depending on it as well. Therefore,

$$\partial F/\partial\delta\xi^i_\alpha \quad \text{as well as} \quad \partial F/\partial\xi^i_\alpha$$

must be taken into account. In other words, the detached configuration is required to be stable in spite of their intervention. The concerned informational aspects constitute an <u>error-correcting code system</u> notwithstanding the variation of F to F + δF.

Under this circumstance, the conditions for intrinsicality are afforded by a modification to involve 2M operators. A remarkable feature has been discovered in this regard to be summarized in what one may call the

MODIFIED KAWAGUCHI-HOMBU THEOREM. <u>In order for a neighbourhood of a properly defined line element to be simultaneously detached with it without loosing intrinsicality it is necessary and sufficient to satisfy the equations of the modified Zermelo conditions which are 2M in number.</u>

9. Mutual Restriction between Dimensionality and Order

The fundamental function F supporting the properly defined line element of $K_N^{(M)}$ is restricted by a number of conditions of the foregoing. But we are concerned more with the generalized line element including the neighbourhood where the Kawaguchi-Hombu Theorem is also modified in the foregoing manner. It is subject to

i) expoint transformation, ii) line element definition, iii) significant modified Zermelo equations, iv) uniformity condition

totaling $1 + (\Lambda + 1) + 4MN + N$ where $\Lambda = M$ if $M \leq 2$ and $\Lambda = 3$ if $M \geq 3$.

In order to satisfy them, we have ξ^i, $\xi^i_{(r)}$ and ξ^a, $\xi^a_{(s)}$ totaling $1 + 4MN + 2N$ so that

$$M + 1 \leq N \quad \text{if} \quad M = 1, 2. \tag{9.1}$$

Obviously for the simplest physics of the detached line element we have

$$N = 3$$

so that

$$M \leq 3 - 1 = 2; \tag{9.2}$$

viz. the highest significant order for the detached line element with the neighbourhood is 2.

Therefore, for this sort of disintegration, which is thought to be the simplest kind practically, the significant higher order features are restricted in $K_{3,3}^{\{2\}}$ from which $K_3^{(2)}$ is detached. The field equation for this is reduced to fourth order.

IV. MEANING OF THE THEORY OF YIELDING

10. Boundary Condition and Strain Tensor

The general solution of the field equation (6.1) is given by the linear combination of a particular solution and a general solution of the homogeneous equation obtained therefrom by replacing the non-homogeneous term k by 0. For isotropic materials, the homogeneous equation is reduced to

$$C^{\alpha\beta}\xi_{\alpha\beta} + B^{\alpha\beta\rho\sigma}\xi_{\alpha\beta\pi\sigma} = 0. \tag{10.1}$$

It has been proven that this is equivalent to the variational criterion:

$$\iiint \delta\xi \left(\tfrac{1}{2}C^{\alpha\beta}\xi_{\alpha\beta} + \tfrac{1}{2}B^{\alpha\beta\rho\sigma}\xi_{\alpha\beta\rho\sigma}\right) du^1 du^2 du^3 = 0 \tag{10.2}$$

with the subsidiary condition on the boundary

$$\iint \{\delta\xi[C^{\alpha(\nu)}\xi_{\alpha} - (B^{\alpha\beta(\nu)\sigma}\xi_{\alpha\beta})_{\sigma}] + \delta\xi_{\rho} B^{\alpha\beta\rho(\nu)}\xi_{\alpha\beta}\} dS_{(\nu)} = 0 \tag{10.3}$$

where $dS_{(\nu)}$ implies the surface element given with the surface normal direction.

The ξ being a displacement outside the average flat space of (u^1, u^2, u^3),

$$e_{\alpha\beta} = \frac{1}{2}\delta_{ij}\xi^i_\alpha\xi^j_\beta, \text{ abbreviated to } \frac{1}{2}\xi_\alpha\xi_\beta,$$

is given the character of strain tensor as can easily be checked.

We have argued that the homogeneous feature so restricted is just in correspondence with the yielding of materials. Regarding this viewpoint we may only follow what we have developed in these thirty years.

11. Solution under Free Boundary Condition

The subsidiary condition (10.3) provides boundary conditions generally relevant. In particular, $\delta\xi$ and $\delta\xi_\alpha$ can be unrestricted, i.e., free if and only if

$$C^{\alpha(\nu)}\xi - (B^{\alpha\beta(\nu)\sigma}\xi_{\alpha\beta})_\sigma = 0, \quad B^{\alpha\beta\pi(\nu)}\xi_{\alpha\beta} = 0. \tag{11.1}$$

These are therefore the boundary conditions relevant for free surfaces.

The fourth rank tensor $B^{\alpha\beta\pi\sigma}$ is reduced to only two independent components B and κB (similar to the case of the elastic tensor) by the assumption of isotropy for materials. The field equation is thensimplified to

$$B\Delta\Delta\xi + C^{\alpha\beta}\xi_{\alpha\beta} = 0, \quad (\alpha, \beta = 1,2,3). \tag{11.2}$$

where Δ is the Laplacian operator.

Solutions of (11.2) and (11.1) have been studied by us (in particular [2]). For the flat free surface condition, we have obtained

$$\left|\frac{\sigma_x - \sigma_y}{B/\lambda^2}\right| = \vartheta(2\pi)^2, \quad \left|\frac{\sigma_y - \sigma_z}{B/\lambda^2}\right| = \left|\frac{\sigma_z - \sigma_x}{B/\lambda^2}\right| = \vartheta'(2\pi)^2 \tag{11.3}$$

where σ_x, σ_y, σ_z are principal stresses, σ_z occuring in the direction normal to the surface, λ is a small characteristic length of the disturbance distribution that is practically periodic and ϑ and ϑ' depend on certain characteristic combinations of σ_x, σ_y, σ_z; λ and κ.

We shall not go into details of the eigencondition. For a practically plausible range we have

$$\vartheta, \vartheta' \doteqdot 1 \tag{11.4}$$

so that (11.3) agrees with the criterion of maximum shear associated with Tresca's name. Elaboration to more detailed formulae such as the Mohr-Coulomb and/or Lade-Duncan Criteria [3] can also be feasible by further scrutinizing the suggestions included in the formulae (11.3) and the deviation from (11.4).

12. Equivalent Hydrostatic Pressure

A remarkable feature of (11.3) is that all hydrostatic (isotropic) stresses, being excluded in this criterion, become comparable with some internal physics inherent in the material. In particular, from dimensional considerations, B is equivalent to a hydrostatic stress, say p, multiplied by an area:

$$B = pb^2$$

where b is a linear dimension comparable with the foregoing λ.

The originally discrete material elements are tied together into a continuum whether by a preexisting interaction or an impressed force between them. Even if there is no external pressure apparently observed, there needs to be a certain internal stress tying together the elements. This is the construction of the quantity B/b^2 reinterpreted as a pressure (or isotropic attraction).

There is a limiting case in which no preexisting cohesive interaction can be observed between the elements, such as the mass of cohesionless sand grains, where the tying forces must be given by an apparent external pressure imposed upon the field by something like a hydrostatic device. Therefore, experiments upon sand are particularly equipped to reveal the constitution that entails yield phenomena.

13. An Entropy Criterion for a Soil-mechanical Illustration

Regarding the sites of sand particles, not more than their probabilities are known a particle being hardly distinguished from another. If the probability density is p, a measure for the indistinction is afforded by the information-theoretical entropy defined by $\eta = p \log p - p$. Its increment

$$d\eta = dp \log p$$

for a change in a more probable direction $dp > 0$ contributes a positive or negative amount to the indistinction measure according as $p >$ or < 1.

The more the distribution is concentrated so that $\log p$ has a large positive value, the more number of sand particles have probability to come into the same spot apparently. More positive contributions to η imply more indistinguishability and more negative contributions more distinguishability. The neutral condition occurring at

$$p = 1, \text{ where } d\eta = 0,$$

should be just the limit beyond which the continuum character would fail, individual particles becoming at any rate mutually distinguished.

Material elements being distinguished from one another at yielding by slips,

$$\text{i) } p > 1 \ (d\eta > 0) \quad \text{and} \quad \text{ii) } p < 1 \ (d\eta < 0)$$

imply configurations i) without and ii) beyond yielding respectively.

A Gaussian distribution may be assumed. If the variance of the number of just noticeable differences or distances, comparable with the number of λ, is restricted to

$$2\pi\theta^2 < 1, \tag{13.1}$$

where θ is the standard deviation, the condition for ii) will be fulfilled for all conceavable values of p so entailed.

The upper bound $\theta = \sqrt{1/2\pi}$ imposed upon the variance may be adopted as the mean value of the foregoing characteristic length b divided by the j.n.d. which may be assumed to be just a half wave length $\lambda/2$ of the nearly sinusoidal disturbances recognized in our analysis of yielding (cf. [2]). Substituting this in the formula (11.3), we have

$$|\Delta P/p| = \vartheta\pi/2 \geq \pi/2 \quad (\vartheta \geq 1) \tag{13.2}$$

where ΔP is the difference between principal stresses.

In Figure 2, experimentally obtained data for $\Delta P/p$ with sand specimens are plotted against v/p (where v is the volumetric strain). The relation is parabolic below the yield point as normally required. The experimental yield region agrees remarkably well with the limit at 1.57 on the horizontal axis in conformance with the foregoing theoretical analysis.[5]

[5] For information regarding these data, referred to also elsewhere by us the author owes to those who engaged in the experiment in the University of Tokyo. Several years since more materials have been published in [4] to confirm our viewpoint further.

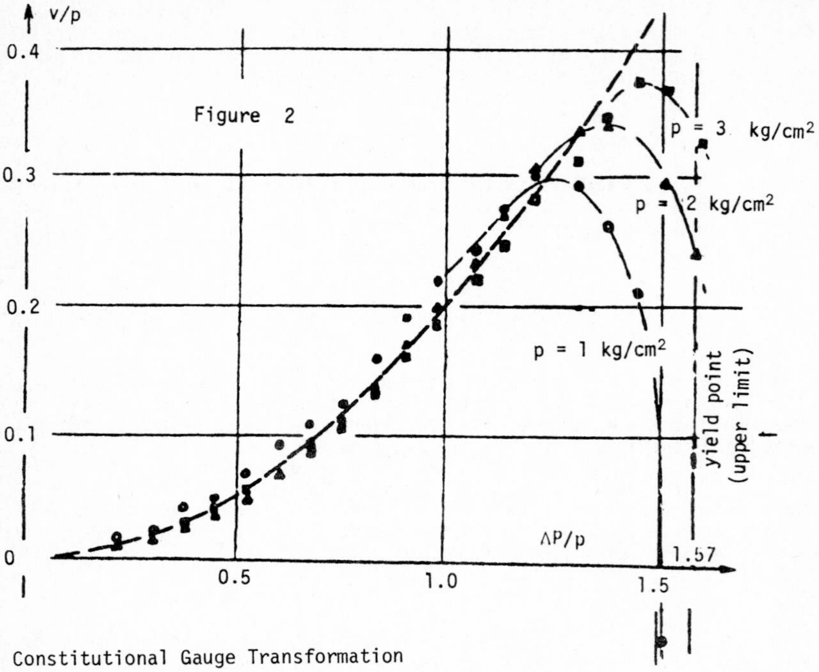

14. Constitutional Gauge Transformation

Returning to the general standpoint, we may point out that the higher order features can be added linearly to the Lagrangian formalism modified as suggested without altering the required intrinsicality. Here needs to originate a class of gauge transformations.[6] Notice that this is based on the constitutional compromise between the discretum background and the continuum picture.

[6] A similar gauge transformation was noted by O.Brulin and S.Hjalmars [5]. As regards our interpretation, see [6] and [7].

References

[1] M.KAWAGUCHI, An Introduction to the Theory of Higher Order Spaces, I: The Theory of Kawaguchi Spaces; II: Higher Order Spaces in Multiple Parameters. *RAAG Memoirs Vol.III* (1962), 718-734; *Vol.IV* (1968), 577-591.

[2] K.KONDO, Mathematical Analysis of Yield Points of an Isotropic Material. *ibid., Vol.I,* D-IV (1955), 495-521.

[3] P.V.LADE and J.M.DUNCAN, Elastoplastic Stress-Strain Theory for Cohesionless Soil. *Journal of the Geotechnical Engineering Division, GT10* (1975), 1037-1053.

[4] Y.YAMADA, Deformation Characteristics of Sand in Three Dimensional Stress State. *Continuum Mechanics and Statistical Approach in the Mechanics of Granular Materials* (edited by C.COWIN and M.SATAKE), 1978, 91-97.

[5] O.BRULIN and S.HJALMARS, Grade-Consistent Micropolar Theory. *PREVENIENT NATURAL PHILOSOPHY (POST-RAAG REPORTS), No.110* (June 1980).

[6] K.KONDO, Constitutional Construction of External Force by Osculation of Higher Order. *ibid., No.112* (September 1980).

[7] K.KONDO, Constitutional Restriction on Micromultipolar Disturbances. *ibid., No.113* (September 1980).

Continuum Models of Discrete Systems 4
eds. O. Brulin and R.K.T. Hsieh

PROJECTION OPERATOR METHOD IN CONTINUUM MECHANICS

I. A. Kunin

Department of Mechanical Engineering
University of Houston
Houston, Texas 77004, USA

1. INTRODUCTION

We suggest a new mathematical formulation of linear field theories in terms of projection operators which are closely related to Green's functions but have certain advantages. Here the method and certain applications will be described in the scope of elastostatics and elastodynamics.

Note that the projection operators were first introduced in [1]. Their detailed investigation, finite-dimensional representations and connection with Green's operators for strain and stress are given in [2-4].

2. STRESSES DUE TO BODY FORCES AND DISLOCATIONS

Let us consider an unbounded elastic medium with the elastic moduli tensor $C(x)$ where $x = x(x_1, x_2, x_3)$ is a point of the three-dimensional Eucledian space E_3. As is well known, the displacement $u(x)$ and the body force $f(x)$ satisfy the equation of elastostatics

$$- \partial C \partial u = f \tag{2.1}$$

which is equivalent to the following system of equations for strain $\varepsilon(x)$ and stress $\sigma(x)$

$$\begin{aligned} &\mathrm{Rot}\ \varepsilon = 0, \qquad \varepsilon = \mathrm{def}\ u\ , \\ &\sigma = C\varepsilon\ , \\ &\mathrm{div}\ \sigma = -f\ . \end{aligned} \tag{2.2}$$

Here def is the symmetrized gradient and Rot = curl curl^T is the operator in the Saint-Venant compatability equations.

The equations (2.1) or (2.2) will be understood in the sense of generalized functions. This permits one to incorporate boundary value problems into the equations by viewing $C(x)$ or $B(x) = C^{-1}(x)$ as piecewise smooth functions [5]

In parallel, we can consider the equations for internal stresses due to dislocations (or quasi-dislocations)

$$\begin{aligned} &\mathrm{Rot}\ \varepsilon = \eta \\ &\sigma = C\varepsilon \\ &\mathrm{div}\ \sigma = 0 \end{aligned} \tag{2.3}$$

where $\eta(x)$ is related to a dislocation density.

A general state of stress is a superposition of stresses due to body forces and dislocations. This permits one to introduce projection operators which split the stress function space into two corresponding subspaces. For this purpose, let us define the new tensor field

$$\tau = B'\sigma = C'\varepsilon \qquad (2.4)$$

where B' and C' are the uniquely defined symmetric square roots from $B(x)$ and $C(x)$. Then under certain additional conditions, we have the unique decomposition

$$\tau = \Pi\tau + \Theta\tau \qquad (2.5)$$

where the projection operators Π and Θ are related to body forces and dislocations, respectively.

Our aim is to reformulate the theory of elasticity in terms of these operators and to show that this formalism is convenient for treating a class of problems including media with inhomogeneities.

3. INTEGRAL EQUATIONS FOR STRESS AND STRAIN

Let

$$C(x) = C_o + C_1(x) \quad , \quad B(x) = B_o + B_1(x) \qquad (3.1)$$

where C_o and B_o are constant tensors and C_1, B_1 are perturbations vanishing rapidly enough at infinity.

Let G_o be the Green's operator of the infinite homogeneous medium associated with the elastic tensors C_o or B_o. Applying def G_o to both sides of (2.1) we have

$$\varepsilon + K_o C_1 \varepsilon = \varepsilon_o \quad , \qquad (3.2)$$

$$\sigma + S_o B_1 \sigma = \sigma_o \quad , \qquad (3.3)$$

where

$$K_o = -\operatorname{def} G_o \operatorname{def} \quad , \quad S_o = C_o - C_o K_o C_o \quad , \qquad (3.4)$$

$$\varepsilon_o = \operatorname{def} G_o f, \qquad \sigma_o = C_o \varepsilon_o \; . \qquad (3.5)$$

We shall call ε_o and σ_o the external fields, i.e. the fields that would be present in the homogeneous medium under the action of the force f vanishing at infinity. Under our assumptions, ε_o and σ_o must vanish at infinity. In reality, it can be shown that the equations (3.2) and (3.3) are valid for external fields ε_o, σ_o with arbitrary conditions at infinity, the same conditions at infinity being valid for ε, σ. Moreover, ε_o, σ_o can arise from both body forces and dislocations. In what follows, we shall consider the external fields as given.

The cases of a cavity and a rigid inclusion need special precautions. We consider these cases as limiting ones: for a cavity V

$$C_1(x) \to -C_o \quad , \quad B_1(x) \to \infty \text{ for } x \in V \qquad (3.6)$$

and for a rigid inclusion V

$$C_1(x) \to \infty \quad , \quad B_1(x) \to -B_o \text{ for } x \in V \; . \qquad (3.7)$$

It follows that the equations (3.2) and (3.3) are not equivalent for these special cases; (3.2) being valid for a cavity and (3.3) for a rigid inclusion.

4. GREEN'S OPERATORS FOR STRAIN AND STRESS

Let G be the Green's operator corresponding to $C(x)$, i.e.

$$G\,(-\partial C \partial) = I \tag{4.1}$$

where I is the identity operator. Analogously to (3.4) we define operators

$$K = -\,\mathrm{def}G\mathrm{def}, \qquad S = C - CKC \tag{4.2}$$

which are symmetric with respect to a usual scalar product. It can be shown [4] that K admits an interpretation as the Green's operator for strain due to body forces and S is the Green's operator for internal stress. It is readily seen that K and S are in a one-to-one correspondence with G, but they are more efficient for problems where external fields are considered as given.

Let us indicate the basic identities for K and S. From the definition (4.2) we immediately obtain

$$\mathrm{Rot}\,K = 0 \quad , \qquad \mathrm{div}\,S = 0 \; . \tag{4.3}$$

Using (4.1) and the symmetry of $C(x)$ we have

$$\begin{aligned} KCK &= \mathrm{def}G\partial C\partial G\mathrm{def} = \\ &= -\,\mathrm{def}G\mathrm{def}, \end{aligned} \tag{4.4}$$

and from (4.2) we find

$$KCK = K \quad , \qquad SBS = S. \tag{4.5}$$

Now let us rewrite equations (3.2) and (3.3), replacing C_o by C and vice versa (take into account that $C_1 \to -\,C_1$ and $B_1 \to -\,B_1$)

$$\varepsilon_o - KC_1\varepsilon_o = \varepsilon \quad , \tag{4.6}$$

$$\sigma_o - SB_1\sigma_o = \sigma \; . \tag{4.7}$$

Substituting ε and σ in (3.2) and (3.3) we find the identities

$$K - K_o + KC_1K_o = 0 \quad , \tag{4.8}$$

$$S - S_o + SB_1S_o = 0 \; . \tag{4.9}$$

Note that if we formally put $C = 2C_o$, we obtain the identities (4.5) as a particular case.

Other properties of the Green's operators K and S, as well as explicit formulas for K_o and S_o, are investigated in [2-4].

It is important to indicate that the identities (4.8), (4.9) can be considered as equations which uniquely determine K and S if K_o, S_o and C_1, B_1 are given. Solutions ε, σ of (3.2), (3.3) are then expressed through K and S by (4.6), (4.7).

5. PROJECTION OPERATORS

The Green's operators K and S admit an elegant description in terms of associated projection operators. Let us introduce the operators

$$\Pi = C'KC' \quad , \quad \Theta = B' \cdot B'S. \tag{5.1}$$

Using (4.2), (4.5), (4.8) and (4.9) we find a complete system of equations defining Π and Θ if C and Π_o, Θ_o are given:

$$\Pi^2 = \Pi \, , \quad \Theta^2 = \Theta \quad , \quad \Pi + \Theta = I, \tag{5.2}$$

$$\Theta_o C_o' B' \Pi = 0 \quad , \quad \Pi_o B_o' C' \Theta = 0 \quad . \tag{5.3}$$

The last two equations are equivalent when C is strictly positive.

As we see from (5.2), Π and Θ are projection operators (i.e., projections). It follows from the symmetry of C', B' and K, S that Π and Θ are orthogonal projections. The corresponding Hilbert space is thus split into two orthogonal subspaces of solutions generated by body forces and dislocations. From (4.6), (4.7) and (5.1) it is seen that Π, Θ determine the solution of equations (3.2), (3.3).

The most interesting question is the nature of dependence Π and Θ on C or B. We shall indicate certain results in this direction.

Let an operator A be represented in the following block matrix form generated by the projections Π_o, Θ_o

$$A = \begin{pmatrix} A_{11} & A_{12} \\ A_{21} & A_{22} \end{pmatrix} \quad . \tag{5.4}$$

Assuming A_{11}^{-1} exists, we put

$$[A^{-1}] = \begin{pmatrix} A_{11}^{-1} & 0 \\ 0 & 0 \end{pmatrix} \quad . \tag{5.5}$$

It can be shown that the following formulas are valid

$$\Pi = C' \, [C^{-1}] C' \quad , \quad \Theta = B' \, [B^{-1}] B' \quad . \tag{5.6}$$

Note that finding C_{11}^{-1} and B_{11}^{-1} is a nontrivial operation.

We indicate different and more convenient representation for Π and Θ. Under slight restrictions on C the projections Π, Θ are "orthogonaly" equivalent to Π_o, Θ_o respectively, i.e.

$$\Pi = U \Pi_o U^+ \quad , \quad \Theta = U \Theta_o U^+ \tag{5.7}$$

where U is an orthogonal operator. Note that U and UV_o, where V_o commutes with Π_o, define the same Π, i.e. U is defined up to an orthogonal multiplier V_o from the right.

We put

$$U = e^A \tag{5.8}$$

where A is a skew - symmetric operator and assume for simplicity that $C_o = I$. Then from (5.3), (5.7) and (5.8) we obtain the following equation for A

$$\Pi_o C' e^A \Theta_o = 0 \; . \tag{5.9}$$

Let us introduce a one-parameter orthogonal group

$$U_t = e^{tA} \tag{5.10}$$

with the generator A.

Let C_t be the corresponding family of positive operators satisfying the equation

$$\Pi_o C_t' e^{tA} \theta_o = 0 \quad (5.11)$$

and the obvious conditions $C_{t=o} = I$ and $C_{t=1} = C$. Then solutions of this equation describe a class of projections Π_t, θ_t corresponding to C_t.

It is readily seen that Π_t satisfies the Heisenberg type equation

$$\frac{d}{dt} \Pi_t = [A, \Pi_t] \quad , \quad \Pi_{t=o} = \Pi_o \ . \quad (5.12)$$

We give an example of a solution of this equation. Let

$$A = \begin{pmatrix} o & a \\ -a^+ & o \end{pmatrix} \quad , \quad D = \begin{pmatrix} o & a \\ a^+ & o \end{pmatrix} \quad , \quad (5.13)$$

where a is an orbitrary operator. Then

$$C_t = I - \sin(2tD) \quad (5.14)$$

satisfies (5.11) and

$$\Pi_t = e^{tA} \Pi_o e^{-tA} \quad . \quad (5.15)$$

Let us note that explicit formulas for Π_o, θ_o and their properties are given in [3,4].

6. LOCAL PROJECTION OPERATORS

Let two elastic media with tensors of elastic moduli C_o and C be joined. We assume that, on the (sufficiently smooth) boundary Ω between them, the usual conditions of continuity of displacement and the normal stress vector

$$u_o = u \quad , \quad n.\sigma_o = n.\sigma \quad (6.1)$$

are satisfied. Here n is a normal to Ω at a given point x.

Let ε_o, ε and σ_o, σ be the limiting values of strain and stress from the both sides of Ω at the point x. It was shown in [2] that the strains and stresses satisfy the equations

$$\varepsilon + K_o(n)\, C_1 \varepsilon = \varepsilon_o \ , \quad (6.2)$$

$$\sigma + S_o(n)\, B_1 \sigma = \sigma_o \quad (6.3)$$

or

$$\varepsilon_o - K(n) C_1 \varepsilon_o = \varepsilon \quad , \quad (6.4)$$

$$\sigma_o - S(n) B_1 \sigma_o = \sigma \quad . \quad (6.5)$$

Here C_1, B_1 are defined as before,

$$K(n) = n.G(n).n, \quad (6.6)$$

$$S(n) = C - CK(n)C, \quad (6.7)$$

$$G(n) = (n.C.n)^{-1}, \tag{6.8}$$

and the corresponding expressions for $K_o(n), S_o(n)$.

These local equations have exactly the same form as the global ones considered above. The tensor functions K(n), S(n) satisfy equations which are exact analogies of (4.5), (4.8), (4.9). This permits one to introduce the local projection operators

$$\Pi(n) = C'K(n)C', \qquad \theta(n) = B'S(n)\,B' \tag{6.9}$$

that satisfy equations analogous to (5.2), (5.3). Thus, we have a remarkable finite-dimensional representation of the global Green's and projection operators by local ones.

We indicate the relationship between the global and local operators. Let $\Pi(x,x')$ be the kernel of the global operator Π. Assuming that C-const, i.e Π is translationally invariant, we have $\Pi(x,x') = \Pi(x-x')$. We denote by $\Pi(k)$ the Fourier transform of the tensor function $\Pi(x)$. It can be shown that $\Pi(k)$ is a homogeneous function of k of zero order, i.e.

$$\Pi(k) = \Pi(n) \quad \text{where} \quad n = k/|k|. \tag{6.10}$$

The function $\Pi(n)$ is identical with the local projection operator $\Pi(n)$, where n is the normal to Ω at x. The same results hold for the local Green's operators K(n) and S(n).

The local projection operators admit the invariant decomposition

$$\Pi(n) = \Pi'(n) + \Pi''(n), \tag{6.11}$$

$$\theta(n) = \theta'(n) + \theta''(n), \tag{6.12}$$

where Π', θ' project the six-dimensional space of tensors τ onto one-dimensional subspaces, and Π'', θ'' project the space onto two-dimensional subspaces of traceless tensors [3]. The corresponding decomposition is also valid for the global projection operators. Thus, the Hilbert space of solutions is decomposed into four invariant orthogonal subspaces.

7. APPLICATION

As a first example of application of the operators Π, θ and K, S we consider an ellipsoidal inhomogeneity. Let the ellipsoid Ω be given by the equation

$$x a^{-2} x = 1 \tag{7.1}$$

where a is a tensor whose eigen-values are the semi-axes. The formulas

$$x = \frac{a^2 n}{\sqrt{n a^2 n}}, \qquad n = \frac{a^{-2} x}{\sqrt{x a^{-4} x}}, \tag{7.2}$$

where n is a normal to Ω at a point x establish a one-to-one correspondence between a function $f(x)$ on Ω and a function f(n) on the unit sphere S.

In particular,

$$\rho(x) = (x a^{-4} x)^{-1/2} \quad \leftrightarrow \quad \rho(n) = (n a^2 n)^{-1/2} \tag{7.3}$$

We define

$$\langle f(x)\rangle = \frac{1}{4\pi \det a} \int_{\Omega} \frac{f(x)}{\rho(x)}\, dx =$$
$$= \langle f(n)\rangle = \frac{\det a}{4\pi} \int_{S} f(n)\rho^3(n)\,dn \qquad (7.3)$$

as the mean value of $f(x)$ over Ω or of $f(n)$ over S.

Let an ellipsoidal inhomogeneity have elastic moduli C = const inside Ω and C_o = const outside Ω. In particular, C = 0 and C = ∞ correspond to a cavity and a rigid inclusion respectively.

We assume that a constant stress σ_o is applied at infinity. Then the stress $\sigma(x)$ at Ω or $\sigma(n)$ at S are linear functions of σ_o,

$$\sigma(n) = F(n)\sigma_o \ , \qquad (7.4)$$

where F(n) is a fourth order tensor. It was shown [2] that F(n) can be expressed explicitly through the local projection or Green's operators. Assuming, for simplicity, that C = 0 and C_o = I, we have

$$F(n) = \theta_o(n) \langle\theta_o(n)\rangle^{-1} \ . \qquad (7.5)$$

As a result

$$\langle F(n)\rangle = I, \qquad \langle\sigma(n)\rangle = \sigma_o \ . \qquad (7.6)$$

The formula (7.5) permits one to obtain easily stress concentration factors for limiting cases of ellipsoidal crack and needle.

As another example we indicate a problem of arbitrary curvilinear cracks in an anisotropic three-dimensional elastic medium. Using the Green's operator S_o, the problem can be reduced to a pseudo-differential equation on the crack surface [6].

The operators K_o, S_o are particularly convenient in treating problems connected with stochastic media. We indicate, for example, the recent works [7, 8, 4].

8. ELASTODYNAMICS

Let us briefly indicate a generalization appropriate to the dynamical case. We start from the equations of motion

$$\rho\partial_t^2 u - \partial C\partial u = f \qquad (8.1)$$

and rewrite them in the form

$$\bar{\partial}^+\bar{C}\bar{\partial}u = f, \qquad (8.2)$$

where

$$\bar{C} = \begin{pmatrix}\rho & o\\ o & C\end{pmatrix}, \qquad \bar{\partial} = \begin{pmatrix}\partial_t\\ \partial\end{pmatrix}, \quad \bar{\partial}^+ = \begin{pmatrix}\partial_t\\ -\partial\end{pmatrix} . \qquad (8.3)$$

We introduce the dynamical "strain" and "stress"

$$\bar{\varepsilon} = \begin{pmatrix}\partial_t u\\ \varepsilon\end{pmatrix}, \quad \bar{\sigma} = \begin{pmatrix}\rho\partial_t u\\ \sigma\end{pmatrix} \qquad (8.4)$$

and dynamical "Hook's law"

$$\bar{\sigma} = \bar{C}\ \bar{\varepsilon}\ . \tag{8.5}$$

Now, following the general scheme described above, we can define the dynamical Green's operators $\bar{K}$, $\bar{S}$ and projection operators $\bar{\Pi}$, $\bar{\Theta}$. In this case the kernels of these operators will depend also on the time variables, for example

$$\bar{\Pi} \rightarrow \begin{cases} \bar{\Pi}(x,x', t - t') \\ \bar{\Pi}(k,k', \omega) \end{cases} .$$

New and interesting problems arise in elastodynamics, in particular, direct and inverse scattering problems. The method of solution of these problems usually based on the scattering matrix or S-matrix. It can be shown [4], that the S-matrix is closely related to Green's functions or projection operators. Roughly speaking

$$S(k) \sim \bar{\Pi}\ (k,k,\omega(k))\ , \tag{8.6}$$

where $\omega(k)$ is the corresponding dispersion law. Thus S(k) is defined by kernels of Green's operators at the "diagonal" k = k!

9 CONCLUSIONS

Let us summarize certain advantages of the projection operator method.

1. The equations for projection operators are universal for different field theories.

2. The equations and their solutions have a geometrical significance.

3. Projection operators have a particularly simple spectrum.

4. There is a remarkable relationship between global and local projection operators.

5. Projection operators are especially appropriate to problems connected with inhomogeneities, cracks, dislocations, stochastic media, etc. when external fields are given.

6. Projection operators are effective in obtaining approximate solutions by using perturbation and variational methods.

REFERENCES

[1] Kunin, I. A., *The Green's tensor for an anisotropic elastic medium with sources of internal stress*, Dokl. AN SSSR, 157 (1965) 76-79

[2] Kunin, I. A., and Sosnina, E. G., *The stress concentration due to an ellipsoidal inhomogeneity in an anisotropic elastic medium*, Appl. Math. and Mech. 37 (1973) 306-312

[3] Kunin, I. A., *An algebra of tensor operators and its applications to elasticity*, J. Eng. Sci. (to appear).

[4] Kunin, I. A., *Elastic media with microstructure*, (Springer-Verlag, to appear), vol. 1, 2.

[5] Gel'fand, I. M., *Generalized functions* (Academic Press, 1964), vol. 1,4.

[6] Kunin, I. A., and Shlyapobersky, J., *The spatial crack problem in elasticity*, J. Eng. Sci. (to appear).

[7] Willis, J. R., *Bounds and self-consistant estimates for the overall properties of composites*, J. Mech. Phys. Solids 25 (1977) 185-195

[8] Kröner, E., *Graded and perfect disorder in random media elasticity*, J. of the Engineering Mechanics Division, ASCE, 106 (1980) 889-914.

Continuum Models of Discrete Systems 4
eds. O. Brulin and R.K.T. Hsieh

CONTINUUM THEORIES OF BLOOD FLOW

Richard Skalak and Aydin Tözeren

Bioengineering Institute
Department of Civil Engineering and Engineering Mechanics
Columbia University
New York, New York 10027

Blood is a suspension of discrete cells in a Newtonian fluid. The red blood cells are the most numerous and behave like small elastic sacks filled with a Newtonian fluid. Depending on the scale involved it is necessary to treat blood as a discrete suspension or as a continuum. The bulk rheological properties of blood are nonlinear and viscoelastic. The possibilities of applying rheology of suspensions to blood flow are discussed. Continuum fluids with microstructure are also considered.

INTRODUCTION

The viscosity of blood is a subject which has a long history. Somewhere in antiquity the opinion was formed that blood is thicker than water. This is correct, but numerical values and the reasons for its behavior have only become clear in the last two decades. Analysis from first principles is further behind and still quite incomplete today.

One of the earliest estimates of blood viscosity was made in 1809 by Thomas Young [1] in his Croonian lecture on the functions of the heart and arteries. He estimated from measurements of the arterial pressure drop that the resistance to the motion of blood must be about 4 times greater than that of water. This is close to the typical value accepted today of 3.5 at high shear rates. In 1840 Poiseuille [2] undertook precise measurements of the pressure drop in fine glass capillaries with the intent to understand the laws governing the flow of blood. However due to troubles with clotting he made his tests on water, alcohol and mercury. His measurements were quite accurate for these liquids but did not establish the viscosity of blood itself. Later tests, mostly on larger diameter tubes, showed that the viscosity of normal blood at high shear rates was of the order of 3 to 4 centpoise.

The variation of the apparent viscosity of blood with shear rate has been extensively explored only in the last two decades. It is shown to exhibit a shear thinning behavior attributable both to the flexibility of the cells and the tendency of the cells to adhere to each other at low shear rates [3,4]. Under oscillatory flow it has been shown in the last decade that blood exhibits a certain viscoelasticity. While these effects are understood in the sense of the blood cell properties responsible for them, derivations from first principles are still at a semi-empirical stage.

In the last few years the properties of blood cells have been more clearly defined. They may be regarded as viscoelastic membranes filled with a Newtonian fluid. This model has some success in the analysis of capillary blood flow in which cells pass in single file. In this range there does not appear to be much hope to derive a continuum model of the discrete system. In many capillaries the diameter of the blood vessel is smaller than the diameter of the red blood cells when they are at rest. The blood cells are readily deformed and may undergo very large deformations during normal blood flow.

In the next section the properties of red blood cells and of blood as a suspension are described in more detail. In the subsequent sections the pertinent aspects of the theory of suspensions and of the theory of fluids with microstructure are outlined to the extent they may be applicable to blood flow. It will be seen that the transition from known properties of the constituents to a continuum theory for blood flow is far from complete.

RHEOLOGY OF BLOOD

Blood is a suspension of particulate cells usually well dispersed in plasma which is a Newtonian fluid. Red blood cells make up more than 99% of the volume of particulate matter and their mechanical behavior is the most important factor determining the bulk rheology of blood. The volumetric percentage of red cells in whole blood is called the hematocrit and ranges between 40% and 50% in normal blood. Human red blood cells are biconcave, disk-shaped particles with an average diameter of about 7.6 μm and thickness 2.8 μm.

Red blood cells consist of a thin membrane, about 100 Å thick, filled with a hemoglobin solution that is a Newtonian fluid with a viscosity of about 6 cp. The membrane is a viscoelastic solid, and possesses unusual properties compared to ordinary materials like rubber [5]. The red cell membrane has a shear modulus of about 0.004 dyn/cm (for the entire thickness) but has an areal modulus of about 200 dyn/cm at steady state deformation. It thus strongly resists any change of area due to isotropic tension and may be considered as a constant area membrane [6,7]. Large deformations are readily produced at constant area due to the low shear modulus. The red cell membrane also has a bending stiffness of about 10^{-12} dyn/cm which may be neglected in many situations in which the deformations are large [8].

The red blood cell membrane also has a definite viscosity in shear, of about 10^{-3} dyn sec/cm, for the entire thickness [9,10]. The fact that the red cell membrane has some viscosity and maintains nearly constant area has been used to develop an alternate model of a viscoelastic, incompressible, two-dimensional fluid. The membrane consists of two layers of phospholipid molecules and has been discussed in terms of a layer of liquid crystal fluid [11].

Among the other particulate matter in blood, white blood cells are the largest. They are more or less spherical, about 8 μm in diameter. They are stiffer but much less numerous than red cells. White blood cells have a viscoelastic interior which makes them about two orders of magnitude stiffer than red blood cells under rapid deformations [12]. The cell membrane is also a bilipid layer, but is thought to play a much smaller role in the rheology of the cell than that of the red blood cell membrane. In the red cell, all elastic properties derive from the cell membrane.

Blood contains platelets which are smaller ellipsoidal particles with a length of about 2.5 μm. They play an important role in blood clotting but are not important rheologically in normal blood because of their small numbers.

The specific gravity of red blood cells is about 1.10 and that of plasma is 1.03 so red blood cells tend to settle slowly when at rest in plasma. However, the normal motion of the blood disperses the cells quite uniformly except near rigid boundaries where there may be a thin layer deficient in red cells. The coefficient of viscosity of plasma is 1.2 cp at 37°C [3]. In large blood vessels and at high shear rates in a viscometer, blood behaves as a Newtonian fluid with a constant coefficient of viscosity which depends on the hematocrit (Fig. 1). For normal blood, the high shear rate viscosity is about 3.5 cp. At low shear rates, the apparent viscosity may be an

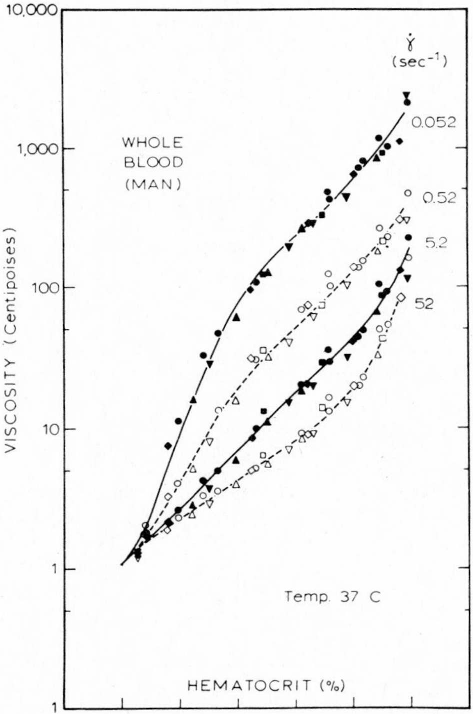

FIG. 1. Relationship between the logarithm of viscosity at four different shear rates and the hematocrit of whole blood. Closed symbols represent the data at shear rates of 0.052 and 5.2 sec^{-1} and open symbols depict the results at shear rates of 0.52 and 52 sec^{-1}. Symbols of different shapes are used to indicate results on each of five subjects. (From Chien et al. [3] .)

order of magnitude greater [3,4,13]. As the shear rate increases, normal blood exhibits a shear-thinning behavior as shown in Fig. 2.

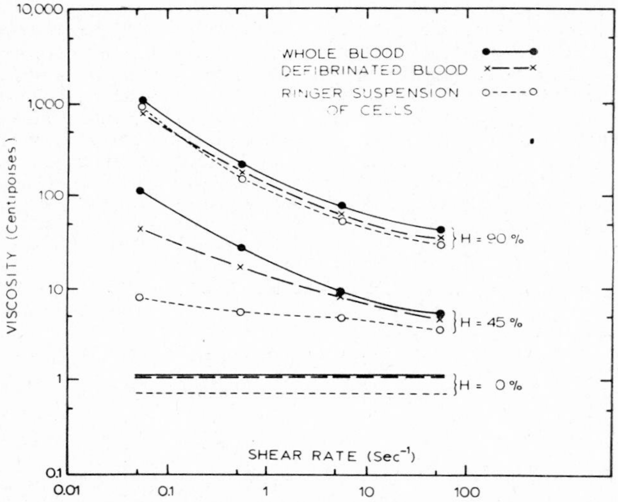

FIG. 2. Relationship between the logarithm of viscosity and the logarithm of shear rate for whole blood, defibrinated blood, and Ringer suspensions of cells at hematocrits (H) of 90 and 45%. (From Chien et al. [37].)

This is due to two effects. First, red cells form stacks, called rouleaux in the presence of the fibrinogen and globulin in the plasma. These long chain molecules bind the neighboring cells together like a weak adhesive. The rouleaux are gradually broken up as the shear rate increases and as a result the viscosity decreases. Secondly, the higher shear rates deform the flexible red blood cells and tend to align them in the direction of flow which further reduces the viscosity.

The above discussion is based on measurements in viscometers in which the red cell diameter is small compared to the typical gap dimension of the instrument. In the circulation the ratio of the red cell diameter to the blood vessel diameter runs the full gamut from the order of one in the capillaries to a very small value in large arteries. In studies of wave propagation and of local flow patterns at bends and branch points in the major arteries, blood is usually assumed to be a Newtonian fluid with the high shear rate value of the viscosity [14]. This is a valid approximation for most purposes because the viscous effects in wave propagation in the large arteries are relatively small.

At the other extreme, in capillary blood vessels, the cell diameter is about equal to or less than the capillary diameter. The cells travel in single file and may fill most of the lumen, leaving only a thin lubricating layer of plasma between the cell and the vessel wall. There have been a number of analysis of capillary blood flow treated as a two phase flow consisting of the discrete elastic particles and the Newtonian plasma [15,16,17]. Asymptotic expansions, lubrication theory and numerical methods have been applied to various cases (see Skalak and Chien [15]). In capillary flow there is not any theory purporting to represent the discrete system as a continuum, although the pressure drop is conveniently summarized in terms of an apparent viscosity of the equivalent Newtonian fluid as far as discharge is concerned. Such apparent viscosities are functions of the diameter ratio, velocity, hematocrit and elastic properties of the blood cells.

The intermediate range of blood vessel diameters from about 12 μm to 500 μm is the most interesting from the standpoint of utility of developing a continuum theory for the discrete medium. The Reynolds number is sufficiently small so that viscous behavior of blood dominates the flow. The diameter ratio of the red blood cells to the blood vessels is small but finite and the rheology of whole blood depends strongly on this length ratio. There have been extensive experimental studies on the flow of concentrated suspensions of various kinds through cylindrical tubes in this range [4,16,17].

In blood vessels apparent viscosity of blood decreases with decreasing vessel diameter. This is known as the Fåhraeus-Lindqvist [19] effect. It has been observed also in the flow of other suspensions such as oil in oil emulsions. The Fåhraeus-Lindqvist effect in blood is closely related to the flow behavior and deformability of red cells. It is observed experimentally that deformable particles in a suspension near a rigid boundary migrate radially away from the wall (16,19,20). It has been shown analytically that rigid particles in a slow viscous tube flow do not migrate perpendicular to streamlines, but that flexible particles may migrate provided the stresses induced are sufficient to deform the particles. As the Reynolds number of the flow increases a radial migration may result from inertial effects also, even in the case of rigid particles [21]. This is the basis of the Segre-Silberberg [22] results which show a maximum particle concentration at about 0.6 of the radius of a circular tube with a low concentration of rigid particles.

In blood flow the Reynolds numbers are sufficiently small so that the effects of deformability predominate radial migration. At the same time, the hematocrit is sufficiently high so that axial accumulation of the cells is limited by cell collisions. The net effect is the development of a cell-depleted layer near the vessel walls. The existence of a cell-depleted layer at the wall has two results which affect the bulk flow properties of the suspension. Firstly, it reduces the hematocrit in the tube compared to that which is observed in a collecting reservoir into which the blood is flowing. This decrease in tube hematocrit with the decreasing tube size was first observed by Fåhraeus [23] in 1929. More recent experiments have confirmed and extended these observations [24]. Their data show that the Fåhraeus-Lindqvist effect of decreasing apparent viscosity with decreasing tube diameter may be largely attributed to the decreased hematocrit in the tube. Secondly, the cell depleted layer at the wall has a lower viscosity than the bulk suspension. This causes, in effect, an apparent slip velocity which further contributes to the Fåhraeus-Lindqvist effect. The cell depleted layer is always present at least to the extent that exclusion by the wall occurs, but it becomes more significant when radial migration of the cells takes place. It is generally not more than two or three cell diameters in thickness.

Experimental velocity profiles were obtained by Goldsmith and Marlow [4] for ghost blood cell and other suspensions flowing in clyindrical tubes. As shown in Fig. 3 the velocity profile in cell suspen-

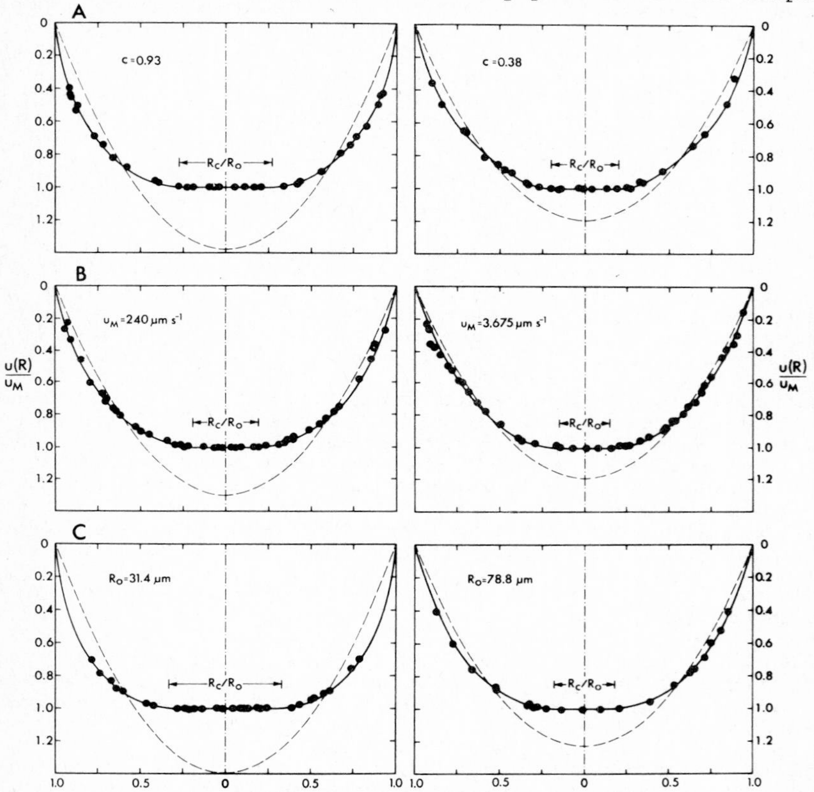

FIG. 3. Dimensionless plots of the tracer rbc velocity distributions showing the effects of (A) concentration, (B) flow rate, and (C) rube radius. (A) R_o = 41.7 μm; Q = 8.82×10^{-4} (left) and 8.83×10^{-4} mm^3 sec^{-1} (right). (B) R_o = 42.6 μm; c = 0.430; Q = 8.82×10^{-4} (left) and 5.37×10^{-3} mm^3 sec^{-1} (right). (C) c = 0.600; Q = 5.66×10^{-4} (left) and 6.89×10^{-3} mm^3 sec^{-1} (right); G = 11/7 (left) and 9.0 sec^{-1} (right). (From Goldsmith & Marlow [4].)

sions is not parabolic; as in Poiseuille flow, but is more blunted in suspensions when the volumetric particle concentration is greater than twenty percent. Then there exists a core in the center of the tube in which an approximately uniform particle velocity is observed. The centerline velocity is smaller than that for a Newtonian fluid at the same flow rate. The diameter of the core region decreases with increasing flow rate and increases with increasing concentration of red blood cells. The dependence of the velocity profile on the flow rate is an indication of non-Newtonian behavior and is also observed in concentrated oil-in-oil emulsions [25]. On the other hand, for a suspension of rigid spheres the velocity profile is independent of the flow rate and pressure drop is proportional to the volumetric flow rate [26]. The degree of blunting of the velocity profile decreases with increasing deformability of the suspending particles.

Comparison of relative viscosity of the human blood with that of concentrated oil-in-water emulsions indicate that red blood cells are much more easily deformed than liquid droplets. The biconcave shape of the red cell has a large area in relation to its volume which enables red cells to deform and move past each other easily in crowded particle configurations. Goldsmith and Maslow [4] observed that particle paths in ghost cell suspensions exhibit random lateral movements due to the frequent encounters of a particle with neighboring cells. Viscometric and microrheological observations of whole blood indicate that interaction effects and the aggregation of cells are important in determining the behavior at low rates of shear ($\mathring{\gamma} < 50\ sec^{-1}$), while the deformability of red cells and red cell aggregates become a dominant factor for higher shear rates.

Blood exhibits viscoelastic properties when tested under oscillatory shear or transient conditions. The viscoelastic behavior of blood is associated with the elastic properties of the red cell membrane and the viscosity of the membrane and of the internal and external fluids. The surface forces which produce aggregation may be important in establishing an elastic network of cells. The viscoelasticity of blood cell suspensions has been demonstrated in oscillatory tests using cone and plate [27] and tube flow [28] geometries. However, there is no comprehensive theory that allows prediction of bulk viscoelastic behavior starting from the known blood cell properties.

THE SUSPENSION THEORY OF BLOOD FLOW

Insofar as blood is a suspension, progress in the theory of suspensions should be useful in discussing the rheology of blood. The vocabulary and general concepts developed in suspension theory are directly applicable to blood flow to a large extent. However as the red blood cells are flexible and usually have a high concentration, very little has been accomplished in the way of specific formulas for the viscosity or other aspects of behavior of blood from first principles. Progress in analytical work has been confined mostly to dilute suspensions. The pertinent literature has been reviewed by Brenner [29], Batchelor [30], Jeffrey and Acrivos [31], Herczynski and Pienkowska [32] and others.

The most appropriate way to discuss blood rheology in relation to the theory of suspensions would seem to be to adopt the notion of an ensemble average q_{ij} which is defined as the average over values of q_{ij} at a fixed point $(\underset{\sim}{x},t)$ occurring in a large number of different realizations of a flow under the same macroscopic boundary conditions. Equations of continuity and linear momentum are obtained by taking ensemble averages over the corresponding equations of any realization. The equation of continuity for an incompressible suspension is then

$$\partial U_i/\partial_{xi} = 0 \qquad (1)$$

where $U_i = \langle u_i \rangle$ denotes the mean velocity defined as the ensemble average. The momentum equation for the mean flow can be written as

$$\rho(\frac{\partial U_i}{\partial t} + U_j \frac{\partial U_i}{\partial x_j}) = \frac{\partial \Sigma_{ij}}{\partial x_j} + F_i \qquad (2)$$

where by definition the bulk stress is

$$\Sigma_{ji} = \langle \sigma_{ji} - \rho\ (u_i - U_i)(u_j - U_j) \rangle \qquad (3)$$

In the above equations ρ is the density which is assumed to be constant both for the fluid and suspended particles, F_i is the body force which is also to be uniform over the fluid and particulate phase. The stress tensor Σ_{ij} is the bulk stress of the suspension defined by Batchelor [33]. In many cases it is convenient to make an ergodic principle assumption, namely that a suitable volume average will be equal to the ensemble average. This will be generally valid for blood flow in larger vessels but in the intermediate range where the Fåhraeus Lindqvist effect occurs the important region close to the wall will not be homogeneous.

For suspensions of solid or liquid particles in an incompressible viscous fluid Batchelor [30] expressed the bulk stress in the following form:

$$\Sigma_{ij} = -p\delta_{ij} + 2\mu E_{ij} + (1/V)\sum_{m=1}^{n} [\int_{A_o} [\sigma_{ik} x_j n_k - \frac{1}{3}\sigma_{\ell k} x_\ell n_k \delta_{ij}$$
$$- \mu(u_i n_j + u_j n_i)] dA \qquad (4)$$

where A_o is the surface area of the typical particle, σ_{ij} is the hydrodynamic stress, E_{ij} is the mean rate of strain, p is the mean pressure and μ is the viscosity of the suspending fluid. The summation in equation (3.4) is over all the particles in the averaging volume, V. The surface integrals lead to a component which may be called the particle stress, denoted by S_{ij}. In writing equation (3.4) the terms due to the fluid and particle inertia are neglected, which is a valid assumption for blood flow. The effect of inertia on the bulk stress of the dilute suspensions was recently considered by Lin et al. [34] and Ryskin [35]. The resulting stress system is non-Newtonian and depends on the Reynolds number based on the particle radius and the mean velocity gradient. The equation (3.4) indicates that the bulk stress is symmetric in the absence of externally applied couples on the particles.

For dilute suspensions the classical way of proceeding is to approximate the flow in the vicinity of each particle as that of a single particle in an unbounded viscous flow. Results are available for various rigid particles shapes including spheres, ellipsoids and rod-like particles [33,36]. Liquid and spherical viscoelastic particles have also been considered. A realistic analysis for the elastic red blood cell has not been carried out. A model of a spherical elastic capsule is the closest model that has been treated to some extent [37-39].

Progress has been made in recent years on the computation of the bulk stress for more concentrated suspensions, but these analyses have not been carried over to a definitive theory of blood rheology. In the approach of Batchelor and Green [40] the bulk stress in a suspension of rigid spherical particles was studied to order c^2 (where c is the volumetric concentration) by considering two particle interactions in the suspending medium. The basic ingredient is the flow field in the vicinity of two rigid particles which approach each other in a shear flow. The resulting bulk stress expression is in general of non-Newtonian form due to the dependence of the statistical properties of the suspension on the bulk flow. However in the case of pure straining motion of the suspension bulk stress can be represented in terms of a bulk viscosity which was estimated by Batchelor and Green [40] as:

$$\mu = \mu_f(1 + 2.5c + 7.6c^2) \quad (5)$$

where μ_f is the viscosity of the suspending fluid. The consideration of two particle interactions and inertial effects show the complicated nature of the constitutive equation for bulk stress even for dilute suspensions of spherical particles. Recent experiments by Gadala-Maria and Acrivos [41] imply that concentrated suspensions of rigid spheres cannot be modeled as isotropic fluids.

In the wall region where the flow is characterized by particle-wall interactions as well as particle-particle interactions, there has been some progress in the treatment of rigid particle suspensions. But again the carry-over to rheology of blood in a definitive manner is not yet forthcoming. In the wall region the mean particle velocity and particle concentration of the suspension vary on a scale comparable to the particle size. This means that a volume average cannot be utilized to replace the ensemble average. However assuming flow properties are statistically uniform in directions parallel to the wall, surface averages over planes parallel to the wall may be used to replace ensemble averages. On this basis. Tözeren and Skalak [42] analysed the wall layer with the further assumption of a dilute suspension. This allows computation of the effect of the wall on each particle using the solution of uniform shear flow past a plane wall with a particle present by the method of Dean and O'Neill [43]. For any assumed distribution of particle centers the bulk stress component Σ_{21} in the direction of flow at a distance x_2 from the plane wall was expressed as:

$$\Sigma_{21} = 2\mu(1+c°f(x))E_{21} \quad (6)$$

where c° is the volumetric concentration of the particles far from the wall, and E_{21} is the ensemble average over a plane parallel to the wall. The function f(x) can be computed for particular particle distributions. The asymptotic value of f(x) is known for dilute suspensions of rigid spheres far from the wall including the case where external moments are exerted on the particles. This is the result for an infinite shear flow given by Brenner [44] in the form

$$f(x_2) = 2.5+M_3/4\mu E_{21}V_o. \quad (7)$$

At the wall the function f(x) is zero because only fluid particles are present at the wall, and points of contact between the particles and the wall being negligible. Between the wall and the far field where equation (7) is valid, the function f(x) varies depending on the particle density distribution. In any case wall exclusion of the particles results in the lower concentration near the wall itself. As a result, the velocity gradient at the wall is always greater than the uniform gradient far from the wall. This in turn causes an apparent slip velocity which is always present if the velocity profile outside the wall layer is extrapolated linearly to the boundary. Similar macroscopic results may be obtained if the wall layer is replaced by a particle-free layer of suspending fluid outside of which the properties of the suspension at infinity are assumed [42]. This is a common model but since the dependence of the statistical particle distribution in the wall region for a given bulk flow is not yet known the thickness of the cell free layer cannot be determined theoretically at this time.

The more precise analysis of the wall layer will require solution of the problem of creeping motion in which more than one particle is present so that both the particle-particle and particle-wall interactions can be evaluated simultaneously. Recent boundary collocation methods developed, for example, by Ganatos et al. [45,46] allows consideration of such problems. This type of analysis may be useful to approach the behavior of more concentrated suspensions in the vicinity of the wall.

An interesting and important type of motion that red blood cells may undergo involves a so-called tank treading in which the shape of the cells is more or less constant while the membrane steadily rotates around the stationary form (Schmid-Schönbein and Wells [47], Fischer and Schmid-Schönbein [48]. Depending on the several parameters involved, a single cell in an infinite shear flow may exhibit a tank-treading motion or may flip end over end as a rigid body. A theoretical analysis by Keller and Skalak [49] confirmed the presence of either tank-treading or flipping motion depending on dimensionless parameters which describe the shape of the cell, the properties of the cell membrane and the ratio of internal to external fluid viscosities.

The stationary tank treading form results most readily when the viscosity of the external fluid is greater than that of the fluid inside the cell. Tank-treading has also been demonstrated in capillary flow. It occurs when there is non-axisymmetric flow and may enhance mass transport in and out of the cell.

For a high concentration of cells as found in normal blood, a given red blood cell will undergo a certain amount of tank-tread motion since the remainder of the suspension surrounding it acts as a fluid of somewhat increased viscosity. A certain degree of tank-treading will persist even for very high cell concentrations. An analysis is given by Secomb and Skalak [50] for packed cell concentrations of the order of 90% which shows agreement with experimental data for such red blood cell suspensions at high shear rates. For spherical particles at concentrations of the order of 50%, various approaches have been proposed. See Goddard [51], Frankel and Acrivos [52] and Brennan [37]. For realistic, flexible red blood cells such theories have not yet been developed.

The governing equations of a suspension flow reduce to those of an ordinary continuum as the average particle spacing to the typical length of the bulk flow approaches zero. However in blood flow through vessels of smaller than about 300 μm this ratio is small but finite. A perturbation with respect to the length ratio may be in order for the bulk equations governing suspensions. Continuum theories with microstructure may be of interest in this regard. All such fluid models involve an intrinsic length scale ℓ which is derived from the coefficients of viscosity. These continuum models may be useful when the ratio of ℓ to the length scale L_o of the macroscopic flow is small but finite. They will be considered in the next section.

FLUIDS WITH MICROSTRUCTURE AS MODELS OF BLOOD FLOW

Theories of fluids with microstructure have been extensively considdered as models of suspension flows. In particular these theories have been applied to blood flow by a number of authors (Kline, Allen and De Silva [53], Valanis and Sun [54], Ariman [55], Cowin [56], Popel et al. [57], Kang and Eringen [58], Chaturani and Upadhya [59, 60]. Surveys of the various theories and applications have been given by Eringen [61], Ariman et al. [62], and Cowin [63].

The application of theories of fluids with microstructure to blood flow is attractive because it promises to incorporate some of the features of the particulate nature of blood in a continuum theory. However the transition from the particulate suspension to the microstructure theory is still not clear. The new kinematic variable introduced in micropolar theory is the particle angular velocity vector ν_i. If an energy approach is used to derive balance laws, the particle angular velocity vector enters in the expression for the kinetic energy of the form [63]:

$$K = \frac{1}{2} \int_B \rho(v_i v_i + i_{k\ell}\nu_k\nu_\ell)\,dV \tag{8}$$

where v_i is the usual velocity vector at any point, ρ is the density of the fluid and $i_{k\ell}$ is the micro moment of inertia. Similarly the mechanical power supplied to a micropolar fluid is assumed to be of the form

$$P = \int_{\delta B} (t_{ji}\nu_i + m_{ji}\nu_i)n_j\,dA + \int_B \rho(f_i V_i + C_k\nu_k)\,dV \tag{9}$$

where t_{ji} is the stress tensor, m_{ji} is the couple stress tensor, f_i is the body force to unit volume and C_i is the body couple per unit volume. The surface of the volume B is denoted by δB. The actual correspondence of the couple stress m_{ij}, micro inertia $i_{k\ell}$ and microspin ν_i to some mean variables of the suspension calculated as ensemble averages is not clearly established. It was shown by Tözeren and Skalak [64] that micropolar kinetic energy may be a good model for the kinetic energy of highly concentrated suspensions of rigid spheres. This does not mean however that the linear constitutive equation of the micropolar fluid would be valid for this type of suspension. In the case of red cell suspensions of high concentration, the expectation would be that the microspin vector would be related to the tank-treading part of the cell motion.

Using the conservation of energy equation incorporating the above forms and requiring invariance under any combination of rigid body translation and rotation allows derivation of equations of continuity, linear momentum and angular momentum for an incompressible micropolar fluid as follows:

$$\frac{\partial v_i}{\partial x_i} = 0 \tag{10}$$

$$\rho\left(\frac{\partial v_i}{\partial t} + v_j \frac{\partial v_i}{\partial x_j}\right) = t_{ji,j} + \rho f_i \tag{11}$$

$$\rho i_{k\ell}\nu_k = \rho C_\ell + e_{\ell jk}t_{jk} + m_{k\ell,k} \tag{12}$$

where the e_{ijk} is the permutation tensor. The constitutive equations of micropolar fluids are most often assumed to be linear. Then the stress t_{ij} and couple stress m_{ij} are related to the kinematical variables by

$$t_{ji} = -p\delta_{ji} + \mu(v_{i,j} + v_{j,i}) + 2\tau e_{kij}\hat{H}_k \tag{13}$$

$$m_{ji} = (\alpha\nu_{\ell,\ell})\ \delta_{ij} + (\beta+\gamma)\nu_{i,j} + (\beta-\gamma)\nu_{j,i} \tag{14}$$

where $\hat{H}_i$ is called the relative angular velocity and is defined by

$$\hat{H}_i = \nu_i - \frac{1}{2} e_{ijk} V_{k,j} \tag{14}$$

Viscometric flows of blood are usually considered to compute the additional material coefficients of the miscrostructure fluid by curve fitting to experimental velocity profiles, particle rotations, etc. There is no established way to compute the material coefficients τ, α, β, and γ from the known properties of blood cells and plasma.

A micropolar fluid has a material characteristic length defined by:

$$\ell = [(\beta+\gamma)/\mu]^{1/2} \tag{15}$$

This length ℓ is assumed to be related to and of the order of the particle size. Similar characteristic lengths can also be defined for other microstructure fluids such as couple stress fluids and micromorphic fluids. It is also convenient to introduce a dimensionless length parameter L and a coupling number N defined by

$$L = L_o/\ell, \qquad N = [\tau/(\mu+\tau)]^{1/2} \tag{16}$$

where L_o is the characteristic geometric length of the macroscopic flow and the parameter N characterizes the coupling between linear and angular momentum equations.

Various boundary value problems including flow between parallel plates and pipe flows have been solved for micropolar fluids (see Ariman et al. [62] and Cowin [63]). It is usually assumed that the fluid velocity must be equal to the velocity of any rigid boundary. The microspin is set equal to zero at the boundary or it may be assigned a finite value which is less than the fluid rotation at the boundary. The restrictions on the boundary conditions are discussed by Atkin and Cowin [65]. Under the usual boundary conditions it is found that the apparent viscosity of the micropolar fluid increases with decreasing length ratio L. This is just the opposite of the Fåhraeus-Lindqvist effect observed in blood.

The Fåhraeus-Lindqvist effect may be produced analytically by introducing a thin layer of fluid at the boundary which is different from the suspension in the rest of the tube. Cases of the core fluid representing the suspension are a Newtonian fluid of higher viscosity, a non-Newtonian fluid, couple-stress fluid and micropolar fluid [59,60]. These solutions are able to reproduce the Fåhraeus-Lindqvist effect and features such as blunting of the velocity profile. However no analytical method is known to calculate the thickness and properties of the wall layer if only the mean properties of the original suspension are specified. Another equivalent procedure is to allow a slip of the fluid at the boundary as suggested by Brunn [67]. In this approach the microstructure theories represent only the core fluid and that the boundary wall layer is not representable by a continuum theory in detail. Another suggestion has been developed recently by Tözeren [68] in which the no-slip condition is retained at the rigid boundary. The additional boundary condition is obtained by setting the tangential surface traction t_{ij} of the microstructure fluid at the wall equal to the corresponding mean wall stress of the suspension, that is:

$$t_{ji}n_j = (-\rho\delta_{ji} + 2\mu_f E_{ji})n_j \qquad (17)$$

where μ_f is the viscosity of the suspending fluid in the suspension and E_{ij} denotes the mean rate of strain at the boundary. The velocity vector v_i of the microstructure fluid is considered to represent the mean velocity in the suspension defined as an ensemble average. Using the boundary condition (17) the Fåhraeus-Lindqvist effect and blunt velocity profiles can be reproduced however the microspin vector ν_i does not reproduce the mean particle rotation accurately.

It can be shown that the boundary condition (17) requires the existence of couple stress at the wall. The existence of a couple stress in the microstructure theories may have a counterpart in the usual theory of suspensions due to the detailed pressure distributions on the boundary in the vicinity of a particle. Generally speaking the pressure distribution in the vicinity of a particle close to a wall has a positive peak ahead of the particle center and a negative peak in the downstream area. This gives rise to a local couple which might be interpreted as the origin of a couple stress at the wall in a microstructure theory. However for a single particle in a infinite shear flow past a plane wall it can be shown that the moment on the wall integrated over the entire plane is zero. This may be altered by the presence of the other particles in the case of a suspension flow, but at this time the existence of such boundary couples has not been rigorously demonstrated. The boundary condition (17) or the introduction of a slip velocity at the wall may result in additional work being done on the fluid by the wall. This difficulty may be resolved if the boundary condition is considered to replace the net effect of the entire wall layer.

The literature presented above deals with homogeneous continuum theories in which particle concentration c is not a kinematic variable. In a related field of granular media Goodman and Cowin [69] presented a continuum theory in which the solid volume fraction of the media is introduced as an independent variable. To attack the variable concentration suspension problem Quemada [70,72] assumed a minimum principle for viscous dissipation of suspensions (or blood) and obtained equations for the velocity field v and concentration c as Euler-Lagrange equations. If blood is to be modelled by a microstructure fluid it may be reasonable to take local concentration as an internal variable. A logical step in this direction is to consider continuum microstructure theories in which the material constants are a function of position in order to reflect the variable particle concentration in an actual suspension. A model of this type was suggested by Popel [57]. The variable concentration was assumed to be established by a local flux balance involving Brownian motion and the gradient $\nu_{i,j}$ of the spin vector and the relative microspin H_i. Kang and Eringen [58] have considered a micromorphic continuum which includes deformations of the suspended particles in shear and stretching in addition to the rotation field. In this case an additional boundary condition on the particle distortion is necessary to make the equations sufficient and a large number of material coefficients must be introduced. Finally some rule governing the distribution of the particle concentration and the relation of the coefficients to the particle concentration must be assumed. It is in the basics of these laws that the utilization of these theories is unclear. The results show that it is possible to model some of the experimental observations of the velocity and concentration profiles by adjusting the constants and boundary conditions involved.

SUMMARY AND CONCLUSIONS

The mechanical properties of red blood cells and their suspending plasma are quite well established. The bulk properties of blood and its behavior in flow through a tube are also well known experimentally. However the theoretical derivation of the bulk properties of blood starting from the known cellular properties still is not complete. Difficulties remain both in the theory of blood flow as a bulk fluid and in the detailed description of the wall layer near a rigid boundary. For the bulk flow, non-linear large deformation analysis of the individual red blood cell motion would appear to be necessary to reproduce the observed behavior of blood with respect to shear rate and concentration effects. For the wall layer the basic problem to be solved involves the simultaneous interaction of particles near a boundary, with each other, and with the wall. The flexibility of the particle is important in studying the migration of the particles away from the wall.

The possible application of theories of fluids with microstructure to blood flow remains an intriguing possibility but involves unanswered questions both as to the interpretation of the macroscopic equations and the appropriate boundary conditions.

From the standpoint of practical utility it might be said that there are not pressing medical problems which will be alleviated if the above gaps in theory were filled in. The situation is rather one of building up a complete explanation of observed and qualitatively understood effects. This may eventually lead to improved diagnosis and treatment of some circulatory and hematological diseases.

Acknowledgment: This research was supported by U.S. National Heart, Lung and Blood Institute through Grant HL-16851, Dr. Shu Chien, Director.

REFERENCES

[1] Young, T., On the functions of the heart and arteries, Philosophical Transactions of the Royal Society of London, 99 (1809) 1-31.

[2] Poiseuille, J.L.M., Recherches experimentales sur le mouvement des liquides dans les tubes de tres petits diametres, Academie des Sciences, Compter Rendes, 11, (1840) 961-967, 1040-1048; 12, 112-125; 15, 1167-1186.

[3] Chien, S., Usami, S., Taylor, H.M., Lundberg, J.L. and Gregersen, M.I., Effects of hematocrit and plasma proteins on human blood rheology at low shear rates, Jrnl. of Applied Physiology, Vol. 21, No. 1 (1966) 81-87.

[4] Goldsmith, H.L. and Marlow, J., Flow behavior of erythrocytes, Jrnl. of Colloid and Interface Sci., Vol. 71, No. 2 (1979) 383-407.

[5] Tözeren, A., Skalak, R., Sung, K.L.P. and Chien, S. Viscoelastic behavior of erythrocyte membranes. (Submitted to Biophysical Journal, 1981).

[6] Skalak, R., Tözeren, A., Zarda, R.P. and Chien, S. Strain energy function of red blood cell membranes, Biophys. Jrnl., 13 (1973) 245-264.

[7] Evans, E.A., A new material concept for the red cell membrane. Biophys. Jrnl., 13 (1973) 926-940.

[8] Zarda, P.R., Chien, S. and Skalak, R., Interaction of viscous incompressible fluid with an elastic body, in: Belytschko, T. and Geers, T.L. (eds.), Computational Methods for Fluid-Structure Interaction Problems, ASME-AMD Vol. 26 (1977) 65-82.

[9] Evans, E.A. and Hochmuth, R., Membrane viscoelasticity, Biophys. Jrnl., 16 (1973) 1-11.

[10] Chien, S., Sung, P.K-L., Skalak, R., Usami, S., and Tözeren, A. Theoretical and experimental studies on viscoelastic properties of erythrocyte membrane, Biophys. Jrnl., 24 (1978) 463-482.

[11] Helfrick, W. and Dueling, H.J., Some theoretical shapes of red blood cells, Jrnl. Physique, 36, in: Proceedings of the Vth International Liquid Crystal Conference (1975) Colloque 1-327.

[12] Schmid-Schönbein, G.W., Sung, P.K-L., Tözeren, H., Chien, S. and Skalak, R. The passive mechanical properties of human leukocytes. Biophys. Jrnl. (1981, in press).

[13] Chien, S., Usami, S., Jan, K-M., and Skalak, R., Chapt. 2 in: Gabelnick, H.G. and Litt, M. (eds.), Rheology of Biological Systems (Charles C. Thomas, Springfield, Ill., 1973).

[14] Pedley, T.J., The Fluid Mechanics of Large Blood Vessels (Cambridge Univ. Press, N.Y., 1980).

[15] Skalak, R. and Chien, S., Capillary blood flow: history, experiments and theory, Biorheology, 18 (in press, 1981).

[16] Goldsmith, H.L., Mason, S.G., Bibl. Anat., 10 (Karger, Basle and N.Y., 1969) 1-8.

[17] Wells, R., Schmid-Schönbein, H., J. Appl. Physiol., 27 (1969) 213-217.

[18] Fåhraeus, R., Lindqvist, T., Am. J. Physiol., 96 (1931) 562-568.

[19] Goldsmith, H.L. and Mason, S.G., Jrnl. of Colloid Sci, 17 (1962) 448.

[20] Karnis, A., Mason, S.G., J. Colloid Interface Sci., 24 (1967) 164-169.

[21] Cox, R.G., Mason, S.G., Ann. Rev. Fluid Mech., 3 (1971) 291-316.

[22] Segre, S., and Silberberg, A., Fluid Mech. 14 (1962) 136.

[23] Fåhraeus, R. Physiol. Rev., 9 (1929) 241-274.

[24] Barbee, J.H. and Cokelet, G.R., Prediction of blood flow in tubes with diameter as small as 20 microns, Microvascular Res., 3, (1971) 17-21.

[25] Gauthier, F.J., Goldsmith, H.L. and Mason S.G., Biorheology, 9 (1972), 205.

[26] Karnis, Mason, and Goldsmith, Jrnl. Colloid Interface Sci., 22 (1966b) 531-533.

[27] Chien, S., King, R.G., Schuessler, G.B., Skalak, R., Tözeren, A., Usami, A., and Copley, A.L., AIChE Symp. Series No. 182, 74 (1978) 56-60.

[28] Thurston, G.B., Rheological parameters for the viscosity, viscoelasticity and thixotropy of blood, Biorheology, 16 (1979) 149-162.

[29] Brenner, H., Rheology of a dilute suspension of axisymmetrical Brownian particles, Int. Jrnl. Multiphase Flow, 1 (1974) 195-341.

[30] Batchelor, G.K., Transport properties of two-phase materials with random structure, Ann. Rev. Fluid Mech., 6 (1977) 227.

[31] Jeffrey, D.J. and Acrivos, A., The rheological properties of suspension of rigid particles, AIChE Jrnl. 22 (1976) 417-432.

[32] Herczynski, R. and Pienkowska, I., Toward a statistical theory of suspension, Ann. Rev. Flúid Mech., 12 (1980) 237-269.

[33] Batchelor, G.K., The stress system in a suspension of force-free particles, Jrnl. Fluid Mech., 41 (1970) 545.

[34] Lin, C.J., Lee, K.J. and Sather, N.J., Slow motion of two spheres in a shear field, Jrnl. Fluid Mechanics, 43 (1970) 35.

[35] Ryskin, G., Extensional viscosity of a dilute suspension of spherical particles at intermediate microscale Reynolds numbers, Jrnl. Fluid Mechanics, 99 (1980) 513-530.

[36] Batchelor, G.K., Slender-body theory for particles of arbitrary cross-section in Stokes flow, Jrnl. Fluid Mech., 44, part 3 (1970) 419-440.

[37] Brennen, C., A concentrated suspension model for the rheological properties of blood, The Canadian Jrnl. of Chem. Eng., 53 (Apr., 1975) 126-133.

[38] Barthes-Biesel, Motion of a spherical microcapsule freely suspended in linear shear flow, Jrnl. Fluid Mech.,100 (1980)831-855.

[39] Brunn, P., On the rheology of viscous drops surrounded by an elastic shell, Biorheology, 17 (1980) 419-430.

[40] Batchelor, G.K. and Green, J.T., The determination of the bulk stress in a suspension of spherical particles to order c^2, Jrnl. Fluid Mech., 56 (1972b) 401.

[41] Gadala-Maria, F. and Acrivos, A., Shear-induced structure in a concentrated suspension of solid spheres, Jrnl. of Rheology, 24 (1980) 799-814.

[42] Tözeren, A. and Skalak, R., Stress in a suspension near rigid boundaries, Jrnl. Fluid Mech., 82 (1977) 289-337.

[43] Dean, W.R. and O'Neill, M.E., A slow motion of viscous liquid caused by the rotation of a solid sphere, Mathematika, 10 (1963) 13-24.

[44] Brenner, H., Dynamics of neutrally buoyant particles in low Reynolds number flows, in: Metsroni, G.(ed.), Progress in Heat and Mass Transfer, vol. 6 (Pergamon Press, 1972) 509-574.

[45] Ganatos, P., Pfeffer, R., and Weinbaum, S., A numerical-solution technique for three-dimensional Stokes flow, with application to the motion of strongly interacting spheres in a plane, Jrnl. Fluid Mech., 84 (1978) 79-111.

[46] Ganatos, P., Pfeffer, R., and Weinbaum, S., A strong interaction theory for the creeping motion of asphere between plane parallel boundaries. Part 2: Parallel motion, Jrnl. Fluid Mech. 99 (1980) 755-783.

[47] Schmid-Schönbein, H., Wells, R., in: Hartet, H.H. and Copley, A. (Eds.), Theory and Clin. Hemorheology (Springer-Verlag, Berlin, 1971) 348.

[48] Fischer, T. and Schmid-Schönbein, H., Tank-tread motion of red cell membranes in viscometric flow: behavior of intracellular and extracellular markers (with film), Blood Cells, 3 (1977) 351-365.

[49] Keller, S.R. and Skalak, R., Motion of tank treading ellipsoidal particles in shear flows (Submitted to Jrnl. of Fluid Mech.).

[50] Secomb, T.W. and Skalak, R., Rheology of highly concentrated red blood cell suspensions, ASME Advances in Bioengineering, 1981.

[51] Goddard J.D., An elastohydrodynamic theory for the rheology of concentrated suspensions of deformable particles, J. Non-Newtonian Fluid Mech., 2 (1977) 169-189.

[52] Frankel, N.A. and Acrivos, A., On the viscosity of concentrated suspension of solid spheres, Chem. Engin, Sci., 22 (1967) 847-853.

[53] Kline, K.A., Allen, S.J., and De Silva, C.N., A continuum approach to blood flow, Biorheology, 5 (1968) 111-118.

[54] Valanis, K.C. and Sun, C.T., Poiseuille flow of a fluid with couple stress with applications to blood flow, Biorheology, 6, (1969) 85-97.

[55] Ariman, T., On the analysis of blood flow, J. Biomech, 4 (1971) 185-192.

[56] Cowin, S.C., On the polar fluid as a model for blood flow in tubes, Biorheology, 9 (1972) 23-25.

[57] Popel, A.S., Regirer, S.A., and Usick, P.I., A continuum model of blood flow, Biorheology, 11 (1974) 427-437.

[58] Kang, C.K. and Eringen, A.C., The effect of microstructure on the rheological properties of blood, Bulletin of Math. Bio., 38 (1976) 38.

[59] Chaturani, P., and Upadhya, V.S., A two fluid model for blood flow through small diameter tubes, Biorheology, 16 (1979) 109-118.

[60] Chaturani, P., Upadhya, V.S., On micropolar fluid model for blood flow through narrow tubes, Biorheology, 16 (1979) 419-428.

[61] Eringen, A.C., On bridging the gap between macroscopic and microscopic physics. Presented at the 10th Annual Meeting, Soc. Engng. Sci., Raleigh, N.C., 1975.

[62] Ariman, T., Turk, N.A. and Sylvester, N.D., Microcontinuum fluid mechanics - a Review, Int. J. Eng. Sci, 11 (1973) 905-930.

[63] Cowin, S.C., The theory of polar fluids, reprinted from: Advances in Applied Mecahnics, vo. 14 (Academic Press, Inc., N.Y., 1974) 279-347.

[64] Tözeren, A. and Skalak, R., Micropolar fluids as models for suspensions of rigid spheres, Int. Jrnl. Engng. Sci., 15 (1977) 511-523.

[65] Atkin, R.J., On boundaryconditions of polar materials, Jrnl. of Applied Mathematics and Physics, 28 (1977).

[66] Brunn, P., The velocity slip of polar fluids, Rheol. Acta, 14, (1975) 1039-1054.

[67] Brunn, P., The general solution to the equations of creeping motion of a micropolar fluid and its applications, Int. Jrnl. of Engin. Sci. (in press).

[68] Tözeren, A., Couple stress in a suspension near rigid boundaries (submitted for publication).

[69] Goodman, M.A. and Cowin S.C., A continuum theory for granular materials, Arch. Rational Mech. Anal., 44 (1971) 249-266.

[70] Quemada, D., Rheology of concentrated disperse systems and minimum energy principle. I. Viscosity concentration relationship, Rheol. Acta, 16 (1977) 82-94.

[71] Quemada, D, Rheology of concentrated disperse systems. II. A model for non-Newtonian shear viscosity in steady flows, Rheol. Acta, 17 (1978) 632-642.

[72] Quemada, D., Rheology of concentrated disperse systems. III. General features of the proposed non-Newtonian model. Comparison with experimental data, Rheol. Acta, 17 (1978) 643-653.

Part 5
MICROSTRUCTURE, FRACTURE, FATIGUE

Continuum Models of Discrete Systems 4
eds. O. Brulin and R.K.T. Hsieh

STABILITY CONDITIONS IN CONTINUUM MODELS OF DISCRETE SYSTEMS

Olof Brulin and Stig Hjalmars

Department of Mechanics
The Royal Institute of Technology
Stockholm, Sweden

Stability in conventional continuum theory requires that the free energy density is positive definite. In the linear theory this requirement gives the well known bounds of e. g. the Lamé constants. The same arguments are generally used in all cases of continuum models of discrete systems. Such a model is however valid only down to a typical length L, and volume elements smaller than L^3 have no physical meaning in the continuum model. Stability requires thus a positive energy content in volume elements of the order of magnitude L^3 or larger only. The implications for the bounds of the material constants will be discussed.

The crucial problem in producing continuum models of discrete systems is the dimension at which the model is satisfactorily valid. If the dimensions of the system to be described, statically or dynamically, are much larger than the dimensions, which characterize the internal structure of the discrete system, a satisfactory model of the system can be given by the well known, classical continuum theory. The specific internal properties of the discrete system enforce modifications on the continuum model essentially only if the dimensions of the system to be described approach the internal dimensions of the system. Simultaneously the problem arises if the continuum description of the system is then meaningful. Roughly speaking, for obtaining non-classical continuum effect the relative change $lf_{,k}/f$ of a continuum variable f for a typical internal length l of the discrete system should be smaller, but not too much smaller than 1.

Mathematically a continuum theory is supposed to be valid for all dimensions, and mathematical limit procedures are assumed to be permitted. Evidently, problems of validity of such procedures arise particularly in the region described above. It is conceivable that some aspects of the continuum model as e.g. a description of the motion by differential equations may give a satisfactory description of the corresponding aspects of the discrete system, whereas in other aspects the limit procedure is not allowed.

The conditions of stability in continuum models of discrete systems seem to be a good example of this type of problem. As well known the conventional method of obtaining the conditions of stability in linear elasticity is to consider the consequences of a supposedly positive definite potential energy density. This means of course that the limit passage of a positive energy content in a finite volume element, divided by the volume, to a positive energy density, is allowable i.e. physically meaningful. The authors will discuss this problem in a typical continuum model where the internal properties of the discrete system are important.

Such a theory is the linear grade-consistent micropolar theory [1], of which the conventional micropolar theory is a special case. In [2] the magnitude of the polar effects are estimated from a simple micromodel. The continuum equations may be obtained from a potential energy density ρU where

$$\begin{aligned}\bar{U} &= \frac{\bar{\mu}+\bar{\alpha}}{2}\,\bar{u}_{k,\bar{l}}\bar{u}_{k,\bar{l}} + \frac{\bar{\mu}-\bar{\alpha}+\bar{\lambda}}{2}\,\bar{u}_{k,\bar{k}}\,\bar{u}_{l,\bar{l}} + \\ &+ 2\,\bar{\alpha}\,\varepsilon_{klm}\,\bar{u}_{k,\bar{l}}\,\varphi_m + 2\,\bar{\alpha}\,\varphi_k\varphi_k + \\ &+ \frac{\bar{\gamma}+\bar{\varepsilon}}{2}\,\varphi_{k,\bar{l}}\,\varphi_{k,\bar{l}} + \frac{\bar{\gamma}-\bar{\varepsilon}+\bar{\beta}}{2}\,\varphi_{k,\bar{k}}\,\varphi_{l,\bar{l}} + \\ &+ \bar{\zeta}\,\varepsilon_{klm}\,\bar{u}_{k,\bar{l}\bar{n}}\,\varphi_{m,\bar{n}} + \\ &+ \frac{\bar{\xi}}{2}\,\bar{u}_{k,\bar{l}\bar{l}}\,\bar{u}_{k,\bar{m}\bar{m}} + \frac{\bar{\eta}}{2}\,u_{k,\bar{k}\bar{m}}\,u_{l,\bar{l}\bar{m}}\,. \end{aligned} \tag{1}$$

The bars denote dimensionless quantities, all lengths divided by the micro length l. We recognize $\mu = \bar{\mu}\,\mu$ ($\bar{\mu} = 1$) and $\lambda = \bar{\lambda}\,\mu$ as the ordinary Lamé constants. $\alpha = \bar{\alpha}\,\mu$ and $(\beta, \gamma, \varepsilon) = \mu l^2\,(\bar{\beta}, \bar{\gamma}, \bar{\varepsilon})$ are the conventional micropolar constants and $(\xi, \eta, \zeta) = \mu l^2\,(\bar{\xi}, \bar{\eta}, \bar{\zeta})$ are the constants, introduced through the hypothesis of grade-consistency.

From the conventional assumption of $\int_V \rho U dV$ as a positive definite quantity, follows by the usual limit procedure, $V \to 0$, the well known stability conditions

$$\bar{\mu} > 0,\ 3\,\bar{\lambda} + 2\,\bar{\mu} > 0,\ \bar{\alpha} > 0, \tag{2}$$

$$\bar{\gamma} > 0,\ 3\,\bar{\beta} + 2\,\bar{\gamma} > 0,\ \bar{\varepsilon} > 0 \tag{3}$$

and for the new constants

$$\bar{\xi} > 0,\ \bar{\eta} > 0,\ |\,\bar{\zeta}\,| < \sqrt{(\bar{\gamma}+\bar{\varepsilon})\ \bar{\xi}}. \tag{4}$$

However, as discussed above the continuum model can be valid only down to some macro length L much larger (about 10 times say) than some characteristic length l of the discrete system. Volume elements smaller than L^3 have no physical meaning in the continuum models. In consequence stability requires a positive energy content in volume elements only of the order of magnitude L^3 or larger.

At first from the fact that it is always possible to regard systems with a characteristic macro-length L so large that l/L is negligibly small, follows immediately that the inequalities (2) for the macro-constants μ, λ and α must be valid. Further, as l/L is a quantity that cannot become large it is reasonable to expect that relations characterizing the material micro-constants also have to contain the macro-constants. The inequalities (3) and (4) are probably too strong in a physically realistic continuum model.

Let us thus regard a volume element of magnitude L^3 with an isolated energy content. Then there is no work flow across the boundary and, consequently the first second and third order derivatives of the macro-displacement field $\bar{u}_k$ must be zero on the boundary as well as the micro-rotation field φ and its first and second order derivatives. Further the total energy inside this volume element must be positive. As the form of the element is not essential we shall make the simplest possible assumption, i.e.discuss a cube with edge L. We make a Fourier expansion of the fields $\bar{u}_k$ and φ_k with the boundary conditions

$$\bar{u}_{k,\bar{l}} = \bar{u}_{k,\bar{l}\bar{m}} = \bar{u}_{k,\bar{l}\bar{m}\bar{n}} = 0\,,$$

$$\varphi_k = \varphi_{k,\bar{l}} = \varphi_{k,\bar{l}\bar{m}} = 0$$
$$\text{on } \bar{x}_s = 0,\ \bar{x}_s = L/l\ . \tag{5}$$

The lowest Fourier components fulfilling these conditions can be given as

$$\begin{aligned} f_2\ (x_s) &= 2 \sin g_s - \sin 2\, g_s\ , \\ f_4\ (x_s) &= 5 \sin g_s - 4 \sin 2\, g_s + \sin 3\, g_s\ , \\ g_s &= \frac{2\pi l}{L}\ \bar{x}_s\ , \end{aligned} \tag{6}$$

which vanish and have two or four vanishing derivatives on $\bar{x}_s = 0,\ L/l$.
From the functions (6) we can obtain different types of functions $\bar{u}_k$, φ_k e.g.

$$\begin{aligned} \bar{u}_k &= A_k F\ (\bar{x}_1, \bar{x}_2, \bar{x}_3), \\ \bar{u}_k &= B\ F_{,\bar{k}}\ (\bar{x}_1, \bar{x}_2, \bar{x}_3) + \varepsilon_{klm}\ C_m\ F_{,\bar{l}}\ (\bar{x}_1\ , \bar{x}_2, \bar{x}_3)\ , \\ \varphi_k &= D_k F\ (\bar{x}_1, \bar{x}_2, \bar{x}_3)\ , \\ \varphi_k &= E\ F_{,\bar{k}}\ (\bar{x}_1, \bar{x}_2, \bar{x}_3) + \varepsilon_{klm}\ G_m\ F_{,\bar{l}}\ (\bar{x}_1, \bar{x}_2, \bar{x}_3)\ , \end{aligned} \tag{7}$$

where

$$F(\bar{x}_1, \bar{x}_2, \bar{x}_3) = f\ (\bar{x}_1)\ f(\bar{x}_2)\ f\ (\bar{x}_3) \tag{8}$$

and

$$A_k,\ B,\ C_k,\ D_k,\ E,\ G_k,\ k = 1,\ 2,\ 3 \tag{9}$$

are constants.

As functions $f\ (\bar{x}_s)$ we choose the function (6) in such a way that the conditions (5) are fulfilled. If we choose the constants in (9) suitably and introduce (7) into the energy (1) and integrate over the cube L^3 we obtain new inequalities, which the material constants have to fulfil. As the numerical coefficients in these inequalities are rather complicated we give the following reasonable approximations of the results:

$$\begin{aligned} &\bar{\gamma} + \bar{\varepsilon} > - \left(\frac{L}{2\pi l}\right)^2\ \frac{2}{3}\ \bar{\alpha}, \\ &2\ \bar{\gamma} + \bar{\varepsilon} + \frac{\bar{\beta}}{2} > - \frac{5}{4}\left(\frac{L}{2\pi l}\right)^2\ \bar{\alpha}, \\ &\bar{\gamma} + \frac{\bar{\beta}}{2} > - \left(\frac{L}{2\pi l}\right)^2\ \frac{2\ \bar{\alpha}}{3}, \\ &\bar{\xi} > - \left(\frac{L}{2\pi l}\right)^2\ \frac{\bar{\mu} + \bar{\alpha}}{12}, \\ &\bar{\xi} - \frac{\bar{\eta}}{4} > - \left(\frac{L}{2\pi l}\right)^2\ \frac{1}{12}\left(\bar{\mu} + \frac{\bar{\alpha}}{2} + \frac{\bar{\lambda}}{4}\right), \\ &\left(\bar{\gamma} + \bar{\varepsilon}\right)\bar{\xi} - \ \bar{\zeta}^2 > 0. \end{aligned} \tag{10}$$

The inequalities (2) are still valid.

In [2] table 1 the material constitutive constants are estimated from a structural micromodel. It is seen that these estimates satisfy the stability conditions (2) for the macro constants but not the too strong conditions (3), (4) for the micro constants. However, with the smallest admissible value of $L \approx 2\pi l$ the weaker stability conditions (10) are seen to be fulfilled by the estimates. It may be noted that the satisfaction of the stability conditions by the estimates ceases to be valid just about the continuum limit.

Inserting the smallest value $L = 2\pi l$ of the continuum limit, the approximate stability conditions on the dimensionless material constants (2), (10) will be given as

$$\bar{\mu} > 0,\ 3\bar{\lambda} + 2\bar{\mu} > 0,\ \bar{\alpha} > 0 \tag{11}$$

$$\bar{\gamma} + \bar{\varepsilon} + \frac{2}{3}\bar{\alpha} > 0,\quad 4\bar{\gamma} + 2\bar{\varepsilon} + \bar{\beta} + \frac{5}{2}\bar{\alpha} > 0,$$

$$2\bar{\gamma} + \bar{\beta} + \frac{4}{3}\bar{\alpha} > 0, \tag{12}$$

$$\bar{\xi} + \frac{1}{12}(\bar{\mu} + \bar{\alpha}) > 0,\quad 4\bar{\xi} + \bar{\eta} + \frac{1}{12}(4\bar{\mu} + 2\bar{\alpha} + \bar{\lambda}) > 0,$$

$$(\bar{\gamma} + \bar{\varepsilon})\,\bar{\xi} - \bar{\zeta}^2 > 0. \tag{13}$$

REFERENCES

[1] Brulin, O. and Hjalmars, S., *Linear grade-consistent micropolar theory*, Int.J. Engng. Sci. (1981) in press

[2] Fischer-Hjalmars, I. and Hjalmars. S,. *The magnitude of polar effects estimated from a simple structural micromodel*, Proceedings of Continuum models of Discrete Systems (CMDS 4), (North-Holland, Amsterdam, 1981). (This volume)

Continuum Models of Discrete Systems 4
eds. O. Brulin and R.K.T. Hsieh

THE MAGNITUDE OF POLAR EFFECTS ESTIMATED FROM A SIMPLE STRUCTURAL MICROMODEL

Inga Fischer-Hjalmars
Department of Theoretical Physics
University of Stockholm
Stockholm
Sweden

Stig Hjalmars
Department of Mechanics
Royal Institute of Technology
Stockholm
Sweden

An estimate is made of the order of magnitude of the micropolar constitutive constants by means of a cubic, orthotropic, four-fold rotation-symmetric micromodel. By passing to the continuum limit by means of a systematic expansion with respect to the dimensionless wave number, the constitutive constants are then determined as functions of the micro-interaction constants. The grade-consistent micropolar constants, foreseen by Brulin and Hjalmars are seen to enter on the same level as the conventional ones. The contribution of the different constants to the dispersion near the continuum limit is also determined.

1. INTRODUCTION AND SUMMARY

The constitutive constants of the isotropic,micropolar, elastic medium are given by the field equations

$$\mu^{-1} l^2 \rho \ddot{\bar{u}}_k = (\bar{\mu} + \bar{\alpha})\, \bar{u}_{k,\bar{l}\bar{l}} + (\bar{\mu} + \bar{\lambda} - \bar{\alpha})\, \bar{u}_{l,\bar{l}\bar{k}} + 2\bar{\alpha}\varepsilon_{klm}\varphi_{m,\bar{l}} -$$

$$- \bar{\xi}\bar{u}_{k,\bar{l}\bar{l}\bar{m}\bar{m}} - \bar{\eta}\bar{u}_{l,\bar{l}\bar{k}\bar{m}\bar{m}} - \bar{\zeta}\varepsilon_{klm}\varphi_{m,\bar{l}\bar{r}\bar{r}} , \qquad (1.1)$$

$$\mu^{-1} l^2 \rho \bar{j} \ddot{\varphi}_k = 2\bar{\alpha}(\varepsilon_{klm}\bar{u}_{m,\bar{l}} - 2\varphi_k) +$$

$$+ (\bar{\varepsilon} + \bar{\gamma})\,\varphi_{k,\bar{l}\bar{l}} + (\bar{\beta} + \bar{\gamma} - \bar{\varepsilon})\,\varphi_{l,\bar{l}\bar{k}} - \bar{\zeta}\varepsilon_{klm}\bar{u}_{m,\bar{l}\bar{r}\bar{r}} . \qquad (1.2)$$

Here the bars denote dimensionless quantities, all lengths being divided by the micro length l. We recognize easily $\mu = \bar{\mu}\mu$ ($\bar{\mu} = 1$) and $\lambda = \bar{\lambda}\mu$ as the ordinary Lamé constants, $\alpha = \bar{\alpha}\mu$ and $(\beta,\gamma,\varepsilon) = \mu l^2(\bar{\beta},\bar{\gamma},\bar{\varepsilon})$ as the conventional micropolar constants, and $(\xi,\eta,\zeta) = \mu l^2(\bar{\xi},\bar{\eta},\bar{\zeta})$ as the constants of the grade-consistent terms,argued by Brulin and Hjalmars [1,2] to be of the same order of magnitude as the conventional ones.

The aim of the present work is to estimate the constants of (1.1) and (1.2) and their individual contributions to the wave dispersion by means of a quantitatively fairly realistic micromodel of equal spherical micropoles in a cubic lattice with the least possible anisotropy.

The difference equations of motion are converted to continuum differential equations, (2.23), (2.24), by means of a systematic expansion with respect to the dimensionless wave number. By comparing the equations, adapted to plane waves along the axes, estimates of the constants of (1.1) and (1.2) as functions of the micro force constants are read off, (3.10-17). The contributions to the dispersion in ω^2 as given in [2] are calculated (3.18-21). Numerical results for a bar model are finally given in Tables 1 and 2. The different trends in the constants and contributions for vanishing size of the micropoles are pointed out.

2. THE DIFFERENTIAL FIELD EQUATIONS

We assume the same configuration as Tauchert [3], i.e. a periodic cubic lattice of equal spherical, rigid particles with mass m , moment of inertia J , and central distance l along the cubic edges. We also label the particles by integer index triples $(l \pm p\ , m \pm q\ , n \pm r)$, which denotes a particle with the coordinates

$$x_1^{l\pm p} = x_1^{l} \pm pl\ , \quad x_2^{m\pm q} = x_2^{m} \pm ql\ , \quad x_3^{n\pm r} = x_3^{n} \pm rl\ , \tag{2.1}$$

as in Fig. 2 of [3].

As in [3] we assume bonds only between nearest and next nearest neighbors, i.e. with lengths l along the cubic edges and $l\sqrt{2}$ along the diagonals of the square surfaces of the cube. The bonds parallel to the k-axis are denoted kk , and the bonds along the diagonals in the kl coordinate plane are denoted kl in the 1:st and 3:rd, and lk in the 2:nd and 4:th quadrant, as shown in Fig. 3 of [3].

Now, consider a bond of length l_{ik} between two particles, denoted (i) and (k) , and assume that this bond acts with the force $t_l(ik)$ and the moment $m_l^{(ik)}$ on particle (i) and $t_l(ki)$ and $m_l(ki)$ on particle (k) . Introducing a local coordinate system with x_3 as (i) → (k) and denoting the particle displacement and rotation vector components by $u_l(i)$, $\varphi_l(i)$, and $u_l(k)$, $\varphi_l(k)$, we have for bonds, assumed to be symmetric between (i) and (k) :

$$t_3^{(ik)} = -a\left(u_3^i - u_3^k\right)\ , \tag{2.2}$$

$$t_1^{(ik)} = -b\left(u_1^i - u_1^k\right) - \frac{1}{2}\, l_{ik}\, b\left(\varphi_2^i + \varphi_2^k\right)\ , \tag{2.3}$$

$$t_2^{(ik)} = -b\left(u_2^i - u_2^k\right) + \frac{1}{2}\, l_{ik}\, b\left(\varphi_1^i + \varphi_1^k\right)\ , \tag{2.4}$$

$$t_3^{(ki)} = a\left(u_3^i - u_3^k\right)\ , \tag{2.5}$$

$$t_1^{(ki)} = b\left(u_1^i - u_1^k\right) + \frac{1}{2}\, l_{ik}\, b\left(\varphi_2^i + \varphi_2^k\right) \tag{2.6}$$

$$t_2^{(ki)} = b\left(u_2^i - u_2^k\right) - \frac{1}{2}\, l_{ik}\, b\left(\varphi_1^i + \varphi_1^k\right)\ , \tag{2.7}$$

$$m_3^{(ik)} = -c\left(\varphi_3^i - \varphi_3^k\right)\ , \tag{2.8}$$

$$m_1^{(ik)} = \frac{1}{2}\, l_{ik}\, b\left(u_2^i - u_2^k\right) - \frac{d}{2}\left(h\varphi_1^i + \varphi_1^k\right)\ , \tag{2.9}$$

$$m_2^{(ik)} = -\frac{1}{2}\, l_{ik}\, b\left(u_1^i - u_1^k\right) - \frac{d}{2}\left(h\varphi_2^i + \varphi_2^k\right)\ , \tag{2.10}$$

$$m_3^{(ki)} = c\left(\varphi_3^i - \varphi_3^k\right) , \tag{2.11}$$

$$m_1^{(ki)} = \frac{1}{2} l_{ik}\, b\left(u_2^i - u_2^k\right) - \frac{d}{2}\left(\varphi_1^i + h\varphi_1^k\right) , \tag{2.12}$$

$$m_2^{(ki)} = -\frac{1}{2} l_{ik}\, b\left(u_1^i - u_1^k\right) - \frac{d}{2}\left(\varphi_2^i + h\varphi_2^k\right) , \tag{2.13}$$

where

$$d = l_{ik}^2 \frac{1}{1+h} b , \tag{2.14}$$

and where the constants a, b, c, d and h are parameters, which characterize the bond, and which also should be thought of as labelled by the index ik. The expressions (2.2) - (2.14) are the same as those, given by Tauchert in the equation (A1) of the Appendix in [3], only that they are generalized by the introduction of the coefficients h, which Tauchert puts equal to 2, according to his assumption that the bonds are given by elastic cylindrical bars. The only assumption made by us in (2.2) - (2.14) is that the bonds are symmetric between (i) and (k), and in this case we have four independent constants a, b, c and h, characterizing the bond.

Using the assumptions (2.2) - (2.14) to construct the forces and moments on particle (l, m, n) from the nearest and next nearest neighbors, we obtain as $m\ddot{u}$- and $J\ddot{\varphi}$-equations of motion for this particle the same difference equations as given by Tauchert on pages 329-332 of [3], only that we insert the generalization $2 \rightarrow h$ on proper places, and that we do not generally assume the uniform bar relations (A2) of [3] but only the relations

$$d_{ij} = l^2\, 2\,(1 + h_{ij})^{-1}\, b_{ij} , \quad (i \neq j) , \tag{2.15}$$

$$d_{ii} = l^2\,(1 + h_{ii})^{-1}\, b_{ii} , \tag{2.16}$$

derived from (2.14) by means of $l_{ii} = l$ and $l_{ij} = l\sqrt{2}$, $(i \neq j)$.

Up to this point our treatment coincides substantially with that of Tauchert. However when turning to the conversion of the difference equations to differential continuum equations, it is necessary to make essential amendments. In order to perform the conversion we have to expand the translations and rotations in the neighboring lattice points in Taylor series about the origin as

$$u^{l\pm1,m,n} = \left(1 \pm l\partial_1 + \frac{l^2}{2!}\partial_1^2 \pm \ldots\right) u^{l,m,n} , \text{ etc.,} \tag{2.17}$$

$$u^{l+1,m\pm1,n} = \left(1 + l\partial_1 \pm l\partial_2 + \frac{l^2}{2!}\partial_1^2 \pm l^2\partial_1\partial_2 + \frac{l^2}{2!}\partial_2^2\right) u^{l,m,n} , \text{ etc., (2.18)} \tag{2.18}$$

which of course means expanding each Fourier-component in a power series after the dimensionless wave number $\bar{q}$:

$$\bar{q} = ql \quad \text{with} \quad q = 2\pi/\lambda , \tag{2.19}$$

$\bar{q}$ being < 1 in the proper continuum.

Now Tauchert performs the expansion up to $\bar{q}^2$ in both the u- and φ- equations, thus obtaining derivative terms of the kind entering in conventional micropolar theories, i.e. $u_{k,lm}$- and $\varphi_{k,l}$-terms in the $\ddot{u}_k$-equation and $u_{k,l}$- , φ_k- and $\varphi_{k,lm}$-terms in the $\ddot{\varphi}_k$- equation. However, by arguments drawn from general dimensional considerations as well as from the general close coupling between φ_k and $u_{k,l}$, Brulin and Hjalmars [1 , 2] have argued that higher derivatives should consistently appear in the field equations, as shown in (1.1). Moreover, they have also shown [2], that the coefficients ξ , η and ζ for the higher derivatives enter on the same level of approximation in the dispersion relations as the conventional polar constitutive constants ε , γ , β and j . They must thus also be estimated at the same time as the conventional ones. In order to obtain all terms, which are of the same order of magnitude, we have thus to perform the expansions (2.17), (2.18) up to 4:th derivatives, i.e. up to $\bar{q}^4$, in the $\ddot{u}_k$ - equation, and up to 3:rd derivatives, i.e. up to $\bar{q}^3$, in the $\ddot{\varphi}_k$-equation. In doing so, we have also, in order to get the orders of magnitude straight, written all equations in dimensionless form, introducing the following dimensionless quantities, denoted by a bar:

$$\bar{x}_k = l^{-1} x_k \,, \quad \partial f/\partial \bar{x}_k = f_{,\bar{k}} = l\,\partial f/\partial x_k = l f_{,k} \,, \quad \bar{j} = j\, l^{-2} \,, \tag{2.20}$$

$$\bar{a}_{ik} = \mu^{-1} a_{ik} \,, \quad \bar{b}_{ik} = \mu^{-1} b_{ik} \,, \tag{2.21}$$

$$\bar{c}_{ik} = \mu^{-1} l^{-2} c_{ik} \,, \quad \bar{d}_{ik} = \mu^{-1} l^{-2} d_{ik} \,. \tag{2.22}$$

Under these assumption we obtain the following field equations for $\ddot{\bar{u}}_k$ and $\ddot{\varphi}_k$:

$$\mu^{-1} l^2 \rho \ddot{\bar{u}}_1 = \bar{a}_{11}\left(\bar{u}_{1,\bar{1}\bar{1}} + \frac{1}{12}\bar{u}_{1,\bar{1}\bar{1}\bar{1}\bar{1}}\right) +$$

$$+ \bar{b}_{22}\left(\bar{u}_{1,\bar{2}\bar{2}} + \frac{1}{12}\bar{u}_{1,\bar{2}\bar{2}\bar{2}\bar{2}}\right) + \bar{b}_{33}\left(\bar{u}_{1,\bar{3}\bar{3}} + \frac{1}{12}\bar{u}_{1,\bar{3}\bar{3}\bar{3}\bar{3}}\right) +$$

$$+ \frac{1}{2}\left(\bar{a}_{12} + \bar{a}_{21} + \bar{b}_{12} + \bar{b}_{21}\right)\left[\bar{u}_{1,\bar{1}\bar{1}} + \bar{u}_{1,\bar{2}\bar{2}} + \frac{1}{12}\left(\bar{u}_{1,\bar{1}\bar{1}\bar{1}\bar{1}} + \bar{u}_{1,\bar{2}\bar{2}\bar{2}\bar{2}}\right) + \frac{1}{2}\bar{u}_{1,\bar{1}\bar{1}\bar{2}\bar{2}}\right] +$$

$$+ \frac{1}{2}\left(\bar{a}_{12} - \bar{a}_{21} + \bar{b}_{12} - \bar{b}_{21}\right)\left[2\bar{u}_{1,\bar{1}\bar{2}} + \frac{1}{3}\left(\bar{u}_{1,\bar{1}\bar{2}\bar{2}\bar{2}} + \bar{u}_{1,\bar{2}\bar{1}\bar{1}\bar{1}}\right)\right] +$$

$$+ \frac{1}{2}\left(\bar{a}_{31} + \bar{a}_{13} + \bar{b}_{31} + \bar{b}_{13}\right)\left[\bar{u}_{1,\bar{1}\bar{1}} + \bar{u}_{1,\bar{3}\bar{3}} + \frac{1}{12}\left(\bar{u}_{1,\bar{1}\bar{1}\bar{1}\bar{1}} + \bar{u}_{1,\bar{3}\bar{3}\bar{3}\bar{3}}\right) + \frac{1}{2}\bar{u}_{1,\bar{1}\bar{1}\bar{3}\bar{3}}\right] +$$

$$+ \frac{1}{2}\left(\bar{a}_{31} - \bar{a}_{13} + \bar{b}_{31} - \bar{b}_{13}\right)\left[2\bar{u}_{1,\bar{1}\bar{3}} + \frac{1}{3}\left(\bar{u}_{1,\bar{1}\bar{3}\bar{3}\bar{3}} + \bar{u}_{1,\bar{3}\bar{1}\bar{1}\bar{1}}\right)\right] +$$

$$+ \left(\bar{b}_{23} + \bar{b}_{32}\right)\left[\bar{u}_{1,\bar{2}\bar{2}} + \bar{u}_{1,\bar{3}\bar{3}} + \frac{1}{12}\left(\bar{u}_{1,\bar{2}\bar{2}\bar{2}\bar{2}} + \bar{u}_{1,\bar{3}\bar{3}\bar{3}\bar{3}}\right) + \frac{1}{2}\bar{u}_{1,\bar{2}\bar{2}\bar{3}\bar{3}}\right] +$$

$$
\begin{aligned}
&+ \left(\bar{b}_{23} - \bar{b}_{32}\right)\left[2\bar{u}_{1,\bar{2}\bar{3}} + \frac{1}{3}\left(\bar{u}_{1,\bar{2}\bar{3}\bar{3}\bar{3}} + \bar{u}_{1,\bar{3}\bar{2}\bar{2}\bar{2}}\right)\right] + \\
&+ \frac{1}{2}\left(\bar{a}_{12} - \bar{a}_{21} - \bar{b}_{12} + \bar{b}_{21}\right)\left[\bar{u}_{2,\bar{1}\bar{1}} + \bar{u}_{2,\bar{2}\bar{2}} + \frac{1}{12}\left(\bar{u}_{2,\bar{1}\bar{1}\bar{1}\bar{1}} + \bar{u}_{2,\bar{2}\bar{2}\bar{2}\bar{2}}\right) + \frac{1}{2}\bar{u}_{2,\bar{1}\bar{1}\bar{2}\bar{2}}\right] + \\
&+ \frac{1}{2}\left(\bar{a}_{12} + \bar{a}_{21} - \bar{b}_{12} - \bar{b}_{21}\right)\left[2\bar{u}_{2,\bar{1}\bar{2}} + \frac{1}{3}\left(\bar{u}_{2,\bar{1}\bar{2}\bar{2}\bar{2}} + \bar{u}_{2,\bar{2}\bar{1}\bar{1}\bar{1}}\right)\right] + \\
&+ \frac{1}{2}\left(\bar{a}_{31} - \bar{a}_{13} - \bar{b}_{31} + \bar{b}_{13}\right)\left[\bar{u}_{3,\bar{1}\bar{1}} + \bar{u}_{3,\bar{3}\bar{3}} + \frac{1}{12}\left(\bar{u}_{3,\bar{1}\bar{1}\bar{1}\bar{1}} + \bar{u}_{3,\bar{3}\bar{3}\bar{3}\bar{3}}\right) + \frac{1}{2}\bar{u}_{3,\bar{1}\bar{1}\bar{3}\bar{3}}\right] + \\
&+ \frac{1}{2}\left(\bar{a}_{31} + \bar{a}_{13} - \bar{b}_{31} - \bar{b}_{13}\right)\left[2\bar{u}_{3,\bar{1}\bar{3}} + \frac{1}{3}\left(\bar{u}_{3,\bar{1}\bar{3}\bar{3}\bar{3}} + \bar{u}_{3,\bar{3}\bar{1}\bar{1}\bar{1}}\right)\right] - \\
&- \left(\bar{b}_{33} + \bar{b}_{13} + \bar{b}_{31} + \bar{b}_{23} + \bar{b}_{32}\right)\left(\varphi_{2,\bar{3}} + \frac{1}{6}\varphi_{2,\bar{3}\bar{3}\bar{3}}\right) + \\
&+ \left(\bar{b}_{22} + \bar{b}_{12} + \bar{b}_{21} + \bar{b}_{23} + \bar{b}_{32}\right)\left(\varphi_{3,\bar{2}} + \frac{1}{6}\varphi_{3,\bar{2}\bar{2}\bar{2}}\right) + \\
&+ \left(\bar{b}_{32} - \bar{b}_{23}\right)\left[\varphi_{2,\bar{2}} - \varphi_{3,\bar{3}} + \frac{1}{6}\left(\varphi_{2,\bar{2}\bar{2}\bar{2}} - \varphi_{3,\bar{3}\bar{3}\bar{3}}\right) + \frac{1}{2}\left(\varphi_{2,\bar{2}\bar{3}\bar{3}} - \varphi_{3,\bar{3}\bar{2}\bar{2}}\right)\right] + \\
&+ \left(\bar{b}_{13} - \bar{b}_{31}\right)\left(\varphi_{2,\bar{1}} + \frac{1}{6}\varphi_{2,\bar{1}\bar{1}\bar{1}} + \frac{1}{2}\varphi_{2,\bar{1}\bar{3}\bar{3}}\right) + \\
&+ \left(\bar{b}_{12} - \bar{b}_{21}\right)\left(\varphi_{3,\bar{1}} + \frac{1}{6}\varphi_{3,\bar{1}\bar{1}\bar{1}} + \frac{1}{2}\varphi_{3,\bar{1}\bar{2}\bar{2}}\right) - \\
&- \frac{1}{2}\left(\bar{b}_{23} + \bar{b}_{32}\right)\left(\varphi_{2,\bar{3}\bar{2}\bar{2}} - \varphi_{3,\bar{2}\bar{3}\bar{3}}\right) + \\
&+ \frac{1}{2}\left(\bar{b}_{12} + \bar{b}_{21}\right)\varphi_{3,\bar{2}\bar{1}\bar{1}} - \frac{1}{2}\left(\bar{b}_{13} + \bar{b}_{31}\right)\varphi_{2,\bar{3}\bar{1}\bar{1}} \, , \qquad (2.23)
\end{aligned}
$$

$$\mu^{-1} l^2 \rho \bar{j} \ddot{\varphi}_1 =$$

$$= -\left(\bar{b}_{12} + \bar{b}_{21} + \bar{b}_{13} + \bar{b}_{31} + 2\bar{b}_{23} + 2\bar{b}_{32} + \bar{b}_{22} + \bar{b}_{33}\right) \varphi_1 +$$

$$+ \left[\bar{c}_{11} + \frac{1}{2}\left(\bar{c}_{12} + \bar{c}_{21} + \bar{c}_{13} + \bar{c}_{31}\right) - \frac{1}{4}\left(\bar{d}_{12} + \bar{d}_{21} + \bar{d}_{13} + \bar{d}_{31}\right)\right] \varphi_{1,\bar{1}\bar{1}} +$$

$$+ \frac{1}{2}\left[\bar{c}_{12} + \bar{c}_{21} - \frac{1}{2}\left(\bar{d}_{12} + \bar{d}_{21}\right) - \bar{d}_{23} - \bar{d}_{32} - \bar{d}_{22}\right] \varphi_{1,\bar{2}\bar{2}} +$$

$$+ \frac{1}{2}\left[\bar{c}_{13} + \bar{c}_{31} - \frac{1}{2}\left(\bar{d}_{13} + \bar{d}_{31}\right) - \bar{d}_{23} - \bar{d}_{32} - \bar{d}_{33}\right] \varphi_{1,\bar{3}\bar{3}} +$$

$$+ \left[\bar{c}_{12} - \bar{c}_{21} - \frac{1}{2}\left(\bar{d}_{12} - \bar{d}_{21}\right)\right] \varphi_{1,\bar{1}\bar{2}} +$$

$$+ \left[\bar{c}_{31} - \bar{c}_{13} - \frac{1}{2}\left(\bar{d}_{31} - \bar{d}_{13}\right)\right] \varphi_{1,\bar{1}\bar{3}} - \left(\bar{d}_{23} - \bar{d}_{32}\right) \varphi_{1,\bar{2}\bar{3}} +$$

$$+ \left(\bar{b}_{12} - \bar{b}_{21}\right) \varphi_2 +$$

$$+ \frac{1}{2}\left[\bar{c}_{12} - \bar{c}_{21} + \frac{1}{2}\left(\bar{d}_{12} - \bar{d}_{21}\right)\right] \left(\varphi_{2,\bar{1}\bar{1}} + \varphi_{2,\bar{2}\bar{2}}\right) +$$

$$+ \left[\bar{c}_{12} + \bar{c}_{21} + \frac{1}{2}\left(\bar{d}_{12} + \bar{d}_{21}\right)\right] \varphi_{2,\bar{1}\bar{2}} +$$

$$+ \left(\bar{b}_{31} - \bar{b}_{13}\right) \varphi_3 +$$

$$+ \frac{1}{2}\left[\bar{c}_{31} - \bar{c}_{13} + \frac{1}{2}\left(\bar{d}_{31} - \bar{d}_{13}\right)\right] \left(\varphi_{3,\bar{1}\bar{1}} + \varphi_{3,\bar{3}\bar{3}}\right) +$$

$$+ \left[\bar{c}_{31} + \bar{c}_{13} + \frac{1}{2}\left(\bar{d}_{31} + \bar{d}_{13}\right)\right] \varphi_{3,\bar{1}\bar{3}} -$$

$$- \left(\bar{b}_{33} + \bar{b}_{31} + \bar{b}_{13} + \bar{b}_{23} + \bar{b}_{32}\right) \left(\bar{u}_{2,\bar{3}} + \frac{1}{6} \bar{u}_{2,\bar{3}\bar{3}\bar{3}}\right) +$$

$$+ \left(\bar{b}_{22} + \bar{b}_{12} + \bar{b}_{21} + \bar{b}_{23} + \bar{b}_{32}\right) \left(\bar{u}_{3,\bar{2}} + \frac{1}{6} \bar{u}_{3,\bar{2}\bar{2}\bar{2}}\right) +$$

$$+ \left(\bar{b}_{13} - \bar{b}_{31}\right) \left(\bar{u}_{2,\bar{1}} + \frac{1}{6} \bar{u}_{2,\bar{1}\bar{1}\bar{1}} + \frac{1}{2} \bar{u}_{2,\bar{1}\bar{3}\bar{3}}\right) +$$

$$+\left(\bar b_{32}-\bar b_{23}\right)\left(\bar u_{2,\bar 2}+\frac{1}{6}\bar u_{2,\bar 2\bar 2\bar 2}+\frac{1}{2}\bar u_{2,\bar 2\bar 3\bar 3}\right)+$$

$$+\left(\bar b_{12}-\bar b_{21}\right)\left(\bar u_{3,\bar 1}+\frac{1}{6}\bar u_{3,\bar 1\bar 1\bar 1}+\frac{1}{2}\bar u_{3,\bar 1\bar 2\bar 2}\right)+$$

$$+\left(\bar b_{23}-\bar b_{32}\right)\left(\bar u_{3,\bar 3}+\frac{1}{6}\bar u_{3,\bar 3\bar 3\bar 3}+\frac{1}{2}\bar u_{3,\bar 3\bar 2\bar 2}\right)+$$

$$+\frac{1}{2}\left(\bar b_{12}+\bar b_{21}\right)\bar u_{3,\bar 2\bar 1\bar 1}-\frac{1}{2}\left(\bar b_{31}+\bar b_{13}\right)\bar u_{2,\bar 3\bar 1\bar 1}-\frac{1}{2}\left(\bar b_{23}+\bar b_{32}\right)\left(\bar u_{2,\bar 3\bar 2\bar 2}-\bar u_{3,\bar 2\bar 3\bar 3}\right), \tag{2.24}$$

where we have used

$$\bar d_{kk}=(1+h_{kk})^{-1}\bar b_{kk}, \quad \bar d_{ik}=2(1+h_{ik})^{-1}\bar b_{ik}, \quad (i\neq k). \tag{2.25}$$

Specializing (2.23), (2.24) to an orthotropic lattice, i.e. with orthogonal planes of symmetry giving $\bar a_{ij}=\bar a_{ji}$, and moreover to the case of four-fold rotation symmetry, giving

$$\bar a_{ii}=\bar a. \quad \bar b_{ii}=\bar b, \quad \bar c_{ii}=\bar c, \quad \bar d_{ii}=\bar d, \quad h_{ii}=h \tag{2.26}$$

$$\bar a_{ij}=\bar a', \quad \bar b_{ij}=\bar b', \quad \bar c_{ij}=\bar c', \quad \bar d_{ij}=\bar d', \quad h_{ij}=h', \quad (i\neq j), \tag{2.27}$$

we obtain

$$\mu^{-1}l^2\rho\ddot{\bar u}_1=(\bar a+2\bar a'+2\bar b')\left(\bar u_{1,\bar 1\bar 1}+\frac{1}{12}\bar u_{1,\bar 1\bar 1\bar 1\bar 1}\right)+$$

$$+(\bar b+\bar a'+3\bar b')\left[\bar u_{1,\bar 2\bar 2}+\bar u_{1,\bar 3\bar 3}+\frac{1}{12}\left(\bar u_{1,\bar 2\bar 2\bar 2\bar 2}+\bar u_{1,\bar 3\bar 3\bar 3\bar 3}\right)\right]+$$

$$+2(\bar a'-\bar b')\left[\bar u_{2,\bar 1\bar 2}+\frac{1}{6}\left(\bar u_{2,\bar 1\bar 2\bar 2\bar 2}+\bar u_{2,\bar 2\bar 1\bar 1\bar 1}\right)\right]+$$

$$+2(\bar a'-\bar b')\left[\bar u_{3,\bar 1\bar 3}+\frac{1}{6}\left(\bar u_{3,\bar 1\bar 3\bar 3\bar 3}+\bar u_{3,\bar 3\bar 1\bar 1\bar 1}\right)\right]+$$

$$+\frac{1}{2}(\bar a'+\bar b')\left(\bar u_{1,\bar 1\bar 1\bar 2\bar 2}+\bar u_{1,\bar 1\bar 1\bar 3\bar 3}\right)+\bar b'\,\bar u_{1,\bar 2\bar 2\bar 3\bar 3}-$$

$$-(\bar b+4\bar b')\left[\varphi_{2,\bar 3}-\varphi_{3,\bar 2}+\frac{1}{6}\left(\varphi_{2,\bar 3\bar 3\bar 3}-\varphi_{3,\bar 2\bar 2\bar 2}\right)\right]-$$

$$-\bar b'\left(\varphi_{2,\bar 3\bar 2\bar 2}-\varphi_{3,\bar 2\bar 3\bar 3}+\varphi_{2,\bar 3\bar 1\bar 1}-\varphi_{3,\bar 2\bar 1\bar 1}\right), \tag{2.28}$$

$$\mu^{-1} l^2 \rho \bar{j} \ddot{\varphi}_1 = -2(\bar{b} + 4\bar{b}')\varphi_1 +$$

$$+ \left(\bar{c} + 2\bar{c}' - \bar{d}'\right) \varphi_{1,\bar{1}\bar{1}} + (\bar{c}' - 2\bar{d}') \left(\varphi_{1,\bar{2}\bar{2}} + \varphi_{1,\bar{3}\bar{3}}\right) +$$

$$+ (2\bar{c}' + \bar{d}') \left(\varphi_{2,\bar{1}\bar{2}} + \varphi_{3,\bar{1}\bar{3}}\right) -$$

$$- (\bar{b} + 4\bar{b}') \left[\bar{u}_{2,\bar{3}} - \bar{u}_{3,\bar{2}} + \frac{1}{6} (\bar{u}_{2,\bar{3}\bar{3}\bar{3}} - \bar{u}_{3,\bar{2}\bar{2}\bar{2}})\right] -$$

$$- \bar{b}' \left(\bar{u}_{2,\bar{3}\bar{1}\bar{1}} - \bar{u}_{3,\bar{2}\bar{1}\bar{1}} + \bar{u}_{2,\bar{3}\bar{2}\bar{2}} - \bar{u}_{3,\bar{2}\bar{3}\bar{3}}\right) , \qquad (2.29)$$

with

$$\bar{d} = (1 + h)^{-1} \bar{b} , \quad \bar{d}' = 2(1 + h')^{-1} \bar{b}' . \qquad (2.30)$$

3. THE PRINCIPAL WAVE MODES

Evidently the wave equations (2.28), (2.29) for the highest symmetry, attainable in a cubic lattice with bonds only between nearest and next nearest neighbors, are still anisotropic. Full isotropy can be attained only by introducing still higher symmetry in the micro-structure, e.g. by having bonds also between next next nearest neighbors. The authors intend to investigate this possibility at a later occasion. However, the anisotropy as it is here must be judged to be rather low, which means that the most symmetric wave modes, i.e. those along the symmetry axes, should give a fairly good estimate of the constants in the fully isotropic case, and give good information about how the isotropic constants depend in general on the constants of the bonds in the micro-structure.

Assuming a wave along the x_3-axis

$$\bar{u}_k = \bar{U}_k \exp[i(\bar{q}\bar{x}_3 - \omega t)] , \quad \varphi_k = \Phi_k \exp[i(\bar{q}\bar{x}_3 - \omega t)] , \qquad (3.1)$$

we obtain from (2.28), (2.29) :

$$\mu^{-1} l^2 \rho \ddot{\bar{u}}_1 = (\bar{b} + \bar{a}' + 3\bar{b}') \left(\bar{u}_{1,\bar{3}\bar{3}} + \frac{1}{12} \bar{u}_{1,\bar{3}\bar{3}\bar{3}\bar{3}}\right) -$$

$$- (\bar{b} + 4\bar{b}') \left(\varphi_{2,\bar{3}} + \frac{1}{6} \varphi_{2,\bar{3}\bar{3}\bar{3}}\right) , \qquad (3.2)$$

$$\mu^{-1} l^2 \rho \bar{j} \ddot{\varphi}_2 = -2(\bar{b} + 4\bar{b}')\varphi_2 +$$

$$+ (\bar{b} + 4\bar{b}') \left(\bar{u}_{1,\bar{3}} + \frac{1}{6} \bar{u}_{1,\bar{3}\bar{3}\bar{3}}\right) + (\bar{c}' - 2\bar{d}')\varphi_{2,\bar{3}\bar{3}} , \qquad (3.3)$$

$$\mu^{-1} l^2 \rho \ddot{\bar{u}}_3 = (\bar{a} + 2\bar{a}' + 2\bar{b}')\left(\bar{u}_{3,\bar{3}\bar{3}} + \frac{1}{12}\bar{u}_{3,\bar{3}\bar{3}\bar{3}\bar{3}}\right) , \tag{3.4}$$

$$\mu^{-1} l^2 \rho \bar{j} \ddot{\varphi}_3 = -2(\bar{b} + 4\bar{b}')\varphi_3 + (\bar{c} + 2\bar{c}' - \bar{d}')\varphi_{3,\bar{3}\bar{3}} . \tag{3.5}$$

Comparing (3.1) - (3.5) with the corresponding equations, obtained from (1.1), (1.2), i.e.

$$\mu^{-1} l^2 \rho \ddot{\bar{u}}_1 = (\bar{\mu} + \bar{\alpha})\bar{u}_{1,\bar{3}\bar{3}} - 2\bar{\alpha}\varphi_{2,\bar{3}} - \bar{\xi}\bar{u}_{1,\bar{3}\bar{3}\bar{3}\bar{3}} + \bar{\zeta}\varphi_{2,\bar{3}\bar{3}\bar{3}} , \tag{3.6}$$

$$\mu^{-1} l^2 \rho \bar{j} \ddot{\varphi}_2 = -4\bar{\alpha}(\varphi_2 - \frac{1}{2}\bar{u}_{1,\bar{3}}) + (\bar{\varepsilon} + \bar{\gamma})\varphi_{2,\bar{3}\bar{3}} - \bar{\zeta}\bar{u}_{1,\bar{3}\bar{3}\bar{3}} , \tag{3.7}$$

$$\mu^{-1} l^2 \rho \ddot{\bar{u}}_3 = (2\bar{\mu} + \bar{\lambda})\bar{u}_{3,\bar{3}\bar{3}} - (\bar{\xi} + \bar{\eta})\bar{u}_{3,\bar{3}\bar{3}\bar{3}\bar{3}} , \tag{3.8}$$

$$\mu^{-1} l^2 \rho \bar{j} \ddot{\varphi}_3 = -4\bar{\alpha}\varphi_3 + (\bar{\beta} + 2\bar{\gamma})\varphi_{3,\bar{3}\bar{3}} , \tag{3.9}$$

we obtain by means of (2.30) the following estimates of the constants $\bar{\mu} = 1$, $\bar{\lambda}$, $\bar{\alpha}$, $\bar{\beta}$, $\bar{\gamma}$, $\bar{\varepsilon}$, $\bar{\xi}$, $\bar{\eta}$, $\bar{\zeta}$:

$$\bar{\mu} = \frac{1}{2}(2\bar{a}' + \bar{b} + 2\bar{b}') = 1 , \tag{3.10}$$

$$\bar{\lambda} = \bar{a} - \bar{b} , \tag{3.11}$$

$$\bar{\alpha} = \frac{1}{2}(\bar{b} + 4\bar{b}') , \tag{3.12}$$

$$\bar{\varepsilon} + \bar{\gamma} = -4(1 + h')^{-1}\bar{b}' + \bar{c}' , \tag{3.13}$$

$$2\bar{\gamma} + \bar{\beta} = -2\bar{b}'(1 + h')^{-1} + \bar{c} + 2\bar{c}' , \tag{3.14}$$

$$\bar{\xi} = -\frac{1}{12}(\bar{a}' + \bar{b} + 3\bar{b}') , \tag{3.15}$$

$$\bar{\eta} = -\frac{1}{12}(\bar{a} + \bar{a}' - \bar{b} - \bar{b}') , \tag{3.16}$$

$$\bar{\zeta} = -\frac{1}{6}(\bar{b} + 4\bar{b}') . \tag{3.17}$$

Expressing the dispersion relations for the wave modes along the principal axes, as calculated by Brulin and Hjalmars [2] in the micro-constants by means of (3.10) - (3.17), we obtain:

The transversal displacement-rotation mode:

$$\omega^2_{t,u\varphi} = q^2 \frac{\mu}{\rho}\left\{1 - \frac{1}{12}\left[1 - \frac{3}{2}\bar{b} - 3\bar{c}' - 6\bar{b}'(h' - 1)/(h' + 1) + 3\bar{j}\right](q^2 l^2)\right\} , \tag{3.18}$$

The transversal rotation-displacement mode:

$$\omega^2_{t,\varphi u} = q^2 \frac{\mu}{\rho} \frac{4\bar{\alpha}}{\bar{j}} (q^2 l^2)^{-1} \times$$

$$\times \left\{ 1 - \frac{1}{2} \left[\left(4(1 + h')^{-1} \bar{b}' - \bar{c}' \right) \left(\bar{b} + 4\bar{b}' \right)^{-1} - \frac{1}{2} \bar{j} \right] (q^2 l^2) \right\} , \qquad (3.19)$$

The longitudinal displacement mode:

$$\omega^2_{l,u} = q^2 \frac{\mu}{\rho} (2 + \bar{\lambda}) \left[1 - \frac{1}{12} (q^2 l^2) \right] , \qquad (3.20)$$

The longitudinal rotation mode:

$$\omega^2_{l,\varphi} = q^2 \frac{\mu}{\rho} \frac{4\bar{\alpha}}{\bar{j}} (q^2 l^2)^{-1} \times \left\{ 1 - \frac{1}{2} \left(2\bar{b}'(1 + h')^{-1} - \bar{c} - 2\bar{c}' \right) \left(\bar{b} + 4\bar{b}' \right)^{-1} (q^2 l^2) \right\} . \qquad (3.21)$$

Normal non-dispersion in these modes is given by the first term 1 in the main parentheses, whereas the effect of dispersion is given by the $(q^2 l^2)$-term .

4. APPLICATION AND DISCUSSION

As a first application we assume the bonds to be given by cylindrical bars ($h_{ij} = 2$) of radius r between spherical particles of radius R . Choosing $\nu_{ij} = 1/3$ we then have:

$$a_{ij} = \frac{E_{ij}\pi r^2}{11_{ij}} , \quad b_{ij} = \frac{3E_{ij}\pi r^4}{11^3_{ij}} , \quad c_{ij} = \frac{3E_{ij}\pi r^4}{16 1_{ij}} , \quad j = \frac{2}{5} R^2 . \qquad (4.1)$$

Under these assumptions we obtain the values of the constitutive constants (3.10) - (3.17), as well as of the dispersion in the wave modes (3.18) - (3.21), as given in Tables 1 and 2.

One may note that $\bar{\varepsilon} + \bar{\gamma}$ comes out negative, in apparent contradiction to the conventional conditions of stability for micropolar continua, which require $\gamma > 0$, $\varepsilon > 0$. This negative sign is however a general feature, and as discussed by Brulin and Hjalmars [4] the dilemma is solved by realizing that the conventional stability conditions are in fact too strong.

As can be seen, the constants $\bar{b}$, $\bar{b}'$, $\bar{c}$, $\bar{c}'$ and $\bar{d}$, $\bar{d}'$, and thus $\bar{\alpha}$, $\bar{\varepsilon} + \bar{\gamma}$, $2\bar{\gamma} + \bar{\beta}$ and $\bar{\zeta}$, as well as $\bar{j}$, vanish with the square of the polar parameter r/l , whereas $\bar{a}$, $\bar{a}'$ and thus $\bar{\mu}$, $\bar{\lambda}$, $\bar{\xi}$, and $\bar{\eta}$ remain finite. Moreover. there is in all four wave modes a remaining dispersion of up to 4 % for $r/l = 0$.

This application, which of course is immediately valid for gridworks, should also give reasonable estimates of the order of magnitudes and trends for granular composites, as well as for molecular crystals with van der Waals' interactions. For a more detailed description of the composites, espescially when closely packed, the bar lengths should of course be diminished by about 2R . The crystal case, on the other hand, seems to require a bigger difference between $\bar{a}$ and $\bar{a}'$, i.e. a bigger difference in the properties of the bars in different directions.

Table 1. Constitutive constants given by cylindrical bars of radius r between solid spheres, with radius R, cf. (4.1), (3.10-17).

Constant \ r/l	0.5	0.3	0.2	0.1	0
$\bar{\mu}$	1.00	1.00	1.00	1.00	1.00
$\bar{\lambda}$	0.19	0.78	1.09	1.32	1.41
$\bar{\alpha}$	0.67	0.35	0.18	0.05	0
$\bar{\varepsilon} + \bar{\gamma}$	- 0.24	- 0.12	- 0.06	- 0.02	0
$\bar{\beta} + 2\bar{\gamma}$	- 0.05	- 0.02	- 0.01	- 0.00	0
$\bar{\xi}$	- 0.14	- 0.11	- 0.10	- 0.09	- 0.08
$\bar{\eta}$	- 0.04	- 0.12	- 0.16	- 0.19	- 0.20
$\bar{\zeta}$	- 0.22	- 0.12	- 0.06	- 0.02	0
$\bar{j}(r/R)^2$	0.10	0.04	0.02	0.00	0

Table 2. Percent dispersion in ω at the continuum limit $(\bar{q} = ql = 1)$, cf. (3.18-21), with constitutive constants given in Table 1, and $R = r$.

Mode \ r/l	0.5	0.3	0.2	0.1	0
t,uφ	0	- 2	- 3	- 4	- 4
t,φu	- 3	- 4	- 4	- 4	- 4
l,u	- 4	- 4	- 4	- 4	- 4
l,φ	- 1	- 1	- 1	- 1	- 1

REFERENCES

[1] Brulin, O. and Hjalmars, S., Grade consistent micropolar theory and equivalent media, TRITA-MEK-79-15, and Grade consistent micropolar theory, Post-RAAG Reports (ed. K. Kondo) No. 110 (1981).

[2] Brulin, O. and Hjalmars, S., Linear grade consistent micropolar theory, Int. J. Engng. Sci. (In press).

[3] Tauchert, T.R., A lattice theory for representation of thermoelastic composite materials, Recent advances in engineering science, Vol. 5.I, 325 (1970).

[4] Brulin, O. and Hjalmars, S., Stability conditions in continuum models of discrete systems, Proceedings of Continuum Models of Discrete Systems (CMDS 4), (North-Holland, Amsterdam, 1981). (This volume).

Continuum Models of Discrete Systems 4
eds. O. Brulin and R.K.T. Hsieh

REEXAMINATION OF MICROPOLAR WAVE EXPERIMENT

Richard D. Gauthier

Basic Engineering Department
Colorado School of Mines
Golden, Colorado
U.S.A.

William E. Jahsman

Department 256
Intel Corporation
Santa Clara, California
U.S.A.

The experimental determination of micropolar elastic constants through wave propagation measurements was reported in an earlier paper. A Rayleigh-Lamb infinite plate analysis predicted microrotational wave modes which were detected when a micropolar specimen was tested. By means of a wave equation, two micropolar constants were measured. In the present work an infinite cylinder analysis is used as being a better model for the disk-shaped specimens. The classical Pochhammer-Chree frequency solution is recovered as a special case. An additional microrotational frequency equation results which was earlier obtained by Nowacki and Nowacki. This equation when used to interpret the data from the original experiments provides the information required for the measurement of three micropolar elastic constants.

INTRODUCTION

The authors [1] reported a dynamic measurement technique at CMDS-3 in which wave propagation properties through classical and micropolar elastic media are compared. The method utilizes a compressional Kolsky apparatus (split Hopkinson pressure bar) which is designed to deliver essentially one-dimensional waves into a specimen. When the specimen is a homogeneous classical material, any changes in wave form of the response can be attributed to multiple reflections at the interface surfaces between specimen and input and output bars. However, a two-dimensional effect is observed in a micropolar material. Although all specimens are short cylinders with a diameter to match that of the pressure bars, the interpretation of this observation was made utilizing the micropolar counterpart to the Rayleigh-Lamb problem of the vibration of an infinite plate in plane strain. The plate geometry was selected in preference to the cylinder because the equations are mathematically more tractable. The experiments were conducted on a specially fabricated micropolar material comprising a composite of rigid spherical aluminum shot embedded in an elastic epoxy matrix. Analysis of the data obtained enabled the determination of two micropolar constants for this material in addition to the classical moduli.

In the present work the authors reexamine the data in the context of the micropolar solution to the infinite cylinder problem. Two papers already in the literature deal with certain aspects of the problem. Smith [2] obtained the frequency equation in determinantal form but did not proceed further to investigate the nature of the micropolar modes. Nowacki and Nowacki [3] obtained the classical Pochhammer-Chree solution as a special case, but also an additional frequency equation which is related to a microrotational wave traveling in the axial direction. This last equation contains three micropolar constants wherein the micropolar plate wave equation contains but two. The Nowacki equation will be used in this work to obtain micropolar constants from the original experimental data, and the results compared to those in which the Rayleigh-Lamb plate analysis was applied.

ANALYSIS

Micropolar media are characterized by six elastic moduli; λ, μ, κ, α, β, and γ. The first two are the Lamé constants of classical elasticity. We introduce five wave speeds and a microrotation frequency which are defined in terms of the micropolar constants as follows:

$$c_1^2 = (\lambda+2\mu+\kappa)/\rho, \quad c_2^2 = (\mu+\kappa)/\rho, \quad c_3^2 = \kappa/\rho,$$
$$c_4^2 = \gamma/\rho j, \quad c_5^2 = \beta/\rho j, \quad \omega_0^2 = \kappa/\rho j = c_3^2/j \tag{1}$$

In the above ρ is the density and ρj is the microinertia. The notation used here differs from that of Nowacki and Nowacki [3]. The field equations in cylindrical (r,θ,z) coordinates for the axisymmetric case can be written (no summation on repeated indices)

$$c_1^2\,[r^{-1}(ru_r)_{,r}]_{,r} + c_2^2 u_{r,zz} + (c_1^2-c_2^2)u_{z,rz} - c_3^2\phi_{\theta,z} - \dot{v}_r = 0$$
$$c_2^2 r^{-1}(ru_{z,r})_{,r} + (c_1^2-c_2^2)r^{-1}(ru_{r,z})_{,r} + c_1^2 u_{z,zz} + c_3^2 r^{-1}(r\phi_\theta)_{,r} - \dot{v}_z = 0 \tag{2}$$
$$c_4^2\,\{[r^{-1}(r\phi_\theta)_{,r}]_{,r} + \phi_{\theta,zz}\} - \omega_0^2(u_{z,r}-u_{r,z}+2\phi_\theta) - \dot{\nu}_\theta = 0$$

where the displacement and velocity components are given by u and v, and the microrotation and microgyration by ϕ and ν, and $\dot{v}=\ddot{u}$, $\dot{\nu}=\ddot{\phi}$.

Traction- and couple-free boundary conditions on the free boundary $r=a$ are prescribed to give

$$\rho^{-1}t_{rr}(a) = c_1^2 u_{r,r} + (c_1^2 - 2c_2^2 + c_3^2)(u_{z,z} + a^{-1}u_r) = 0$$

$$\rho^{-1}t_{rz}(a) = c_2^2 u_{z,r} + (c_2^2 - c_3^2)u_{r,z} + c_3^2\phi_\theta = 0 \tag{3}$$

$$(\rho j)^{-1}m_{r\theta} = c_4^2\phi_{\theta,r} - c_5^2 a^{-1}\phi_\theta = 0$$

where t and m are the stress and couple stress, respectively.

Solutions to the field equations are assumed in the form

$$u_r = RJ_1(\eta r)e^{i\xi z}e^{i\omega t}$$

$$u_z = ZJ_0(\eta r)e^{i\xi z}e^{i\omega t} \tag{4}$$

$$\phi_\theta = \Phi J_1(\eta r)e^{i\xi z}e^{i\omega t}$$

in which the proper symmetry is assured. Introduced are the wave numbers η and ξ in the r- and z-directions, respectively, the circular frequency ω, and time t. R, Z, and Φ are amplitudes.

The procedure for obtaining the frequency equation is identical to that employed by Jahsman and Gauthier [1] for the infinite plate. Omitting the details, we present the result where it is assumed that $c_3^2 \ll c_2^2, c_4^2$

$$\{c_4^2(c_4^2+c_5^2)^{-1}\eta_3 J_0(\eta_3 a) - a^{-1}J_1(\eta_3 a)\}\ \{(2c_2^2)^{-1}\eta_1 a^{-1}\omega^2$$
$$\cdot J_1(\eta_1 a)J_1(\eta_2 a) - [(2c_2^2)^{-1}\omega^2 - \xi^2]^2 J_0(\eta_1 a)J_1(\eta_2 a) \tag{5}$$
$$-\xi^2\eta_1\eta_2 J_1(\eta_1 a)J_0(\eta_2 a)\} = 0$$

This frequency equation is the Nowacki result referred to earlier. The factor contained in the second set of braces is the classical Pochhammer-Chree solution when the micropolar constants vanish. The first set of braces contains a strictly micropolar solution which does not appear in classical elasticity,

$$c_4^2\eta a J_0(\eta a) - (c_4^2 + c_5^2)J_1(\eta a) = 0 \tag{6}$$

in which the subscript on η has been dropped. An intermediate result, omitted here but derived in [1], provides an explicit relation between ω and ξ in which the modal $\overset{\circ}{\eta}_n$ are found from the zeros of equation (6)(see [4], p. 414).

$$\omega_n^2 = 2\omega_0^2 + c_4^2(\xi^2 + \overset{\circ}{\eta}_n^2) \tag{7}$$

This will be called the micropolar frequency equation. Of particular interest is the case where $\xi=0$ which defines the modal cutoff frequency $\omega_{cn} = \sqrt{2\omega_0^2+c_4^2\mathring{\eta}_n^2}$ below which the longitudinal wave number ξ is imaginary and waves become nontraveling localized vibrations.

EXPERIMENT

The specimens used with the Kolsky apparatus were disks 38.1 mm in diameter and 12.7 mm thick. The micropolar material comprised 60% by volume of 1.4 mm diameter spherical aluminum shot close-packed in an epoxy matrix. Homogeneous specimens were made also of aluminum and of epoxy for purposes of standardization. When homogeneous materials are tested the signal response to an essentially bell-shaped input stress wave can be accurately predicted on the basis of Young's modulus and density. In the case of the micropolar material there is a marked deviation from predicted response which takes the form of ripples superposed on the calculated wave form. The details of the experimental investigation and data reduction are given in [1] and will not be repeated here except to show the input and output signal variance in Figure 1 which was obtained with the micropolar specimen.

The presence of these ripple frequencies in the composite material is attributed to the excitation of microrotational wave modes by the longitudinal wave pulse. A detailed discussion and rationale leading to this conclusion is given by Jahsman and Gauthier [5]. Together, the micropolar frequency and wave equations (6) and (7) contain three unknowns c_4, c_5, and ω_0. The identification of three micropolar (ripple) frequencies in Figure 1 in conjunction with the measurement of Young's modulus and Poisson's ratio in a conventional tensile or compression test will permit the evaluation of the five moduli λ, μ, κ, β, and γ.

Harmonic analysis of the data shown in Figure 1 reveals two dominant waves with periods of 20 and 10 μs. We assume these periods to be associated with the long wave length cutoff frequencies of the first two microrotational modes. Conditions of thermodynamic stability impose the bounds $-c_4^2 \leq c_5^2 \leq c_4^2$ where $c_4^2 \geq 0$. For real c_5 the third mode must have a period within the narrow range 6.55 to 6.64 μs. No such wave can be identified from the data. If we discount an imaginary c_5, we conclude $c_5=0$. The roots of equation (6) then are $\mathring{\eta}_1 a=1.84$ and $\mathring{\eta}_2 a=5.33$ where the specimen radius $a=19.1$ mm. Equations (7) can be inverted to obtain solutions for c_4 and ω_0.

$$
\begin{aligned}
c_4^2 &= a^2(\omega_{c2}^2-\omega_{c1}^2)[(\mathring{\eta}_2 a)^2 - (\mathring{\eta}_1 a)^2]^{-1} \\
\omega_0^2 &= \frac{1}{2}[(\mathring{\eta}_2 a\omega_{c1})^2 - (\mathring{\eta}_1 a\omega_{c2})^2][(\mathring{\eta}_2 a)^2-(\mathring{\eta}_1 a)^2]^{-1}
\end{aligned}
\qquad (8)
$$

The density of the composite material was measured to be $\rho = 2.19 \times 10^{-3}$ g mm^{-3}. The microinertia is assumed to be the product of the density and the radius of gyration squared of the 1.4 mm diameter aluminum spheres, $j = 0.196$ mm^2. We may now use equations (1)

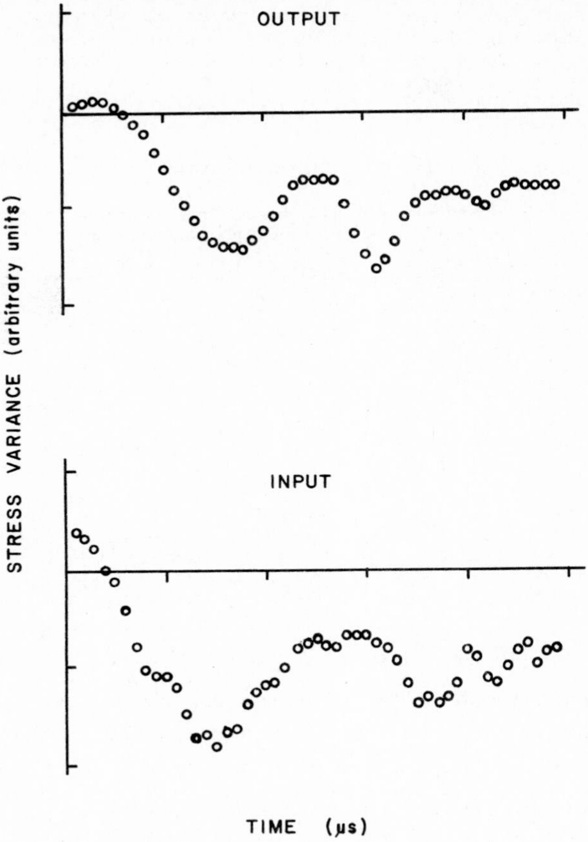

Figure 1. Stress Variance Histories for Input and Output Bars with Aluminum-Epoxy Composite

to calculate the micropolar moduli γ, κ, and β. The results are listed in Table 1 for comparison with those obtained in [1] using the Rayleigh-Lamb plate analysis. Included are the calculated values of c_3 and c_2, the latter obtained from the relation

$$2c_2^{\ 2} = E(1+\nu)^{-1}\rho^{-1}+c_3^{\ 2}$$

where $E=5.31$ GPa and $\nu=0.4$, the measured Young's modulus and Poisson's ratio.

TABLE 1
PROPERTIES OF MICROPOLAR MATERIAL
COMPARISON BETWEEN CYLINDER AND PLATE ANALYSES

PROPERTY	CYLINDER	PLATE
c_2	0.93 km s^{-1}	0.93
c_3	75.7 m s^{-1}	77.9
c_4	2.08 km s^{-1}	2.33
c_5	0	--
ω_0	171 k rad s^{-1}	176
γ	1.85 k N	2.33
κ	12.6 M Pa	13.3
β	0	--

It was assumed in the derivation of the frequency equation that c_3 is small compared to c_2 and c_4. We see from Table 1 that c_3 is only 8 percent of c_2 and less than 4 percent of c_4.

The possibility remains that $c_5=ic_4$ ($\beta=-\gamma$) in the limit, and its associated nontraveling wave would not be detected by the experiment. In this case γ would be less than 2 percent higher than the value shown in Table 1, but κ would be 50 percent less.

DISCUSSION

The infinite cylinder solution is advanced as a more realistic model of the disk-shaped specimens used in the experiments than the Rayleigh-Lamb infinite plate analysis of [1]. The cylinder analysis has the further advantage of containing an extra wave speed with its information about another micropolar elastic modulus. Although c_5 was not detected for the particular micropolar material tested, it is believed that the experimental method is sufficiently sensitive to identify a real c_5.

REFERENCES

1. Jahsman, W.E. and Gauthier, R.D., Dynamic measurements of micropolar elastic constants, in: Kröner, E. and Anthony, K.-H. (eds.), Continuum Models of Discrete Systems (CMDS3), SM Study No. 15 (University of Waterloo Press, Waterloo, Ontario, Canada, 1980).

2. Smith, A.C., Torsion and vibrations of cylinders of a micro-polar elastic solid, in: Eringen, A.C. (ed.), Recent Advances in Engineering Science, Vol. 5, Part II (Gordon and Breach, New York, 1970).

3. Nowacki, W. and Nowacki, W.K., Propagation of elastic waves in a micropolar cylinder. I, Bulletin de l'Academie Polonaise des Sciences, Serie des Sciences Techniques, Vol. XVII, No. 1 (1969) 39-47.

4. Handbook of Mathematical Functions, Abramowitz, A. and Stegun, I.A. (eds.), National Bureau of Standards Applied Mathematics Series 55, U.S. Dept. of Commerce (1964).

5. Jahsman, W.E. and Gauthier, R.D., Dynamic measurements of micropolar elastic constants, University of Oxford, Dept. of Engineering Science, Report O.U.E.L. 1291/79 (1979).

Continuum Models of Discrete Systems 4
eds. O. Brulin and R.K.T. Hsieh

BALANCE EQUATIONS OF THE COSSERAT-CONTINUUM

Gerhard Kluge

Sektion Physik, Friedrich-Schiller-Universität
J e n a
DDR-6900 Jena

The equations of the Cosserat-continuum are solved with the help of stress functions. The gauge transformationa of the stress functions are connected with the conservation laws of the dislocation and disclination densities. Furthermore the balance equations of energy, momentum and angular momentum are calculated, and the forces and torques, acting on the dislocations and disclinations, are discussed.

1. Introduction

Nowadays the Cosserat-continuum is well-known and many papers deal with it. But it may be of interest to give some further insight into the structure of the corresponding theory, especially of the theory of a solid Cosserat-continuum with incompabilities (dislocations and disclinations). Therefore, as common in field theories, we derive balance equations for energy, momentum and angular momentum and obtain, for instance, the forces and angular momenta acting an the incompabilities. Furthermore, we find an interesting relation between the gauge transformation of the stress functions and the conservation laws of the incompabilities. The methods used in this paper can be applied to continua with more complicated internal structures.

2. The Cosserat-continuum without incompabilities

The state of a Cosserat-continuum can be described by the stress σ_{ik} and the momentum stress tensor μ_{ik} or by the distorsion tensor β_{ik} and the curvature tensor $\varkappa_{ik}$. β_{ik} and $\varkappa_{ik}$ can be derived from the displacement field s_i and the rotation field d_i (linear theory)

$$\varkappa_{ik} = d_{k,i} \quad , \quad \beta_{ik} = s_{k,i} - \varepsilon_{ikl} d_l \, , \tag{1}$$

and they fulfil the compatibility conditions (or integrability conditions)

$$\varepsilon_{jri}\varkappa_{ik,r} = 0\,,$$
$$\varepsilon_{jri}(\beta_{ik,r} + \varepsilon_{krl}\varkappa_{il}) = 0\,. \tag{2}$$

The differential equations of the Cosserat-continuum follow, e.g., from the principle of Hamilton

$$\delta \int \mathcal{L}\, dV dt = 0 \tag{3}$$

with the Lagrange density

$$\mathcal{L} = T - U + s_i f_i + d_i m_i\,. \tag{4}$$

Here $T = \frac{1}{2}\varrho \dot{s}_i \dot{s}_i + \frac{1}{2}\Theta_{ik}\dot{d}_i\dot{d}_k$ is the density of kinetic energy, ϱ the mass density, Θ_{ik} the microinertia tensor, $U = U(\beta_{ik}, \varkappa_{ik})$ the elastic potential, from which we can derive σ_{ik} and μ_{ik}

$$\sigma_{ik} = \frac{\partial U}{\partial \beta_{ik}}\,, \qquad \mu_{ik} = \frac{\partial U}{\partial \varkappa_{ik}}\,, \tag{5}$$

f_i an external force density and m_i an external torque density. By variation of s_i and d_i the equations

$$\sigma_{ik,i} + f_k = \varrho \ddot{s}_k = \varrho \dot{v}_k\,,$$
$$\mu_{ik,i} + \varepsilon_{krs}\sigma_{rs} + m_k = \Theta_{ik}\ddot{d}_i = \Theta_{ik}\dot{w}_i \tag{6}$$

follow from the principle of Hamilton (3). For the calculation of the deformation state we have to integrate the eq. (8) considering the initial and boundary conditions. Thereby state equations are needed Often the linearized form of these equations are used (for simplicity without cross effects)

$$\sigma_{ik} = C_{iklm}\beta_{lm}\,, \quad \mu_{ik} = A_{iklm}\varkappa_{lm}\,. \tag{7}$$

Here C_{iklm} and A_{iklm} are elastic modules.

3. The Cosserat-continuum with incompabilities

3.1. The field equations

In the following the conditions (2) may not be valid for the elastic distorsion β^E_{ik} and curvature $\varkappa^E_{ik}$. Then β^E_{ik} and $\varkappa^E_{ik}$ cannot be derived from a displacement field s_i and a rotation field d_i. But we postulate that besides the elastic parts β^E_{ik}, $\varkappa^E_{ik}$ there also exist plastic parts β^P_{ik}, $\varkappa^P_{ik}$ with

$$\varkappa^E_{ik} + \varkappa^P_{ik} = \varkappa_{ik} = d_{k,i} \; , \quad \beta^E_{ik} + \beta^P_{ik} = \beta_{ik} = s_{k,i} - \varepsilon_{ikr} d_r \, . \tag{8}$$

To obtain the corresponding field equations we use the following method which is well-known in the field of electrodynamics. First we solve eq. (6) with the help of stress functions

$$\begin{aligned}
\sigma_{ik} &= \varepsilon_{irs}\varphi_{sk,r} + \dot{\psi}_{ik} + \Omega_{k,i} \, , \\
\mu_{ik} &= \varepsilon_{irs}(\phi_{sk,r} + \varepsilon_{kr\ell}\varphi_{s\ell}) + \dot{\omega}_{ik} + X_{k,i} + \varepsilon_{ik\ell}\Omega_\ell \, , \\
\rho v_k &= \psi_{ik,i} \, , \\
\Theta_{ik} w_i &= \omega_{ik,i} + \varepsilon_{krs}\psi_{rs} \, ,
\end{aligned} \tag{9}$$

with

$$\Omega_{k,\ell,\ell} = f_k \; , \qquad X_{k,\ell,\ell} = m_k \, .$$

For simplicity, we neglect f_k and m_k, then $\Omega_k = 0$ and $X_k = 0$ are valid.

Now we ascribe, to each stress function an incompability tensor (D_{ik}, B_{ik}, J_{ik}, S_{ik}) and extend the Lagrange density in the following way:

$$\begin{aligned}
\mathcal{L} &= \mathcal{L}^E + \varphi_{ik} D_{ik} + \phi_{ik} B_{ik} + \psi_{ik} J_{ik} + \omega_{ik} S_{ik} \, , \\
\mathcal{L}^E &= T - \mathcal{U}(\beta^E_{ik}, \varkappa^E_{ik}) \, .
\end{aligned} \tag{10}$$

The field equations are obtained if we consider $\mathcal{L}$ as a function of the stress function and vary only the stress functions in the principle of Hamilton. Then the field equations follow

$$\varepsilon_{irs} \overset{E}{\beta}_{sk,r} + \varepsilon_{lir} \varepsilon_{lks} \overset{E}{\varkappa}_{rs} = \mathfrak{D}_{ik} , \tag{11}$$

$$\varepsilon_{irs} \overset{E}{\varkappa}_{sk,r} = \mathfrak{B}_{ik} , \tag{12}$$

$$v_{k,i} - \dot{\overset{E}{\beta}}_{ik} - \varepsilon_{ikr} w_r = \mathfrak{J}_{ik} , \tag{13}$$

$$w_{k,i} - \dot{\overset{E}{\varkappa}}_{ik} = S_{ik} . \tag{14}$$

From these equations we see that $\mathfrak{D}_{ik}$ is the dislocation density and $\mathfrak{B}_{ik}$ the disclination density and that

$$\mathfrak{J}_{ik} = \dot{\overset{P}{\beta}}_{ik} , \qquad S_{ik} = \dot{\overset{P}{\varkappa}}_{ik} \tag{15}$$

is valid. $\mathfrak{J}_{ik}$ is the dislocation and S_{ik} the disclination current density. The method applied here was first used by Holländer /1/ and Kosevich /2/.

3.2. The gauge transformations of the stress functions and the laws of conservation of dislocation and disclination density

As in elastostatics where we can add the tensor of the zero-stress functions $a_{k,i}$ to the stress function φ_{ik}, we also have the possibility in our theory to transform the stress functions without changing the physical quantities σ_{ik}, μ_{ik}, ϱv_k, $\Theta_{ik} w_i$. These gauge transformations are as follow

$$\begin{aligned} \varphi'_{ik} &= \varphi_{ik} + a_{k,i} - \dot{b}_{ik} , \\ \phi'_{ik} &= \phi_{ik} - \varepsilon_{ikl} a_l + g_{k,l} - \dot{h}_{ik} , \\ \psi'_{ik} &= \psi_{ik} + \varepsilon_{irl} b_{lk,r} , \\ \omega'_{ik} &= \omega_{ik} + \varepsilon_{irl} h_{lk,r} + \varepsilon_{irl} \varepsilon_{krs} b_{ls} . \end{aligned} \tag{16}$$

The field equations must be invariant in these gauge transformations. That means, the Lagrange density is allowed to change only by a divergence term and a term which can be described as a time derivative. This demand yields the well-known conservation laws of the dislocation and disclination density:

$$
\begin{aligned}
&D_{ik,i} + \varepsilon_{krs} B_{rs} = 0 , \\
&\dot{D}_{ik} + \varepsilon_{irs} (J_{rk,s} + \varepsilon_{kls} S_{rl}) = 0 , \\
&B_{ik,i} = 0 , \\
&\dot{B}_{ik} + \varepsilon_{irs} S_{rk,s} = 0 .
\end{aligned}
\tag{17}
$$

The connection between gauge transformations and conservation laws entirely corresponds to the relations in the electrodynamics, where the conservation law of the electric charge follows from the invariance of the Maxwell equations in gauge transformations of the electrodynamical potentials.

3.3. The balances of energy, momentum and angular momentum

With the aid of Noethers theorem we can obtain the balance equations of energy, momentum and angular momentum from the invariance in time translations, space translations and rotations. But it is possible to derive these equations from the Lagrange density directly by calculating, e.g., $\frac{\partial \mathcal{L}}{\partial t}$ or $\frac{\partial \mathcal{L}}{\partial x_i}$ /3/ and thereby using the field equations. In this way we obtain the balance equations in their canonical form (we use the abbreviations $\varphi_A \sim (\varphi_{ik}$, ϕ_{ik}, ψ_{ik}, ω_{ik}) and $D_A \sim (D_{ik}$, B_{ik}, J_{ik}, S_{ik})):

Balance of energy

$$
\frac{\partial}{\partial t} \left\{ \frac{\partial \mathcal{L}}{\partial \dot{\varphi}_A} \dot{\varphi}_A - \mathcal{L} \right\} + \left\{ \frac{\partial \mathcal{L}}{\partial \varphi_{A,i}} \dot{\varphi}_A \right\}_{,i} = \varphi_A \dot{D}_A , \tag{18}
$$

balance of momentum

$$
\frac{\partial}{\partial t} \left\{ \frac{\partial \mathcal{L}}{\partial \dot{\varphi}_A} \varphi_{A,k} \right\} + \left\{ \frac{\partial \mathcal{L}}{\partial \varphi_{A,i}} \varphi_{A,k} - \mathcal{L} \delta_{ik} \right\}_{,i} = - \frac{\partial \mathcal{L}}{\partial D_A} D_{A,k} , \tag{19}
$$

balance of angular momentum

$$\frac{\partial}{\partial t}\left\{ \frac{\partial \mathcal{L}}{\partial \dot{\varphi}_A} (S_{rl})_{AB} \varphi_B + \overset{can}{P}_l x_r - \overset{can}{P}_r x_l \right\} +$$

$$+ \left\{ \frac{\partial \mathcal{L}}{\partial \varphi_{A,k}} (S_{rl})_{AB} \varphi_B + \overset{can}{T}_{kl} x_r - \overset{can}{T}_{kr} x_l \right\}_{,k} = \tag{20}$$

$$= \overset{can}{k}_r x_l - \overset{can}{k}_l x_r - \frac{\partial \mathcal{L}}{\partial D_A} (S_{rl})_{AB} D_B .$$

Here $(S_{rl})_{AB}$ are the infinitesimal generators of the space rotations. Further we used eq. (19) in the form of

$$\dot{\overset{can}{P}}_k + \overset{can}{T}_{ik,i} = \overset{can}{k}_k \tag{21}$$

Now the stress function should be eliminated in eq. (18) to (20). With regard to eq. (9), (11) to (14) and (17) we obtain after tremendous calculations:

Balance of energy

$$\frac{\partial}{\partial t}(T + U) - (\sigma_{ik} v_k + \mu_{ik} w_k)_{,i} = -\sigma_{ik} J_{ik} - \mu_{ik} S_{ik} , \tag{22}$$

balance of momentum

$$\frac{\partial}{\partial t}\left\{ -\varrho v_k (\overset{E}{\beta}_{rk} + \varepsilon_{rkl} d_l) - \Theta_{il} w_i \overset{E}{\varkappa}_{rl} \right\} + \left\{ \sigma_{lk} (\overset{E}{\beta}_{rk} + \varepsilon_{rkt} d_t) + \right.$$

$$\left. + \mu_{lk} \overset{E}{\varkappa}_{rk} + \overset{E}{\mathcal{L}} \delta_{rl} \right\}_{,l} = \varepsilon_{rkl} \sigma_{lt} D_{kt} + \varepsilon_{rkl} \mu_{lt} B_{kt} + \tag{23}$$

$$+ \varrho v_k J_{rk} + \Theta_{ik} w_i S_{rk} + \varepsilon_{rkl} \sigma_{tk} \overset{P}{\varkappa}_{tl} ,$$

balance of angular momentum

$$\frac{\partial}{\partial t}\left\{ \varepsilon_{lrj} (P_j x_r - \varrho v_j s_r - \Theta_{ij} w_i d_r) \right\} + \left\{ \varepsilon_{lrj} (\sigma_{kj} s_r + \mu_{kj} d_r - \right.$$

$$\left. - T_{kj} x_r) \right\}_{,k} = \varepsilon_{ljr} k_j x_r + \varepsilon_{ljr} (\sigma_{kj} \overset{P}{\beta}_{kr} + \mu_{kj} \overset{P}{\varkappa}_{kr}) . \tag{24}$$

Here we used eq. (23) in the form of $\dot{P}_k + T_{ik,i} = k_k$. The physical interpretation of the balance equations is not difficult. The time derivatives contain the density of energy, field momentum and field angular momentum, and the divergence terms contain the corresponding current densities. The most interesting terms are the source terms on the right sides of the equations. Here $-\sigma_{ik} J_{ik} - \mu_{ik} S_{ik}$ is the energy dissipated during the motion of the dislocations and disclinations. The source term of the balance of momentum describes the force density acting on the imcompabilities. It contains the Peach-Köhler force $\varepsilon_{rik} \sigma_{kl} D_{il}$, the Nabarro force $\varrho v_k J_{rk}$ and the analogous forces connected with the disclinations. Furthermore, there is a force density $\varepsilon_{rkl} \sigma_{tk} \varkappa^P_{tl}$ which arises because forces also act on these disclinations in a stress field through the plastic curvature $\varkappa^P_{ik}$ connected with the disclinations. The source term in the balance of angular momentum describes the torque momentum density acting on the incompabilities. They are split in two parts, the orbit and the spin part. The spin part only depends on the plastic deformation and curvature and not on the dislocation and disclination densities.

3.4. Dislocation and disclination polarisation

To describe point defect densities, it is possible to use dislocation loops /4/ which are connected with the dislocation polarisation P_{ik} . We extend this ides to the Cosserat-continuum and also define a disclination polarisation Q_{ik} . In this case one has to replace in all equations

$$
\begin{aligned}
&B_{ik} \quad \text{through} \quad B_{ik} + \varepsilon_{irs} Q_{sk,r} ,\\
&D_{ik} \quad \text{through} \quad D_{ik} + \varepsilon_{irs}\left(P_{sk,r} + \varepsilon_{krl} Q_{sl}\right), \\
&S_{ik} \quad \text{through} \quad S_{ik} + \dot{Q}_{ik} \\
&J_{ik} \quad \text{through} \quad J_{ik} + \dot{P}_{ik}
\end{aligned}
\tag{25}
$$

As a special example we consider a singular point defect described by

$$P_{ik} = M_{ik}\, \delta(x_l) , \quad M_{ik} = M_{ki} , \quad Q_{ik} = 0 .$$

In a given stress field σ_{ik} ($\mu_{ik} = 0$) the force

$$K_r = \int k_r \, dV = \int \varepsilon_{ril} \varepsilon_{its} \, \sigma_{lk} \, M_{sk} \, \delta(x_a)_{,t} \, dV = \sigma_{sk,r}(0) \, M_{sk}$$

and the torque momentum

$$M_l = \int \varepsilon_{ljr} \, \sigma_{kj} \, M_{kr} \, \delta(x_i) \, dV = \varepsilon_{ljr} \, \sigma_{kj}(0) \, M_{kr}$$

act on this point defect. Both terms are well-known /5/. Corresponding terms can easily be obtained for the forces and torque momenta acting on infinitesimal disclination loops.

References:

/1/ Holländer, E.F., Czech. J. Phys. B10 (1960) 409, 479, 511.

/2/ Kosevich, A.M., J. exp. theor. Phys. 42 (1962) 152,
43 (1962) 637.

/3/ Kluge, G., Int. J. Engng. Sci. 7 (1969) 169.
Landau, L.D., Lifschitz, E.M., Lehrbuch d. Theor. Phys., Bd. II., Klass. Feldtheorie (Akad.-Verlag, Berlin (1964)).

/4/ Kroupa, F., Dislocation Loops, in: Theory of Crystal Defects (Academia, Prag (1966))
Landau, L.D., Lifschitz, E.M., Lehrbuch d. Theor. Phys., Bd. VII, Elastizitätstheorie (Akad.-Verlag, Berlin (1965)).

/5/ Kröner, E., Arch. Rat. Mech. Anal. 4 (196) 273.

Continuum Models of Discrete Systems 4
eds. O. Brulin and R.K.T. Hsieh

CYCLIC PLASTICITY OF MATRIX AND PERSISTENT SLIP BANDS IN FATIGUED METALS

Haël Mughrabi

Max-Planck-Institut für Metallforschung,
Institut für Physik, Heisenbergstr. 1,
7000 Stuttgart 80
F. R. Germany

Dedicated to Prof. Dr.-Ing. Dr. rer. nat. h. c. Ulrich Dehlinger on the occasion of his 80th birthday on 6th July 1981

Current problems related to persistent slip bands (PSBs) in fatigued metals are surveyed. A two-component Masing model is developed in order to describe the responses of the matrix and the PSBs during cyclic stressing in saturation. The dislocation-rich dipolar veins in the matrix and walls in the PSBs are considered as hard components, separated by softer dislocation-poor channels. The composite model is supplemented by micromechanical considerations, based on the details of the actually observed dislocation arrangements, and evaluated quantitatively. A satisfactory characterization of the cyclic plasticity of the matrix and the PSBs in saturation is obtained, including as main features the development of internal stresses and lattice misorientations, the shapes of the hysteresis loops, the saturation stresses and an understanding of the processes leading to the nucleation of PSBs.

INTRODUCTION

Cyclic strain localization in persistent slip bands (PSBs) is one of the most widely studied features of metal fatigue. In the range of intermediate- to high-cycle fatigue, localized cyclic strain in PSBs leads to characteristic surface roughening in marked bands which hence act as the dominant crack initiation sites. PSBs are a general phenomenon in a large class of materials such as, e.g., low-carbon steel, precipitation-hardened alloys and pure face-centred cubic (fcc) or hexagonal close-packed metals, for reviews see [1-8]. Although PSBs differ from material to material with respect to structural details, they always display the property of strain localization causing the initiation of fatigue cracks. Much of the work on PSBs has been guided by the natural primary interest in the crack initiation aspect. In addition, however, the study of the properties of PSBs, especially in fatigued fcc single crystals, has become a rewarding task by itself and has increased our understanding of the relation between dislocation microstructures, as observed by transmission electron microscopy (TEM), and the mechanical behaviour of materials. Here we shall confine ourselves to fcc monocrystals, in particular of copper, fatigued at room temperature, whose dislocation arrangements have been thoroughly studied, as described in the above-mentioned reviews. Fig. 1 shows a three-dimensional TEM montage of the dislocation pattern in the bulk of a fatigued copper crystal. Two types of heterogeneous dislocation arrangments can be recognized: The so-called vein or matrix (M) structure and the so-called wall- or ladder-structure in PSB-lamellae which are about 1 μm thick and which lie parallel to the primary glide plane (111) with primary Burgers vector $\underline{b} = [\bar{1}01]/2$. The PSBs can traverse the entire cross section in the case of single crystals. At the sites of their emergence they give rise to the detrimental surface topography, as shown in Fig. 2.

The most pertinent information on PSBs has been obtained on crystals orientated for single slip which were deformed cyclically in push-pull at constant plastic resol-

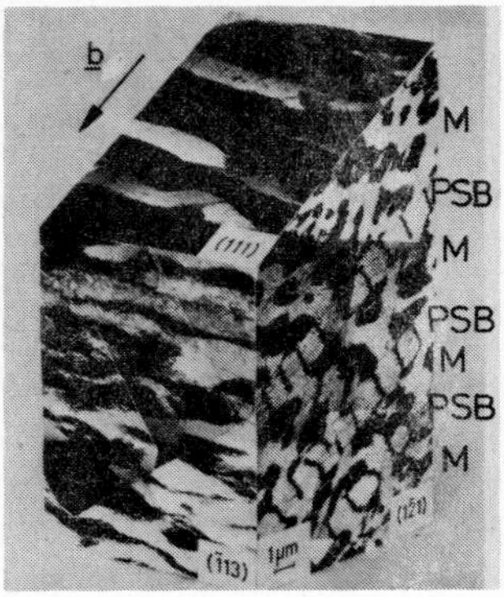

Fig. 1. Three-dimensional view of dislocation arrangement in fatigued copper single crystal. $\gamma_{pl} = 1.5 \times 10^{-3}$. From [4]. Stress axis can be considered as vertical.

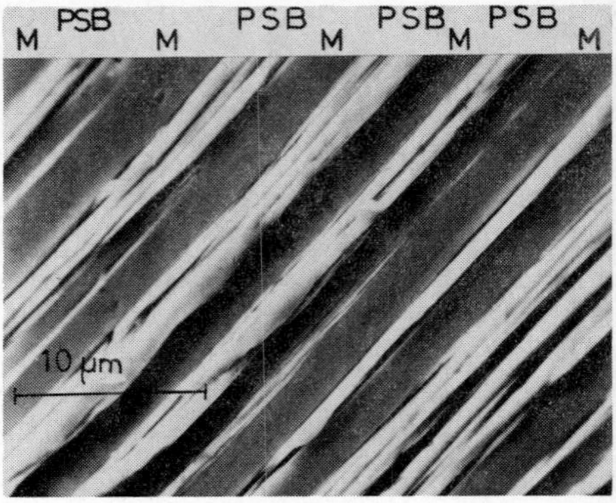

Fig. 2. Scanning electron microscope observation of surface topography at emerging PSBs in the central part of a grain in a fatigued copper polycrystal.

ved shear strain amplitude γ_{pl}. The data are displayed most appropriately in the so-called cyclic stress-strain (css) curve in which the peak resolved shear stress τ_s in cyclic saturation is plotted against γ_{pl}, as shown in Fig. 3 for copper crystals. It is found that below $\gamma_{pl} \simeq 6 \times 10^{-5}$ (regime A) τ_s increases with increasing γ_{pl} [9] and the dislocation arrangements in saturation consist entirely of the vein structure [10]. At $\gamma_{pl} \simeq 6 \times 10^{-5}$ the first discrete PSBs form in the cyclically hardened matrix, acccompanied by slight cyclic softening before saturation finally occurs and by a change in the hysteresis loops from a pointed to a more rectangular shape [9-12]. A cumulative plastic shear strain $\gamma_{pl,cum} = 4\,N\cdot\gamma_{pl}$ (N : number of cycles) of typically 10 to 20 is required till the PSBs are fully developed [9]. Subsequently they carry most of the cyclic strain [12,13]. Above $\gamma_{pl} \simeq 6 \times 10^{-5}$ the volume fraction f_{PSB} occupied by PSBs simply increases approximately linearly with γ_{pl} [9,12,13] and at $\gamma_{pl} \simeq 7.5 \times 10^{-3}$ the PSBs occupy the whole gauge length. In thi entire regime B (cf. Fig. 3) the saturation stress τ_s remains approximately constant (≃28 MPa) and corresponds to the stress required for the localized deformatio in the relatively soft PSBs which are strained in series with the harder matrix. Above $\gamma_{pl} \simeq 7.5 \times 10^{-3}$ (regime C) τ_s increases again with γ_{pl} and a new structure different from that formed in regime B evolves. The occurrence of PSBs is confined to the plateau regime B (Fig. 3) i which the imposed plastic strain amplitude is roughly partitioned between matrix and PSBs according to Winter's [12] two-phase model :

$$\gamma_{pl} = \gamma_{pl,PSB}\cdot f_{PSB} + \gamma_{pl,M}(1-f_{PSB}). \quad (1$$

Here $\gamma_{pl,M}$ and $\gamma_{pl,PSB}$ denote the local values of γ_{pl} in the matrix and in the PSBs, respectively, and can be equated to the values $\gamma_{pl} = 6 \times 10^{-5}$ at the beginning and $\gamma_{pl} = 7.5 \times 10^{-3}$ at the end of the plateau. Winter [12] and Brown [3 have pointed out certain analogies to simple two-phase systems which are documented in Fig. 4 which is almost self-

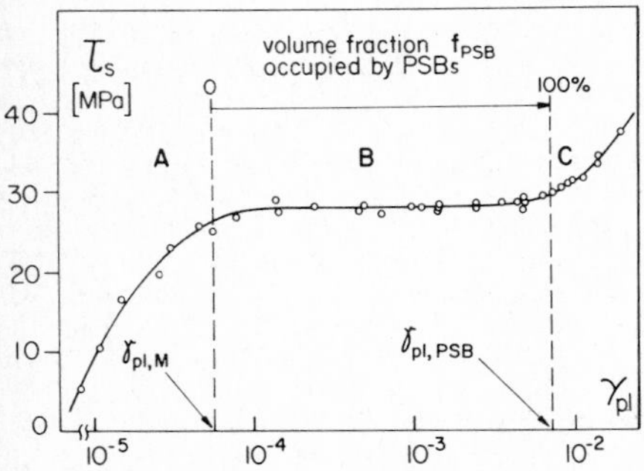

Fig. 3. Cyclic stress-strain curve of copper crystals. After [9].

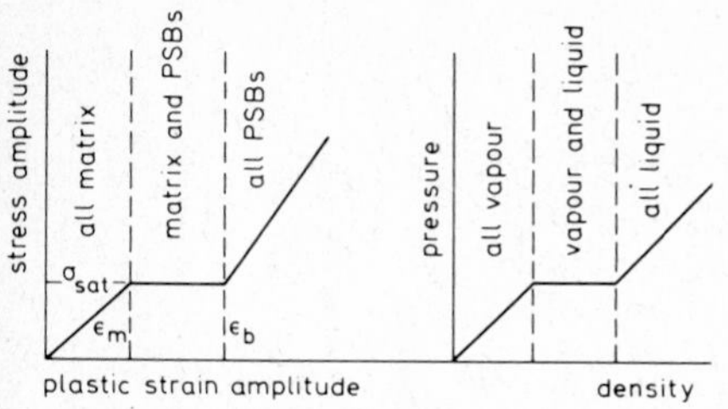

Fatigued copper	Liquid/vapour system
Plastic strain amplitude in PSB's (ε_b)	Density of liquid
Plastic strain amplitude in matrix (ϵ_m)	Density of vapour
Plastic strain amplitude of fatigue test (ϵ_p)	Mean density of mixture
Stress amplitude	Pressure
Saturation stress for $\varepsilon_m < \varepsilon_p < \varepsilon_b(\sigma_{sat})$	Saturation vapour pressure
Motion of dislocations during fatigue test	Brownian motion of molecules
Some function of frequency and temperature	Temperature

Fig. 4. Analogies between fatigued copper crystals and liquid/vapour system. Please note different symbols in figure and accompanying table, as compared to eq. (1). From [12]. Courtesy of A.T. Winter and Taylor & Francis Ltd.

explanatory. A serious limitation, discussed by the authors, is that the formation of PSBs is irreversible in the sense that they do not vanish when γ_{pl} is reduced.

FUNDAMENTAL QUESTIONS

The detailed experimental observations, cf. [1-16], have stimulated theoretical work in several areas. Some of the central questions are related to the following :

a) Heterogeneity of dislocation distribution. Both the matrix and the PSBs consist of alternating high- and low-dislocation-density regions on a microscale, as has been emphasized repeatedly in the literature, cf. in particular [1-8,17-20]. More generally, this is the problem of dislocation cell formation in deformed materials. An adequate theory would have to yield both the ranges of stability in terms of the imposed deformation variables and the characteristic dimensions of the different structures. The work of Burmeister and Hermann [18] seems to be the only attempt made so far in this direction. An existing theory of cell formation [21] pays no attention to the details of the deformation process.
b) Compatibility of deformation of heterogeneous structures. The necessity for strain compatibility in regions of high and low dislocation densities has been pointed out some time ago [1,15]. More detailed considerations have been made only recently [6,19,20].
c) Break-down of matrix, nucleation of PSBs. This "phase transition" is of central interest and several rather different explanations have been suggested [2-9,11, 22-24]. It appears doubtful whether the intriguing but incomplete analogy to two-phase systems indicated in Fig. 4 and in the accompanying Table can be developed further. A successful model must explain how the instability spreads along the glide plane and why the nucleation stress decreases with increasing γ_{pl} from approximately the plateau stress at the beginning of the plateau to significantly lower stresses at higher γ_{pl}, cf. [4,9,24].
d) Reversibility of slip in matrix and PSBs. It has been discussed in more detail elsewhere [4,13,19,25,26] that forward and reverse dislocation glide occur largely reversibly in the matrix but not in the PSBs where cyclic saturation is the result of a dynamic equilibrium between dislocation multiplication and annihilation. The latter slip irreversibility is responsible for the surface roughening at emerging PSBs (Fig. 2). It is largely due to the mutual annihilation of screw dislocations of opposite sign by cross slip [1,4,6-8,27,28], as first recognized by Mott [29]. More recent work has shown that mutual annihilation of close unlike edge dislocations must be considered as well [6,7,25,26,28].
e) Cyclic strain localization in PSBs. Phenomenologically, the origin of the strain localization is clear : at the stress level $\tau_s \simeq 28$ MPa (in copper) the softer PSBs yield plastically to a much larger extent than the harder matrix. In order to explain this behaviour, an understanding of the following is required :
f) Mechanical hysteresis of matrix and PSBs.

g) Internal stresses arising from the inhomogeneous deformation due to the heterogeneity of the dislocation distribution.
h) Saturation stresses of the matrix and the PSBs.

The present study will consider the above topics with the exception of the fundamental topic a) which needs further work and of the question of slip irreversibility, cf. references given under d). The experimentally established dislocation patterns will be taken as given and will be analyzed further with emphasis on certain features that have so far not been widely appreciated and on details that are only observable, if the dislocations are pinned in the stress-applied state (or after unloading) in order to prevent dislocation rearrangement [30,31], cf. [1,4,27,31] for earlier accounts of this work. This characterization of the structure on a microscale will be combined with the simplest possible continuum mechanics considerations in an attempt to provide a basis for the understanding of the above topics. At the same time, it is hoped that this simple-minded experimentalist approach will be able to convey to more theoretically-minded colleagues such as those present at this symposium some of the fascination of the basic questions and to stimulate more rigorous theoretical work in the future.

CHARACTERISTIC FEATURES OF DISLOCATION ARRANGEMENTS

In regime B the dislocation patterns consist largely of dense arrays of segmented primary edge dislocation dipoles clustered into "veins" and "walls" in the matrix and in the PSBs, repectively, separated by "channels" of much lower dislocation density, cf. Fig. 1. Veins, walls and channels are aligned roughly perpendicularly to the primary Burgers vector $\underline{b}$ and extend over several ten µm in the direction $[1\bar{2}1]$ of the primary edge dislocations [14]. The section parallel to the primary glide plane (111) in Fig. 1 shows the dipolar veins of the matrix. In the channels a few weakly curved dislocations of predominant screw character can be recognized in the unloaded state [10]. Inspection of the (111) plane alone suggests erroneously that the mean channel width is smaller than the mean width of the veins, cf. Fig. 1. This result [22] is caused by the projection effects that occur in TEM-specimens having typical thicknesses of 0.3 µm. The correct dimensions are obtained by measuring the widths of channels, d_c, and of veins, d_v, parallel to $\underline{b}$ on TEM-micrographs of $(1\bar{2}1)$-sections of which an example is shown in Fig. 5. In this case the mean widths of channels and veins, $\bar{d}_c$ and $\bar{d}_v$, respectively, are both found to be $\simeq 1.2$ µm (for $\gamma_{pl} = 1.5 \times 10^{-3}$). If one measures pairs of d_c- and d_v-values pertaining to pairs of adjacent channels and veins, one obtains the spatial distributions of d_c and d_v. For our purpose, the result is appropriately plotted as a frequency distribution $Z(f_v)$ of $f_v = d_v/(d_c + d_v)$, i.e. of the area fraction of the veins in a section parallel to (111) which is identical to the volume fraction because of

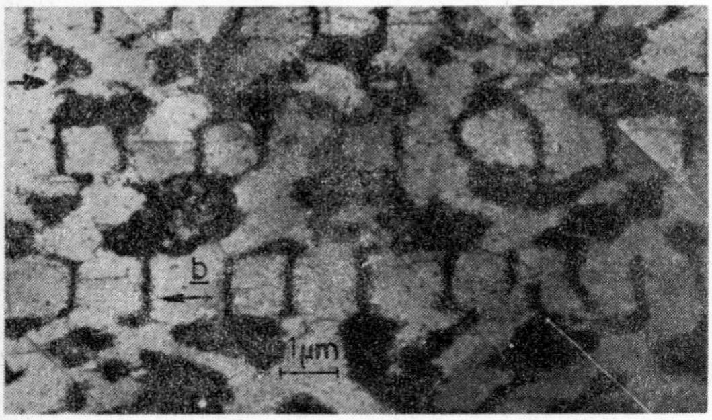

Fig. 5. TEM section parallel to $(1\bar{2}1)$, fatigued copper crystal, $\gamma_{pl} = 1.5 \times 10^{-3}$. Note PSB ladder structures and embryonic PSB marked by arrow. Courtesy of F. Ackermann.

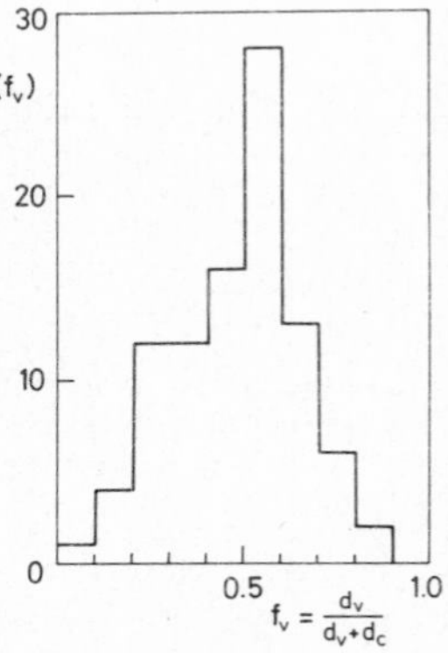

Fig. 6. Frequency distribution of f_v in matrix ($\gamma_{pl} = 1.5 \times 10^{-3}$).

the extended length of the veins in [1$\bar{2}$1]. The result, shown in Fig. 6, indicates a large spatial variation of f_v and hence also of f_c, the volume fraction of the channels. The average volume fractions $\bar{f}_v$ and $\bar{f}_c$ are comparable, i.e. $\bar{f}_v \approx \bar{f}_c \approx 0.5$.

We now turn to the walls of the PSBs. The mean wall spacings are 1.3-1.4 µm [1-14, 22], the mean channel widths, determined from (1$\bar{2}$1)-sections as in the case of the matrix, are $d_c \approx 1.2$ µm [10], i.e. similar to those in the matrix. In view of the widespread belief that the PSB-wall spacings are rather constant [8,22], we point out that, in reality, a wide distribution exists in the range 0.9 µm $< d_c <$ 1.7 µm. Since the wall width d_w lies between 80 and 150 nm, the average volume fraction occupied by the walls in the PSBs is $\bar{f}_w \approx 0.1$. From TEM-micrographs of sections nearly parallel to (111), Winter [22] has concluded that the walls of the PSBs merge into the edges of the veins. It has been pointed out earlier, cf. discussion in [4], that (1$\bar{2}$1)-sections which do not suffer from projection effects in this regard show that many but by no means all of the walls do indeed merge into the edges of the veins, cf. also Fig. 5. It was suggested then that the final positions of the walls in saturation are presumably not indicative of the nucleation process as has been claimed [22,23] but result from gradual adjustments at a later stage, affecting a reduction of the strain energy by minimizing boundary effects.

Regarding the glide processes in the PSBs, the most valuable information is gained from (11$\bar{1}$)-sections, preferably of specimens in which the dislocations have been pinned under load by fast-neutron irradiation, cf. also [1,27,31]. Figs. 7a and 7b show examples. There are three basic dislocation processes occurring in the channels : glide of screw dislocation segments which bow out between adjacent walls and drag out edge dislocations at the wall-channel interfaces (a), encounters of these screw dislocations with screw dislocations of opposite sign (b) and, finally, bowing-out of edge dislocation segments from the walls (c), whereby two new screw dislocations are generated after the edge segment has traversed the channel [1]. In cyclic saturation, screw dislocations of opposite sign on glide planes separated by less than $y_{\odot} \approx 50$ nm [4] annihilate mutually by cross slip and are replaced by the above dislocation multiplication process [1,4,7,25]. It is important, however, that because of the small segment lengths of ≈200 nm, overcritical bowing-out of edge dislocations requires the aid of high internal stresses [1,4]. Evidence for the presence of such internal stresses in the vicinity of the walls has been presented earlier [1,4] but has not found consideration in recent theoretical models [2,3,5,8].

Curved dislocation segments which are well preserved by pinning by fast-neutron irradiation [1,4,27,31,32] represent ideal probes for the determination of local shear stresses τ_{loc} on a microscale, since

$$\tau_{loc} = T_i / b \cdot r_i \quad , \tag{2}$$

where T_i and r_i represent the line tensions and radii of curvature of dislocations of character i, respectively. In our evaluations, we have generally employed experi-

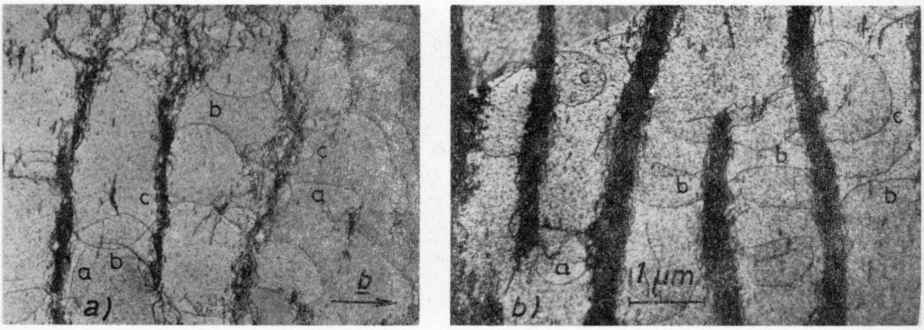

Fig. 7. Dislocation glide in the channels between PSB-walls, stress-applied state. a) Early saturation. b) Saturation. Mechanisms (a), (b) and (c), cf. text.

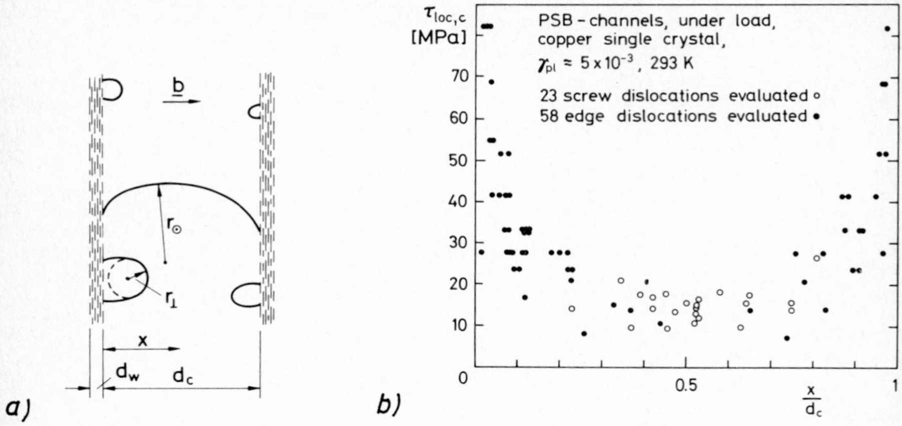

Fig. 8. Local stresses in channels between PSB-walls in the stress-applied state ($\tau_S \simeq 28$ MPa). a) Principle of evaluation. b) Measured stresses.

mentally determined values of line tensions [27] rather than theoretical values of different authors which are neither mutually consistent nor in agreement with experimental data [32]. Using $T_\perp$ = 1 nN and $T_\odot$ = 2.5 nN for edge and screw dislocations in copper, respectively, local stresses $\tau_{loc,c}$ in the channels of PSBs in the stress-applied state were determined from curved edge and screw dislocation segments having the configurations indicated schematically in Fig. 8a . The result is shown in Fig. 8b in which the distance x introduced in Fig. 8a was normalized with respect to d_c in order to combine the data obtained for channels of widely differing widths. The significant fluctuations of $\tau_{loc,c}$ for given values of x/d_c reflect partly fluctuations of local stresses along the channels in [1$\bar{2}$1] and, to the other part, differences related to the different channel widths. The result is clear: $\tau_{loc,c}$ is significantly smaller than the applied stress τ (here $\tau = \tau_S$) in the centre of the channels and considerably larger in the vicinity of the walls, where one finds a large number of short, strongly curved edge dislocation segments, cf. Fig. 7. In the range $0.3 < x/d_c < 0.7$ we find $\overline{\tau}_{loc,c}$ = 14.55 MPa. A furthergoing evaluation is possible by application of Albenga's law [33] which is, in slightly modified form :

$$\frac{\int_V \tau_{loc}\, dV}{V} = 0 \qquad \text{unloaded state} \qquad (3a)$$

$$\frac{\int_V \tau_{loc}\, dV}{V} = \tau, \qquad \text{stress-applied state} \qquad (3b)$$

where V denotes the volume. We now assume that, in the walls, an average stress $\overline{\tau}_{loc,w}$ prevails and that $\tau_{loc,c}$ in the channels extrapolates to this value at the wall-channel interface. With f_w = 0.1, we then obtain with the aid of eq. (3b) the self-consistent values $\overline{\tau}_{loc,c} \simeq 20.55$ MPa as the average local stress in the channels and $\overline{\tau}_{loc,w} \simeq 95$ MPa. Mecke and Messerschmidt [34] have obtained by TEM in-situ deformation work on specimens from a fatigued nickel single crystal a qualitatively similar stress distribution in PSB-channels as that shown in Fig. 8b. In quantitative terms, however, their results appear to be in error, since their values of $\tau_{loc,c}$ exceed τ_S which is $\simeq 50$ MPa in nickel [3,4,11] in the entire range of x/d_c and hence violate eq. (3b) , unless one assumes $\overline{\tau}_{loc,w} \approx -580$ MPa which appears highly unrealistic. The reasons for the discrepancy lie presumably in the use of erroneous theoretical values of the line tension.

MASING MODEL OF CYCLIC PLASTICITY OF TWO-COMPONENT MATERIALS

It appears likely that somebody not familiar with dislocations would consider the structures shown in Fig. 5 as composites consisting of extended dark rod- or plate-like components (veins, walls) separated by light rod-like components (channels). If he then proceeded to consider the compatible shear deformation in the direction of b̲ of such composites, he would realize that this corresponds to deforming the components in parallel, i.e. under the condition that the total shear strain γ_{tot} be constant at any point. This is the basic idea underlying Masing's model of polycrystal plasticity, where the different grains are considered as a large number of components which have different flow stresses [35]. In our case, it suffices in a first approximation to consider just two component elements : the soft dislocation-poor channels and the hard dislocation-rich veins (matrix) and walls (PSBs), as suggested earlier [4,6,36], and to confine ourselves to the particularly simple Masing model of two-component materials [37].

We now consider the transverse shearing of bonded arrays of alternating hard and soft rod-like elements, denoted 1 and 2, having volume fractions f_1 and f_2 and which are assumed to exhibit ideal elastic-plastic behaviour with flow stresses $\hat{\tau}_1$ and $\hat{\tau}_2$ respectively. The shear moduli of both components are assumed to be both equal to G. Upon shearing, the total shear strains in both components must be equal to the imposed total shear strain γ_{tot} and will in general consist of elastic and plastic shear strains, γ_{el} and γ_{pl}, respectively, so that

$$\gamma_{tot} = \gamma_{el,k} + \gamma_{pl,k} . \quad (k = 1 \text{ or } 2) \tag{4}$$

Following initial elastic straining, the softer elements 2 will yield plastically first, when $\gamma_{tot} = \gamma_{el,2} = \hat{\tau}_2/G$ and when the shear stress τ is equal to $\hat{\tau}_2$. The total stress τ at any point will be $\tau = f_1\tau_1(\gamma_{tot}) + f_2\tau_2(\gamma_{tot})$, where τ_1 and τ_2 are the current stresses in 1 and 2, respectively. After yielding of elements 2, τ will increase further with an apparent shear modulus f_1G till the harder elements 1 flow plastically at $\gamma_{tot} = \gamma_{el,1} = \hat{\tau}_1/G$. At that point elements 2 have already undergone a plastic shear strain $\gamma_{pl,2} = (\hat{\tau}_1 - \hat{\tau}_2)/G$. Subsequently τ remains constant and is now equal to the macroscopic flow stress of the composite which is obviously :

$$\hat{\tau} = f_1 \hat{\tau}_1 + f_2 \hat{\tau}_2 . \tag{5}$$

This behaviour is shown in Fig. 9. At the flow stress $\hat{\tau}$ we have for the plastic shear strains :

$$\gamma_{pl,k} = \gamma_{tot} - \hat{\tau}_k/G . \tag{6}$$

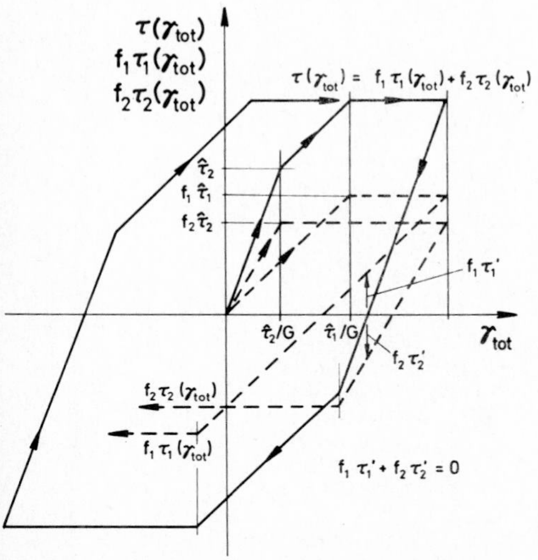

Fig. 9. Stresses, strains and hysteresis in two-component Masing model.

The plastic shear strain γ_{pl} of the composite, on the other hand, is the shear strain remaining after unloading from $\tau = \hat{\tau}$ to $\tau = 0$. Two cases must be distinguished :

$$\gamma_{pl} = \gamma_{tot} - (f_1\hat{\tau}_1 + f_2\hat{\tau}_2)/G \tag{7a}$$

for $\hat{\tau} \leqslant 2\hat{\tau}_2$ and

$$\gamma_{pl} = \gamma_{tot} - (\hat{\tau} - 2f_2\hat{\tau}_2)/f_1G \tag{7b}$$

for $\hat{\tau} \geqslant 2\hat{\tau}_2$.

The first case corresponds to the situation sketched in Fig. 9. In the second case the softer elements 2 undergo reverse plastic yielding while τ is still positive. Hence γ_{pl} is smaller in the second case. It is clear from eqs. (5) to (7) and also evident from Fig. 9 that γ_{pl} is larger than $\gamma_{pl,1}$ and smaller than $\gamma_{pl,2}$. From eqs. (5) and (6) and with the subsidiary condition $f_1 + f_2 = 1$ we obtain for $\tau = \hat{\tau}$:

$$\tau_1 = \hat{\tau}_1 = \hat{\tau} + f_2(\hat{\tau}_1 - \hat{\tau}_2)$$
$$= \hat{\tau} + G(\gamma_{pl,2} - \gamma_{pl,1})f_2 \tag{8}$$

and $$\tau_2 = \hat{\tau}_2 = \hat{\tau} - f_1(\hat{\tau}_1 - \hat{\tau}_2) = \hat{\tau} - G(\gamma_{pl,2} - \gamma_{pl,1})f_1 . \quad (9)$$

In other words, the deformation discussed is affected by a redistribution of the applied stress on a local scale, i.e. by a stress enhancement $\Delta\tau_1 = G(\gamma_{pl,2} - \gamma_{pl,1})f_2$ in elements 1 at the expense of a stress reduction $\Delta\tau_2 = -G(\gamma_{pl,2} - \gamma_{pl,1})f_1$ in elements 2. Upon unloading, these internal stresses remain "frozen in" as residual or eigen-stresses. For the case $\hat{\tau} \leqslant 2\hat{\tau}_2$ to which Fig. 9 refers we find that, after unloading, inspite of τ being zero, the harder elements are still strained elastically in the original sense of deformation, whereas the softer elements are already strained elastically in the reverse sense, so that the respective residual stresses τ_1' and τ_2' are given by

$$f_1\tau_1' + f_2\tau_2' = \tau = 0. \quad (10)$$

Eq. (10) is equivalent to Albenga's law, i.e. to eq. (3a). It is easy to show, since the reverse elastic strain of the composite is $\hat{\tau}/G$, that for $\hat{\tau} \leqslant 2\hat{\tau}_2$:

$$\tau_1' = f_2(\hat{\tau}_1 - \hat{\tau}_2) = \Delta\tau_1 \quad (11)$$

and $$\tau_2' = -f_1(\hat{\tau}_1 - \hat{\tau}_2) = \Delta\tau_2 . \quad (12)$$

For $\hat{\tau} \geqslant 2\hat{\tau}_2$ the softer elements 2 undergo reverse plastic yielding during unloading. Hence $\tau_2' = -\hat{\tau}_2$ and, because of eq. (10), $\tau_1' = f_2\hat{\tau}_2/f_1$. In the case of two-component materials strained to $\hat{\tau}$, reverse straining to the point where the softer elements 2 have undergone a reverse plastic strain equal to the former excess plastic strain $\gamma_{pl,2} - \gamma_{pl,1} = (\hat{\tau}_1 - \hat{\tau}_2)/G$ leads to a situation where, after subsequent unloading to $\tau = 0$, $\tau_1' = \tau_2' = 0$. It can easily be verified that this situation obtains after straining by γ_{tot} into the range $\tau = \hat{\tau}$, followed by reverse straining to $\tau = -\hat{\tau}_2$, whereupon subsequent unloading results in $\tau_1' = \tau_2' = 0$ and in a final plastic strain of the composite

$$\gamma_{pl} = \gamma_{tot} - \hat{\tau}_1/G . \quad (13)$$

The differences between γ_{pl} in eqs.(7a) and (7b) and γ_{pl} in eq. (13) correspond to the extra reverse plastic shear strains of the composite after τ has become negative during the first unloading for the cases $\hat{\tau} \leqslant 2\hat{\tau}_2$ and $\hat{\tau} \geqslant 2\hat{\tau}_2$, respectively. In general multi-component materials (k > 2), similar unloading situations where all τ_k' become zero do not exist, but the conditions to minimize $\sqrt{\langle\tau_k'^2\rangle}$ can be stated [20].

The complete hysteresis loop for fully reversed cyclic deformation is of the form shown in Fig. 9 and exhibits a Bauschinger effect in correspondence with Masing's original results [35,37]. In general the plastic strain is of particular interest and the hysteresis loop is therefore frequently reconstructed by subtracting the elastic from the total strain of the composite. Hereby branches of the hysteresis loop having slopes S in the τ vs. γ_{tot} plot assume slopes S* in the τ vs. γ_{pl} plot,

whereby $$S^* = G\cdot S/(G - S) . \quad (14)$$

In this presentation the elastic line becomes infinitely steep, i.e. vertical.

Next we wish to consider what rôle dislocations play in the model outlined above. In Fig. 10 the shear deformations of the free elements 1 and 2 up to the elastic limit $\gamma_{el,1} = \hat{\tau}_1/G$ are illustrated. The plastic shear strain in the softer element is brought about by the glide of dislocations which leave the element through its surfaces. In the unloaded state element 1 returns to its original shape, whereas element 2 retains the plastic shear strain $(\hat{\tau}_1 - \hat{\tau}_2)/G$. We now consider the same kind of deformation applied to an array of alternating elements 1 and 2 bonded at their interfaces (Fig. 10b). Now the dislocations responsible for the plastic shear deformation in the softer elements 2 get stuck at the interfaces and, in the unloaded state, elements 1 remain slightly strained elastically in the original sense of deformation, whereas elements 2 undergo some reverse elastic straining in addition to that they would have experienced in the unbonded state, so that the requirements of eq. (3a) are fulfilled. We thus realize that the edge dislocations at the interfaces are the sources of the internal stresses discussed before, cf. eqs. (11) and (12), and provide a further basis for Masing's model. It is also apparent from Fig. 10b that, in the unloaded state, these interface dislocations give rise to a to-and-fro bending of the lattice glide planes about the line direction of the

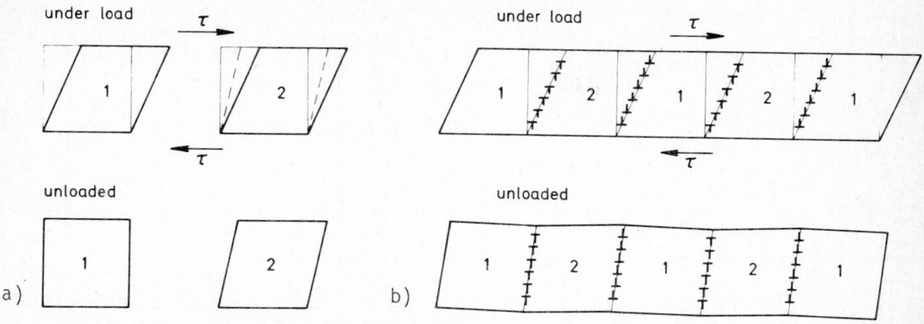

Fig. 10. Shearing of hard (1) and soft (2) elements. Faint lines indicate initial shape. a) unbonded elements. Dashed lines indicate shape of element 2 after elastic shearing. b) Bonded elements. Note shapes after unloading.

edge dislocations by an angle $\tan^{-1}(\hat{\tau}_1 - \hat{\tau}_2)/G$ or $\tan^{-1}(\hat{\tau}_2(f_2 - f_1)/f_1G)$ for the cases $\hat{\tau} \leqslant 2\hat{\tau}_2$ and $\hat{\tau} \geqslant 2\hat{\tau}_2$, respectively. At the same time the lattice plane spacings other than those of the glide planes and of the planes parallel to the plane of Fig. 10 become slightly larger or smaller, depending on the orientations of the planes relative to the glide plane. In general, it will also be necessary to consider that the edge dislocations are parts of closed dislocation loops and have in fact been drawn out along the interfaces by screw dislocations extending across the soft elements 2 very much like the screw segments in the channels in Figs. 7 and 8a. The effect of these screw segments on the internal stresses is expected to be less significant, since they, on the average, alternate in sign, but polarization effects of grouped arrays of screw dislocations [1,27] presumably play a rôle[17].

A comparison with existing theories of dispersion hardening [38-40] shows a complete analogy up to deformations $\gamma_{tot} = \hat{\tau}_1/G$, i.e. up to the point where the hard elements 1 yield plastically, as opposed to, for example, SiO_2 or Al_2O_3 particles (embedded in a soft matrix) which are not considered to flow plastically. Specifically, one finds in these theories which are more rigorously developed than our simple model the internal stresses $\Delta\tau_1 = G\Gamma\gamma_{pl,2}f_2$ and $\Delta\hat{\tau}_2 = -G\Gamma\gamma_{pl,2}f_1$ (with $\gamma_{pl,1} = 0$). Here Γ is the accommodation coefficient introduced by Eshelby [41] which depends on the shape of the particles and the mode of deformation and varies between 0 and 1. This result coincides formally with our evaluation of the two-component Masing model for $\Gamma = 1$. The latter Γ-value is larger than those calculated for particles of various shapes [42] . A more appropriate comparison would be with Γ-values of periodic arrays of transversely sheared closely spaced wall- or vein-like rods which are not available.

In the related work of Asaro [43] and of Holste and Burmeister [20], Masing's early work is also used as a starting-point. Asaro distinguishes between different types of kinematic hardening. The model presented above corresponds to his case of kinematic hardening type I. Holste and Burmeister deal with the general case of multi-component materials in which the flow stresses of the individual components are continuously distributed and show how the eigen-stresses after unloading from any point of the hysteresis loop can be obtained. The two-component model used here lacks similar generality but has the merit of simplicity and appears ideally suited to model the behaviour of matrix and PSBs. If required, different elastic moduli of the components and additional features characteristic of dislocation glide such as non-ideal elastic-plastic behaviour can be incorporated readily. In the following, the dislocation processes in the matrix and in the PSBs, as inferred from TEM, will be analyzed and incorporated into the model in extension of our earlier work [6,36].

BASIC DISLOCATION MECHANISMS IN THE MATRIX AND IN THE PSBS

In general, the flow stress $\hat{\tau}$ consists of an athermal component $\hat{\tau}_G$ and an effective thermal component $\hat{\tau}^*$ [44] . In our case we assume that both $\hat{\tau}_G$ and $\hat{\tau}^*$ are given by relations of the form of eq. (5) with the index 1 referring to veins or walls and the index 2 to the channels. $\hat{\tau}^*$ acts like a friction stress and arises from thermally activated glide processes such as interaction of dislocations with point defect clusters and jog-dragging [4,15,45,46] . Little is known about the components of $\hat{\tau}^*$ in the veins or walls and in the channels. The data of Piqueras et al. [45] suffer from the fact that the specimens were not well saturated and that the evaluation considered thermally activated glide to occur in <u>either</u> elements 1 or 2 but not simultaneously in both, as required by our model and as considered in general terms by Burmeister and Holste [19]. We have determined the effective stress $\hat{\tau}^*_c$ in the channels of the PSBs as follows. As described earlier and as shown in Fig. 8b, the mean local stress in the centre of the channels in the stress-applied state is $\overline{\tau_{loc,c}} \simeq 14.55$ MPa. In a similar experiment the corresponding stress in the unloaded state has been determined as $\overline{\tau^{\mathsf{T}}_{loc,c}} \simeq -6.6$ MPa. Since $2\hat{\tau}^*_c = \hat{\tau}_s - (\overline{\tau_{loc,c}} - \overline{\tau^{\mathsf{T}}_{loc,c}})$, the measured stresses in the channels being athermal stresses, we find $\hat{\tau}^*_c \simeq 4.0$ MPa (with $\hat{\tau}_s \simeq 28$ MPa). At the same time we are able to deduce $\hat{\tau}^*$ from the stress level at which reverse microyielding occurs and obtain $\hat{\tau}^* \approx 4.5$ MPa. We obtain finally, using a relation like eq. (5), $\hat{\tau}^*_w \simeq 10$ MPa for the effective stress in the walls. With regard to the matrix, we assume a similar value of $\hat{\tau}^*_c$ in the channels but neglect $\hat{\tau}^*_v$ in the veins, since the small displacements of the edge dislocations will hardly allow them to sample the obstacles responsible for any stress $\hat{\tau}^*_v$.

In the following the athermal stress contributions in the channels and in the veins and walls will be evaluated. The glide of the screw dislocations in the channels is envisaged to proceed by the sequence indicated in Fig. 11 which refers to the PSB-wall-structure but can also serve to illustrate the situation in the matrix. Under an increasing stress the dislocations will first bow out to the semi-elliptical shape corresponding to the critical Orowan or Frank-Read configuration 1 at a stress

$$\tau_{FR,\odot} \simeq 1.5\ Gb/d_c \quad . \tag{15}$$

Eq. (15) refers to ellipses with an axial ratio of $\simeq 1.4$, cf. [27,32,47], with $G = \sqrt{c_{44}(c_{11} - c_{12})/2} = 42000$ MPa in copper (c_{11}, c_{12} and c_{44} are the elastic moduli). We find, with $\bar{d}_c \simeq 1.2$ µm, $\tau_{FR,\odot} \simeq 13$ MPa. At the configuration 1, the plastic shear strain in the channels is easily shown to be

$$\gamma^1_{pl,c} \simeq 0.1 \cdot \pi \cdot \rho_\odot \cdot b \cdot d_c \quad , \tag{16}$$

where $\rho_\odot$ denotes the density of the screw dislocations. Subsequently the screw dislocations continue to glide at the stress given by eq. (15) and draw out edge dislocations at the interfaces between channels and walls (or veins), cf. [1,2,4,8]. As a consequence of the plastic strain in the channels, internal stresses build up, cf. eqs. (8) and (9), until the local stress in the walls attains the value of the flow stress of the walls, causing the latter to yield. Beyond that point (configuration 2) the excess edge dislocations formed can penetrate into the walls and the internal stress saturates. Steady-state in the walls is maintained by the annihilation of the excess edge dislocations by the collapse of dipoles with heights $y_\perp \lesssim 1.6$ nm [4,6,25]. As will be shown subsequently, the dislocation displacements in saturation are so small in the matrix that they remain confined to a configuration just prior to 2 in Fig. 11, cf. [6], whereas they are more extensive in the PSBs because of the larger strains and go through the full sequence from 0 to 3, cf. Fig. 11.

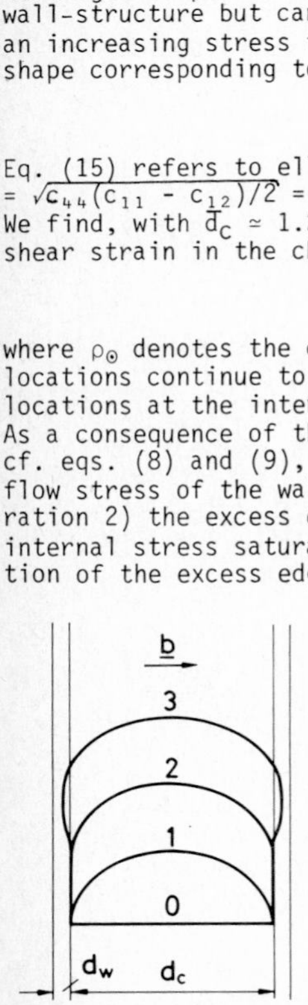

Fig. 11. Glide sequence of screw dislocations.

Next we consider the interaction resulting from encounters between curved screw dislocations of opposite sign. The stress required for the gliding of two screw dislocations of opposite sign past each other is the dipole passing stress

$$\tau_{dip,\odot} = Gb/4\pi y_\odot \quad , \tag{17}$$

where $y_\odot$ denotes the spacing between the two glide planes. In the matrix $\rho_\odot \simeq 10^{12}$ m^{-2} [10]. Hence $y_\odot \simeq 1/\sqrt{\rho_\odot} \simeq 1$ µm and $\tau_{dip,\odot} \simeq 0.8$ MPa. This stress is insignificant relative to $\tau_{FR,\odot} \approx 13$ MPa and can be ignored in the matrix. In the PSBs, however, $\rho_\odot \simeq 1\text{-}2 \times 10^{13}\,m^{-2}$ and evidence has been provided for extensive mutual annihilation of close screw-dislocation dipoles for $y_\odot < 50$ nm [4,10,25]. Hence the largest value of $\tau_{dip,\odot}$ is given by eq. (17) with $y_\odot \simeq 50$ nm as $\tau_{dip,\odot} \simeq 16.7$ MPa, i.e. it is similar in magnitude to $\tau_{FR,\odot}$. We hence have to consider the problem which has been touched upon before [1,8] of whether and how these stresses superimpose. A simple argument shows that these two stresses do not superimpose, as was erroneously assumed by ouselves earlier [1], but that the larger of the two governs the flow stress: In a crude approximation it suffices to consider only the mutual interaction in the centre of the channels, where the available stress is lowest (Fig. 8b) and where the two segments first meet. If $\tau_{dip,\odot} > \tau_{FR,\odot}$, then the dislocations trap each other at $\tau_{FR,\odot}$ and the stress must increase, causing a partial straightening-out of the segments in the centre of the channel, accompanied by a reduction of the retarding stress due to the line tension. When the segments are completely straightened out over a certain fraction of their length, they can glide past each other at just the stress $\tau_{dip,\odot}$. A similar consideration shows that, for $\tau_{dip,\odot} < \tau_{FR,\odot}$, the dislocations can pass at the stress $\tau_{FR,\odot}$. These arguments provide, as a by-product, an important hint to one of the factors governing the channel width and hence the wall spacing. Equating $\tau_{FR,\odot}$ to $\tau_{dip,\odot}$ we find that, for $d_c < 6\pi y_\odot$, the athermal flow stress in the channel would have to be larger than $\simeq 16.7$ MPa and would be $\tau_{FR,\odot}$.

To summarize, the glide of the screw dislocations in the PSB-channels is determined largely by the dipole-interaction stress with the exception of some of the narrower channels in which $\tau_{FR,\odot}$ dominates. The mean local stress of $\simeq 14.55$ MPa available in the centre of the channels according to Fig. 8b is very compatible with the stresses of the processes discussed. In addition, we have in the case of the channels of the PSBs (but not of the matrix) the bowing-out of edge-dislocation segments from the walls and across the channels. This process benefits from the high internal stresses in the vicinity of the walls, resulting from the drawn-out edge dislocations, cf. Figs. 8b and 11, and becomes effective gradually as the local stress in the walls rises with increasing plastic flow to the flow stress level of the walls.

In the final part of this section we consider the processes in the veins and in the walls in which the local density of edge dislocations in the form of narrow dipoles, $\rho_\perp$, is in excess of $\simeq 10^{15}\,m^{-2}$ [1-8,24,48,49]. The passing stress for (unlike) edge dislocations on glide planes separated by a spacing $y_\perp$ is

$$\tau_{dip,\perp} = Gb/8\pi(1-\nu)y_\perp \quad , \tag{18}$$

where ν is Poisson's number ($\nu = 0.35$ in copper). One possibility to evaluate eq. (18) is to use a mean value $y_\perp \simeq 1/\sqrt{2\rho_\perp}$ corresponding roughly to a square array (similar to a Taylor lattice) of alternating positive and negative dislocations [1]. Then, using $\rho_{\perp,v} \simeq 2 \times 10^{15}\,m^{-2}$ and $\rho_{\perp,w} \simeq 5\times10^{15} m^{-2}$ as best average values for the veins and the PSB-walls, respectively, one obtains corresponding values $\tau_{dip,\perp,v} \simeq 40.7$ MPa and $\tau_{dip,\perp,w} \simeq 64.3$ MPa. However, if one considers that the individual dipoles do not form a Taylor lattice but are well separated, cf. [24,50], and that, according to Antonopoulos et al. [50], the dipole heights lie in the range 3 nm $< y_{\perp,v} < 15$ nm, a distribution of passing stresses with 43 MPa $\lesssim \tau_{dip,\perp,v} < 214$ MPa, is found to exist in the veins. We assume that in the PSB-walls a similar distribution exists, presumably peaked to somewhat smaller $y_{\perp,w}$, since $\rho_{\perp,w} > \rho_{\perp,v}$.

Because of the small microstrains in the matrix, some thought must be given to the question whether the high dislocation density gives rise to a significant reduction in shear modulus. As has been discussed earlier [1,51], the elastic polarization of the dislocation dipoles in the veins will certainly cause a shear modulus reduction ΔG. According to Kuhlmann-Wilsdorf [51], this effect should be as large as $\Delta G/G \simeq -50$ % for a Taylor lattice. However, internal friction measurements [52] indicate that fatigued copper exhibits a characteristic ΔG-effect of only $\simeq -10$ %. It has been indicated above and before [24] that a Taylor lattice of infinitely extended edge dislocations is unrealistic. In reality, the areas swept out by the displacements of the dipole dislocations are reduced by the fact that the segments must bow out between the dipole ends. In addition, if the edge-dislocation dipoles corresponding to $\rho_{\perp,v} \simeq 2 \times 10^{15}\,m^{-2}$ and having a mean dipole height of $\simeq 5$ nm [50] were arranged in a Taylor lattice, they would only occupy a small fraction of the veins, leaving the rest volume unaffected. This is reflected in the fact that in [51] it has to be assumed that the Taylor lattice is composed of dipoles with a height as large 19 nm ! An estimate shows that the two effects discussed reduce the ΔG-effect in the veins, as calculated in [51], by more than one order of magnitude. We therefore feel justified to ignore the ΔG-effect in the veins (and walls) and point out that the contribution from the displacements of the screw dislocations in the channels to the reduction of the shear modulus is at least as important and that the two effects together appear to be consistent with the observations [52].

HYSTERESIS LOOPS OF MATRIX AND PSBS IN CYCLIC SATURATION

Under the assumption that τ_s is constant throughout the plateau in Fig. 3 and that f_{PSB} depends on γ_{pl} precisely as described by eq. (1), the implicit assumption being that the properties of matrix and PSBs are independent of $\dot{\gamma}_{pl}$, the hysteresis loops pertaining to the beginning and the end of the plateau, respectively, are of particular interest. With the knowledge of the hysteresis loops for $\gamma_{pl} = \gamma_{pl,M}$ and for $\gamma_{pl} = \gamma_{pl,PSB}$, all other hysteresis loops between these two extremes corresponding to "only matrix" or "only PSBs", respectively, follow in principle with the aid of eq. (1). Fig. 12 shows on the left experimentally observed hysteresis loops for the two limiting situations in cyclic saturation, drawn slightly displaced upwards in order to suppress the commonly observed small difference between tensile and compressive stresses. The saturation stresses at the two extremes are found to differ by ≃ 4 MPa, indicating a deviation from the idealized model (Fig. 4). Based on the dislocation processes discussed above, the forward and reverse yielding of matrix and PSBs were calculated according to the principles of the two-component Masing model.

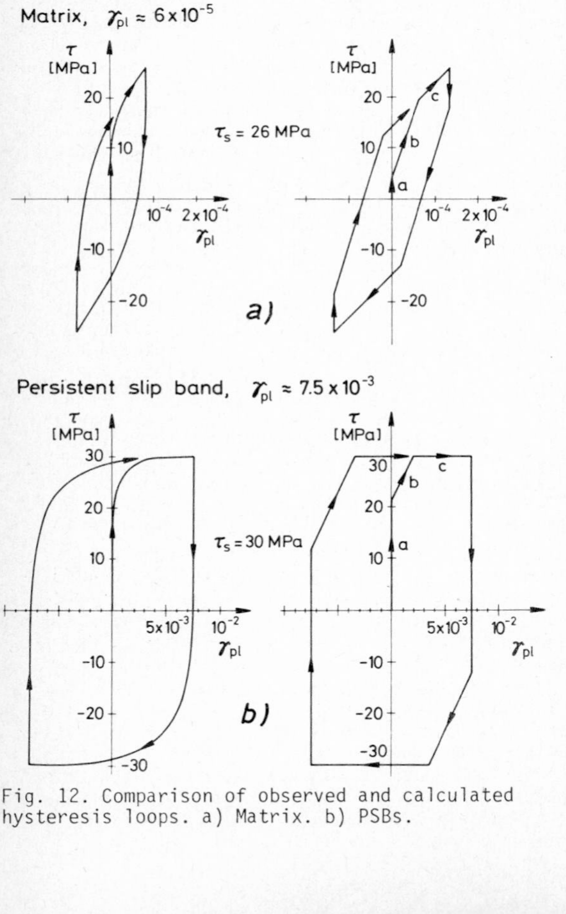

Fig. 12. Comparison of observed and calculated hysteresis loops. a) Matrix. b) PSBs.

In the case of the matrix, the <u>mean</u> numerical values derived above and which pertain to $\gamma_{pl} \simeq 10^{-3}$ were used, since no detailed data exist for $\gamma_{pl} \simeq 6\times10^{-5}$. The hysteresis loop, approximated by a polygon, follows from the branches a, b and c of the calculated <u>microyielding curve</u> (Fig. 12a). In branch a overall elastic straining occurs till τ reaches the friction stress $\hat{\tau}^*_c$ in the channels. Subsequently the screw dislocations bow into the semi-elliptical configuration in branch b (corresponding to $\gamma^1_{pl,c} \simeq 10^{-4}$) and continue to glide in branch c while the veins are still strained only elastically. In the model, deformation was reversed after τ had attained the experimentally observed value $\tau_s \simeq 26$ MPa. At this point the channels had undergone an elastic strain of $\simeq 4\times10^{-4}$ and a slightly larger plastic strain and the veins had been strained only elastically to $\simeq 9\times10^{-4}$.

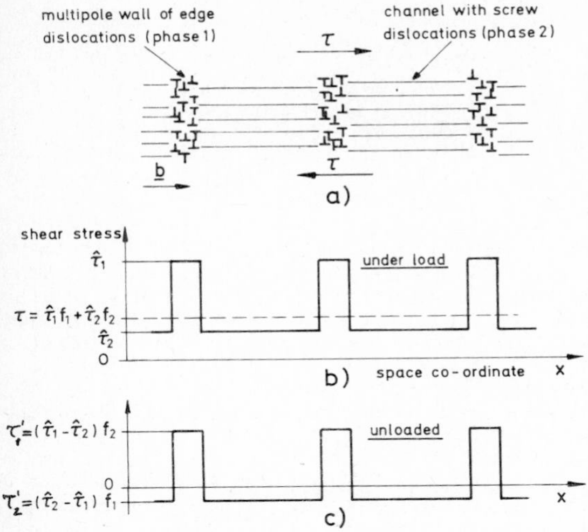

Fig. 13. Model of glide in PSB wall-structure. a) Dislocation distribution. b) and c) show local stress distributions in the stress-applied and unloaded states, respectively. From [6].

At $\tau = \tau_S$, the mean local stress in the veins, calculated according to eq. (8) is found to be almost 40 MPa. This indicates that the available stress is close to the stress at which the widest dipoles flip. Thus the veins are strained approximately to the elastic limit of their weakest parts but do not yield plastically on the whole.

Comparison of the calculated and the observed hysteresis loops indicates fair agreement with regard to the main features. It is apparent, however, that, compared with the measured curve, the calculated loop is somewhat broader and less steep. In order to improve the model further, the actual data for $\gamma_{pl} = 6 \times 10^{-5}$ with the real distributions of channel and vein widths should be evaluated. There is, in fact, evidence that $\bar{f}_V$ increases somewhat with decreasing γ_{pl} so that, at $\gamma_{pl} = 6 \times 10^{-5}$, $\bar{f}_V$ could be larger than 0.5[10]. A higher value of $\bar{f}_V$ would make the loop steeper, just like consideration of the fact that very narrow channels with $d_C \simeq 0.7$ µm yield plastically only at appreciably higher strains than those with $d_C \simeq 1.2$ µm.

In the calculation of the hysteresis loop of the PSBs, the contribution from the bowing of the edge dislocation out of the walls was ignored for the sake of simplicity. This amounts to approximating the real local stress distribution (Fig. 8b) by a rectangular stress profile, as shown in Fig. 13 [6]. The mean flow stress in the channels was taken as $\hat{\tau}_C^* + \tau_{dip,\odot}$ (for $y_\odot$ = 50 nm) = 20.7 MPa. The branches a and b in Fig. 12b correspond to elastic straining of the composite to a shear strain of $\simeq 5 \times 10^{-4}$ followed by plastic yielding of the channels and continued elastic straining of the walls. It was assumed that the walls yield in branch c (macroyielding) after τ has reached the observed value $\tau_S \simeq 30$ MPa and that branch c continues till $\gamma_{pl} = 7.5 \times 10^{-3}$ according to eq. (7a). Calculating backwards as in the case of the matrix, we find that, at the beginning of branch c, the channels have been strained elastically by $\simeq 5 \times 10^{-4}$ and plastically by $\simeq 2.2 \times 10^{-3}$, whereas the walls have been strained only elastically by $\simeq 2.7 \times 10^{-3}$. Thus the local stress in the walls (eq.(8)) is found to be $\simeq 113$ MPa. For comparison with the wall stress deduced from Fig. 8b we note that the stress $\hat{\tau}_W^* \simeq 10$ MPa should be subtracted from the above value. This leaves an athermal flow stress of the walls, $\hat{\tau}_{WG} \approx 103$ MPa, as compared to the value $\overline{\tau_{loc,W}} \simeq 95$ MPa determined from Fig. 8b (for $\tau_S \simeq 28$ MPa). This agreement is considered very satisfactory in view of the approximations made regarding the stress distribution (Fig. 13). The above values of $\simeq 100$ MPa for the flow stress of the walls are compatible with passing stresses of dipoles of heights down to $\simeq 6$ nm, i.e. twice as wide as the narrowest dipoles observed [50]. It seems entirely plausible that, since the dipole population changes continuously during steady-state dislocation production and annihilation, homogeneous deformation of the walls proceeds by the breaking-up of such and wider dipoles without the need to dissociate the more tightly bound narrower dipoles. Regarding the shape of the loop, the PSBs, as opposed to the matrix, exhibit unconstrained plastic flow at $\tau = \tau_S$. Strain localization in PSBs follows naturally. The shape of the loop can be improved by allowing some bowing of the screw dislocations before $\tau_{dip,\odot}$ becomes dominant. This would lead to some reverse dislocation glide upon unloading, as is observed [9].

RESIDUAL STRESSES AND LATTICE MISORIENTATIONS

Evidence for the residual stresses and their changes within a cycle has been provided in the case of fatigued nickel polycrystals. Upon unloading from different points on the hysteresis curves, characteristic changes in the coercive force [53] and in the widths of the X-ray line profiles [54] were observed which can be interpreted in the general Masing-type composite model [20]. With regard to lattice misorientations upon unloading, cf. Figs. 10 and 13, we obtain for unloading at γ_{pl} = = 0, cf. Fig. 12, angles of only ≈0.2' and ≈7' for the matrix and the PSBs, respectively. Experimentally, X-ray rocking curve half-widths of ≈20' have been observed on copper crystals containing mainly PSBs, and it has been noted that this rather large value is mainly due to additionally superimposed misorientations with wavelengths of ≈50 μm [24], as evidenced by TEM and X-ray topography [10]. Furthermore, an X-ray rocking curve half-width of ≈3' was observed on a copper crystal that was not cycled into saturation at $\gamma_{pl} \approx 5 \times 10^{-4}$, i.e. an amplitude considerably larger than $\gamma_{pl,M}$. Hence, this larger value (than ≈0.2') is in accord with the picture that, before development of PSBs and accompanying cyclic softening, the plastic strains in the matrix are significantly larger than $\gamma_{pl,M}$ and cause larger misorientations. Finally, we note for the PSBs and for the matrix that, upon unloading from a point compatible with eq. (13), residual stresses and misorientations should be negligible.

THE PLATEAU STRESS

It follows from our earlier considerations that the saturation stress in the presence of PSBs, i.e. the stress level of the plateau, cf. Figs. 3 and 4, is given by

$$\tau_s = \{(\hat{\tau}_c^* + \tau_{dip,\odot})d_c + (\hat{\tau}_w^* + \hat{\tau}_{w,G})d_w\} / (d_c + d_w). \qquad (19)$$

Brown [3,8] has noted empirically that a relation

$$\tau_s = (2 \pm 0.3)Gb/(1-\nu)(d_c + d_w) \qquad (20)$$

holds for different materials cycled in the plateau regime at different testing temperatures. The uncertainty in eq. (20) of ±15% increases to approximately ±20 %, when the temperature dependence of G is taken into account. Hence, eq. (20), inspite of being useful, should not be considered as accurate. For this reason, we do not find it disturbing that eq. (19) does not yield a simple inverse dependence of τ_s on $(d_c + d_w)$. It is interesting to note that τ_s of eq. (19) is expected to increase with decreasing temperature as has been observed [16], not only through the increase of $\hat{\tau}_c^*$ and $\hat{\tau}_w^*$ but also through the increase of $\tau_{dip,\odot}$, since $y_\odot$ will be expected to decrease as the cross slip ability of the screw dislocations diminishes (and $\rho_\odot$ increases). As suggested earlier [25], the dislocation density in the walls is a little larger at low temperatures, mainly because of the decreasing wall width, observable in the work of Basinski et al. [16]. This may enhance $\hat{\tau}_{w,G}$ moderately.

THE THRESHOLD STRESS FOR THE FORMATION OF PSBS

As noted before [4,6], the characteristic difference between the cyclic plasticity of the matrix and the PSBs in saturation is that, whereas the matrix exhibits only microyielding without plastic flow of the veins, the PSBs show macroyielding with plastic flow of the walls. We hence suggest that at $\gamma_{pl} = \gamma_{pl,M}$, the break-down of the matrix and the formation of the first PSBs are triggered by the first significant enforced dislocation glide activity <u>in</u> the veins <u>after</u> these have hardened to the extent that they resist dislocation glide. We envisage that this occurs at a stress slightly in excess of $\tau_s \approx$ 26 MPa, cf. Fig. 12, when γ_{pl} exceeds a threshold value at which the stresses built up in the veins during the yielding of the channels, cf. eq. (8), suffice to initiate (catastrophic) yielding in some weak vein sites. The freed dislocations would then impinge on neighbouring veins and cause the instability to spread along the glide plane very much like in a Neumann-type strain burst [55]. Mecke [11] has presented TEM observations on nickel single crystals fatigued to the point where cyclic softening commenced. As shown in Fig. 14, Mecke found disturbances going through the veins and along the glide plane which he called "cords". We assume with Mecke that these cords represent the initial stage of the rapidly spreading instabilities and develop subsequently in a gradual process of propagation and widening [9] into the final PSBs with the wall structure.

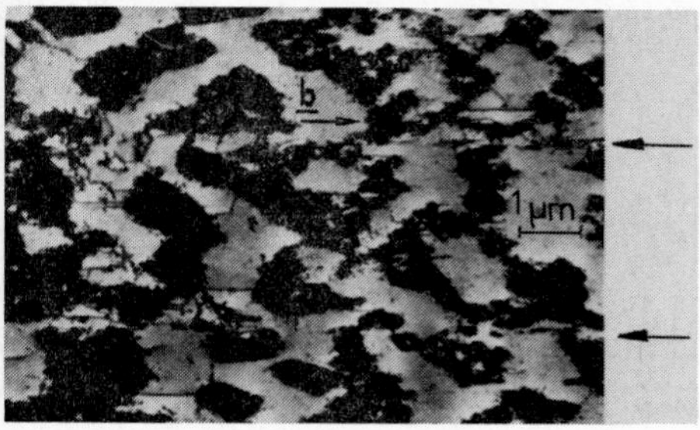

Fig. 14. "Cords" in ≈(1$\bar{2}$1)-section of fatigued nickel crystal. From [11]. Courtesy of Akademie-Verlag and K. Mecke.

This explains why cords are not commonly found in saturation. Contrary to a recent statement [23], remnants of cords are occasionally observed later in saturation. Fig. 5 shows an example of such embryonic PSBs which did not develop further [10]. The final PSB wall-structure does not exhibit similar instabilities as the veins, presumably because the continuously changing steady-state dipole distribution in the walls is not given the opportunity to harden gradually to the point of instability.

Contrary to a widespread belief, the instability does apparently not initiate near the surface but, as shown recently by Yamamoto and Imura [56] and confirmed by ourselves in a preliminary study, in the centre of the crystal where the stress is highest (as in unidirectionally deformed specimens [57,58]) and where the dipoles hence first attain the critically hardened metastable configuration.

In our model the threshold stress for PSB-formation at $\gamma_{pl} = \gamma_{pl,M}$ is given by

$$\tau_{PSB} = \{(\hat{\tau}^*_C + \tau_{FR,\odot})d_C + (\hat{\tau}^*_V + \hat{\tau}_{V,G})d_V\}/(d_C + d_V), \quad (21)$$

in accord with Fig. 12a. At higher γ_{pl} the nucleation of PSBs in copper crystals has been found to occur already at stresses below the final saturation stress [4,9,24]. Recent work on precipitation-hardened alloys [59,60] also indicates that PSBs can nucleate at stresses that are lower than the "plateau stresses" [59,61] of these alloys. In order to explain this behaviour qualitatively, we consider eq. (8). If we replace $\hat{\tau}_1$ and $\hat{\tau}$ by the current stresses τ_1 and τ and set $\gamma_{pl,1} = 0$, corresponding to the situation that the hard component has not yet yielded plastically, then eq. (8) describes the stress in the hard veins or in the precipitates before the instability. In the alloys the precipitates are shearable particles and become unstable during dislocation glide just like the veins. At the instability we have $\tau_1 \rightarrow \hat{\tau}_1$ and $\tau \rightarrow \tau_{PSB}$, so that eq. (8) reads

$$\tau_{PSB} = \hat{\tau}_1 - G\cdot\gamma_{pl,2}\cdot f_2 \quad . \quad (22)$$

In general, $\gamma_{pl,2} > \gamma_{pl}$. Using the data of Fig. 12a ($f_2 = 0.5$, $\gamma_{pl,2} \simeq 5\times10^{-4}$, $\hat{\tau}_1 \simeq 37$ MPa), eq. (22) yields $\tau_{PSB} = \tau_s \simeq 26$ MPa. At very small γ_{pl} ($\simeq 10^{-4}$), the second term on the r.h.s. is significantly smaller than $\hat{\tau}_1$, so that $\tau_{PSB} \lessapprox \hat{\tau}_1$. At higher γ_{pl} ($\gtrsim 10^{-3}$) the second term is non-negligible and, very approximately, $\gamma_{pl} \simeq \gamma_{pl,2}$, so that $\tau_{PSB} \simeq \hat{\tau}_1 - G\cdot\gamma_{pl}\cdot f_2$. According to this approximate relation, τ_{PSB} will in general decrease at higher γ_{pl}. In the case of pure fcc metals, a quantitative evaluation appears premature without knowing how τ_1 builds up during cycling at different γ_{pl}. In precipitation-hardened alloys with shearable particles, however, eq. (22) should be applicable, since $\tau_1 = \hat{\tau}_1$ is a given property of the particles. The shape and hardening mechanism of the particles can be considered more explicitely by replacing G by the appropriate GΓ and by defining f_1 as an effective volume fraction determined not so much by the actual volumes of the particles but more by the effective volumes in which the hardening mechanism is operative.

CONCLUDING REMARKS

In the present work the ideas of Masing have been extended and evaluated for a particularly simple case by taking into account the details of dislocation glide in fatigued fcc metal crystals which deform by single slip and which contain well-characterized heterogeneous dislocation distributions. With the exception of the closely related work of the Dresden group [17-20,53,54], similar ideas involving internal stresses due to deformation inhomogeneities have so far not been widely applied in the study of cyclic micro- and macroyielding of single-phase materials containing heterogeneous dislocation distributions, whereas this approach is quite common in work on two-phase materials [38-42]. Regarding the details of the dislocation glide processes, this study represents an extension and, in several respects, a modification of an earlier paper with Grosskreutz [1]. Reference to the extensive literature [2,3,5,8,13,16,22,23,46,51] shows that the dislocation mechanisms considered here correspond in many aspects to those discussed by other authors. Still,we point out the rather different emphasis on the contributions of the different mechanisms to stress and strain and on the processes responsible for steady-state deformation of matrix and PSBs and for the nucleation of PSBs.

ACKNOWLEDGMENTS

It is with pleasure and in gratitude that I dedicate this paper to my academic teacher, Professor U. Dehlinger, who always provided an incentive to try harder. Furthermore, I thank sincerely Dipl. Phys. F. Ackermann and Dr. K. Mecke for kindly providing original micrographs, Dr. U. Essmann for continuing help and criticism and Professors E.W. Hart and C.Y. Li, Cornell University, for several useful discussions.

REFERENCES

[1] Grosskreutz, J.C. and Mughrabi, H., in : A.S. Argon (editor), Constitutive Equations in Plasticity (MIT Press, Cambridge, Mass. & London, 1975)251-326.
[2] Kuhlmann-Wilsdorf, D. and Laird, C., Mat. Sci. Eng. 27 (1977) 137-156.
[3] Brown, L.M., Metal Sci. 11 (1977) 315-320.
[4] Mughrabi, H., Ackermann, F. and Herz, K., in : Fatigue Mechanisms, Proc. of an ASTM-NBS-NSF Symposium, Kansas City, 1978, J.T. Fong (editor), ASTM-STP 675 (American Society for Testing and Materials, Philadelphia, 1979) 69-105.
[5] Laird, C., in: Fatigue and Microstructure, papers presented at the 1978 ASM Materials Science Seminar, St. Louis, Missouri (American Society for Metals, Metals Park, Ohio,1979) 149-203.
[6] Mughrabi, H., in : P. Haasen, V. Gerold and G. Kostorz (eds.), Proceedings of 5th Int. Conf. on the Strength of Metals and Alloys, Aachen, 1979, (Pergamon Press, Oxford and New York, 1980), Vol. 3, 1615-1638.
[7] Mughrabi, H. and Wang, R., in: Proc. of Int. Symp. "Defects and Fracture", Tuczno, Poland, Oct. 1980, (Sijthoff & Noordhoff, Leyden), in press.
[8] Brown, L.M., in: Proc. of Int. Conf. on Dislocation Modelling of Physical Systems, Gainesville, Florida, June 1980, Acta/Scripta Metallurgica, in press.
[9] Mughrabi, H., Mat. Sci. Eng. 33 (1978) 207-223.
[10] Ackermann, F., Diplomarbeit, Universität Stuttgart (1976).
[11] Mecke, K., phys. stat. sol. (a) 25 (1974) K93-K96.
[12] Winter, A.T., Phil. Mag. 30 (1974) 719-738.
[13] Finney, J.M. and Laird, C., Phil. Mag. 31 (1975) 339-366.
[14] Woods, P.J., Phil. Mag. 28 (1973) 155-191.
[15] Basinski, S.J., Basinski, Z.S. and Howie, A., Phil. Mag. 19 (1969) 899-924.
[16] Basinski, Z.S., Korbel, A.S. and Basinski, S.J., Acta Met. 28 (1980) 191-207.
[17] Holste, C., Schmidt, G.K. and Torber, R., phys. stat. sol. (a) 54 (1979)305-314
[18] Burmeister, H.-J. and Hermann, H., phys. stat. sol. (a) 54 (1979) K59-K61.
[19] Burmeister, H.-J. and Holste, C., phys. stat. sol. (a) 64 (1981) 611-624.
[20] Holste, C. and Burmeister, H.-J., phys. stat. sol. (a) 57 (1980) 269-280.
[21] Holt, D.L., J. Appl. Phys. 41 (1970) 3197-3201.
[22] Winter, A.T., Phil. Mag. 37 (1978) 457-463.
[23] Kuhlmann-Wilsdorf, D. and Laird, C., Mat. Sci. Eng. 46 (1980) 209-219.
[24] Wilkens, M., Herz, K. and Mughrabi, H., Z. Metallk. 71 (1980) 376-384.
[25] Essmann, U. and Mughrabi, H., Phil. Mag. A 40 (1979) 731-756.
[26] Essmann, U., Phil. Mag., in press.
[27] Mughrabi, H., in: Proc. of Third Int. Conf. on the Strength of Metals and Alloys, Cambridge (1973), Vol. 1, 407-410.
[28] Essmann, U., Goesele, U. and Mughrabi, H., Phil. Mag., in press.
[29] Mott, N.F., Acta Met. 6 (1958) 195-197.
[30] Essmann, U., phys. stat. sol. 3 (1963) 932-949.
[31] Mughrabi, H., J. Microsc. Spectrosc. Electron. 1 (1976) 571-584.

[32] Mughrabi, H., to be published.
[33] Albenga, G., Atti Accad. Sci., Torino, Cl. Sci. fis. mat. natur. 54 (1918/19) 864-868.
[34] Mecke, K. and Messerschmidt, U., Kristall und Technik 14 (1979) 1319-1323.
[35] Masing, G., Wissenschaftl. Veröffentl. a. d. Siemens-Konzern 3 (1923) 231-239.
[36] Mughrabi, H., Verhandl. DPG (VI) 15 (1980) 347.
[37] Masing, G., Zeitschr. f. techn. Physik 6 (1925) 569-573.
[38] Tanaka, K. and Mori, T., Acta Met. 18 (1970) 931-941.
[39] Ashby, M.F., Phil. Mag. 21 (1970) 399-424.
[40] Brown, L.M. and Stobbs, W.M., Phil. Mag. 23 (1971) 1185-1199.
[41] Eshelby, J.D., Proc. Roy. Soc. London (A), 241 (1957) 376-396.
[42] Brown, L.M. and Clarke, D.R., Acta Met. 23 (1975) 821-830.
[43] Asaro, R.J., Acta Met. 23 (1975) 1255-1265.
[44] Seeger, A., Phil. Mag., Ser. 7, 45 (1954) 771-773.
[45] Piqueras, J., Grosskreutz, J.C. and Frank, W., phys. stat. sol. (a) 11 (1972) 567-580.
[46] Kuhlmann-Wilsdorf, D. and Laird, C., Mat. Sci. Eng. 37 (1979) 111-120.
[47] Mughrabi, H., in: A.S. Argon (editor), Constitutive Equations in Plasticity (MIT Press, Cambridge, Mass. & London, 1975) 199-250.
[48] Antonopoulos, J.G. and Winter, A.T., Phil. Mag. 33 (1976) 87-95.
[49] Rapps, P., Katerbau, K.-H., Mughrabi, H., Urban, K. and Wilkens, M., phys. stat. sol. (a) 47 (1978) 479-488.
[50] Antonopoulos, J.G., Brown, L.M. and Winter, A.T., Phil. Mag. 34 (1976) 549-563.
[51] Kuhlmann-Wilsdorf, D., Mat. Sci. Eng. 39 (1979) 231-245.
[52] Den Buurman, R. and Snoep, A.P., Acta Met. 20 (1972) 407-413.
[53] Schumann, H.-D., phys. stat. sol. (a) 18 (1973) K27-K29.
[54] Weile, K.-H. and Schmidt, G.K., phys. stat. sol. (a) 31 (1975) K155-K158.
[55] Neumann, P., Z. Metallk. 59 (1968) 927-934.
[56] Yamamoto, A. and Imura, T., in: Proc. of Int. Conf. on Dislocation Modelling of Physical Sytems, Gainesville, Florida, June 1980, Acta/Scripta Metallurgica, in press.
[57] Fourie, J.T., Phil. Mag. 17 (1968) 735-756.
[58] Mughrabi, H., in: Proc. of 4th Int. Conf. on the Strength of Metals and Alloys Nancy (1976), Vol. 3, 1404-1408.
[59] Sinnig, H.-R. and Haasen, P., Scripta Met. 15 (1981) 85-90.
[60] Steiner, D., private communication.
[61] Wilhelm, M., Mat. Sci. Eng. 48 (1981) 91-106.

Continuum Models of Discrete Systems 4
eds. O. Brulin and R.K.T. Hsieh

A MICROSTRUCTURAL FATIGUE INTERPRETATION OF THE SCATTER IN FATIGUE DATA

J.W. Provan

Mechanical Engineering Department
McGill University
817 Sherbrooke St. West
Montreal, Quebec H3A 2K6 Canada

In this paper the probabilistic and stochastic method presently being employed to quantify the known microstructural processes leading to the eventual fatigue failure of polycrystalline metals are reviewed. These analytic and numerical procedures monitor the time or cycle evolution of the various probability distributions involved until a measure of the damage accumulation or material degradation is transformed by a simple mathematical procedure into a measure of the scatter in fatigue results. This new and exciting fatigue reliability law, resulting from the stochastic microstructural analysis, is then discussed along with its possible application to other metal deterioration processes and the paper concludes with a review of the experimental techniques currently being pursued in order to validate this law.

1. INTRODUCTION

As is well known the possibility of experimentally determining fatigue reliability data for large and expensive engineering components is often not feasible. As a result, emphasis has been placed, starting with the work of Freudenthal [1] and Weibull [2], on estimating the fatigue reliability of structures on the basis of both surface and flaw characteristics, size effects and the fracture and fatigue behavior of the materials involved. This approach has indeed resulted in reliability estimates [3,4] and various aspects of the analyses involved have been incorporated into modern design codes.

The current status of the micromechanics approach to further our understanding and description of this problem constitutes the contents of this paper. Since it is based on concepts and analyses carried out over the last few years and detailed elsewhere, a very brief review along with the appropriate references is given in Section 2. In Section 3 a new fatigue reliability law is derived on the basis of microstructural information and its characteristics are discussed in Section 4. In Section 5 both the extension of this approach to the stochastic description of the corrosion degradation process and the experimental procedures currently being formulated are discussed.

2. THE MICROMECHANICS INTERPRETATION OF FATIGUE

The micromechanics approach to fatigue is integrally based on the premise that a meaningful description of both the fatigue crack initiation and propagation processes can be formulated in terms of probabilistic and stochastic terminology. Specifically, in [5] an analytic and numerical method of determining the number of alternating stress cycles, N_0, required to initiate a fatigue crack in metals with a microstructure was developed on the basis of the statistical interference between distribution densities describing the variations in stress, yield strength and ultimate strength of grains/monocrystals existing near the surface of a metallic specimen or component. The technique took into consideration the influence of residual stresses, strain hardening and the cycle-by-cycle accumulation of microstructural damage and indeed led to a nonarbitrary definition of an initiated fatigue crack. For its application to engineering materials the basic experimentally to be determined micromechanic surface condition and material parameter required in this model is the coefficient of variation of the surface microstress distribution, denoted by C.O.V. ($\xi°$). With the specification of a component's C.O.V. ($\xi°$), characterizing the surface finish and material characteristics, the number of cycles at any stress level (including that of random loading [5]) can be specified.

Subsequent to the study of fatigue crack initiation, a linear pure birth Markov stochastic process was utilized in [6] to describe the number of cycles, Np, involved in the propagation of a fatigue crack from its initiation to its catastrophic failure. At any cycle, i, this analysis gives, as its description of crack growth, expressions for the mean, μ_a, and variance, V_a, of the crack length, a considered here as a stochastic process. Specifically, this pair is given by:

$$(\mu_a;V_a) = (\mu_{a_0}e^{\lambda i};\mu_{a_0}\Delta x_1 e^{\lambda i}(e^{\lambda i} - 1)) \quad , \quad 0 \le i \le N_p \ , \tag{2.1}$$

where:

$$\lambda = \phi\frac{\mu_{a_0}}{N_0\Delta x_1} = \text{the crack growth transition intensity} \tag{2.2}$$

μ_{a_0} = the mean of the initiated crack length, and

Δx_1 = the experimental accuracy of crack measuring technique.

In this crack propagation model the basic experimentally to be determined micromechanic crack propagation material parameter is the ϕ indicated in (2.2). Once it is known for any specific material the crack growth characteristics can be determined.

The initial experimental results [7,8] checking on the validity of the pair (2.1) showed good agreement as far as the mean of the crack growth distribution is concerned, namely (2.1)i, but clearly indicated that the expression for the variance, relation (2.1)ii especially near the zone where final fracture occurred, overestimated the scatter in the crack growth. What was observed experimentally was that the variance appeared to stabilize to an approximately constant value as the crack penetrated the polycrystalline metal. As a result, the form of (2.1) used in the about to be described reliability law was changed to:

$$(\mu_a;V_a) = (\mu_{a_0}e^{\lambda i};V_{ac}) \ , \ 0 \le i \le N_p, \tag{2.3}$$

with V_{ac} representing this stabilized variance.

At present, while the value of V_{ac} has been experimentally determined [7,8], its validity is being subjected to a theoretical investigation using Markov stochastic processes other than the pure linear birth process currently being utilized. The main disadvantage in using this particular Markov process has emerged to be that it does not properly account for the spatial correlation of material points along the propagating crack front.

3. THE MICROMECHANICS INTERPRETATION OF RELIABILITY

With the specific interpretation of the mean and variance of the crack growth given by the relation (2.3), the distribution density of crack penetrations translates in a manner, depicted in Figure 1, as an exponentially increasing function of i. This is in sharp contrast to the crack growth characteristics assumed without experimental justification by Birnbaum and Saunders [9] whose reliability law is based on the condition that "if the load ℓ is applied at the ith oscillation, then the incremental crack extension X_i is a random variable with a probability distribution depending only upon ℓ". Furthermore, in this Figure and in the subsequent analysis, the density of crack lengths described by (2.3) is assumed to be Gaussian. This has been partially confirmed experimentally, [8], but is in fact not a limitation since other distributions utilizing ($\mu_a;V_a$) as their basic variables can be used. Naturally, however, the mathematics would be more involved. The form of the reliability law determined in this section does, however, depend on the distribution of crack penetrations being Gaussian.

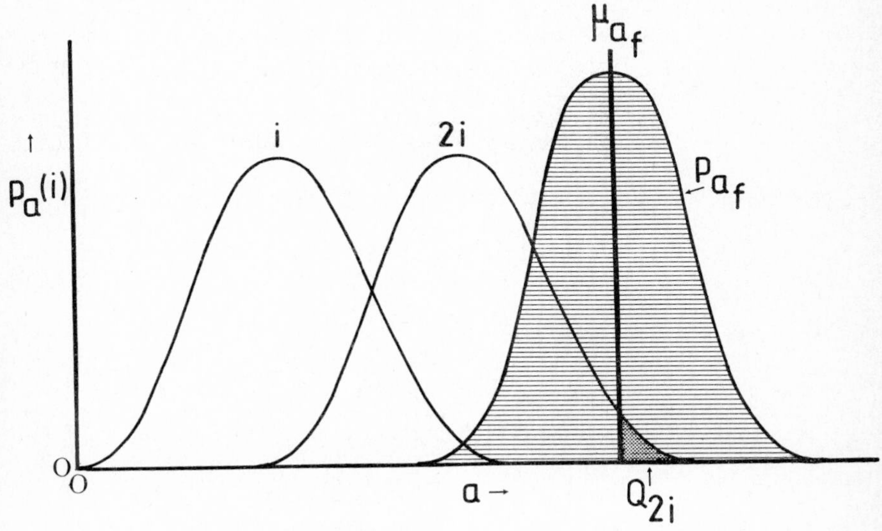

Fig. 1 Schematic of Model Leading to Reliability Law

This distribution translates until it starts to interfere in a reliability sense, [10], with the density of critical crack lengths, described by the pair ($\mu_{a_f};V_{a_f}$), which is again a material characteristic. The monitoring of how much ($\mu_a;V_a$) interferes with ($\mu_{a_f};V_{a_f}$) as i increases constitutes the basis of the new fatigue reliability law. Carrying out the analysis for the case where the distribution of critical crack lengths is chosen to a delta function at μ_{a_f}, see [11], the expression for the fraction of crack lengths at i being larger than μ_{a_f} is:

$$Q_i = \int_{\mu_{a_f}}^{\infty} p_a(i)da = \frac{1}{\sqrt{2\pi}} \int_{Z_i}^{\infty} \exp\left[-\frac{s_i^2}{2}\right] ds_i \quad , \tag{3.1}$$

where:

$$s_i = \frac{a - \mu_{a_i}}{\sqrt{V_{ac}}} \quad , \text{ and} \tag{3.2}$$

$$Z_i = + \frac{\mu_{a_f} - \mu_{a_i}}{\sqrt{Vac}} = \text{the failure coupling coefficient.} \tag{3.3}$$

The fatigue reliability expression at i then simply becomes:

$$P_i = 1 - Q_i = \frac{1}{\sqrt{2\pi}} \int_{-\infty}^{Z_i} \exp\,[-\frac{s_i^2}{2}]\, ds_i \quad , \tag{3.4}$$

which is a monotonically decreasing function of i. Both Q_i and P_i are solely dependent on the nature of Z_i which itself is normally distributed with mean 0 and variance 1 [12]. Hence, since $z_i = (\mu_{a_f} - \mu_{a_i})/\sqrt{Vac}$ is a known function of i, through the choice of $p_a(i)$ being Gaussian and the pair being given by (2.3), it follows from the fundamental relationship:

$$p_{N_p}(i) = p_{Z_i}(z_i)\left|\frac{dz_i}{di}\right| \tag{3.5}$$

that the required reliability density function, $p_{N_p}(i)$, becomes:

$$p_{N_p}(i) = \frac{\mu_{a_0}}{\sqrt{2\pi Vac}} \exp\,\{\lambda i - \frac{(\mu_{a_0}\exp\,[\lambda i] - \mu_{a_f})^2}{2Vac}\} \quad , \tag{3.6}$$

while its associated cumulative form is:

$$P_{N_p}(j) = \int_{-\infty}^{j} p_{N_p}(i)di = \frac{1}{2}\,\{\text{erf}\,(\frac{\mu_{a_0}e^{\lambda j} - \mu_{a_f}}{\sqrt{2Vac}}) - \text{erf}\,(\frac{\mu_{a_f}}{\sqrt{2Vac}})\} \; . \tag{3.7}$$

4. DISCUSSION

The first point to be made concerning the reliability law derived in the previous section is that it strictly pertains to the scatter in the number of fatigue cycles involved in propagating cracks from their initiated dimension, μ_{a_0}, to their critical size, μ_{a_f}. Exactly how to incorporate the initiation model, reviewed in Section 2, into the formulation of this reliability distribution is still, as yet, uncertain.

The second comment is that this failure/reliability law is based solely on information gleaned from the microstructural fatigue processes taking place when a polycrystalline material is being fatigue failed. For example, in the case where OFHC brand copper was examined using a combined experimental program involving an MTS material testing facility and a fractographic analysis, [7,8,11] the pertinent information for implementing (3.6) and/or (3.7) was determined as:

$$\mu_{a_0}(m) = 0.000145\,; \quad \mu_{a_f}(m) = 0.009070\,; \quad Vac(m^2) = 722 \times 10^{-12}\,; \begin{cases} \lambda_{exp}(1/i) = 0.00249\,; \\ \text{or} \\ \lambda_{th}\,(1/i) = 0.00049\,. \end{cases} \tag{4.1}$$

The graphs resulting from these numbers were shown in [11] to be essentially the same as those obtained using a Gaussian distribution. The reason for this is the small experimentally determined value of Vac as indicated in (4.1). In order to highlight, in a more succinct manner, the characteristics of this reliability distribution a value of $Vac = 7 \times 10^{-8} m^2$ was chosen, the remaining variables being unchanged, and the resulting data plotted on probability paper as shown in

Figure 2, (λ_{exp} given in (4.1) was utilized).

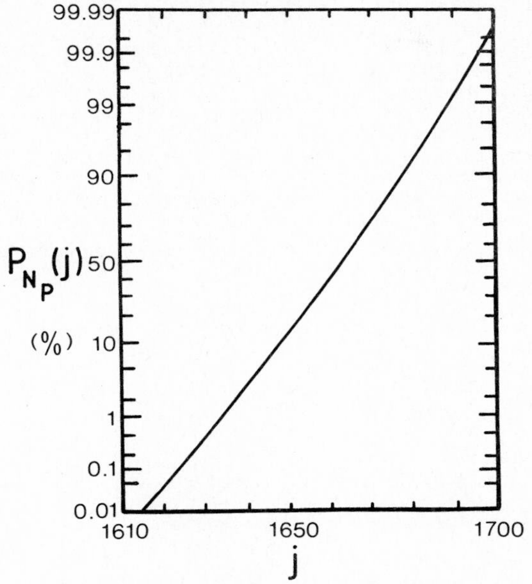

Fig. 2 A Probability Plot of the Proposed Reliability Law
(μ_{a_0} = 0.000145 m ; μ_{a_f} = 0.00907 m ; Vac = 7 x $10^{-8}m^2$; λexp = 0.00249)

This plot clearly indicates the upward swing in the probability curve as is expected from an exponentialnormal cumulative distribution function. This is an attractive feature of this function since it describes the commonly occurring situation where component or specimen failures occur at an ever increasing rate and where there is a tendency for a relatively small number of units surviving for a larger number of fatigue cycles. This latter characteristic is in contrast with other reliability distributions such as the lognormal, normal or Weibull.

5. CONCLUDING REMARKS

At this stage in the development of this new reliability law it is of importance to check its validity with an experimental investigation. One is at present underway where the microstructural information is being obtained from fractographic SEM investigations and the global reliability information is being determined by fatigue failing a large number of ASTM-ANSI E 466 [13] specimens at a constant preset strain amplitude range using an MTS facility. In this manner the first corroborating evidence of this reliability law's value will be obtained. In the long term, if this microstructurally based reliability law proves to be meaningful in industrial applications then it will aid immeasurably in the setting up of inspection intervals, in estimating a component's life expectancy and will also serve to increase the safety of large industrial complexes.

A further advantage of this reliability law is that it can be used, without reference to any microstructural information, to empirically describe the scatter that exists in any fatigue data. Simply by curve fitting existing data the

the constants involved in the exponentialnormal expression can be estimated. From a researcher's point of view this would be an unfortunate application of this law but from an engineer's it may prove to be an invaluable tool.

Finally, based on the projected advantages of viewing material degradation processes from the point of view expressed in this paper, a theoretical and experimental program has been instigated to characterize the corrosion problem related to polycrystalline metals in hostile environments. The philosophy of the approach detailed in this paper is currently being applied to the description of the corrosion stochastic process. At present a bivariate Markov theory and a companion experimental investigation are being developed. Its projected applications are numerous with researchers in the gas pipeline, paper, thermal power generation and nuclear industries already showing some interest in this approach.

REFERENCES

[1] Freudenthal, A.M., The safety of structures, Trans. Am. Soc. C.E., 112 (1947) 125-180.

[2] Weibull, W., The effect of size and stress history on fatigue crack initiation and propagation, in: Symposium on Crack Propagation, Cranfield (1962) 271.

[3] Heckel, K., Ziebart, W., A new method for calculation of notch and size effect in fatigue, in: Taplin, D.M.R. (ed.), Fracture 1977 2 (Waterloo, 1977) 937-941.

[4] Schütz, W. and Oberparleiter, W., VI Experimental techniques for determining fracture toughness values, in: Liebowitz, H. (ed.), Fracture Mechanics of Aircraft Structures, AGARD, 176 (1974) 370-394.

[5] Provan, J.W., A model for fatigue crack initiation in polycrystalline solids, in: Taplin, D.M.R. (ed.), Fracture 1977 2 (Waterloo 1977), 1169-1176.

[6] Provan, J.W. and Ghonem, H., Probabilistic descriptions of microstructural fatigue failure, in: Provan, J.W. and Leipholtz, H.E. (eds.), Continuum Models of Discrete Systems (Waterloo, 1977) 407-430.

[7] Provan, J.W. and Mbanugo, C.C.I., Stochastic fatigue crack growth - an experimental study, RES MECHANICA, the Int. Jrnl. of Struct. Mechs. and Matl. Sci. 2 (1981) 53-72.

[8] Provan, J.W., The micromechanics approach to the fatigue failure of polycrystalline metals, in: Gittus, J. (ed.), Voids, Cavities and Cracks in Metallic Alloys (Elsevier's Applied Science, 1981).

[9] Birnbaum, Z.W. and Saunders, S.C., A probabilistic interpretation of Miner's rule, SIAM J. Appl. Math. 16 (1968) 637-652.

[10] Haugen, E.B., Probabilistic Mechanical Design (Wiley-Interscience, 1980).

[11] Provan, J.W., A fatigue reliability distribution based on probabilistic micromechanics, in: Sih, G.C. and Matczynski, M. (eds.) Proc. Defects and Fracture Symp., Tuczno, Poland, Oct. 1980.

[12] Mann, N.R., Schafer, R.E. and Singpurwalla, N.D., Methods for statistical analysis of reliability and life data (John Wiley, 1974).

[13] 1980 Annual Book of ASTM Standards, 10, American National Standard, ANSI/ASTM E 466-76, 614-619.

Continuum Models of Discrete Systems 4
eds. O. Brulin and R.K.T. Hsieh

IMPROVED LATTICE SUMMATION TECHNIQUE: APPLICATION TO PSEUDOPOTENTIAL COMPUTATION OF CRYSTAL ELASTICITY

Daniel J. Rasky and Frederick Milstein

Department of Mechanical and Environmental Engineering
University of California
Santa Barbara, California 93106 USA

The use of lattice summations in the computation of physical properties of crystals (e.g. binding energy or elastic moduli) is widespread. In the present paper, an improved technique for computing general lattice summations is presented. A single computer algorithm performs lattice summations on general crystals (including polyatomic and homogeneously deformed crystals). As an example, the technique is used to compute elastic moduli and domains of stability of cubic crystals (face centered, body centered, and simple cubic) under hydrostatic loading, within the framework of a pseudopotential model for copper. The definition of elastic moduli and the stability criteria are those of Milstein and Hill [J. Mech. Phys. Solids 25, 457 (1977); 26, 213 (1978); 27, 255 (1979)]. The pseudopotential model is that of Thomas [Phys. Rev. B 7, 2385 (1973)]. Comparisons are made with other computations of lattice stability based upon the same stability criteria.

I. INTRODUCTION

Lattice summations are used in theoretical computations of a variety of crystal properties. For example, typical computations of crystal binding energies or elastic moduli involve summations of interactions between a reference atom and other atoms in the crystal and/or summations over reciprocal lattice space of a Fourier series expansion. In semi-empirical, central force models (e.g. Morse, Lennard-Jones), the binding energy E_{bind} is computed from a pairwise summation (over atom sites $\vec{r}_\alpha$) of the form $\Sigma\, E_{str}(\vec{r}_\alpha)$. In the more sophisticated pseudopotential models, the crystal binding energy of a monatomic crystal can be written as a combination of a real space and reciprocal space part,

$$E_{bind} = E_{real} + E_{reciprocal} \tag{1}$$

Both the real and reciprocal space parts are of the form

$$E = E_{vol}(\Omega) + \sum_{\vec{r}_\alpha} E_{str}(\Omega, \vec{r}_\alpha), \tag{2}$$

where E_{vol} is a volume dependent term, E_{str} depends explicitly upon the atomic volume Ω and the relative positions of the interacting atoms (i.e. upon volume and "structure"), and $\vec{r}_\alpha$ is either a vector connecting two atomic sites in the crystal (for E_{real}) or a reciprocal lattice vector (for $E_{reciprocal}$).

In the present paper we are concerned with the numerical technique for evaluating lattice summations (such as those appearing in the various theoretical expressions for E_{bind}) as well as with the results of some specific elasticity computations. In particular, a pseudopotential model for copper is used to compute the bulk modulus κ and the shear moduli μ and μ' (as defined in Section III) of the three cubic Bravais crystals (i.e. fcc, bcc, and sc) under arbitrary hydrostatic pressure; the moduli, in turn, are used to determine the domains of elastic stability of the crystals. Previously, analogous studies were carried out for the complete family of Morse-function fcc, bcc, and sc crystals [1-3]. The Morse function computations yielded an extensive collection of useful, semi-empirical data as well as valuable insights into the nature of crystal elasticity under hydrostatic pressure. However, a basic limitation of the Morse model is that of the Cauchy symmetry [4] (which is of course inherent in the elastic moduli of crystals described by all such central force models). It is thus of particular interest to extend and generalize the computations to the pseudopotential models (since the volume dependencies of the energy in the pseudopotential models remove Cauchy symmetry from among the moduli). On the other hand, the computational complexity (and corresponding computing time) is much greater in the pseudopotential than in the semi-empirical models [particularly for crystals that have been subjected to general homogeneous deformations (e.g. under uniaxial or shearing loads) or that are of low symmetry]. Since we expect (in due course) to carry out pseudopotential computations of crystals under a variety of loading conditions, the foregoing considerations prompted the development of an improved lattice summation technique which is applicable to either monatomic or polyatomic crystals of arbitrary symmetry. In this paper we give prominence to the presentation of the lattice summation technique (see Section II); although results of the elasticity and stability computations will be summarized, a detailed presentation of computational results will be given elsewhere [5].

II. SUMMATION METHOD

Generally there are two important computational considerations associated with evaluating lattice summations. These are to minimize computing costs and to include a sufficiently large number of terms (or "summation sites") to obtain convergence (to the desired numerical accuracy) without introducing significant computer "round-off" or "truncation" errors. More slowly convergent summations tend to be more susceptible to such errors and more costly to evaluate. (As perhaps a somewhat extreme example, in order to study the behavior of the Morse model, as the continuum limit was approached, Milstein and Hill [1] reported doing lattice summations over about 10^6 atomic sites.) With regard to computing costs, since the arguments of the summations often depend on $|\vec{r}_\alpha|$ rather than on $\vec{r}_\alpha$, significant savings can be made if the locations of all summation sites that are equidistant from the origin are determined. Thus, with this determination, instead of having to evaluate a function of $|\vec{r}_\alpha|$ at all lattice sites $\vec{r}_\alpha$, the function need be evaluated but once for all $\vec{r}_\alpha$ that have the same magnitude (assuming that the $\vec{r}_\alpha$ are occupied by like atoms). As will be seen below, such a determination can also be used in reducing computational error.

Cumulative round-off or truncation errors can be minimized, in principle, by first ordering the terms to be summed according to magnitude and then summing in the order smallest first, largest last. However, for the practical purposes of effectively reducing such potential errors and avoiding the necessity of storing and ordering a very large number of terms, the following procedure can be used. Start the computations at the most distant summation sites that are to be included (where $|\vec{r}_\alpha| \equiv R_L$, say) and compute and add the arguments of the summation at those sites; then move to the next most distant sites (at $|\vec{r}_\alpha| \equiv R_{L-1}$, say) and compute the arguments of the summation at those sites and add them to the "running" summation; repeat the process until all summation sites for which $|\vec{r}_\alpha| \leq R_L$ have been included. (The particular value of R_L depends on the crystal model, the property of interest, and the desired number of significant figures.) In order to carry out the above procedure, it is necessary to be able to determine (i) the locations of all summation sites at positions $\vec{r}_\alpha$ on a sphere of radius $R_M = |\vec{r}_\alpha|$ (call this the M-sphere) and (ii) the radius R_{M-1} of the next largest sphere (call this the (M-1)-sphere) containing summation sites (or equivalently, the largest radius R_{M-1} of a sphere containing summation sites, such that $R_{M-1} < R_M$). We now consider a procedure for making such determinations for arbitrary crystals.

Owing to computer round-off or truncation, the numerical (i.e. computed) values of $|\vec{r}_\alpha|$, for summation sites that are in fact equidistant from the origin, can differ very slightly. Thus, for computational purposes, in order to locate all of the summation sites that are actually <u>on a sphere</u> of radius R_M, it is convenient to locate all sites <u>within a shell</u> of mean radium R_M and of thickness 2ε, where $\varepsilon \ll \ldots \ldots \ll R_M$ and ε is considerably greater than any possible round-off or truncation error (to insure that all sites on the M-sphere are contained in the shell); typically, $\varepsilon \sim 10^{-12} R_M$ in our computations. In other words, all $\vec{r}_\alpha$ that satisfy

$$R_M - \varepsilon \leq |\vec{r}_\alpha| \leq R_M + \varepsilon \tag{3}$$

are to be found; R_M and ε are always positive algebraically. Equivalently, to first order in ε,

$$2R_M\varepsilon \geq \beta_\alpha \geq -2R_M\varepsilon, \tag{4}$$

or

$$|\beta_\alpha| \leq 2R_M\varepsilon \quad , \tag{5}$$

where

$$\beta_\alpha \equiv R_M^2 - |\vec{r}_\alpha|^2 = R_M^2 - r_\alpha^2. \tag{6}$$

Thus, the problem of locating the $\vec{r}_\alpha$ that satisfy (3) is equivalent to finding the $\vec{r}_\alpha$ which have associated values of β_α that satisfy (5). Figure 1 illustrates the values of β_α for three summation sites (at $\vec{r}_a$, $\vec{r}_b$, and $\vec{r}_c$); here, the summation sites at $\vec{r}_a$ and $\vec{r}_b$ satisfy (5), and are therefore considered to lie on the M-sphere, while $\beta_c > 2R_M\varepsilon$, and thus the site at $\vec{r}_c$ lies on the N-sphere, say, where $R_N < R_M$. Thus, given the M-sphere, the problem of determing the (M-1)-sphere (of radius R_{M-1}) is equivalent to determining the smallest value of

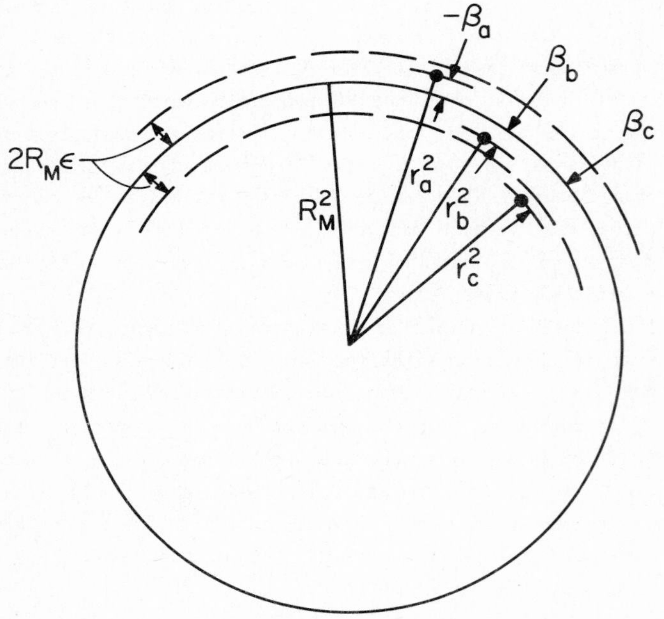

Figure 1

Illustration of the use of the quantity β_α; if $|\beta_\alpha| \leq 2R_M\varepsilon$, then the summation site at $\vec{r}_\alpha$ is in the shell of mean radius R_M and thickness 2ε.

$\beta_\alpha = \beta_t$, say, such that $\beta_t > 2R_M\varepsilon$; the value of R_{M-1}^2 can then be taken as $R_M^2 - \beta_t - 2R_M\varepsilon$. We now consider the procedure for finding all of the β_α that satisfy (5) for the M-sphere and for finding the next smallest or (M-1)-sphere. (The terms "M-sphere" and "sphere M" are used interchangeably).

A standard description of a crystal consists of a Bravais lattice and a basis containing one or more atoms. The Bravais lattice sites are generated by translations of the form $m_1\vec{b}_1 + m_2\vec{b}_2 + m_3\vec{b}_3 = m_i\vec{b}_i$ (summation convention), where the $\vec{b}_i$ are the primitive translation vectors of the Bravais lattice and the m_i are all integers (positive, negative, and zero). An identical basis or grouping of atoms is associated with each lattice site; the atoms in the basis can be connected to their associated Bravais lattice site by basis vectors $\vec{h}_s$ (s = 1,2,...S for the case of S atoms in the basis; in practice, $\vec{h}_1$ can conveniently be taken as zero). The location of any given atom in the crystal (with respect to an origin atom) is then written as

$$\vec{r}_\alpha = m_{\alpha i}\,\vec{b}_i + \vec{h}_s \quad , \tag{7}$$

where the $m_{\alpha i}$ comprise a particular set of integers (m_1, m_2, m_3). (For a crystal with one atom in the basis or for a reciprocal lattice, the $\vec{h}_s$ are not required.)

The principle that is used in the location of summation sites within the spherical shell M (i.e. the shell of mean radius R_M and thickness 2ε) is to determine whether or not, for a given pair ($m_{\alpha 2}$, $m_{\alpha 3}$), there is an $m_{\alpha 1}$ for which relation (5) is satisfied; however, the specific technique is much more efficient than a simple examination of all possible sets ($m_{\alpha 1}$, $m_{\alpha 2}$, $m_{\alpha 3}$). In order to explain the workings of the technique, it is convenient first to consider an expression for r_α^2 in which all functional dependence upon $m_{\alpha 1}$ is shown explicitly, i.e.

$$r_\alpha^2 = am_{\alpha 1}^2 + 2bm_{\alpha 1} + c \quad , \tag{8}$$

where, from (7),

$$a = b_1^2 \quad , \tag{9a}$$

$$b = (m_{\alpha 2}\,\vec{b}_2 + m_{\alpha 3}\,\vec{b}_3 + \vec{h}_s) \cdot \vec{b}_1 = b_{11}\,X_1 + b_{12}\,X_2 + b_{13}\,X_3, \tag{9b}$$

$$c = |m_{\alpha 2}\,\vec{b}_2 + m_{\alpha 3}\,\vec{b}_3 + \vec{h}_s|^2 = X_1^2 + X_2^2 + X_3^3 \ , \tag{9c}$$

with $X_j = m_{\alpha 2}\, b_{2j} + m_{\alpha 3}\, b_{3j} + h_{sj}$ (9d)

and b_{ij} and h_{sj} are the jth cartesian components of $\vec{b}_i$ and $\vec{h}_s$, respectively, referred to the same cartesian axes. Equation (8) can be inverted to yield

$$m_{\alpha 1} = -b/a \pm d^{\frac{1}{2}} \tag{10}$$

where $d = (b/a)^2 - (c - r_\alpha^2)/a$ (11)

Next, consider a line AA' that is parallel to $\vec{b}_1$ and that passes through the summation site located at $m_{\alpha 2}\, \vec{b}_2 + m_{\alpha 3}\, \vec{b}_3 + \vec{h}_s$; this is illustrated in Fig. 2 for the particular case of $|\vec{h}_s| = 0$. We wish to determine whether any summation sites at $\vec{r}_\alpha$ on AA' lie within the spherical shell M (of mean radius R_M and

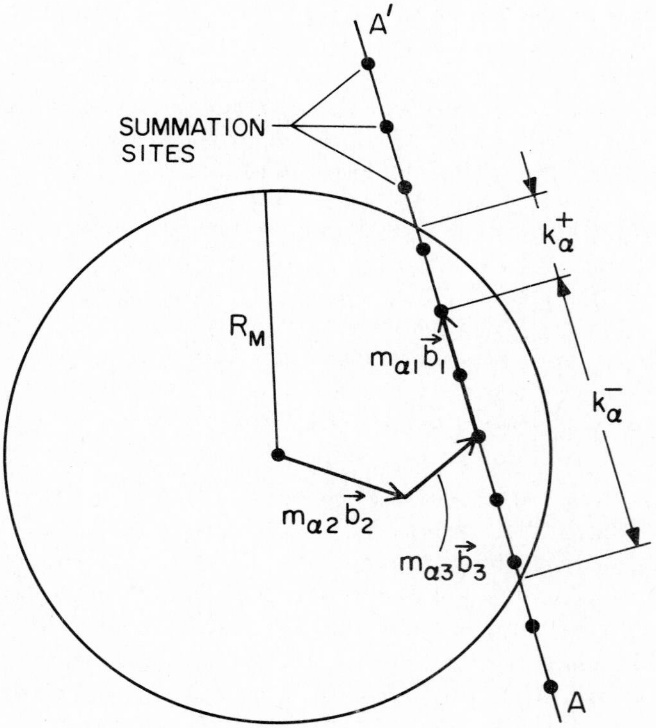

Figure 2

Geometric interpretation of $k_\alpha^\pm = \pm K_\alpha^\pm\, |\vec{b}_1|$, where the $K_\alpha^\pm$ appear in Eqs. (12) and (13).

thickness 2ε). Let $K_\alpha^\pm$ be defined such that the vectors from the summation site at $\vec{r}_\alpha$ to the intersections of AA' with the sphere M are $K_\alpha^\pm \vec{b}_1$; then the distances from $\vec{r}_\alpha$ to the intersections of AA' and sphere M are $k_\alpha^\pm = \pm K_\alpha^\pm |\vec{b}_1|$ (assuming for the moment that AA' does intersect the sphere). Then substituting R_M^2 for r_α^2 and $(m_{\alpha 1} + K_\alpha^\pm)$ for $m_{\alpha 1}$ in (8) gives

$$R_M^2 = a(m_{\alpha 1} + K_\alpha^\pm)^2 + 2b(m_{\alpha 1} + K_\alpha^\pm) + c \; ; \tag{12}$$

inverting (12) yields

$$m_{\alpha 1} + K_\alpha^\pm = - b/a \pm D^{\frac{1}{2}} \tag{13}$$

where $$D = (b/a)^2 - (c - R_M^2)/a. \tag{14}$$

The distance between the two intersections of AA' with sphere M (i.e. the length of the chord in Fig. 2) is $k_\alpha^+ + k_\alpha^- = 2D^{\frac{1}{2}} |\vec{b}_1|$. If $D = 0$, line AA' is tangent to sphere M; if $D < 0$, line AA' does not intersect sphere M. This condition is useful for eliminating values of $(m_{\alpha 2}, m_{\alpha 3})$ which can not correspond to an $\vec{r}_\alpha$ on the sphere M; i.e., when $D < -\varepsilon$, the associated values of $(m_{\alpha 2}, m_{\alpha 3})$ need not be considered. For a given R_M, the summing procedure is thus to determine the pairs $(m_{\alpha 2}, m_{\alpha 3})$ for which $D + \varepsilon \geq 0$. For each such pair, a determination is made of the $\vec{r}_\alpha$ on AA' that are closest to the sphere M; the corresponding β_α are then determined and the summation site is "tested" according to relation (5) to determine whether it is within the shell M. Note that each $\vec{r}_\alpha$ (given by Eq. (7)) has its corresponding $m_{\alpha 1}$ that appears in Eq. (13); for each $\vec{r}_\alpha$ there are two possible values of K_α (i.e. $K_\alpha^\pm$) in Eq. (13); the smaller the value of $|K_\alpha^\pm|$, the closer the summation site (at $\vec{r}_\alpha$) is to the sphere M. In practice, it is sufficient to "test" only those summation sites for which $m_{\alpha 1}$ is given by the two integer values that are closest to $-b/a \pm D^{\frac{1}{2}}$ (i.e. to the right hand side of Eq. (13)), since all other sites on AA' will be too far from the M-sphere to satisfy relation (5). However, in order to determine the radius R_{M-1} of the (M-1)-sphere, two other summation sites on AA' are also considered; these are the next nearest sites (on AA') to the sphere M and are determined by setting $m_{\alpha 1}$ equal to the integer values that are closest to $-b/a \pm D^{\frac{1}{2}} \mp 1$. The values of β_α are computed for the four summation sites so determined (for each considered set $m_{\alpha 2}, m_{\alpha 3}$); these values of β_α (for all considered sets $m_{\alpha 2}, m_{\alpha 3}$) are examined to determine which of them is the smallest, positive value that is greater than $2R_M\varepsilon$; that value defines R_{M-1}, as discussed earlier. If $0 > D \geq -\varepsilon$, then D is set equal to zero for computational purposes, and only one value of $m_{\alpha 1}$ need be considered in the foregoing procedure (i.e. the integer value closest to -b/a).

In summary, the summing procedure is started by defining some sufficiently large value of R_L. Then determinations are made of all of the summation sites in the shell L and of the mean radius R_{L-1} of the next largest shell (with $R_{L-1} < R_L$) by use of the procedure described above. The contributions to the summation (from the sites in the shell L) are evaluated and added. Then the process is repeated for the (L-1), the (L-2), etc., shells; the summation is completed when the origin site is found to be in the "next interior shell."

III. PSEUDOPOTENTIAL COMPUTATIONS OF ELASTIC MODULI AND CRYSTAL STABILITY

Basic work that defines and characterizes the elasticity and stability of cubic crystals under hydrostatic pressure has been carried out by Milstein and Hill [1-3]. Briefly, they adopted the view that the shear moduli $\mu(P)$ and $\mu'(P)$ and bulk modulus $\kappa(P)$ provide a more direct, physically meaningful measure of the elastic response of cubic crystals at nonzero pressure P than other moduli (e.g. the Green moduli) that have been employed by various investigators of crystal elasticity under nonzero loading. Here κ is the bulk modulus and μ and μ' are the shear moduli in the relation between the cubic-axes components of the Cauchy stress increment $\delta\sigma_{ij}$ and the rotationless strain increment $\delta\varepsilon_{ij}$ (reckoned relative to the current configuration under P). Milstein and Hill [3] presented a complete account of the application of the principles of bifurcation analyses for general materials to the particular case of cubic crystals subjected to hydrostatic loading. The analyses are carried out in a manner equivalent to Hill [Ref. 6, chapter III, section C2] but without recourse to the general mathematical apparatus for handling follower-loadings. Their treatment of crystal stability is rigorous and complete; i.e. (i) the loading environment is fully specified, to sufficient order and in both its active and passive modes, and (ii) the potential energy of the system as a whole is examined in all the nearby, possibly inhomogeneous, configurations allowed by the kinematic constraints, if any. Under a hydrostatic pressure that does not vary during any departure from a considered configuration of equilibrium, elastic stability is guaranteed if

$$\kappa(P) > 0 \quad , \quad \mu(P) > 0 \quad , \quad \text{and} \quad \mu'(P) > 0 \qquad (15)$$

simultaneously.

Milstein and Hill [1-3] computed κ, μ, and μ' for the complete Morse family of cubic crystals (fcc, bcc, and sc) under arbitrary hydrostatic pressure; evidently their work presented the first atomistic computations of shear moduli and ranges of crystal stability (for a constant hydrostatic pressure environment). The present paper extends the pressure dependent computations of κ, μ, and μ' (and associated stability ranges) to pseudopotential fcc, bcc and sc crystals that do not obey Cauchy symmetry.

In the pseudopotential formalism, the binding energy per ion E_{bind} can be written as

$$E_{bind} = E_{fe} + E_{es} + E_{bs} + E_{ol} \tag{16}$$

The volume dependent free electron energy, E_{fe}, includes the kinetic, exchange, and correlation energies of a free electron gas, and also the first order perturbation energy term for the pseudopotential. The classical electrostatic energy of an array of positive point charges embedded in a uniform-compensating electron gas is given by E_{es}. The band structure energy, E_{bs}, is the second order perturbation energy for the pseudopotential, and the overlap energy, E_{ol}, represents the exchange energy due to the d-electron states overlapping on the neighboring ion cores (this term is included for d band metals).

The present computations employed the particular pseudopotential formulation used by Thomas [7] to compute third order elastic moduli of copper. [Thomas used Harrison's modified point-ion model potential and the Hubbard-Sham form for the dielectric screening function in calculating the band structure energy. The overlap energy is computed from a real-space summation over Born-Meyer interactions and the electrostatic and band structure energies are computed using a reciprocal-space summation.] Thomas "adjusted" five empirical pseudopotential parameters to obtain agreement between theoretical and experimental values of binding energy, lattice parameter, and elastic moduli κ, μ, and μ' of unstressed fcc copper, extrapolated to 0°K. We employed the same empirical parameters in the present computations.

Some specific computational results are summarized in Table 1. The slight differences in the quantities a, κ, and μ for the fcc Morse and pseudopotential values at $\lambda = 1$ (P=0) are owing to small differences in the experimental values of these quantities used to determine the empirical parameters; the larger differences in μ' are a result of the Cauchy symmetry of the Morse model. [For central force cubic crystals under pressure, $\kappa - 2/3\ \mu = \mu' + 2P$, as given by Milstein and Hill [2]; this can be considered as a generalization of the Cauchy condition $C_{23} = C_{44}$ in the sense that at P = 0, $\kappa - 2/3\ \mu = \mu'$ is equivalent to $C_{23} = C_{44}$.]

Owing to space limitations, additional computational details (including the detailed functional dependencies of the elastic moduli κ, μ, and μ' upon pressure P and stretch λ in the pseudopotential model) will be presented elsewhere

Table 1. Theoretical ranges of elastic stability and computed values of stretch λ, lattice parameter a, pressure P, and the elastic moduli κ, μ and μ' at the points of termination of the stable ranges and at λ = 1. The stretch λ is defined as the current lattice parameter divided by the lattice parameter at P = 0 for the particular crystallographic configuration.

Crystal Model and Stable Ranges	λ	a (Å)	P	κ	μ	μ'
				(10^{12} dyn/cm^2)		
pseudopotential fcc	1.000	3.597	0	1.439	0.270	0.831
λ < 1.145	1.145	4.118	-0.197	0	0.038	0.203
pseudopotential	0.649	1.862	113.9	445.8	0	215.6
bcc	1.000	2.868	0	1.415	-0.057	1.046
λ < 0.649 and	1.088	3.120	-0.179	0.238	0	0.365
1.088 < λ < 1.145	1.145	3.284	-0.194	0	0.012	0.225
pseudopotential sc no stable range	1.000	2.380	0	1.334	1.152	-0.292
Morse potential fcc	1.000	3.615	0	1.420	0.257	1.249
λ < 1.144	1.144	4.134	-0.218	0	0.095	0.372
Morse potential bcc	1.000	2.871	0	1.411	0.065	1.368
λ < 1.143	1.143	3.282	-0.216	0	0.002	0.431
Morse potential sc no stable range	1.000	2.385	0	0.971	2.208	-0.500

[5]. For the present, we note the following general conclusions. (1) The qualitative dependences of the moduli upon pressure and stretch, for the respective lattices, are the same for the Morse and the pseudopotential models; this suggests that the particular crystal symmetries (rather than the details of atomic binding) determine the qualitative characteristics of the responses of the crystals to pressure; these characteristics are examined in great detail in Refs. [1-3]. (2) For both the Morse function and the pseudopotential models of fcc copper, stability is lost in tension where κ = 0. The critical values of stretch, at which κ passes from positive to negative are virtually identical in the Morse and pseudopotential models! (The Morse model fcc copper crystal has the correct (i.e. experimental) values of lattice parameter and moduli κ and μ at zero pressure, extrapolated to 0°K.) (3) Both the Morse and pseudopotential

models of the simple cubic configuration of copper are unstable at all values of stretch (i.e. when the model parameters determined for fcc copper are used to describe sc lattices, the lattices are unstable throughout). (4) Milstein and Hill [3] found that the functions $\mu(\lambda)$ for bcc crystals "exhibit particularly interesting behavior; at sufficiently small λ, the functions are large and positive; with increase in λ, they pass from positive to negative; they then pass, successively, through negative minima, from negative to positive, through positive maxima, and finally approach zero asymptotically from the positive." In the present pseudopotential computations, $\mu(\lambda)$ for the bcc copper configuration behaves similarly. (5) Owing to the existence of the two positive ranges of $\mu(\lambda)$ for bcc configurations, Morse function bcc crystals with very "short range, steep" interatomic interactions have two distinct ranges of λ (one in compression and one tension) over which the crystals are stable; however, when the Morse function parameters of copper are applied to the bcc configuration, there is only one stable range [3] (since $\kappa(\lambda)$ has already passed from positive to negative in the tensile region where $\mu(\lambda) > 0$). For pseudopotential bcc copper, however, the intermediate positive region of $\mu(\lambda)$ occurs at smaller values of λ and as a result, two stable ranges do occur.

ACKNOWLEDGMENT

This work was supported by the National Science Foundation under grant number DMR78-06865.

References

1. Milstein, F. and Hill, R., "Theoretical properties of cubic crystals at arbitrary pressure - I. Density and Bulk Modulus," J. Mech. Phys. Solids, 25, pp. 457-477 (1977).

2. Milstein, F. and Hill, R., "Theoretical properties of cubic crystals at arbitrary pressure - II. Shear Moduli," J. Mech. Phys. Solids, 26, pp. 213-239 (1978).

3. Milstein, F. and Hill, R., "Theoretical properties of cubic crystals at arbitrary pressure - III. Stability," J. Mech. Phys. Solids, 27, pp. 255-279 (1979).

4. Hill, R., "On the elasticity and stability of perfect crystals at finite strain," Mathematical Proceedings of the Cambridge Philosophical Society, 77, pp. 225-240 (1975).

5. Milstein, F. and Rasky, D., to be published.

6. Hill, R., "Aspects of invariance in solid mechanics," Advances in Applied Mechanics, edited by Yih, C.S., 18, Academic Press, New York, p. 1 (1978).

7. Thomas, Jr., J.F., "Pseudopotential calculation of the third-order elastic constants of copper and silver," Phys. Rev. B, 7, pp. 2385-2392 (1973).

Part 6
POROUS MEDIA, POLYMERS, CRYSTAL MECHANICS, AMORPHOUS SOLIDS

Continuum Models of Discrete Systems 4
eds. O. Brulin and R.K.T. Hsieh

DISCRETE AND CONTINUUM MODELS OF POLYMER ELASTICITY

R.G.C. Arridge

H.H. Wills Physics Laboratory
University of Bristol
Royal Fort
Bristol BS8 1TL England

In deciding the size of the representative volume element (RVE) in polymers, above which continuum mechanics is relevant but below which the molecular nature of the polymer must be considered, it is suggested that mechanisms of load transfer be studied.

Using Gaussian statistics, a RVE for amorphous polymers is proposed having the linear dimension $\frac{1}{2} \langle r^2 \rangle^{\frac{1}{2}}$, where $\langle r^2 \rangle$ is the mean square end-to-end distance of a chain.

1. Since the realization in the 1920's that polymers are composed of long chain molecules explanations have been sought in molecular terms of their elastic, thermal and electromagnetic properties. In rubbers, and in other amorphous polymers above their glass transition, the statistics of chain configurations explain very well the entropy-elastic behaviour and orientation-dependent effects such as birefringence, but in the glassy state quantitative theories are less simple (Ferry (1)).

The discovery of crystallization in polymers and its complex nature has encouraged a 'two-phase' interpretation of polymer properties in which each phase contributes by a rule of mixtures law to the property in question. Thus entropy-elastic and energy-elastic mechanical properties can be added according to the respective volume proportions of the amorphous and crystalline phases. Similarly, their respective contributions to dielectric properties, refractive index,density etc. can be calculated.

Together with the interest in the 1960's and 70's in fibre composites and in composites in general there arose also an interest in treating polymers as composite materials and in trying to explain their behaviour quantitatively by this means. I want in this paper to consider this approach both in terms of the scale of the elements involved and also in terms of the validity of using continuum concepts such as stress, strain and modulus when we are considering molecular assemblies such as polymer chains.

2. It seems convenient to trace the origin of models of the behaviour of polymers to the early work of Takayanagi (See Arridge (2) for a description of these models) who recognized that certain polymer blends and mixtures had mechanical properties very well modelled by series or parallel (or combinations of series and parallel) blocks of the constituent materials. Such models were, and are still, used to predict both real and imaginary parts of the dynamic moduli of numerous different polymers over a wide range of temperature. The Takayanagi models are reminiscent of the Voigt and Reuss models used in explaining the properties of polycrystalline metals in terms of individual crystal properties, using the assumptions of uniform strain (Voigt) or stress (Reuss) in the

constituent crystals.

That the Voigt and Reuss calculations of the elastic moduli of a polycrystalline assembly of elastic elements give upper and lower bounds respectively was shown by (Hill (3)). Since that date better (closer) bounds have been found both for isotropic and anisotropic elements (see, for example, Walpole (4) and references therein). Extensions to linear viscoelastic elements have also been published (see for example Laws and McLaughlin (5) and references therein).

A considerable body of rigorous theory therefore exists which can be used to explain the properties of polymer composites if the properties of the respective elements, crystalline and amorphous are known. We need, however, to consider three problems at this stage. First: what are the elements in a polymer considered as a composite ? A polymer crystal or spherulite is itself a composite of lamellar folded chain crystals and less-ordered interlamellar material. We discuss this in §3 below. Second: what is the scale on which we consider the composite approach to be valid ? This involves the determination of the size of a representative volume element (RVE) in the sample as a whole. We consider this in §4. Third: how valid is our use of continuum theory when we think of the mole -cular configurations involved in polymer structures ? We discuss this problem in §5 where we propose a criterion for the RVE in an amorphous polymer.

3. The elements in a polymer. In semi-crystalline polymers X-ray diffraction reveals the presence of crystalline elements with, in general, random orientation and poor perfection. The amount of crystalline material may vary from some 20% to over 90% by volume. In melt-crystallized material the crystals are spherulitic consisting of a substructure of folded chain lamellae which slowly twist along the radial direction of the spherulite. Non-crystalline material is present between lamellar surfaces and between the spherulites.

Mechanical treatments to induce orientation (rolling, drawing, extrusion) may convert the spherulite structure to one of stacked lamellae or to oriented fibrils, depending upon the stress system in operation, the latter showing a very high degree of orientation and improved crystal perfection. While exact analysis of the spherulitic structure in terms of its substructure has not been carried out, both the stacked lamellae and the fibrillar structures have been modelled using classical elasticity. For lamellar stacks (Ward (6)) and his associates used simple shear, assuming homogeneity of stress within and between the lamellae in the stack. This was not justified, as the present author has pointed out (Arridge (7)).

In our laboratories we have modelled the lamellar stack using both a stress function approach and finite element numerical methods (Arridge and Lock (8)). In the case of fibrillar geometry simple shear lag theory has been used by (Arridge and Barham (9)) although its use implies uniformity of stress over a fibre cross section as well as approximations regarding the environment of the assumed fibres.(Halpin and Tsai (10)) have proposed a useful general formulation adjustable to various geometries by a "contiguity factor".

The modelling of a spherulite by an assembly of lamellar elements is possible since the elastic properties of a stack of lamellae under any stress system are exactly calculable. The calculation has not, as far as the author knows, yet been attempted. If it were done, then the properties of spherulites would be predictable in terms of their elements and therefore the properties of assemblies of spherulites in the macroscopic polymer would be in turn predictable.

4. The scale on which the composite approach is valid. This problem is not peculiar to the case of polymers, but exists wherever heterogeneity is found. In the field of petrology illuminating examples of scale effects were first

collected by (Sander (11)) who defined statistical homogeneity as the scale at which the average of the internal constitution of a body is the same for all volume elements with dimensions not smaller than that scale. In 1963 (Hill (12)) defined the representative volume element (RVE), which includes the statistical homogeneity of Sander, as follows.

A representative volume element is a) structurally entirely typical of the whole mixture on average, b) contains a sufficient number of inclusions for the apparent overall moduli to be effectively independent of the surface values of traction and displacement so long as these values are "macroscopically uniform". That is, they fluctuate about a mean with a wavelength small compared with the dimensions of the sample, and the effects of such fluctuations become insignificant within a few wavelengths of the surface.

Using such definitions and remembering that the size of a spherulite in a polymer is on the scale of a few microns, it would seem reasonable to consider a scale of millimetres to treat as the RVE in a bulk polymer, but a smaller scale (to which we shall refer shortly) in amorphous polymers.

5. The validity of continuum theories. We cannot strictly apply continuum theories to assemblies of polymer chains since differentiability is implied in the usual definitions of deformation gradient and stress, yet the smallest dimension definable in a polymer is that of one chain. The same problem is, of course, true in any solid and it was appreciated by Cauchy and Poisson in the early 19th Century. However, in well-ordered crystals a unit cell of atomic dimensions can be found and both deformation and deformation gradient defined in terms of atomic separation. In glasses and in glassy polymers, an averaging process is required which needs to be defined carefully if it is not to lead to confusion. We should seek in any solid a representative volume element above the scale of which contin-uum theories may be applied, but below which we must consider forces and displacements only since stress and strain cannot be defined.

If we consider the way in which loads are transferred through a solid then there is a difference apparent as between crystalline and non-crystalline materials. The size of the unit cell in a crystalline material will in general be smaller than the RVE, for in considering load transfer at an atomic scale, not only nearest neighbour interactions need to be taken into account.

It is true that the range of atomic forces is small, so that 2nd and 3rd nearest neighbours are probably all that need to be considered. We may confidently take the RVE in a crystal as being no more than about three times the linear dimension of the crystal unit cell.

In a glassy solid (ionic or metal) the RVE will be larger because of the requirement of isotropy, but it is unlikely to be very much larger than for the crystal case.

For an amorphous polymer however, we have to consider how load is transferred from chain to chain and we need to answer the question: if we take a unit cube in the polymer how is load transferred from one face to another ? We can attempt to answer this question by two methods a) using configurational statistics b) by model building. I shall use the first of these methods to show that a reasonable size for the RVE in an amorphous polymer is a cube of scale $0.5 < r^2 >^{\frac{1}{2}}$, where $< r^2 >$ is the mean square end-to-end distance of a polymer chain.

6. The RVE in an amorphous polymer. Supposing that all load is tranferred along chains we need to answer the question: how many chains cross from one unit plane to the similar plane displaced by a distance a along their mutual perpendicular ?

(We suppose that the stress system in the continuum i.e. on a scale greater than that of the RVE, is one of principal stresses and that these are tensile). We consider an elementary solution first and a more refined one later. Assuming Gaussian chain statistics the probability, W(x,y,z) dx dy dz of a chain of n links, each of length ℓ, which commences at (0,0,0) finishing within the volume (dx dy dz) at the point (x,y,z) is

$$W(x\,y\,z)\,dx\,dy\,dz = \left(\frac{b}{\sqrt{\pi}}\right)^3 \exp(-b^2 r^2)\,dx\,dy\,dz$$

where

$$b = [3/(2n\ell^2)]^{1/2}$$

The probability therefore of a chain starting within the area 0 < (x,y)< a; z = 0 and finishing in the prism 0 < (x,y)< a; z > a will be

$$W(ab) = \frac{1}{a^2}\left(\frac{b}{\sqrt{\pi}}\right)^3 \int_0^a\!\!\int_0^a \exp[-b^2(x_2-x_1)^2]\,dx_1\,dx_2 \times$$

$$\times \int_0^a\!\!\int_0^a \exp[-b^2(y_2-y_1)^2]\,dy_1\,dy_2 \int_{z=a}^{\infty} \exp(-b^2z^2)\,dz$$

or

$$W(ab) = \frac{1}{a^2}\left(\frac{b}{\sqrt{\pi}}\right)^3 \left\{\int_0^a\!\!\int_0^a \exp[-b^2(x_2-x_1)^2]\,dx_1\,dx_2\right\}^2 \int_a^{\infty} \exp(-b^2z^2)\,dz$$

This is easily found to be given by

$$W(ab) = \frac{1}{2a^2b^2}\,\mathrm{erfc}\,(ab)\left[\frac{\exp[-(ab)^2]-1}{\sqrt{\pi}} + ab\,\mathrm{erf}(ab)\right]^2$$

In terms of $u = ab = a\sqrt{3/2}/(\ell\sqrt{n})$ this has values tabulated below.

u	W(u)
.1	.00488
.2	.01381
.3	.02024
.4	.02154
.5	.01858
.6	.01364
.7	.00876
.8	.00502
.9	.00258
1.0	.00121

Showing a maximum for W(u) at a value of u of about 0.43.

There are certain assumptions we have made however, in the above derivation, such as including only chains starting on the lower Z-face of the cube and traversing it, while neglecting chains starting at $z < 0$ which also traverse it. We should also restrict ourselves to chains wholly confined to the cube while traversing it. An improved, but still not exact, treatment of the problem is as follows. (Haward, Daniels and Treloar (13)) give an exact solution for the one-dimensional Gaussian chain problem for a slab of width a. They define the probability of not spanning the gap a as

$$R = 1 - 8 \sum_{r=1}^{\infty} (-1)^{r-1} \frac{A(rb')}{b' + 2\sqrt{2/\pi}}$$

where

$$b' = a/\sigma = a\sqrt{3}/(l\sqrt{n})$$

$$A(u) = \frac{1}{\sqrt{2\pi}} \exp(-u^2/2) - u[1 - \Phi(u)]$$

and

$$\Phi(u) = \frac{1}{\sqrt{2\pi}} \int_{-\infty}^{u} \exp(-t^2/2)\, dt$$

The probability of spanning the gap is therefore 1-R. In order to use this result for a unit cube, rather than for two planes of infinite extent separated by a distance a, we need to restrict the number of chains crossing the gap in the Z-direction to those which remain in the cube in the other two directions. (If we restrict ourselves to chains which remain within (0,a) for all Z this will be too restrictive. The multiplying factor for this case can be shown to be f^2, where

$$f = \frac{2b'}{b' + 2\sqrt{2/\pi}} - R$$

This reduces the probability of spanning the unit cube from side to side to a value of about .03%).

We may arrive at a more realistic, but still not exact, result using the argument of Haward, Daniels and Treloar once more to arrive at the probability of random walks in x and y which enter but do not span the interval (0,a).

This probability is just R, so that the required probability that chains entering one face (the Z-face) of the cube will leave by the opposite face is $(1-R)R^2$. Note: this argument cannot be completely correct, for we may be including some chains which enter the face $x = 0$ and leave it before traversing the plane $z = a$. Similarly, a chain may be neglected wrongly which spanned $0 < x < a$ but then returned within the cube and left by the face $z = a$. To limit the chains which span the gap in z to those lying within $0 < x < a$ and $0 < y < a$ for all z would be easy, but too restricting.

A proper computation would be subject to the boundary conditions of the cube, but it is felt that the approximate treatment above is probably sufficiently accurate.

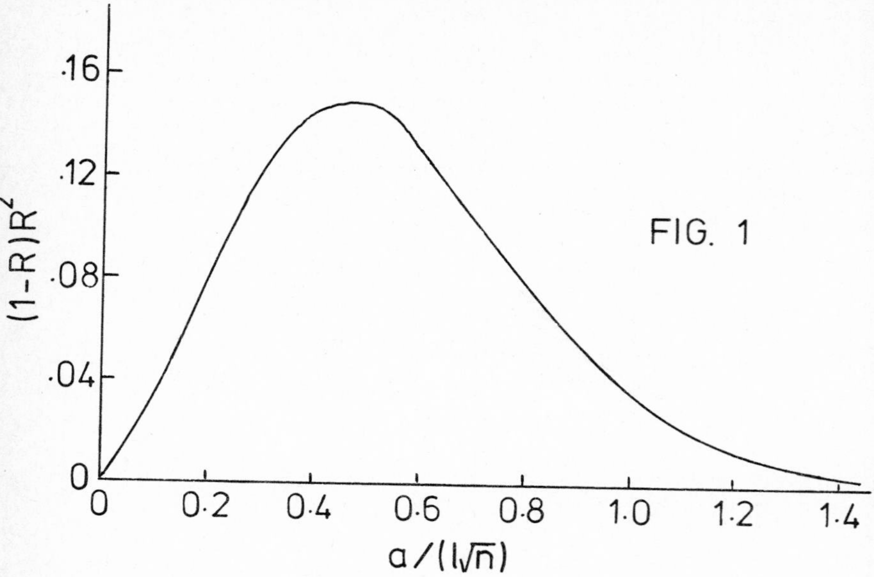

Figure 1 shows a plot of $(1-R)R^2$ against the parameter $a/(\ell\sqrt{n})$ using the figures of Haward, Daniels and Treloar. It indicates a maximum probability of about 0.148 at $a/(\ell\sqrt{n})$ = 0.5. This means that for a cube of side equal to half the RMS end-to-end distance of a chain about one chain in seven will traverse the cube from face to face. If we take a unit cube appreciably smaller than this it is difficult to see how it can be in equilibrium under forces transferred along chains since very few of these will traverse it, whereas for a much larger cube we will have to consider inter chain transfer by friction or entanglements since effectively no chains will ever reach across from face to face.

Let us consider a numerical example. For amorphous polystyrene $\ell\sqrt{n}$ is about 22nm giving a cube size of 11nm. Considerations of chain packing suggest that about 584 chains will cross any area (11nm x 11nm). The same applies to the opposite face of the cube except that some of the 584 chains on each face span to the opposite face.

Figure 2 illustrates the situation for 12 chains in the cube. There are 3 chains which traverse from face to face. On each face there are also 3 chains which enter but do not traverse, passing through an adjacent face. There are 9 of these. In any direction therefore, out of 7 chains which enter the gap between cube faces, 3 chains on each face enter but do not leave and 1 traverses. For the 11nm cube therefore we have 146 chains spanning the gap in, say, the z-direction and three times this number, on each face, that is 438 chains which enter the gap on a z-face, but leave it by an x- or y-face, roughly 110 to each face.

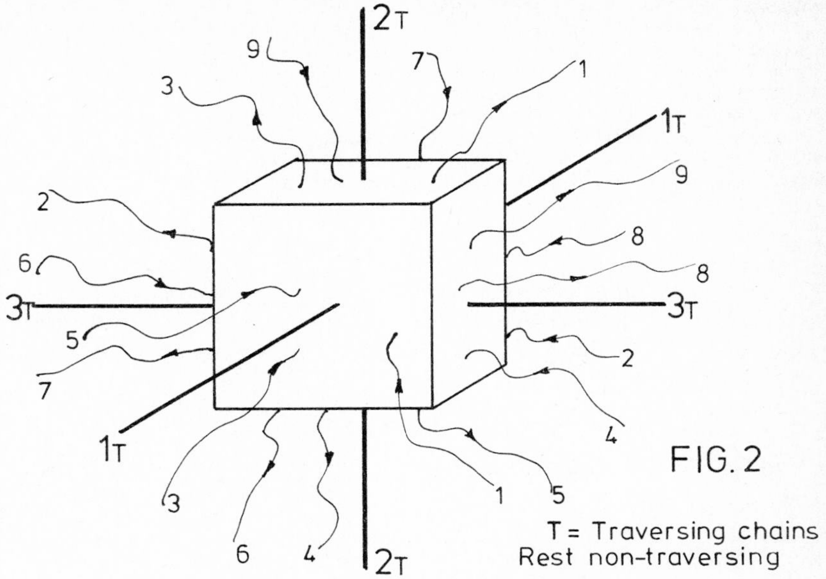

FIG. 2

In a cube of side 11nm there will be approximately 10^6 atoms, far too many for computer modelling. The cube size is, however, not too large for modelling using for example "Gaussian spaghetti", that is flexible cylinders of circular cross section the direction of whose axes at any point follows Gaussian statistics. Studies of such models, using various systems of uniform stress on the cube surfaces would be valuable as models for amorphous polymers.

7. Representative volume elements for semi-crystalline polymers. In these materials various types of crystal structure can occur ranging in size from a few hundred Ångströms (fringed micelle structure) through lamellae with dimensions of about 1 micron to dendrites and spherulites which may be as large as several hundred microns. The size to be taken for an RVE in such materials will therefore depend upon the crystal morphology, itself a function of polymer type and molecular weight and the conditions of crystal formation, whether from solution or from the melt. As in metals, annealing can alter crystal size and, even more than in the case of metals, physical working (drawing, rolling or extruding) will change both the orientation and the nature of the crystal phase. There is therefore no unique RVE proposed for semi-crystalline polymers, its size must be determined by microscopy or other (x-ray diffraction, light scattering) techniques.

At high orientation some polymers may show quasi-perfect crystal orientation and models for their internal structure postulate a crystalline array with defects, rather than an aggregate of crystal and amorphous phases. The appropriate RVE for such a material would be the crystal unit cell.

8. Significance of measures of scale. Some studies have been reported in the literature of the effects of crystal size on mechanical properties (Patel and Phillips (15)) although it is not clear from their paper to what extent the variation (change of modulus with spherulite size in polyethylene) depended upon the sample size.

Where the size of the representative volume element will become important is, of course, when a very small sample is being tested or when properties sensitive to the stress distribution in a very small region are being studied. This occurs when the fracture toughness G_c is being determined using notched specimens for the size of the 'process zone' at the crack tip may be of the order of the RVE for that material. The commonly used Dugdale model in which a stress equal to the yield stress is assumed throughout the process zone may then be inappropriate. A related problem is that of crazing in polymers where the conditions at the crack tip become of importance. Similar problems arise, of course, in composite materials and dispersion hardened metals.

REFERENCES:

1. Ferry, J.D., Viscoelastic properties of polymers (Wiley, New York 1970).

2. Arridge, R.G.C., Mechanics of Polymers (Clarendon Press, Oxford 1975).

3. Hill, R., The elastic behaviour of a crystalline aggregate, Proc. Phys.Soc. A65 (1952), 349-354.

4. Walpole, L.J., On the overall elastic moduli of composite materials, J. Mech. Phys. Solids 17 (1969), 235-251.

5. Laws, N. and McLaughlin R., Self-consistent estimates for the viscoelastic creep compliances of composite materials, Proc. Roy. Soc. A359 (1978) 251-273.

6. Owen, A.J. and Ward, I.M., A simple model for mechanical anisotropy in specially oriented sheets of low density polyethylene, J. Materials Sci. 6 (1971) 485-489.

7. Arridge, R.C.C., Assumption of homogeneous elasticity in theories of lamellar texture of polymers. J. Materials Sci. 9 (1974), 155.

8. Arridge, R.G.C. and Lock, M.W.B., Stresses and displacements in lamellar composites Part II. J. Phys. D: Appl. Phys. 9 (1976), 329-351.

9. Barham, P.J. and Arridge, R.G.C., A fibre composite model of highly oriented polyethylene, J. Polymer Sci. (Poly. Phys. ed.) 15 (1977) 1177-1188.

10. Halpin, J.C., Effects of environmental factors on composite materials, AFML-TR-67-423 (1969).

Halpin, J.C. and Kardos, J.L., The Halpin-Tsai equations: a review, Polymer Eng. and Sci. 16 (1976), 344-352.

11. Sander: Gefügekunde der Gesteine (1911) discussed in Turner F.J. and Weiss L.E., Structural Analysis of Metamorphic Tectonites (McGraw-Hill, New York, 1963).

12. Hill, R., Elastic properties of reinforced solids: some theoretical principles, J. Mech. Phys. Solids, 11 (1963) 357-372.

13. Haward, R.N., Daniels, H.E. and Treloar, L.R.G., Molecular conformation and craze fracture, J. Polymer Sci. (Poly. Phys. Ed.) 16 (1978) 1169-1179.

14. Patel, J. and Phillips, P.J., The Young's modulus of polyethylene, J. Polymer Sci., Polymer Letters ed. 11 (1973), 771-776.

Continuum Models of Discrete Systems 4
eds. O. Brulin and R.K.T. Hsieh

CRYSTALLOGRAPHY OF AMORPHOUS BODIES

Maurice Kléman

Physique des Solides, Bât.510
Université Paris-Sud, 91405 - Orsay Cedex
France

Continuous random networks and dense random packings can be classified as specific mappings of ordered latticesin spaces of constant curvature onto the usual 3-dimensional euclidean space. Mapping of spherical spaces introduces naturally surface defects, while mapping of spaces of negative curvature introduces line defects (disclinations). This classification and the defects it involvesare discussed in light of the known universality properties of glasses near Tg.

UNIVERSALITY IN THE GLASSY STATE

Glasses (sensu stricto, or amorphous metals and alloys, or glassy polymers, or spin glasses, etc...) are characterized experimentally by two sets of universal properties, whose understanding would resolve one of the most exciting challenge in present-day condensed matter physics, viz : the nature of disorder. In this paper, in which we restrict to metallic and covalent simple amorphous systems, we try to shed some light on the origin of universality by putting into evidence the existence of very general models for disordered states ; the notions used are geometrical and topological, and lead to a kind of "crystallography" of amorphous states. The exemples we have in mind are a)amorphous silicon (with or without hydrogen) for covalent systems, which has been extensively studied by x-rays diffraction (recently by electron microscopy) but also by computational methods ; b)amorphous iron for metallic systems, which in a way is a virtual system, since iron yields amorphous phases only in presence of a rather substantial quantity of metalloïd atoms, but has been largely studied with computational methods. However it is a good first approximation to consider the compact arrangement of iron atoms independently of that of metalloïd atoms (which are probably bond by covalent forces).

Universality properties fall broadly into two classes :

a - at very low temperatures ($\sim$ 1 K), one observes an extra heat capacity proportional to T and a heat conductivity proportional to T^2. This is explained by Anderson et al. [1] on the basis of the existence of localized two level centers ; the nature of these centers is still unknown, probably related to some "defects" in the amorphous structure, perhaps the crystallographic defects we shall describe. However this connexion remains of the realm of hypotheses, and we shall not try to relate our results to the low temperature behavior.

b - at high temperature, inverse viscosity, diffusivity, rate of thermal contraction, and in fact all transport rates D_i decrease very rapidly as one approaches the glass transition temperature Tg from above ; they follow the well-known Vogel (Fulcher-Doolittle) law [2]

$$D = \text{const. } \exp - \frac{E}{T - T_o} \qquad (1)$$

where T_0 is experimentally a relatively well defined temperature, generally much below T_g. On the other hand T_g is in a sense a variable temperature, which marks the freezing of the hysteretic phenomena which keep the liquid in a supercooled state, and depends on the way this freezing point is reached ; there are many (metastable) amorphous states, of practically the same energy. But T_0 has a thermodynamic meaning.

The understanding of the nature of T_0 would be a great step in the physics of the liquid state, as emphasized by Anderson [3]. There is a general agreement to say that the transport phenomena which obey Vogel's law happen through some kind of "defects" mechanism, but these defects themselves are objects of dispute:

- in the "free volume" theory of Turnbull and Cohen [4], it is claimed that the thermodynamic coalescence of voids in the dense random packing (DRP) of hard spheres (this is a model for metallic atoms) provide sites for the activated motion of atoms. Mechanical properties below T_g are also explained in the frame of this theory [2].

- annealing in the glassy state can be performed without crystallization, and induces changes in the structure (like densification, increased Young's modulus, embrittlement, reduced diffusivity,...) which are interpreted by Egami [5] as a tendency towards a more amorphous state. According to related computer experiments, the defects which anneal during this "structural relaxation" could be coupled positive (p) and negative (n) local density fluctuations which do not affect short range order [6]. The same computations indicate the existence of strong shear stress defects (τ) with large deviations from spherical symmetry.

- Anderson [3] proposes that the (virtual) transition which takes place at T_0 is of the Kosterlitz - Thouless type, i.e. a second order transition driven by defects [7], between a low temperature non defective state, and a high temperature amorphous state where defects of some sort, with a density $\rho \sim \exp - E/(T_g-T_0)$, exist in equilibrium. The importance of Anderson's suggestion is that these defects must be intrinsic defects, i.e. defined by some topological rules with respect to an ideal state at T_0, as dislocations and disclinations are defined topologically with respect to the crystalline state.

- Rivier [8], in agreement with Anderson, also considers T_0 as a K.- T.transition, and identifies the defects with lines threaded through odd-numbered rings of bonds in a continuous random network (in short, CRN ; this is a model for covalent glasses),avoiding even rings. He has a nice demonstration of the existence of these lines. But his theory relies implicitly on the fact that there is some ideal CRN without lines of defects, which as we shall see is disputable.

In conclusion it appears that the universality of T_0, if true, implies the existence of some sort of ideal amorphous structure, or the understanding in what sense current CRN and DRP models are ideal models. The present author and Sadoc [9] have proposed to define the amorphous state as the state full of defects of a crystal lattice in a curved space (of positive or negative curvature according to the case). It is tempting to define the ideal state at T_0 as this virtual crystal lattice in curved space. We review and develop the point of view of [9] in this paper. Recent electron microscopic observations of amorphous hydrogenated Si have put into evidence a three dimensional lattice of lines[10] ; these observations are perhaps related to our theory.

C R N and D R P

As stated above, there are two types of models for amorphous glasses. Most used are the tetracoordinated CRN and the DRP of tetrahedra. They are dual one from the other. The recent review by Zallen [11] of the present knowledge in this "stochastic" geometry is to be consulted for more details. The DRP model was first proposed by Bernal [12] for liquids and later refined by Finney [13]. The CRN model dates back to Zachariasen [14] and developed by Polk [15].

SMALL AGREGATES OF ATOMS WITH CENTRAL FORCES

Electron diffraction experiments (coupled with computer simulation) on small aggregates of Ar atoms, condensed at low temperature in ultra high vacuum [16] have shown that the first atoms to condense form regular tetrahedra of closed packed atoms ; therefore the initial growth process of these small clusters does not yield a nucleus of f.c.c. phase (which is a regular packing of an equal number of tetrahedral and octahedral arrangements), but icosahedral arrangement of atoms : a central atom is surrounded by 12 others at equal distance from this center ; each of these 12 atoms sees a pentagonal environment, so that the line linking the central atom to any of the 12 surrounding ones is the common edge of 5 equal tetrahedra, which are in fact slightly distorted with respect to perfect tetrahedra (the dihedral edge of a perfect tetrahedron is somewhat smaller than $2\pi/5$). The total cluster of 13 atoms is made of 20 tetrahedra. It is known that one cannot fill space regularly with tetrahedra, so that adding new atoms condensing on the already formed cluster either leads to other (interpenetrating) icosahedra, or to a multilayered distorted large icosahedron ; the largest observed clusters contain up to 1000 atoms. They are stabilized by a balance between the (disfavourable) elastic distortion energy of the tetrahedron and (favourable) surface tension.

Sadoc has advanced the idea that such arrangements should be described in spherical space rather than in euclidean space. More precisely, he states that the ideal arrangement which approaches best the amorphous aggregate is a regular polytope which covers the three-dimensional sphere S_3, and contains 600 tetrahedra (5 of them around each edge of the polytope), 1200 triangles, 720 edges, 120 vertices (i.e. 120 atoms). Each vertex has 12 neighboring vertices. We have to explain what such a sort of geometry means, since it is not familiar to most physicist and mechanicians, and how the (real) amorphous aggregates can be constructed from this polytope: (details can be found in ref.[9].)

A polytope is nothing else but a n-dimensional analog of an ordinary 2-dimensional polyhedron scribed on the ordinary sphere S_2 (here we shall have $n = 3$). For example consider a 2-dimensional planar "crystal" where each atom likes to be surrounded by five others; such an arrangement does not fill the plane regularly. On the other hand the same symmetry conditions imposed to atoms lying on a sphere S_2 yield a regular icosahedron. Therefore one can think that various mappings of this sphere on the plane leads to various amorphous states, amongst which are the real ones, which obviously have to minimize the distortion which occurs in the mapping, and which are of two kinds : distortions in mapping one sphere (or a part of one sphere) on the plane, distortions coming from the addition along the boundaries of repeated mappings of spheres, in order to fill the entire plane. All these processes can be extended to 3-space. The polyhedron is now a 3-polytope, made of vertices, edges, faces and elementary 2-polyhedra. It is always possible, in such a curved space, to define lengths and angles, as on a S_2. Mapping will necessarily introduce distortions of these lengths and angles, and yield an "amorphous" arrangement in R_3. In fact, it is possible to show in a very general way that to any regular ("crystalline" so to speak) arrangement on S_3 corresponds an irregular arrangement on R_3, or, in other words : any symmetry pattern (we considered above the tetrahedral arrangement as an example) which fills in regularly a space of constant positive curvature cannot be regularly reproduced on a flat space. The reciprocal is true. Sadoc has recently made an extensive use of all the possible "space groups ("Schoënflies groups") of S_3 to describe possible amorphous structures [17] possessing the same symmetries, but only on a local scale, than on S_3.

REGULAR LATTICES AND SPACE GROUPS IN CURVED SPACES

Although mappings on R_3 on spheres S_3 regularly filled with atoms yield a number of possibilities of amorphous structures, they do not provide us with an ideal lattice just below T_0, since S_3 is finite. Also, in order to compute the amorphous state, we have to introduce complex interfaces, which can be in many ways relaxed by disclinations. The projection itself of one S_3 on R_3 is not unique. We therefore prefer to restrict the use of S_3 to small aggregates. Note however that Sadoc and Mosseri have done an extensive study of the structure of amorphous tetracoordinated semiconductors starting with lattices on S_3.

Another (more general) possibility is to look at spaces of constant negative curvature, rather than spaces of constant positive curvature (spheres). All spaces of constant curvature have properties which make them useful objects to define ideal states : a)they are invariant under a continuous group G of isometries depending on six parameters ; G depends only on the sign of the curvature ; b)this implies properties of isotropy and homogeneity ; c)therefore such spaces can carry regular lattices of points (with eventually a pattern) invariant under a subgroup H of G (all these properties pertain also to the ordinary flat space R_3 ; G is the euclidean group in this case, and H any of the 230 Schoënflies groups). But, unlike spheres, spaces of constant negative curvatures (also called hyperbolic spaces) are infinite ; if they are simply connected (note them H_3), they are diffeomorphic to R_3. Therefore the mapping of H_3 on R_3 does not require successive steps as above. We shall see in the following paragraph that this mapping, which we want to be as isometric as possible, introduces defects in a very natural way. Let us here indicate briefly how one can construct on H_3 regular honeycombs (i.e. lattices for which the unit cell is a regular polyhedron). The general case (i.e. the denumeration of all space groups of H_3) is to be found in an important paper by Coxeter and Whitrow [18].

Any regular polyhedron can be defined by the symmetry order p of a face (p = 3 for an equilateral triangle, p = 4 for a square) and by the number q of faces around a vertex. This yields the Schläfli symbol {p,q}. For example {3,5} denotes an icosahedron and {5,3} a dodecahedron. It is well known that there are only five regular polyhedra, viz. the tetrahedron {3,3}, the cube {4,3}, the octahedron {3,4} and the two polyhedra already mentionned. In all these cases the total angle span by the q faces at a vertex, i.e. $q\,(1 - \frac{2}{q})\,\pi$, is smaller than 2π ; this reads also

$$(p-2)\,(q-2) < 4 \qquad (2)$$

and this angular condition expresses the fact that the polyhedron is scribed on a sphere. The equality obtains when the faces {p,q} tile regularly a plane (three possibilities {4,4}, {3,6} , {6,3}). This generalization of the meaning of the symbol can also be extended to an hyperbolic plane H_2, for which

$$(p-2)\,(q-2) > 4 \qquad (3)$$

There are in this case an infinite number of possible regular tilings of H_2. H_2 is the two dimensional connected simply connected surface with constant negative curvature.

Now, going to three dimensions, the Schläfli symbol {p,q,r} represents a regular tiling of space by {p,q} cells, each edge being common to r cells.

Here the condition on curvature generalizing eq.2 reads [19] :

$$p - 4/_p + r - 4/_r + 2q - 12 < 0 \qquad (4)$$

for S_3. This leads to the six spherical regular polytopes :

{3,3,3} {3,3,4} {4,3,3} {3,4,3} {3,3,5} {5,3,3}

Notice that the polytope we have been considering above for amorphous metallic aggregates is {3,3,5}.

Equality in eq.(4) yields only one euclidean honeycomb, viz.{4,3,4}, which is a tiling of R_3 with cubes (cells of other symmetry groups in R_3 are not regular polyhedra).

Finally the inequality inverse to (4) yields all the regular honeycombs of H_3. In the sequel, we consider only a few of them. Let us discuss here the honeycomb {6,3,3}.

The first two symbols define a regular tiling of hexagons in the plane, not a regular polyhedron. {6,3} is not scribed on a sphere of H_3, but on a so-called "horosphere" of infinite radius. 3 such horospheres pass through each edge. Each vertex has 4 neighboring vertices. This lattice represents conveniently a covalent arrangement with a coordination number of 4. Since vertices arrange in six-numbered rings, it bears some analogy with the diamond lattice of carbon or silicium atoms, or with the wurtzite lattice, but is in fact much simpler, since it is an unique arrangement.

Apart from representing a covalent network with z = 4, {6,3,3} can as well be interpreted as a close packing of spheres. The last two figures in the Schläfli symbol constitute indeed the Schläfli symbol of a cell whose vertices are the centers of the cells of {6,3,3}. The inspheres of the cells of {6,3,3} pack therefore in tetrahedra {3,3}. The same interpretation can be given to any honeycomb {p,3,3} ; p is then the symmetry order of a line joining two adjacent spheres of the packing. The honeycomb {p,3,3} itself constitutes the set of the Voronoï polyhedra related to the sphere packing. This dual interpretation of {p,3,3} should be kept in mind ; note incidentally that while {6,3,3} can be interpreted as an hyperbolic close packing of spheres, {5,3,3} is a spherical close packing.

MAPPING IN THE DISCLINATION MODE

The general problem of mapping a given manifold M on another one M' is discussed in a qualitative way in the beautiful book of Hilbert and Cohn-Vossen [20], Geometry and the Imagination (which is perhaps the most valuable reading in mathematics one can advise to a non-specialist). For example it is always possible to map M conformally (i.e. conserving angles) on R_3, in a one-to-one mapping ; but such a mapping changes the lengths, except in special cases. A one-to-one geodesic mapping is possible between manifolds of constant curvature each, as in the case we are considering ($M = H_3$, $M' = R_3$) but only the lengths measured along geodesics are conserved ; the other ones are strongly distorted. An isometric mapping (conserving the distances, hence all the angles) would be most interesting, but is impossible. In fact, a general theorem states that isometric mapping between points P and P' of two manifolds M and M' implies that the curvatures in P and P' are equal. For example one can map isometrically any (2-dimensional) developable surface on the plane, but not the sphere on the plane. Let us discuss more fully how curvature and isometry are related in a mapping, since this is at the heart of our problem.

Consider a closed path C on M, and a point P on C. It is always possible to define an euclidean space tangent to M in P along C. (If M is a two dimensional surface, it is the tangent plane). This local euclidean space can be mapped isometrically on R_3, P' being in coïncidence with P. Moreover, it is possible to roll M on R_3 along C, from P to P, along a path C' in R_3 such that lengths along C as well as angles between vectors belonging to the tangent space in P are conserved:

this is a particular case of a general theorem stating that the parallel transport of vectors from one point of a Riemannian manifold to another point along a fixed path C is a linear isometric operation [21]. The reader can visualize the properties we have in mind by considering a closed path on a sphere S_2, and rolling S_2 on a plane Π along this path. This is an unique operation (assuming starting point and starting direction given), which defines local mappings between planes tangent to S_2 along C and Π. Note that C' is an isometric mapping which cannot be extended to all the sphere ; moreover C' is not a closed path. These properties are true for any M ; but we are still able to use this kind of mapping, by a sort of non-analytical extension we discuss now.

The closure failure of C' is in this respect the essential thing to understand ; it contains a failure in rotation and a failure in translation.

- the failure in rotation reveals curvature ; take for C an infinitesimally small closed path, bounding a small area σ in M. when rolling M on R_3 along C, any frame of reference attached to P has rotated by an angle $\Omega_i = \omega_i \sigma$, where

$$\omega_i = K_{ji} \nu_j \tag{5}$$

ν_j are the direction cosines of the normal to σ and K_{ji} the six independant components of the Riemann curvature tensor (see [21], chap.7).

The Riemannian (sectional) curvature in P is :

$$K = \omega_i \nu_i = K_{ij} \nu_i \nu_j \tag{6}$$

and is a constant, independant of P and $\vec{\nu}$, in a space of constant curvature. K is positive for a spherical space S_3, negative for H_3.

The failure in translation reads

$$D \sim K \lambda \sigma \tag{7}$$

where λ is the diameter of C. Clearly, D is a second order quantity compared to $\Omega = |\Omega_i|$, so that we restrict only to the consideration of the rotation in the local mapping $C \to C'$.

To go further in our analysis, we need to interpret Bianchi's identities ([21], chap.8) according to which the sum of the rotations $\int \omega_i \, d\sigma$ relative to a small closed surface is zero. This condition is akin to the node condition for dislocation lines in conventional crystals, and leads us to introduce here sets of disclination lines, in finite or closed, drawn in H_3, whose role is to decurve the hyperbolic lattice into an euclidean arrangement. They are the non-analytic objects we alluded to above. This is best seen by using an approach due to Regge [22] in another context (how to describe the Einstein space of general relativity without using coordinates). The hyperbolic space H_3 can be approximated by a triangulation in which the lengths of the edges of the honeycomb are conserved, but the faces are replaced by euclidean faces. Such a triangulation can be carried for any riemannian manifold, indeed, and in the case of an S_2, it consists in approximating the sphere by a polyhedron whose vertices are on the sphere. This is equivalent to a deformation of S_2 in which one concentrates curvature at the vertices, with a Dirac density equal to the angle deficit at the vertex:

$$2\pi - \sum_i \alpha_i^*$$

where α_i^* is the angle at vertex V of the euclidean face i.

In the case of a 3-dimensional manifold like H_3, the curvature is now concentrated along the euclidean edges of the lattice which approximates the hyperbolic honeycomb. This means that if one moves a vector by parallel transport on a circuit C around an edge i, this vector rotates by an angle $\vec{\Omega}_i^*$ after completing a turn, and $\vec{\Omega}_i^*$ does not depend on the precise location of C as long as it surrounds only one edge. Bianchi's identities tell us that the sum of the rotations $\vec{\Omega}_i^*$ relative to all the edges originating from a common vertex is zero. Therefore one can extract from the edges a 3-dimensional network of infinite lines, along each of them $\vec{\Omega}$ is constant. Then one performs a Volterra process along those lines to map H_3 on R_3 ; by the Volterra process the curvature concentrated along each line is removed.

In order to visualize this operation, consider an analogous 2-dimensional situation : a cone is a manifold where all curvature (positive) is concentrated at the apex. Cut the cone along a generator and develop it on a plane ; then fill with 2-dimensional matter the remaining empty sector. This is the realization of a mapping with a disclination point. Up to now we have not introduced any change in length. Now suppose that the cone carries a regular lattice. The development of the cone does not bring any distortion, but the filling of the void does, if one wants to insure that no singularity subsists along the cut, so that finally one is left with some distortion, small enough if the angle at the vertex of the cone is small. A singularity in the lattice order subsists at the image of the apex.

We can perform similar operations in H_3, not only on the lines extracted from the edges, but on any set of lines. Since the curvature is negative, we shall have to remove matter at the cuts ; if we want that the contact between the lips of the cut be perfect, the angle of the disclination must be an angle of rotation symmetry of the lattice. Finally this mapping leaves us with a lattice arrangement which contains the fewer distortions (with respect to the lengths measured in the hyperbolic honeycomb) as the number of disclination lines is large enough to approach best the curvature of H_3 ; but along these disclination lines the lattice order is drastically changed. Balance of these two contributions determines the density of disclination lines.

Concerning the short range order at the disclinations cores, we have essentially two kinds of situations, which we illustrate on the tetracoordinated or close-packed {6,3,3} pattern.

- assume close-packed cells, and that a disclination line passes along an edge of their Voronoi {6,3,3} honeycomb ; then the Volterra process consists in removing one tetrahedron out of six at the line. The core symmetry becomes pentagonal.

- assume a tetracoordinated pattern, and that the wedge disclination line passes through centers of hexagons ; the Volterra process consists in creating a five-numbered ring around the line, and in fact a sequence of five-numbered rings all along. This result fits with Rivier's analysis [8], at least in spirit.

- if in the tetracoordinated pattern the disclination line follows the edges, the Volterra process consists in creating a line of dangling bonds.

THE DENSITY OF DISCLINATIONS IN THE IDEAL AMORPHOUS STATE

For a {p,q,r} lattice in H_3, supposed for simplicity made of regular cells, we introduce perfect disclinations of angle $\frac{2\pi}{p}$ or $\frac{2\pi}{r}$. The first are certainly of lower energy if they are located along the centers of the faces, the second if they are along the edges. Also for energetic reasons, we consider only wedge lines.

Introducing $2\pi/_p$ disclinations is equivalent to introduce a certain proportion of faces with (p-1) edges, without changing q (number of faces per vertex) and r (number of cells per edge). Similarly $2\pi/_r$ disclinations introduce a certain proportion of edges with r - 1 cells. Since the obtained configuration is in flat euclidean space, with zero curvature, a way of estimating these proportions is to write eq.4 as an equality, as proposed by Coxeter [19].

For example, consider the hyperbolic honeycomb {6,3,3} and introduce $\frac{2\pi}{6}$ disclinations. Locally, we have {5,3,3} configurations, which pertain to a spherical regular honeycomb. The condition (cf.eq.4).

$$I_{(p,q,r)} \equiv p^2 - (12 - 2q - r + \frac{4}{r})\, p - 4 = 0 \tag{8}$$

with q = r = 3 leads to

$$p = (13 + \sqrt{313}) / 6 = 5.115... \tag{9}$$

Therefore we have a large predominance of pentagons over hexagons. Experiments made with a froth [23] composed of 1900 bubbles give similar results. Recent computations by Srolovitz et al.[24] on a cluster of 2067 Fe atoms bound by a Johnson potential give the frequency f_i of occurence of each value of p_i = 3,4,...7. One finds

$$\sum_i p_i f_i / \sum f_i \sim 5.1 \tag{10}$$

which is also in agreement with the value obtained by only geometrical considerations. In the case of Srolovitz et al. the atoms are assumed to occupy the inspheres of the cells {p,3} and to form close-packed tetrahedra {3,3} ; note that the number of faces F of a cell is the number of neighbors z of each atom. For q = 3, one has $z = \frac{12}{6-p}$, which amounts to z = 13.56 with p = 5.115. Matzen's result [23] on froth is z = 13.70. Using as above the relative frequencies obtained in ref.[24], one finds

$$\sum z_i f_i / \sum f_i = 13.6 \tag{11}$$

which is also remarkably close to the geometrical result of Coxeter.

Let us remark that the quantity I(p,q,r) (q=r=3) changes sign between p = 6 and p = 5. This circumstance forbids us to obtain a regular flat (R_3) honeycomb by introduction of $\frac{2\pi}{6}$ disclinations. But the simultaneous introduction of, say $+\frac{2\pi}{6}$ in p position and $-\frac{2\pi}{3}$ in r position can transform {6,3,3} in {4,3,4} which is a regular honeycomb of R_3, made of cubes. This phenomenon might happen in other circumstances, with other lattices ; it might have some relevance to the process of recrystallization of an amorphous body.

A difficulty, not mentionned up to now, occuring with the particular example {6,3,3} is that the cells {6,3} are horospheres, (a special object in hyperbolic geometry), i.e. spheres of infinite radius. This is not bothering in the case of tetracoordinated structures, but yields difficulties when considering close packed structures : the inspheres of {6,3} are infinite. This makes the success of the above reasoning quite surprising. However note that eq.4 is a condition on angles, not lengths. It would then probably be useful, with some refinements involving lengths, to extend these considerations to more general cases.

CONCLUSION

With our picture, an ideal amorphous solid is a disclinated crystal in hyperbolic space with some lattice distortion and a large core energy. The short range order observed in the glassy state, and which we believe locally decreases the energy (as, for example, the tetracoordination along 4 perfectly symmetric directions around a covalent atom) cannot be extended regularly in euclidean space but can be so in hyperbolic space. The disclination density and the choice of disclination strength is not uniquely defined, so that the ideal amorphous solid is not uniquely defined. However one can believe that structural relaxation in the solid state or viscous relaxation in the supercooled state tend to a better disclinated arrangement. This might involve motion, creation or annihilation of disclinations, by a process which remains to be studied, but also eventually other processes, like formation or disappearance of kinks on the lines ; in this case dislocations should play a role through their own stress fields. There is therefore some hope, the real processes being known, to obtain a law of the Vogel's type, as suggested by Anderson [3], but the ideal state at T_0 would be the most defectuous one (in the topological sense), while rising the temperature decreases the disclination content and eventually introduces other defects. As stated above [10], a three-dimensional lattice of lines has been observed in a thin sample of hydrogenated or post-hydrogenated silicon obtained by sputtering ; the mean distance between lines (100 Å) is of the order of the sample thickness. These lines preexist before hydrogenation which implies either there are dangling bonds (lines $2\pi/r$) or that the $2\pi/p$ preexisting (through the faces)move towards the edges under hydrogenation.

Note that our model, although somewhat related in spirit, is incompatible with Rivier's model, which implies loopswhose opposite sides have opposite strengths. In our model, since disclinations are equivalent to curvature, all disclinations have the same sign.

We do not cancel to ourselves the fact that a major difficulty remains with the definition of the corresponding symmetry group in H_3, an amorphous body being given. We have not at our disposal anything like X-ray measurements yielding the "reciprocal" lattice. So that we are left with the necessity of computational methods, and comparison with experimental radial distribution functions (R D F). Also, the knowledge of the symmetry group yields the knowledge of the various classes of topological defects, through their homotopy groups [25] ; hence, electron microscopy observations, of the type first done in [10], should be helpful. Finally one cannot avoid the possibility [26] that the amorphous short range order depends on the method of preparation (for example clusters with icosahedral symmetry by sputtering methods in vacuo), but then a part of structural relaxation process can be due to healing towards a better amorphous arrangement.

REFERENCES

[1] Anderson, P.W., Halperin, B.I. and Varma, C.M., Phil. Mag. 25 (1972) 1.
[2] For a review, see Spaepen, F., in Balian, R., Kléman, M. and Poirier, J.P. (eds.), Les Houches 1980, Physics of Defects (North Holland, Amsterdam, in press).

[3] Anderson, P.W., in Balian, R., Maynard, R. and Toulouse, G. (eds.), Les Houches 1978, Ill-condensed Matter (North-Holland, Amsterdam, 1979).
[4] Cohen, M.H. and Turnbull, D., J. Chem. Phys. 31 (1959) 1164.
[5] Egami, T., Mater. Res. Bull. 13 (1978) 557.
[6] Egami, T., Maeda, K. and Vitek, V., Phil. Mag. 41 (1980) 883 ; see also Srolovitz, D., Maeda, K. et al. Phil Mag. (in press).
[7] Kosterlitz, J.M. and Thouless, D.J., J. Phys. C 6 (1973) 1181.
[8] Rivier, N., Phil. Mag. 40 (1979) 859 ; see also Rivier, N. and Duffy, D.M., in Demongeot, J. (ed.), Springer Series in Synergetics (Springer, Berlin, 1980).
[9] Kléman, M. and Sadoc, J.F., Jour. Phys. Lettres 40 (1979) L.569.
[10] Bourret, A., Communication to be presented at the 9th International Conference on amorphous and liquid semiconductors (Grenoble, 1981).
[11] Zallen, R., in Montroll, E.W. and Lebowitz, J.L. (eds.), Fluctuation Phenomena (North Holland, Amsterdam, 1979).
[12] Bernal, J.D., Proc. Roy. Soc. A280 (1964) 299.
[13] Finney, J.L., Proc. Roy. Soc. A319 (1970) 479.
[14] Zachariasen, W.H., J. Amer. Chem. Soc. 54 (1932) 3841.
[15] Polk, D.E., J. Non-Cryst. Sol. 5 (1971) 365.
[16] Farges, J., de Feraudy, M.F., Raoult, B. and Torchet, G., Jour. Phys. 36 (1975) 62.
[17] Sadoc, J.F., Jour. Phys. C 8-41 (1980) 326.
J. Non Cryst. Solids (1981), in press.
[18] Coxeter, H.S.M. and Whitrow, G.J., Proc. Roy. Soc. A201 (1950) 417.
[19] Coxeter, H.S.M., Illinois J. Math. 2 (1958) 746.
[20] Hilbert, D. and Cohn-Vossen, S., Geometry and the Imagination (Chelsea Pub. Cy., New York, 1952).
[21] Cartan, E. Leçons sur la Géométrie des Espaces de Riemann (Gauthier-Villars, Paris, 1963).
[22] Regge, T., Nuovo Cimento 19 (1961) 558.
[23] Matzke, E.B., Bull. Torrey Botanical Club, 77 (1950) 222, cited in [19].
[24] Srolovitz, D., Maeda, K., Vitek, V. and Egami, T., Phil. Mag. (1981) in press.
[25] Kléman, M., Points, lines and walls (Wiley and Sons, 1981, in press).
[26] Friedel, J., private communication.

Continuum Models of Discrete Systems 4
eds. O. Brulin and R.K.T. Hsieh

MOLECULAR DESCRIPTION OF TEMPORARY POLYMER NETWORKS

E. Kröner and R. Takserman-Krozer

Institut für Theoretische und Angewandte Physik
der Universität Stuttgart
und
Max-Planck-Institut für Metallforschung
Stuttgart, F.R. Germany

Dedicated to Professor Dr. Ulrich Dehlinger, Stuttgart, with the best wishes for his 80th birthday

The objects of this investigation are temporary polymer networks, i.e. networks in which the decay and the formation of a junction causes a viscoelastic component of the flow. These so-called temporary networks are treated by an extension of the well-known spring-bead model. Using a plausible molecular model, a generalized diffusion equation for the fluid at rest and in flow is derived and the many-particle distribution function for the fluid in equilibrium is determined. The diffusion equation contains transition probabilities which can be calculated approximately if the elementary kinetic processes are seldom enough to occur independently. A relaxation time which depends on the velocity gradient is defined with the help of the transition probabilities. Factorizing the many-mode distribution function obtained by the diagonalization of the elastic matrix a set of one-mode differential equation (or integro-differential equations, if hydrodynamic interaction is included) is obtained which can be solved numerically and will later lead us to the quantitative prediction of experiments. In particular, it will be possible to calculate, within the present model, the non-Newtonian viscoelastic constitutive equations and the rheooptical properties of the considered networks.

INTRODUCTION

Condensed amorphous polymer systems such as concentrated polymer solutions and polymer melts are characterized by a large interpenetration of molecular coils and high frequency of intermolecular contacts. As shown by Takserman-Krozer and Ziabicki/1/ and by Dobson and Gordon /2/, the number of such contacts can be considerable. Under suitable conditions of temperature, stress etc. two (or more) molecules coming into contact can be bond together by a chemical or physical (van der Waals) reaction leading to formation of a junction. The bond energy of the junctions lies within the range of 10^{-1} kcal/mole (case of van der Waals forces) and 10^{2} kcal/mole (case of homopolar chemical bonds). It is much less than the bond energy of the skeleton atoms in the macromolecule.

The junctions are the objects of special interest in the present investigation. In certain ranges of temperature and stress they can dissociate provided sufficient energy is supplied to overcome the potential barrier for decay. We assume that such a barrier always exists, i.e. also in the case of van der Waals bonding. The energetic situation is shown in fig. 1. The existence of the barrier implies, that not only the dissociation but also the formation of a junction needs thermal activation. The difference $e_D = |h_{diss} - h_{form}|$ of the two activation energies is the energy of dissolution, also called the bond energy. The energy of activation required for the occurrence of the mentioned processes can be provided both by thermal motion and by a stress acting through the system. A quantitative treatment of this problem was given by Takserman-Krozer et al. /3/.

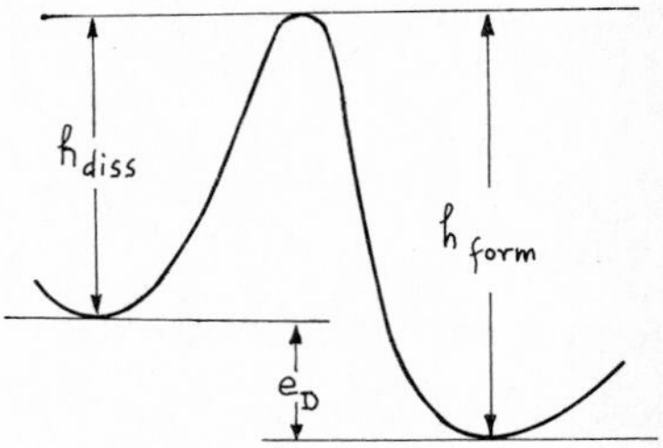

Fig. 1

A polymer system with junctions forms a polymer network. The distribution of junctions within the network can be more or less irregular. Since in the situation considered here the junctions decay after some time and new junctions are formed the network is called temporary. Temporary networks were discussed among others by Green and Tobolky /4/, Scott and Stein /5/, Lodge /6/ and by Yamamoto /7/. Our investigation is meant as a subsequent development of these works.

Beside the contacts which lead to junctions there occur many other contacts which can be described as normal collisions, leading to friction, between elementary units of two (or more) molecules. In their effect these collisions are not basically different from collisions between the molecules of solvent and polymer or those of solvent molecules among themselves. Now the main aim of this work is the calculation of the viscoelastic behaviour of the temporary polymer network. For this purpose we can include the effect of the collisions in the form of a friction parameter, later called ζ , which does not depend on the velocity gradient $\underline{q}$ in the flowing polymer system if hydrodynamical interactions are neglected.

More problematic is a further sort of contacts, namely the so-called entanglements. These are not localized like the junctions, but rather can glide along the macromolecules under the action of Brownian forces or elastic forces transmitted through the chains between junctions. A calculation of the contribution to viscoelasticity of the entanglements appears to be rather complex. Here we shall assume that the effect is sufficiently well described by a contribution to the mentioned friction parameter ζ . A more thorough investigation of this effect is, however, highly desirable. For recent work see Doi and Edwards /8/ and Soong et al. /9/.

In section 2 we explain qualitatively the molecular model of the temporary network treated here. The elementary processes leading to the viscoelasticity of this network are explained in section 3 which includes an approximate calculation of the transition probabilities involved. In section 4 the spring-bead model is applied to the network and a generalized diffusion equation obtained which defines the model quantitatively. The diffusion equation is solved for the case when the fluid is in thermodynamic equilibrium (section 5). In order to describe the network in flow we introduce in section 6 an approach which uses a mean relaxation time. This relaxation time depends on the velocity gradient and can be calculated by means of the transition probabilities of the elementary processes. The diffusion equation simplified in this manner is transformed into a many-mode equation which can be factorized without loosing the information that the beads (junctions) form a network. The resulting one-mode equations are relatively simple. In section 7, finally, the simplifications introduced are listed up and briefly discussed.

2. THE MOLECULAR MODEL

In analogy to Rouse's theory of dilute polymer solutions /10/ the network is treated by a spring-bead model where the beads represent the junctions and are taken as carriers of mass and friction, whereas the springs, assumed mass- and frictionless, interconnect the beads on which they exert elastic forces. Each spring represents a part of a macromolecule and connects two topologically neighboring junctions. Thus the springs model what are usually called the chains of the polymer network.

In our model, the polymer system is a random four-functional network which, in order to simplify the calculations, possesses a Hamilton-Euler path. The latter restriction implies in particular that topological pecularities like molecules forming closed loops etc. are omitted. For details see e.g. Eichinger /11/.

Since the bond energy within the macromolecule is much higher than that of a junction, the decay and formation processes occur in such a manner, that the macromolecules forming the junctions remain intact. This means that the principle units of the temporary network are not only the junctions and chains (beads and springs) but also the primary macromolecules themselves. They are stable structural elements existing independent of the network and of the process of decay and formation.

Each decay and each formation of a junction destroys the enumeration of the junctions and the chains in the network. This leads to great difficulties in the quantitative description of the polymer system. One can, however, imagine situations where decay and formation occur strongly correlated namely such that a junction which decays reforms immediately in such a way that it preserves the same four topologically neighboring junctions. If, for simplicity, we confine ourselves to this situation, then always two correlated processes of decay and formation are described as only one process which takes place within a very short time. We shall assume later that this time is much shorter than the average time between the occurrence of these processes in a certain neighborhood. This situation is similar to that in a gas in which the elementary processes are collisions between particles. There it is often assumed that the "process time" is much shorter than the average time between such processes in a certain neighborhood. The linear dimensions of these neighborhoods are related to the spatial distance over which two processes would occur correlated.

It is instructive to consider the following analogy to the inelastic deformation of metals. In certain cases dislocation loops are locally produced from some source. These loops expand thereby causing an inelastic glide along the plane bounded by the loop. Due to the developping back stress the loops are suddenly stopped and no longer contribute to the inelastic deformation which rather continues at some other place in the body. The analogy between the process "formation and stopping of the dislocation loop" and the process "decay and reformation of the junction" is obvious.

If the polymer system is in equilibrium, i.e. at rest, then there will be no preferred direction for the discussed junction processes. A preferred direction, however, enters, when the system is in flow. The superposition of all viscous elementary steps caused by the junction processes then results in a nonvanishing macroscopic viscous deformation. This is the phenomena of our present interest.

The polymer network can, but need not be embedded in some fluid (solvent). Any solvent considered here is assumed to be a simple Newtonian fluid.

3. THE ELEMENTARY DECAY-FORMATION PROCESS

A fundamental part of the theory is the treatment of the elementary decay-formation process. Consider a junction at point $\underline{r}$ which is surrounded by topological neighbor junctions at points $\underline{r}_\alpha$ (α = a,b,c,d, fig. 2a). The labelling a,b,c,d of these junctions shall be done in such a manner that $\underline{r}_a$ and $\underline{r}_b$ belong to one and $\underline{r}_c$ and $\underline{r}_d$ to the other of the two macromolecules forming the junction. The four chains have contour lengths $l_\alpha = z_\alpha l$ with l as the length of the elementary link (or segment) and z_α the number of the links in the considered chain.

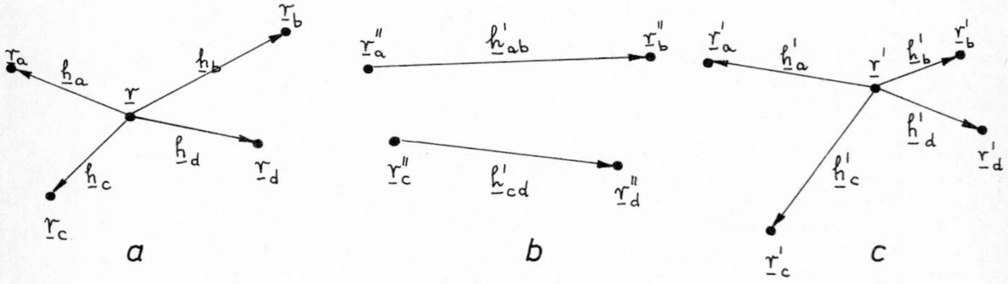

Fig. 2

As discussed in section 1 we assume that the bond state and the dissociated state are separated by an activated state. The activation energies h_{diss} and h_{form} are barriers which have to be overcome by thermal agitation in the processes of decay and formation. The thermal agitation consists of both collisions with the solvent and other polymer molecules (see above) and of the thermal vibrations of the polymer chains. Let the rate constants for the two types of processes be called $\mathring{k}_1$ and $\mathring{k}_2$. They are proportional to $\exp\{-h_{diss}/k_BT\}$ and $\exp\{-h_{form}/k_BT\}$ where k_BT has the usual meaning. The preexponential factors are frequencies which were calculated in detail in ref. /3/, but, for simplicity, can also be taken from experiment. Also h_{diss} and h_{form} are conveniently taken as adjustable parameters so that $\overset{\circ}{k}_1$ and $\overset{\circ}{k}_2$ can be considered as given numbers.

So far we have tacitly assumed that the junction at $\underline{r}$ is stress-free in a macroscopic sense. Since the considered chains participate in the thermal motion they will, judged momentaneously, be under stress. Macroscopically stress free means that if the momentary stress is averaged over a sufficiently long time, it vanishes. It is clear that the junction breaks at the instant when by thermal agitation either the momentary stress at a junction or the action of the solvent molecules or both together are large enough to overcome the energy barrier.

Let us now consider the junction at $\underline{r}$ under a macroscopic stress which is exerted on the junction via the chains ending at $\underline{r}_\alpha$. We can also say that the complex of junctions at $\underline{r}$ and $\underline{r}_\alpha$ (α = a,b,c,d) is in a (macroscopic) stress state which is caused by tractions exerted from the environment of the complex. In response to this stress the mean positions of the junctions at $\underline{r}$ and $\underline{r}_\alpha$ now describe a deformed configuration.

The chains α tear at the junction at $\underline{r}$ with the effect that the activation energy h_{diss} is apparently lowered by the elastic energy stored (in average) in the four chains. If we neglect a possible slight change of the preexponential factor we obtain an apparent rate constant k_1 of the form

$$k_1 = \overset{\circ}{k}_1 \exp\{-(h_{diss} - \Delta f)/k_B T\} \tag{1}$$

with the elastic energy

$$\Delta f = \frac{1}{2}\sum_\alpha (\kappa_\alpha h_\alpha^2 - \overset{\circ}{\kappa}_\alpha \overset{\circ}{h}_\alpha{}^2), \quad \underline{h}_\alpha \equiv \underline{r}_\alpha - \underline{r}, \quad \underline{\overset{\circ}{h}}_\alpha \equiv \underline{\overset{\circ}{r}}_\alpha - \underline{\overset{\circ}{r}} \tag{2}$$

where the "o" refers to the unstressed state. The κ_α and $\overset{\circ}{\kappa}_\alpha$ are the linear or nonlinear elastic spring constants of the chains. The "constant" κ_α, for instance, depends on z_α and, in the nonlinear case, also on $\underline{h}_\alpha$, the end-to-end vector of the chain α. It can be calculated from a small scale statistics like that of Volkenstein and Ptitsyn /12/.

The occurrence of $\underline{\overset{\circ}{h}}_\alpha$ in eq. (2) has to do with the fact that the elastic energy of a chain vanishes only when the end-to-end vector is zero. In equilibrium $\underline{h}_\alpha = \underline{\overset{\circ}{h}}_\alpha$ and the elastic energy $\sum \overset{\circ}{\kappa}_\alpha \overset{\circ}{h}{}_\alpha^2/2$. This energy, however, is by convention included into the activation energy h_{diss} and, therefore, does not contribute to its depression. Thus it has to be subtracted in (2). Once the junction at $\underline{r}$ is decayed the junctions at $\underline{r}_\alpha$ will move into new average positions, say $\underline{r}''_\alpha$, which correspond to the new distribution of the z-numbers. At this moment, there are only two chains instead of the four chains α (fig. 2b). One chain connects $\underline{r}''_a$ with $\underline{r}''_b$ and has $z'_a + z'_b$ links whereas the other connects $\underline{r}''_c$ with $\underline{r}''_d$ and has $z'_c + z'_d$ links. Since the thermal motion enables frequent contacts between these two chains there exists a probability that a new junction is formed between them at some average position, say $\underline{r}'$ (fig. 2c). The relevant part of the elastic energy of the new configuration compared to that of the intermediate (two-chain) state is

$$\Delta f' = \frac{1}{2}\sum_{\alpha}(\kappa'_\alpha h'^2_\alpha - \mathring{\kappa}'_\alpha \mathring{h}'^2_\alpha) - \frac{1}{2}(\kappa'_{ab}h'^2_{ab} - \mathring{\kappa}'_{ab}\mathring{h}'^2_{ab})$$
$$- \frac{1}{2}(\kappa'_{cd}h'^2_{cd} - \mathring{\kappa}'_{cd}\mathring{h}'^2_{cd}) \tag{3}$$

where e.g. the elastic spring constants κ'_{ab} and end-to-end vectors $\underline{h}'_{ab}$ refer to the chain connecting the junctions at $\underline{r}''_a$ and $\underline{r}''_b$. The quantities $\mathring{\kappa}'_{ab}$ and $\mathring{\underline{h}}'_{ab}$ have the analogous meaning but referred to the unstressed state with missing junction at $\underline{r}'$. The energy (3) has to be provided, in addition to the formation energy h_{form}, by the thermal agitation if a new junction is to be formed. The pertinent rate constant is

$$k_2 = \mathring{k}_2 \exp\{-(h_{form} + \Delta f')/k_B T\}. \tag{4}$$

As described in section 2 we shall now assume that each decay is followed by a reformation of the junction within a very short time. We also assume that the reformation process is uncorrelated with the decay process in the sense that at the moment of reformation the system does not remember the state before the decay.

We now define a probability density $p(\underline{h}_\alpha, z_\alpha \rightarrow \underline{h}'_\alpha, z'_\alpha)dt$ (density with respect to $\underline{h}', z'$) for the transition of a junction $(\underline{h}_\alpha, z_\alpha)$ into a junction $(\underline{h}'_\alpha, z'_\alpha)$. This density is proportional to $k_1 k_2$. If we absorb those parts of k_1 and k_2 which do not depend on the velocity gradient $\underline{q}$ into the preexponential frequency, say ν, then we obtain with (1) - (4)

$$p(\underline{h}_\alpha, z_\alpha \rightarrow \underline{h}'_\alpha, z'_\alpha) = \nu \exp\{\frac{1}{2}(\sum[\kappa_\alpha h^2_\alpha - \kappa'_\alpha h'^2_\alpha] + \kappa'_{ab}h'^2_{ab} + \kappa'_{cd}h'^2_{cd}) \quad . \tag{5}$$

This expression depends on $\underline{q}$ via $\underline{h}_\alpha$, $\underline{h}'_\alpha$, $\underline{h}'_{ab}$ and $\underline{h}'_{cd}$. We shall later determine the frequency ν by comparison with experiment.

According to our previous remark, the elastic constants κ'_{ab} and κ'_{cd} in (5) depend on

$$z'_{ab} = z'_a + z'_b = z_a + z_b, \quad z'_{cd} = z'_c + z'_d = z_c + z_d \tag{6}$$

(conservation of links in a molecule) and, in the nonlinear case, on $\underline{h}'_{ab}$ and $\underline{h}'_{cd}$. For a first orientation we may assume that the positions $\underline{r}'_\alpha$ of the neighboring junctions remain unchanged from the intermediate (two-chain) state to the final state. Then

$$\underline{h}'_{ab} = \underline{h}'_a + \underline{h}'_b, \quad \underline{h}'_{cd} = \underline{h}'_c + \underline{h}'_d. \tag{7}$$

With (6) and (7) the "primed" part of the exponential in (5) can be expressed in terms of $\underline{h}'_\alpha, z'_\alpha$ and the unprimed part depends on $\underline{h}_\alpha, \underline{z}_\alpha$. Thus we have found the desired transition probability. Its explicit form depends on the chosen relation between the κ's and z's. This relation follows from the mentioned small scale statistics. In an improved theory eqs. (7) have to be replaced by more accurate relations which unfortunately require a rather extended calculation.

4. THE DIFFUSION EQUATION

The spring-bead model discussed in section 2 will now be treated with the methods of statistical physics, i.e. analogously to the methods often applied to dilute polymer solutions /10,13/. We assume that the bead velocities are relatively low so that inertia effects can be neglected. This implies that the momentum part of the phase space reduces to the configuration space, say Γ. Each point in this space specifies a configuration of the network at a particular instant t. Within our model such a configuration is completely specified by the full set $\underline{R} \equiv \{\underline{r}_i\}$ of the position vectors of the junctions plus the full set $Z \equiv \{z_{ij}\}$ of the link numbers*. These numbers are now labelled with two subscripts in order to indicate which junctions are connected by the considered chain. The numeration of the junctions does not offer a particular problem as long as we confine ourselves to networks with a Hamilton-Euler path. The more general situation is left for later work.

As usual, a probability density $W(\underline{R},Z,t)$ in the configuration space is introduced which describes the probability of finding the various configurations among the polymer systems forming the statistical ensemble. The calculation of W, also called the distribution function, is the primary aim of the theory because its knowledge permits us to calculate the expectation values of physical quantities which are observed in experiments. Some expectation values, for instance that for the developping stress, can even be calculated if only the second moment of W is known. The basic equation for the determination of W serves a similar purpose as the Liouville equation of the general statistical mechanics. It can be derived as follows:

Observe that the time change of the density W in a volume element $d\Gamma = d\underline{R}dZ$ of the configuration space has two distributions: The first one, called $(\partial W/\partial t)_{per}$, can be described as an in- and outflow of systems as it would occur in the permanent network situation where no junctions decay or form under the acting stresses. The second one, $(\partial W/\partial t)_{kin}$, is the contribution of the kinetic processes of decay and formation of junctions. Thus we have

$$\partial W/\partial t = (\partial W/\partial t)_{per} + (\partial W/\partial t)_{kin}. \tag{8}$$

The first term on the right hand side is formally like in dilute polymer solutions, namely

$$(\partial W/\partial t)_{per} = -\sum_i \nabla_i \cdot (\dot{\underline{r}}_i W) \tag{9}$$

where $\dot{\underline{r}}_i$ is the velocity vector of the junction at $\underline{r}_i$. Observe that (9) does not contain $\dot{z}_{ij}$ because decay and formation of junctions are not considered in (9).

The bead velocities $\dot{\underline{r}}_i$ can be calculated from the force balance at the beads like in dilute polymer solutions /10,13/. In this balance we need not regard the inertia forces since these are neglected. Various types of forces should be considered, namely polymer-polymer and polymer-solvent friction forces including hydrodynamic interactions, repulsive forces leading to the excluded volume effect, forces of the Brownian motion of the beads, elastic forces transmitted by the chains and perhaps other forces. In order not to complicate the situation we shall here include the mentioned friction forces without hydrodynamic interactions, the Brownian and the

---

* This statement suffices for the present purpose. More strictly, it is necessary to distinguish between the various molecules because junctions break in such a manner that the molecules remain intact.

elastic forces. If the friction forces exerted on the bead i are taken proportional to the relative velocity between this bead and the surroundings we obtain in a standard calculation

$$\dot{\underline{r}}_i = \underline{q} \cdot \underline{r}_i - \frac{1}{\zeta}\sum_j \kappa_{ij}\underline{r}_j - \frac{k_B T}{\zeta} \nabla_i \ln W \tag{10}$$

where $\underline{q}$, as before, is the macroscopic velocity gradient and $(\kappa_{ij}) = K$ is the elastic matrix which represents the spring constants and also describes the connectivity of the network. In fact, the non-diagonal elements κ_{ij} $(i \neq j)$ are zero if the pertaining junctions are not connected by a chain. In the case of nonlinear spring constants the matrix K depends on $\underline{R}$ and Z whereas it depends on Z only for Gaussian chains. The quantity ζ finally is an effective polymer-polymer and polymer-solvent friction coefficient. We assume that it does not depend on the velocity gradient $\underline{q}$ and is given by experiment. In a more refined theory friction constants ζ_i which are different from bead to bead can be introduced in a manner described by Takserman-Krozer and Ziabicki /14/.

For the kinetic change of the probability density W we can write down a rigorous master equation, namely

$$(\partial W/\partial t)_{kin} = \iint [\, W(\underline{R}',Z')\vec{p} - W(\underline{R},Z)\overleftarrow{p}\,] d\underline{R}' dZ', \tag{11}$$

where

$$\vec{p}dt \equiv p(\underline{R}',Z' \longrightarrow \underline{R},Z)dt, \quad \overleftarrow{p}dt \equiv p(\underline{R},Z \longrightarrow R',Z')dt \tag{12}$$

are the transition probability densities, relative to $\underline{R}$,Z and $\underline{R}'$,Z' respectively, that a configuration $\underline{R}'$,Z' changes to $\underline{R}$,Z and vice versa in the time interval dt. The first integral in eq. (11) describes the fact that the kinetic change of W in a volume element at $\underline{R}$,Z consists of an increase due to processes taking place in the configurations $\underline{R}'$,Z' and of decrease by the processes in the configuration $\underline{R}$,Z itself.

Inserting (9) and (11) into (8) we obtain the required equation which has the form of a generalized diffusion equation

$$\frac{\partial W}{\partial t} + \sum_i \nabla_i \cdot [\, \dot{\underline{r}}_i(\underline{R},Z)W(\underline{R},Z)] = \iint [W(\underline{R}',Z')\vec{p} - W(\underline{R},Z)\overleftarrow{p}\, d\underline{R}'dZ'. \tag{13}$$

Together with eqs. (10) and (12) which define $\dot{\underline{r}}_i$, $\vec{p}$ and $\overleftarrow{p}$ eq. (13) can be considered as defining our model completely. In this sense it is a rigorous equation.

5. THE EQUILIBRIUM STATE

If we want to describe stationary flow then $\partial W/\partial t = 0$ in eqs. (8) and (13). This, however, does not mean that simultaneously

$$(\partial W/\partial t)_{per} = 0, \quad (\partial W/\partial t)_{kin} = 0. \tag{14}$$

In fact, in stationary flow a continuous change of W due to the kinetic processes takes place. This change, however, is compensated by the contribution of the elastic deformation. In the equilibrium state, on the other hand, the fluid is at rest. In such a situation the kinetic processes in the two directions balance so that $(\partial W/\partial t)_{kin} = 0$. Thus both equations (14) must be satisfied in order to have equilibrium. The situation is analogous to that of Boltzmann's kinetic theory (cf. e.g. Huang /15/).

In order to find the equilibrium distribution W_{eq} we start from the hypothesis that all configurations which possess the same energy, say H, the same total link number $L \equiv \sum z_{ij}$ and the same junction number M occur with equal a priori probability. The degree of reliability of this hypothesis corresponds to that of the a priori probability postulate of general statistical mechanics. Now let the considered system which has values H,L and N of the mentioned quantities be part of a very large system of the same sort which also is in thermal equilibrium. Let two subsystems I and II of the smaller system have values H_I, L_I, M_I and H_{II}, L_{II}, M_{II} so that

$$H = H_I + H_{II}, \quad L = L_I + L_{II}, \quad M = M_I + M_{II}. \tag{15}$$

Since the probability densities of the three systems must have the same appearance we can write

$$W(H,L,Z) = \sum W(H_I,L_I,M_I)W(H_{II},L_{II},M_{II}) \tag{16}$$

where the summation extends over all combinations of H_I, H_{II} and L_I, L_{II} and M_I,M_{II} which satisfy (15). It follows that W has the form

$$W = W_{eq} = C'\exp\{-\beta H - \lambda L - \mu M\} \tag{17}$$

where C' is independent of H,L and M and can be determined by normalizing W to unity. Eq. (17) describes a sort of grandcanonical ensemble in which β is identified as $1/k_BT$ and λ and μ are the chemical potentials of the elementary link and of the junction respectively. The chemical potential μ of the junction consists of the energy e_D of dissolution, which is equal to $|h_{diss} - h_{form}|$, and of an entropic part which can be calculated by simple combinatorics. This will be done in later work. Similarly, λ can be obtained. Within our model eq. (17) is a rigorous result.

In the section on elementary processes we have combined the decay and formation processes to one single decay-formation process. If only such processes occur, then the junction number M is constant. It has been calculated in ref. /3/ as a function of the parameters defining the system. If we furthermore restrict ourselves to systems with constant link numbers, then L and M are fixed numbers and we can pass over to the canonical ensemble described by the density

$$W_{eq} = C\exp\{-H/k_BT\} \tag{18}$$

with a new constant C.

For simplicity, let us confine ourselves to linear spring constants so that the elastic energy can be written in the form

$$H = \frac{1}{2}\sum \kappa_{ij}\underline{r}_i \cdot \underline{r}_j \tag{19}$$

where κ_{ij} depends on Z but not on $\underline{R}$. We then obtain

$$W_{eq} = C\exp\left\{\frac{-1}{2k_BT}\sum \kappa_{ij}\underline{r}_i \cdot \underline{r}_j\right\}. \tag{20}$$

Inserting this into (9) with (10) it is easily seen that $(\partial W/\partial t)_{per} = 0$.

We now show that (20) satisfies also the condition that $(\partial W/\partial t)_{kin}$ as given by eqs. (11,12) vanishes. The integration over the configuration space involved can be split into integrations over shells of constant H' and subsequent integration over H'. If for simplicity the value of W_{eq} at the energy H' is denoted by $W_{eq}(H')$ then the shell integration at a constant H' gives

$$\iint\limits_{\text{shell } H'} W_{eq}(H')\vec{p}\,d\underline{R}'dZ' = W_{eq}(H')\iint\limits_{\text{shell } H'} \vec{p}\,d\underline{R}'dZ' \tag{21}$$

because all states with the same energy have the same probability (see eq. (18)). Now in equilibrium all occupied states group themselves very closely around one energy, here called H, so that we may replace $W_{eq}(H')$ by $W_{eq}(H)$. It is shown in general statistical mechanics that this procedure is rigorous in the thermodynamical limit. The shell part of the integration in (11) then is

$$W_{eq}(H)\iint\limits_{\text{shell } H} (\vec{p} - \overleftarrow{p})\,d\underline{R}dZ'. \tag{22}$$

According to the hypothesis of equal a priori probabilities none of the various configurations $\underline{R}$,Z which belong to the same energy is distinguished. Therefore, if $\underline{R}'$,Z' and $\underline{R}$,Z belong to the same energy, it follows that $\vec{p} = \overleftarrow{p}$, thus with (11) $(\partial W/\partial t)_{kin} = 0$. This proves that in the case of linear spring constants (20) gives the rigorous equilibrium distribution function belonging to the diffusion equation (13).

6. RELAXATION TIME APPROACH

Because W_{eq} satisfies the stationary part of (13) it is convenient to introduce $\Phi \equiv W - W_{eq}$ in (13) with (10). The result is

$$\frac{\partial\Phi}{\partial t} + \sum_i \nabla_i \cdot [(\underline{g}\cdot\underline{r}_i - \frac{1}{\zeta}\sum_j \kappa_{ij}\underline{r}_j - \frac{k_BT}{\zeta}\nabla_i)\Phi]$$
$$= \iint[\Phi(\underline{R}',Z')\vec{p} - \Phi(\underline{R},Z)\overleftarrow{p}]\,d\underline{R}'dZ' \tag{23}$$

and Φ obeys the normalization

$$\iint \Phi(\underline{R}',Z')\,d\underline{R}'dZ' = 0. \tag{24}$$

Eq. (23) is a complicated many-particle integro-differential equation which has to be simplified somehow. Most effective is the relaxation time approach at which we arrive by replacing $\vec{p}$ by its average relative to $\underline{R}'$,Z' and, consequently, $\overleftarrow{p}$ by its average relative to $\underline{R}$,Z. We thus substitute

$$\vec{p} \to \iint p(\underline{R}',Z' \to \underline{R},Z)W(\underline{R}',Z')\,d\underline{R}'dZ, \quad \overleftarrow{p} \to \iint p(\underline{R},Z \to \underline{R}',Z')W(\underline{R},Z)\,d\underline{R}dZ. \tag{25}$$

In view of this and the normalization (24) the first integral in (23) vanishes and we are left with

$$\frac{\partial\Phi}{\partial t} + \sum_i \nabla_i \cdot [(\underline{g}\cdot\underline{r}_i - \frac{1}{\zeta}\sum_j \kappa_{ij}\underline{r}_j - \frac{k_BT}{\zeta}\nabla_i)\Phi] = -\overleftarrow{p}\,\Phi \tag{26}$$

where

$$\overline{p} \equiv \iint \iint p(\underline{R},Z \rightarrow \underline{R}',Z')\, W(\underline{R},Z)\, d\underline{R}\,dZ\,d\underline{R}'dZ'. \tag{27}$$

Here $1/\overline{p} = \tau$ has the physical meaning of an average relaxation time. The approximation introduced here is specified by eqs. (25). The motivation for (25) is, of course, the dramatic simplification of the diffusion equation and we expect that the loss in accuracy remains within limits. Of course, a quantitative argument for the applicability of this approach is desirable.

We shall discuss later in this section how τ can approximately be calculated from the elementary processes considered in section 3. Since its calculation from (27) involves the distribution function W which is not known at the beginning, a final determination of τ can only be achieved at the end of the mathematical procedure. In other words, the diffusion equation (26) is still an integro-differential equation. Its solution, however, is naturally reduced to the iterative solution of a pure differential equation where in each step a newly calculated τ can be used. It is clear that τ depends on the velocity gradient $\underline{q}$, namely via W in eq. (27).

Let us turn to the differential equation (26) where $\overline{p} = 1/\tau$ is now considered as a given constant. We still have a complex many-particle equation. Factorization of Φ into one-particle distributions would greatly simplify the situation. This factorization would, however, destroy the most important information, namely that the particle system forms a network. A method which was developpped for dilute polymer solutions /12/ helps to overcome this difficulty: Now the factorization is done not with the many-particle function but with the corresponding many-mode function which is obtained when normal coordinates are introduced. To this end we only need to diagonalize the elastic matrix. Eq. (26) then assumes the form

$$\frac{\partial \Phi}{\partial t} + \sum_{i'} \nabla_{i'} \cdot \left[\left(\underline{q} \cdot \underline{s}_{i'} - \frac{1}{\zeta}\lambda_{i'}\underline{s}_{i'} - \frac{k_B T}{\zeta}\nabla_{i'} \right) \Phi \right] + \frac{\Phi}{\tau} = 0 \tag{28}$$

where Φ is now understood as a function of the normal coordinate vectors $\underline{s}_{i'}$, and $\lambda_{i'}$ are the eigenvalues of the elastic matrix. They depend on Z, but, in the linear spring approximation, not on $\underline{R}$.

We now introduce the factorization

$$\Phi = \varphi^{(1)}(\underline{s}_1)\, \varphi^{(2)}(\underline{s}_2) \ldots \varphi^{(M)}(\underline{s}_M). \tag{29}$$

Inserting this into (28) and integrating over all $s_{i'}$ except one, say $\underline{s}_{1'}$, we obtain with

$$\frac{\partial \varphi^{(1')}}{\partial t} + \nabla_{1'} \cdot \left[\left(\underline{q} \cdot \underline{s}_{1'} - \frac{1}{\zeta}\lambda_{1'}\underline{s}_{1'} - \frac{k_B T}{\zeta}\nabla_{1'} \right) \varphi^{(1')} \right] + \frac{\varphi^{(1')}}{\tau} = 0 \tag{30}$$

the desired set of one-mode equations. Contrary to our former result on dilute solutions the set (30) is uncoupled because here we have not included the hydrodynamic interactions. If they are included one obtains coupled equations.

The next problem concerns the eigenvalues $\lambda_{1'}$ of the elastic matrix. They follow by the classical diagonalization procedure from the matrix elements κ_{ij}. These, however, are expressed by the z_{ij}. Thus we need the values of the z_{ij}. These are random variables and therefore not known. Among various possibilities that are being explored we can use the expectation values which are defined as

$$\bar{z}_{ij} = \iint z_{ij}\, W(\underline{R}, Z)\, d\underline{R}dZ \tag{31}$$

and so depend on the velocity gradient $\underline{q}$ via W.

The procedure of solving (26) is as follows. Since W_{eq} is known we can use it as the starting point of our iteration procedure by calculating τ, z_{ij} and $\lambda_{1'}$ in zeroth approximation. The $\varphi^{(1')}$ follow from (30) in first approximation. From (29) we find Φ, thus W, in first approximation. This can be used to calculate τ, z_{ij} and $\lambda_{1'}$ in first approximation and so on.

At present time a rigorous calculation of the transition probability densities $\vec{p}$ and $\overleftarrow{p}$ is certainly unrealistic. It is therefore necessary to introduce simplifying assumptions for the calculation of τ. A rather effective assumption which may be valid in many situations is that the time distribution of the decay-formation processes is of the Poisson or some similar type which guarantees that the processes can be considered as independent. The calculation of p can then be reduced to integrations over the transition probabilities for the single elementary process discussed in section 3. Since we are still in the preliminary stage of finding an optimum way of doing this we renounce a more explicit description and content ourselves with the statement that we can, at least in a reasonable approximation, calculate τ as a function of $\underline{q}$ where the $\underline{q}$-dependence enters via the appearance of W in eq. (25_2).

7. CONCLUSION

In complex physical situations such as those considered here one cannot hope to derive expectation values of observable quantities without using any simplifications or assumptions. One is usually forced to treat idealized models which, however, should show the main features of the real physical problem. A quantitative definition of the model is highly desirable. The corresponding set of equations is, in a sense, rigorous. In our problem the equations defining the model are (13) with (10) and (12). The model excludes

1) the effects of entanglements except those which can be absorbed in a friction constant which is independent of the velocity gradient,
2) hydrodynamic interactions,
3) inertia forces.

   These three simplifications are of the <u>classification type</u>. They can be removed at the expenses of the simplicity of the model. At the moment we see no sufficiently simple possibility of an inclusion of entanglements. The simplifications 2 and 3 confine us to not too high velocity gradients, in any case to laminar flow.

   Once the model is established one tries to go through the calculation as well as possible. Also in this stage one is usually forced to make simplifications, this time of the <u>approximation type</u>. The following simplifications have been introduced in the derivation of the final equations (29,30):

4) the relaxation time approach,
5) the factorization of the distribution function.

   We have given arguments which indicate that these approximations are still bearable. In particular, the relaxation time approach has been very successful in Boltzmann's kinetic theory. It is an important feature of the simplification 5 that it does not destroy the information according to which our system is a network.

In order to solve eqs. (30) we need to know the eigenvalue of the elastic matrix K as well as the mean relaxation time τ. Their determination is perhaps the most central physical problem within this theory. In order to remain within a manageable frame we propose to use the following approximations:

6) Use of a linearized small scale statistics in order to express the elastic matrix by the link numbers z_{ij}.

7) Replacement of z_{ij} by their expectations $\bar{z}_{ij}$.
We have found by numerical analysis of some formulae contained in ref. /12/ that the error due to the approximation 6 is rather small, typically of the order 1%. The approximation 7 implies a transition from a fine grain equation to a coarse grain equation. Such a transition is needed in all statistical problems involving Liouville's equation because a solution of the original fine grain equation is impossible. Clearly, the simplification 7 is an approximation which, however, should not be too bad. Further investigation on this point is desirable.
Even more difficult is the determination of the mean transition probability $\bar{p}$. At the moment we restrict ourselves to

8) Poisson type distributions of the elementary processes.
The simplification consists in the consequence that it is good enough to calculate transition probabilities for the elementary processes only. This restriction is analogous to that of the Boltzmann kinetic theory where only two-particle collisions are included. We have to use furthermore

9) approximations in calculating the transition probabilities for the elementary processes.
A closer inspection of these approximations shows that they are probably not so drastic.

In addition to the simplifications listed above there are a few minor simplifications which we have not discussed explicitly. They certainly do not invalidate the general conclusions drawn in this section.

Among all the listed classifications and approximations, the number one seems to be the most critical. Others may be serious in certain circumstances, in particular in the range of high velocity gradients. Further studies in this regime are highly desirable. Of special interest is also the temperature range where the transition to full viscosity, for instance in the form of reptation, occurs.

The preparation of the numerical solution of eqs. (30) which leads us to the quantitative prediction of relevant expectation values such as the stress tensor, is now undergoing. We hope to report about the results in a later publication.

ACKNOWLEDGMENT

Financial support by the Deutsche Forschungsgemeinschaft is gratefully acknowledged.

REFERENCES

/1/ R. Takserman-Krozer and A. Ziabicki, Polimery (Poland) 12, 401 (1967)

/2/ G.R. Dobson and M. Gordon, J. Chem. Phys. 41, 2389 (1964)

/3/ R. Takserman-Krozer, S. Krozer and E. Kröner, Colloid and Polymer Sci. 257, 1033 (1979)

/4/ M.S. Green and A.V. Tobolsky, J. Chem. Phys. 14, 80 (1946)

/5/ K.W. Scott and R.S. Stein, J. Chem. Phys. 21, 1281 (1950)

/6/ A.S. Lodge, Trans. Faraday Soc. 52, 1o0 (1956)

/7/ M.Y. Yamamoto, Phys. Soc. Japan 11, 413 (1956), 12, 1148 (1957)

/8/ M. Doi and S.F. Edwards, J.C.S. Faraday II, 14, 1789, 1816 (1978)

/9/ D.S. Soong, M.C. Williams and M. Shen, Structure and Properties of Amorphous Polymers, Proc. 2nd Symp. on Macromolecules, Cleveland, Ohio 1978, A.G. Walton (Ed.) Studies in Physical and Theoretical Chemistry, Vol. 10, Elsevier, Amsterdam 1980

/10/ P.E. Rouse, J. Chem. Phys. 21, 1272 (1953)

/11/ E. Eichinger, Macromolecules 5, 496 (1972)

/12/ M.W. Volkenstein and B.O. Ptitsyn, Rev. Ac. Sci (UdSSR) 91, 1313 (1953)

/13/ E. Kröner and R. Takserman-Krozer, Rheol. Acta 18, 431 (1979)

/14/ R. Takserman-Krozer and A. Ziabicki, J. Polymer Sci., A2, 8, 321 (1970)

/15/ K. Huang, Statistical Mechanics, J. Wiley, New York 1963

Continuum Models of Discrete Systems 4
eds. O. Brulin and R.K.T. Hsieh

WAVE VELOCITIES AND ATTENUATION IN POROUS MEDIA WITH FLUIDS

Amos Nur and William Murphy

The Stanford Rock Physics Program
Department of Geophysics
Stanford University
Stanford, California, U.S.A.

INTRODUCTION

Interest in porous media with fluids is growing very rapidly because of the energy crisis. A great number of technological and geological energy problems are in fact tied to the detection and mobility of fluids in porous and cracked rocks, such as in the recovery of heavy oil in tar sands and oil shales, geothermal energy exploration, in situ liquefaction and gasification of coal, and nuclear waste disposal. The theory of fluids in porous media plays an important role also in the mechanics of earthquakes and is intimately related to earthquake prediction and the prediction of earthquake hazards such as liquefaction potential and slope stability.

Wave propagation--the focus of this review--is one of the more important aspects of rocks with fluids. The main emphasis of this paper is on experimental results, some of which have been obtained for the first time only in the past few years. Simple models are explored to understand the underlying processes. These models are mostly incapable as yet of making quantitative predictions because the topological nature of pore space is usually not sufficiently known.

WAVES IN POROUS MEDIA

The main elements which significantly control waves in porous media with fluids at low temperatures are: (1) mechanical properties of the solid framework; (2) the nature and properties of the fluid in the pores; (3) the configuration of the pore space. The properties of the solid matrix which are of common importance are the size and shape of grains, their cementation and cohesive strength. The relevant fluid aspects are compressibility, pore pressure, viscosity, mixed phases and temperature. The important aspects of the pore space include porosity--the pore volume per unit composite volume--and the shape in which porosity occurs, described by an idealized aspect ratio distribution of ellipsoidal cavities, by the ratio of pore surface area to pore volume, or by other more statistical measures. Often in rocks, the part of pore space associated with equidimensional pores is responsible for the bulk of the pore volume, whereas inter and intra granular cracks are responsible for most of the surface area density. For equal porosities, solids with large surface area density or many cracks are much more compliant than those with small surface area density, or equidimensional pores. Also, the application of confining pressure and pore pressure have larger effects on cracked rocks than on rocks with rounded pores.

In this paper we summarize several experimental results on the dependence of wave propagation in porous solids--body wave velocities and their attenuation factors--on the confining stress applied to the solid matrix; on pressure of the pore fluid; on the compressibility of the pore phase; and on the wave strain amplitude.

THE ROLE OF CRACK POROSITY

Figure 1 shows an example of compressional and shear wave velocities in several rocks as a function of confining pressure, applied to samples which are jacketed in thin copper. Granite is a typical example of rock in which the entire porosity, about 1%, is in the form of thin cracks, without any equi-dimensional pores. As confining pressure increases, both P and S velocities increase very drastically, by 50% or more, due to crack closure under external pressure. The increase of velocities in the dry sandstone is less pronounced, whereas the Solenhofen Limestone exhibits hardly any increase with confining pressure. As verified by microscopic observations, the Massilon Sandstone and Bedford Limestone contain a mix of rounded pores and cracks, whereas the Solenhofen graphic limestone contains only well-rounded pores.

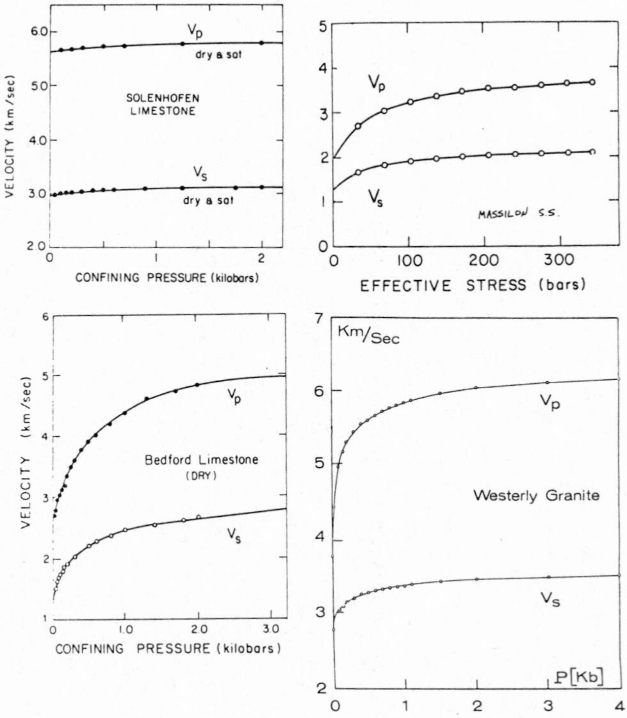

Figure 1
Compressional velocity as a function of confining pressure in four rocks with different pore shape configurations. Pores are dry.

Using models such as by Walsh (1965), we can estimate the fraction of porosity associated with cracks--which tend to close under fairly low confining pressure. In Figure 2 we show the relation between crack porosity and the compressional velocity defect--the difference between velocity without confining pressure and the velocity at 10,000 atm, where slim cracks are closed by pressure. As might be expected, the velocity defect increases with crack porosity, showing that thin cracks tend to control the effective moduli and velocities in rocks. The crack-free Solenhofen Limestone in contrast exhibits almost no velocity defect.

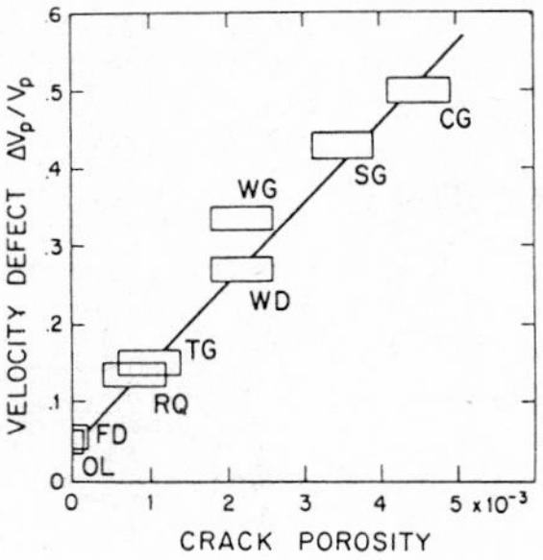

Figure 2
Velocity defect--the normalized reduction of compressional velocity relative to crack-free rock--as a function of crack porosity

A strong test for the role of cracks is provided by cases in which they have preferred orientations leading to velocity anisotropy. Figures 3 and 4 show experimental results of the dependence of velocity anisotropy on direction in a crack-rich granite sample, subject to uniaxial compression. As shown in Figure 3, a clear pattern of velocity anisotropy develops with increasing stress: the P wave and the shear wave velocities show a strong, and similar dependence on direction, whereas the second S wave, polarized in the plane of the applied stress direction, shows a weaker and distinctly different dependence on direction. These patterns due to uniaxial stress are quite typical for crystals with hexagonal symmetry--a symmetry which corresponds to uniaxial stress. Table 1 lists other, more complex stress induced symmetries in cracked solids which possess an initial, unstressed, anisotropic crack distribution.

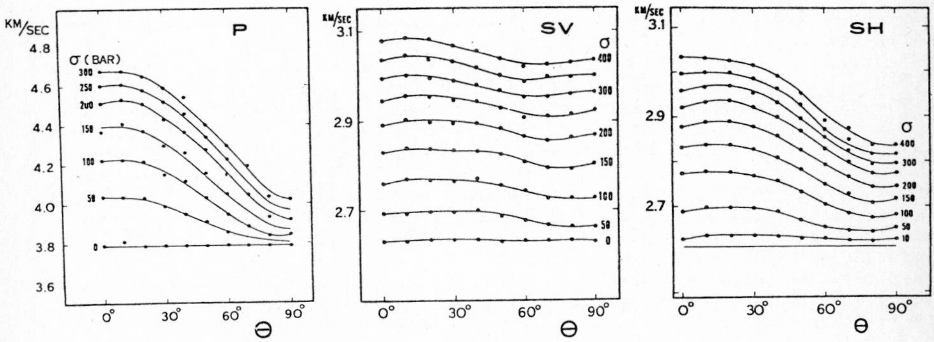

Figure 3
Velocity anisotropy induced by uniaxial stress in granite. The stress is applied in the direction θ = 0°. SH denotes the shear velocity associated with particle motion which is perpendicular to the direction of stress for all θ, and SV denotes the shear velocity associated with particle motion in the plane containing the direction of the stress.

TABLE 1

Dependence of Symmetry of Induced Velocity Anisotrophy on Initial Crack Distribution, Applied Stress, and Its Orientation

Symmetry of Initial Crack Distribution	Applied Stress	Orientation of Applied Stress	Symmetry of Induced Velocity Anisotrophy	Number of Elastic Constants
Random	Hydrostatic		Isotropic	2
	Uniaxial		Axial	5
	Triaxial*		Orthorhombic	9
Axial	Hydrostatic		Axial	5
	Uniaxial	Parallel to axis of symmetry	Axial	5
	Uniaxial	Normal to axis of symmetry	Orthorhombic	9
	Uniaxial	Inclined	Monoclinic	13
	Triaxial*	Parallel to axis of symmetry	Orthorhombic	9
	Triaxial*	Inclined	Monoclinic	13
Orthorhombic	Hydrostatic		Orthorhombic	9
	Uniaxial	Parallel to axis of symmetry	Orthorhombic	9
	Uniaxial	Inclined in plane of symmetry	Monoclinic	13
	Uniaxial	Inclined	Triclinic	21
	Triaxial*	Parallel to axis of symmetry	Orthorhombic	9
	Triaxial*	Inclined in plane of symmetry	Monoclinic	13
	Triaxial*	Inclined	Triclinic	21

* Three generally unequal principal stresses.

Figure 4 shows the development of stress-induced acoustic S wave birefringence, in the direction of propagation perpendicular to the applied stress. With a single pair of favorably oriented shear plate transducers, we notice that two shear pulses, one faster than the other, can be recognized: the first corresponds to the SV wave, the second to the SH wave.

THE ROLE OF FLUID SATURATION ON VELOCITIES

Figure 5 shows the effects on velocities in the rocks of Figure 1, upon saturation with tap water, held at 1 atm. A very large increase in P velocity is observed in the granite, sandstone and the Bedford Limestone--all of which have also shown large velocity increases with confining pressure (e.g. Fig. 1). Solenhofen Limestone exhibits very little change of both P and S velocities upon saturation. These effects can be anticipated from a variety of models for the compressional and shear response of fluid saturated elastic solids with pores and cracks (Walsh, 1968; Nur and Simmons, 1969; O'Connell and Budiansky, 1977). The great decrease in pore compressibility due to the addition of pore water has a first order effect on the compressibility of the composite material, but only a second order effect on shear modulus. Figure 6 shows the influence of saturation on Poisson's ratio ν in dry and water-saturated solids, as a function of confining pressure. At low pressure where cracks are mostly open, ν is anomalously high in saturated rocks, and very low, even negative in dry rocks. This behavior of Poisson's ratio with saturation is widespread, and provides, for example, a good basis for detecting gas or steam pockets in the crust of the earth (DeVilbiss, 1980).

Two subtle but important points can be raised concerning the data for granite and limestone in Figure 5. First, it can be noticed that the shear velocity in saturated rocks at low confining pressure is greater, and at higher confining pressure is less than for the dry samples. We will return to this subtle effect

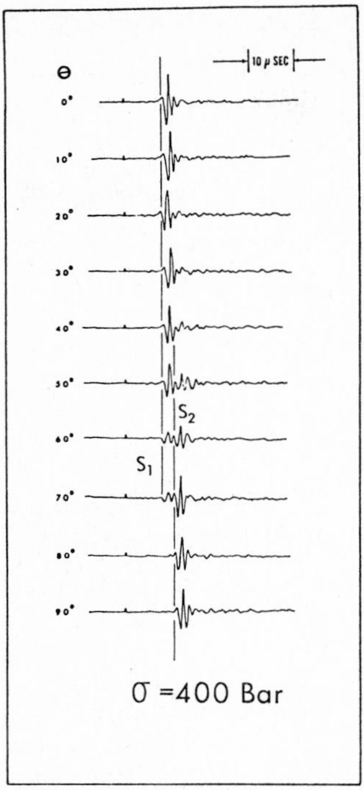

Figure 4
Traces of shear wave arrivals in uniaxially stressed samples as a function of direction of wave propagation θ relative to direction of applied stress. Acoustic birefringence is clearly observed at θ = 50°-70°.

later, to show that it is related to dispersion and attenuation in rocks with fluids, and to the viscosity of the pore fluid. The second point to note is that in order to saturate the samples shown here, the pores must be connected, so that fluid can flow throughout the pore space. This suggests, as we discuss later, that a relationship might exist between the wave period and the hydraulic diffusivity: long period waves could experience lower bulk stiffness than shorter period waves, as flow is more restricted at higher frequencies. We believe in fact, as discussed below, that this kind of wave-induced fluid flow is a major mechanism responsible for wave energy dissipation in saturated and partially saturated porous media.

GRAIN BOUNDARY FRICTION AND WAVE ATTENUATION

Although fluid-related dissipation mechanisms have long been recognized in porous media (e.g. Biot, 1956), a prevalent view has been that wave energy dissipation in rocks is caused by frictional losses at grain boundaries. Support for this notion was derived in part from the fact that confining pressure--which tightens grain contacts--also reduces attenuation. Mavko (1979), for example, has developed a theoretical model for this mechanism--considering crack tip contact sliding. His study leads to the conclusion that the inverse quality factor

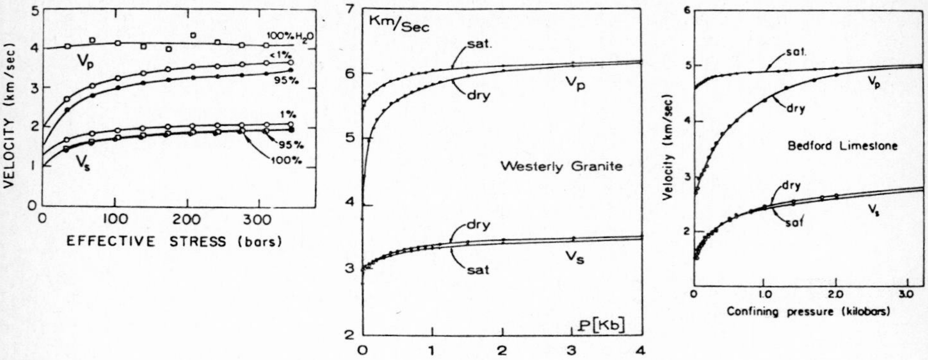

Figure 5
The effect of water saturation on compressional wave velocities in Massilon Sandstone, Westerly Granite and Bedford Limestone. Note the large increases in compressional velocities with saturation, and the small changes in shear.

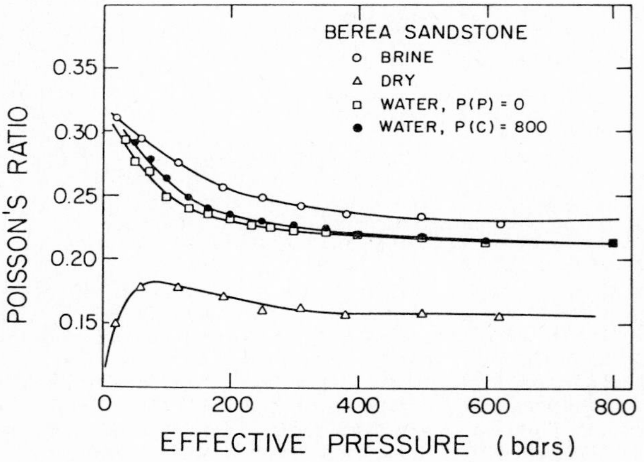

Figure 6
Typical effects of confining pressure and water saturation on Poisson's ratio in rocks. Effective pressure is the difference between confining pressure and pore pressure.

$Q^{-1} = \Delta W/W$ (where ΔW is the energy dissipated and W is the average energy per cycle) should increase roughly linearly with the amplitude of the wave strain, and that the effective moduli, or wave velocities should decrease. Figure 7 shows just these effects for a cemented sandstone and for a loose sand: at sufficiently large strain $\varepsilon \geq 10^{-6}$ or so, Q_S^{-1} increases and the shear modulus decreases with shear wave strain amplitude. Consequently the frictional sliding process is the dominant loss mechanism at high wave strain. However, at small strains $\varepsilon < 10^{-6}$ attenuation must be dominated by some other mechanisms. The most important of these mechanisms in rocks is wave-induced fluid flow, as discussed next.

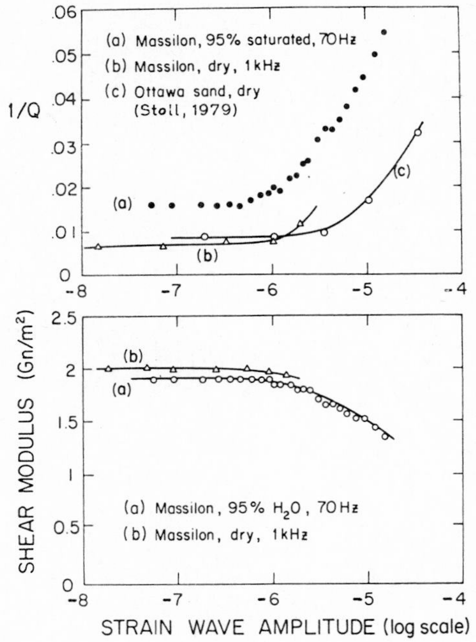

Figure 7
The dependence of the dissipation factor Q^{-1} and the dynamic shear modulus on strain wave amplitude.

THE EFFECTS OF PARTIAL SATURATION ON WAVES

The most compelling evidence for the direct role which fluids play in the dissipation of wave energy in rock with fluids comes from studies in which the amount of fluid in the pores, e.g. water, is varied from 100% down to 1% and less. Figure 8 shows the variation of the compressional and shear velocities and dissipation with partial water saturation in Vycor porous glass, measured at 1 KHz using a resonant bar method. Although systematic, only small changes in velocities are observed. These effects are due to the combined effects of changing sample density and its elastic moduli with water content, with a range of change of about 5% and 3% for the S and P wave velocities, respectively. The velocity ratio V_p/V_s, which is independent of density, shows roughly a linear decrease by about 5% as water content decreases by 80%.

In contrast the dependence of the energy dissipation factor Q^{-1} on saturation is very large, changing by a factor of 5 for Q_p^{-1} with saturation from 100% to 60% and down to 20%. Unlike Q_p^{-1} the S wave attenuation, Q_s^{-1}, barely changes at all. The very large peak in dissipation in bulk compression, with no dissipation in shear, strongly argues for some kind of fluid mechanism associated with

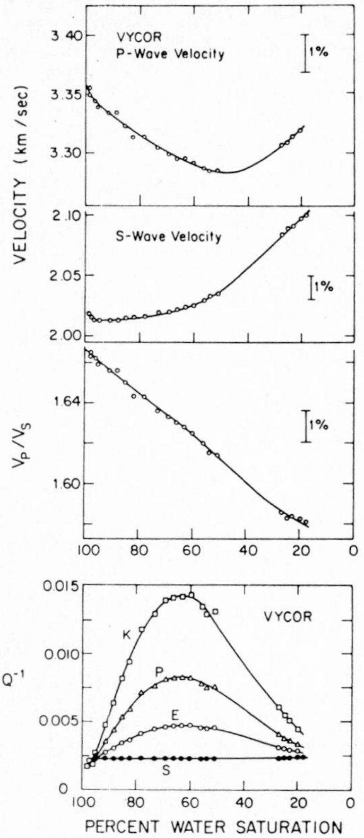

Fig. 8 Velocities and Q^{-1} in Vycor porous glass as a function of the degree of partial water saturation.

compression. Figure 9 shows a simplified conceptual model for the flow mechanism which might be involved. Because pores in Vycor are fairly equidimensional, there is only a second-order volume change of the pore space under shear whereas isotropic compression causes first-order pore volume changes. Part of this deformation is imparted to the fluid phase, which can easily flow in or between pores, which are in turn partly occupied by low viscosity, very compliant air. Thus, a local fluid squirt mechanism may well be responsible: at low saturation the amount of fluid is small with little stored strain energy. Dissipation by flow is small. At high saturation, the flow becomes restricted and strain energy in the fluid is large enough and fluid mobility is still sufficiently unrestricted to maximize the loss Q_p^{-1}.

Figure 10 shows the dependence of wave velocities and energy dissipation as a function of water content in sandstone. Although somewhat different in details from the Vycor behavior, the results show similar general trends. Almost no change in shear velocity is observed between 10% to 98% water saturation, but a significant increase of extensional velocity from 90% up. This increase is most likely due to the rapid stiffening of the pore space as the last air bubbles are driven out of the pores. At very low saturation, below 2% or 1%, large increases of both S and P velocities are observed. These are apparently associated with the removal of the last remains of free water from narrow parts of the pore space. However with the exception of the behavior at very low <2% or very high saturation >90%, wave velocities are insensitive to the degree of partial

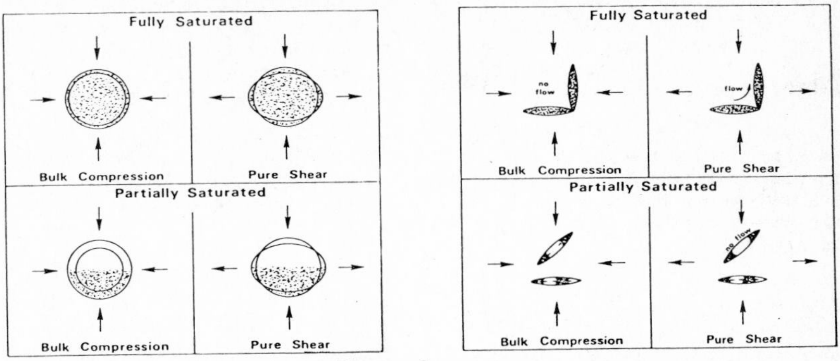

Figure 9
Conceptual models for the fluid flow and squirt mechanisms responsible for wave energy dissipation in partially and fully saturated rocks.

saturations. In contrast the dissipation is strongly dependent on saturation with a strong peak of Q_p^{-1} at ~82% water. The loss Q_s^{-1} is about twice as large in the fully saturated rock than in most of the partial saturated state. Both waves show a small but clear dissipation peak at very low saturation.

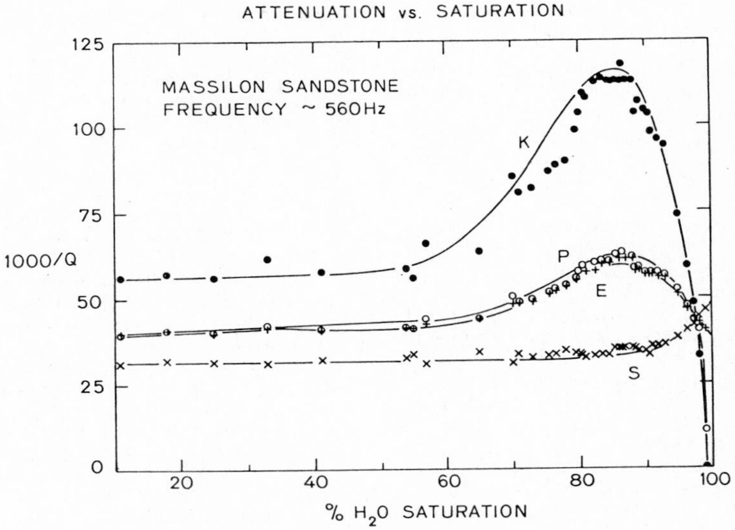

Figure 10
The dependence of the dissipation factor Q^{-1} on the degree of water saturation in Massilon Sandstone at 500 Hz.

A schematic model of the wave-induced fluid flow or squirt in sandstone is shown in Figure 9. Where wave-induced flow occurs in and between cracks, both compression strain and shear strain cause first-order volumetric strains of individual cracks, although some cracks' orientations are not sensitive to shear. In fully saturated rock the hydrostatic component of the wave strain induces pressure changes of the same sign in all pores, so that not much flow is expected. The shear wave strain in contrast dilates cracks normal to the direction of tension, and compresses cracks normal to compression. Squirt flow between these cracks can thus be substantial, leading to substantial energy loss. Cracks at 45° to tension or compression do not contribute to the shear loss.

In partially saturated sandstone, the loss in bulk compression becomes larger than in shear: the compression of the fluid with subsequent squirt into the regions with air or gas occurs in all cracks, whereas shear losses occur mostly in cracks which are favorably oriented. It can be shown (Mavko and Nur, 1978) that this model predicts $Q_p^{-1} \simeq 2Q_s^{-1}$, in fairly reasonable agreement with the measured results.

DEPENDENCE ON FREQUENCY AND WAVE DISPERSION

As mentioned already, the wave-induced fluid squirt concept suggests that dissipation might depend on frequency. Strong dependence is actually observed, as shown in Figure 11, for Vycor: dissipation is small and frequency independent when dry. As pore water is added, a loss peak develops at 7 KHz. At 80% water, the loss at the peak is about a factor of 4 greater than at 1 KHz. Losses in shear at 80% are smaller than in compression. In fully saturated Vycor, a dissipation peak is still present, but now shear losses are greater than in compression.

A similar pattern is found in sandstone (Fig. 12): Data ranging from 20 Hz to 10,000 Hz, show that a sharp peak develops with increasing water content. At 75% saturation shear losses are about half of the losses in compression. The loss peaks here at about 2,000 Hz.

The sharp peak of dissipation at some critical frequency is not yet adequately explained. The simple squirt model suggests that relaxation times of the induced pore pressure might be distributed over a fairly wide range of values, leading to a fairly broad spectrum of Q^{-1}. Although a variety of other mechanisms can be proposed for this frequency dependence, we believe that only careful further experimental data will reveal the nature of the underlying processes.

One of the direct consequences of the strong dissipation in porous rocks with fluids is the associated velocity dispersion. Figure 13 shows the shear velocity V_s as a function of water content in a crack-rich granite, measured at about 1 KHz and 500 KHz. The high frequency velocity is generally greater, and shows a strong dependence on saturation, whereas the low frequency velocity is smaller, and shows only weak dependence on saturation.

Using the formulation of Kjartansson (1980) the 5 KHz V_s values have been used, together with their corresponding values of Q_s^{-1} to compute preducted V_s values at 500 KHz. Assuming that shear dissipation Q_s^{-1} in granite is frequency independent, we obtain a close agreement between the measured and the extrapolated 500 KHz values. This suggests that in granite Q_s^{-1} is not strongly frequency dependent over the band from 1 KHz to 500 MHz. Independent verification of the validity of this assumption is being made at present.

EFFECTS OF PORE FLUID VISCOSITY

Figure 14 shows the dependence of velocities and amplitude in two rocks on the viscosity of the pore fluids, air, water and glycerol at temperatures ranging from -58°C to 100°C. In the two samples shown, the velocity and Q_s^{-1} go through a transition, at which Q_s^{-1} has a clear maximum and velocity increases from low "relaxed" values, to high "unrelaxed" ones. This effect can be explained in two ways: Nur (1972), using the shear relaxation theory of Walsh (1968), suggested

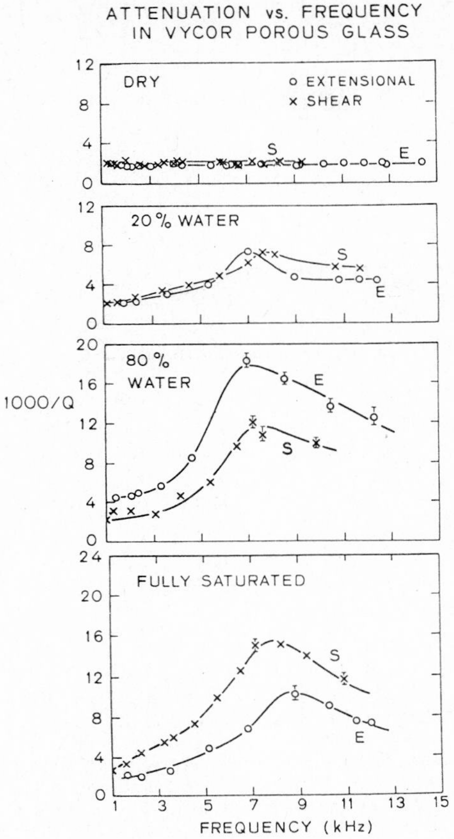

Figure 11
The dependence of Q^{-1} on frequency for various levels of water saturation in Vycor porous glass.

that the transition frequency at which Q_s^{-1} peaks, is consistent with the value $\omega_c \simeq \alpha\mu/\eta$ where α is a typical crack aspect ratio, μ the effective shear modulus of the rock, and η is the fluid viscosity. For the case at hand, $\omega_c \simeq 10^6$ Hz, $\eta = 1$ poise, $\mu = 10^{10}$ dyne/cm^2, we get aspect ratios $\alpha \simeq 10^{-3}$-10^{-4}. O'Connell and Budiansky (1977) prefer instead to match these data using the viscous squirt model.

DeVilbiss (1980) has suggested that the crossover of the shear velocity, in dry and saturated rock (Fig. 5) mentioned earlier is also due to Walsh's viscous relaxation. At low confining pressure, the thinnest cracks with smallest aspect ratios and longest relaxation times are mostly open. As pressure increases, these cracks tend to close, shifting the critical frequency to higher values, shifting the effective elastic modulus to the "relaxed" side of the spectrum. A clear example of this effect is shown in Figure 15. Here a large increase in shear velocity is found in granite at 150°C as steam in the pores, at low pore pressure, is transformed into hot water by raising pore pressure. The effect is

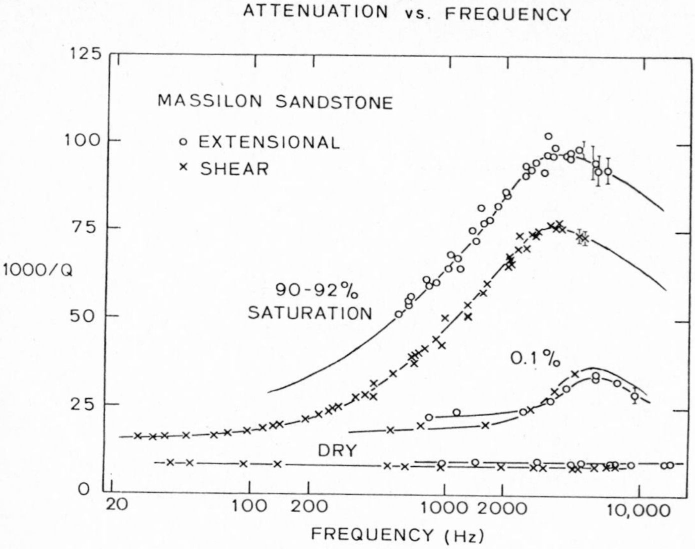

Figure 12
The dependence of Q^{-1} on frequency for various degrees of water saturation in Massilon Sandstone.

small in the porous sandstones, (~20% porosity), where it is also partly masked by the offsetting effect of the sample increasing density as hot water replaces steam. The effect is fairly large (~8%) in low porosity granite (~1% porosity). It is likely that the viscosity of the hot water increases the shear stiffness of the rock at the frequency at which measurements were made--about 300 KHz.

The change in the velocity ratio V_p/V_s at the phase transformation shows a simple increase (Fig. 16)--reflecting the fact that the V_p change is the dominant one, due to the great decrease of pore compressibility when water replaces steam. The attenuation of the P wave (Fig. 17), in contrast, shows a strong minimum at the transition, whereas the S wave attenuation does not. These results are similar to the behavior of rock at room temperature. In both dry or steam-bearing rocks, as well as in fully saturated rock, Q_p^{-1} is relatively small. When both steam and water, or air and water are present together, the dissipation of compressional energy is very large.

SUMMARY

A review of experimental studies of wave propagation in rocks with fluids shows that several variables are of particular importance: the nature of the pore space; the compressibility of the fluid, its viscosity and its pressure; the presence of gas, air, or steam; the state of stress; and the amplitude of waves.

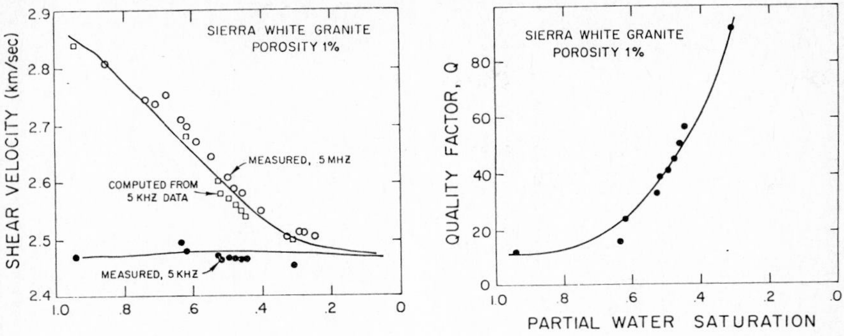

Figure 13
Shear velocity dispersion between 5 KHz and 500 KHz, as a function of quality factor Q.

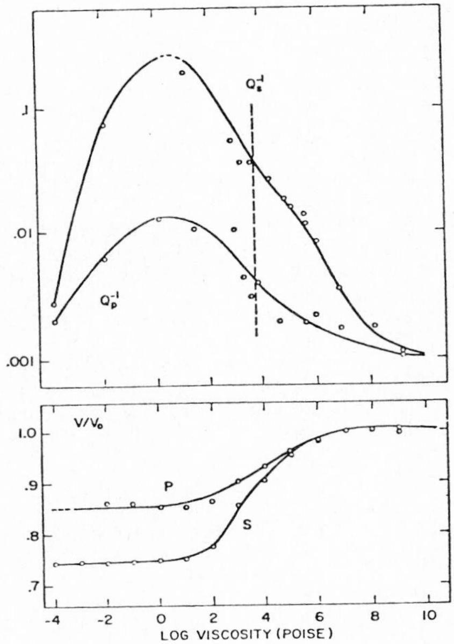

Figure 14
Normalized velocities and relative attenuation in glycerol-saturated Bedford Limestone. The viscosity of the glycerol was controlled by temperature. Values for air ($\eta \simeq 10^{-4}$ poise) and water ($\eta \simeq 10^{-2}$ poise) are also included.

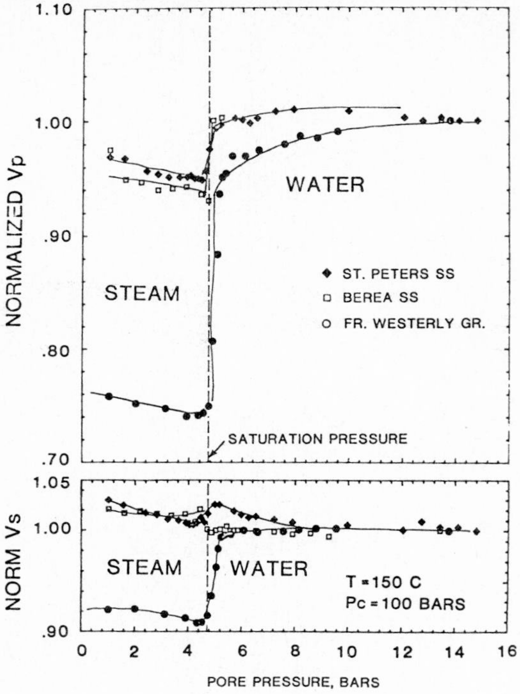

Figure 15
The effect of pore fluid on P and S velocities in sandstone and granite. By changing pore pressure the pore fluid changes from hot water to steam and back.

One of the most important results is that solids with pores or cracks and fluids exhibit strong wave attenuation, which is dependent on frequency and the degree of saturation in an intricate and complex way. Figure 18 shows a three-dimensional presentation of the dependence of Q^{-1} on these two variables. More data and better experiments should not only allow refinement of these results, but provide the understanding of the physics involved also of those effects which are still poorly understood.

Acknowledgement. This report is the result of research by several investigators associated with the Stanford Rock Physics Program. Support for these studies came from the U.S. Department of Energy, and the Geophysics Program, the U.S. National Science Foundation.

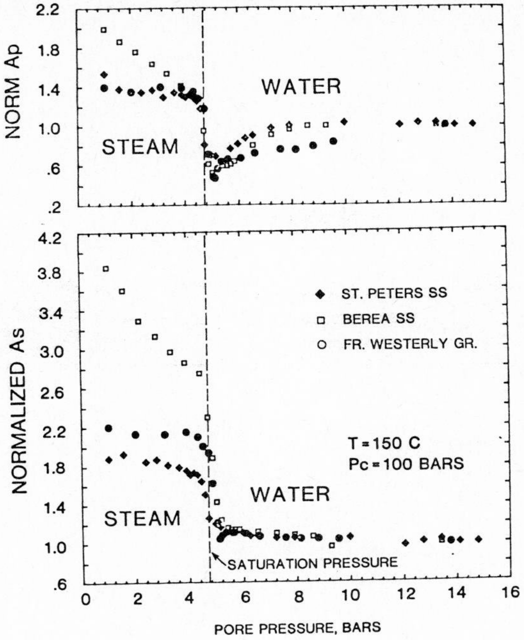

Figure 16
Dependence of V_p/V_s ratio on the nature of the pore fluid.

REFERENCES

Biot, M.A., Theory of propagation of elastic waves in a fluid saturated, porous solid, low-frequency range, J. Acoust. Soc. Am. 28 (1956a) 168-178.

Biot, M.A., Theory of propagation of elastic waves in a fluid saturated, porous solid, II. Higher frequency range, J. Acoust. Soc. Am. 28 (1956b) 179-191.

Born, W.T., The attenuation constant of earth materials, Geophysics 6 (1941) 132-148.

DeVilbiss, J., Wave dispersion and absorption in partially saturated rocks, Ph.D. Thesis, Dept. of Geophysics, Stanford Univ. (Jan. 1980).

Domenico, N.S., Effect of brine-gas mixture on velocity in an unconsolidated sand reservoir, Geophysics 41, 5 (1976) 882-894.

Elliot, S.E. and Wiley, B.F., Compressional velocities of partially saturated unconsolidated sands, Geophysics 40, 6 (1975) 949-954.

Gordon, R.B., Mechanical relaxation spectrum of crystalline rock containing water, J. Geophys. Res. 79 (1974) 2129-2131.

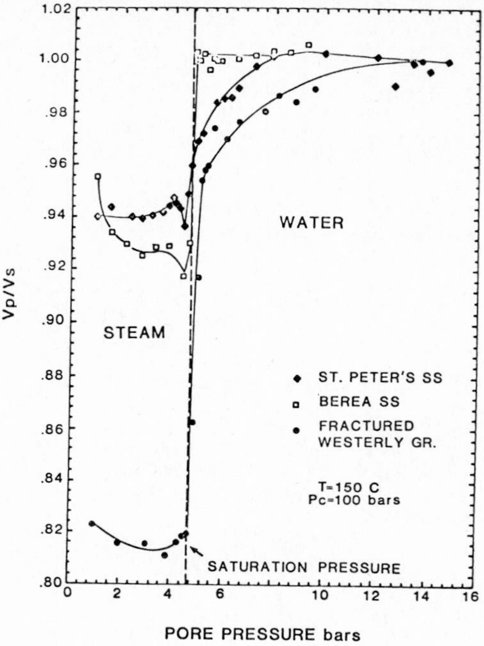

Figure 17
The dependence of wave amplitude on the nature of the pore fluid.

Johnston, D.H., Toksoz, M.N. and Timur, A., Attenuation of seismic waves in dry and saturated rocks, II. Mechanisms, Geophysics 44 (1979) 691-711.

Kjartansson, E., Constant Q wave propagation and attenuation, J. Geophys. Res. 84 (1979) 4737-4752.

Mavko, G., Frictional attenuation: an inherent amplitude dependence, J. Geophys. Res. 84 (1979) 4769-4775.

Mavko, G., Kjartansson, E. and Winkler, K., Seismic wave attenuation in rocks, Rev. Geophys. Space Phys. 17 (1979) 1155-1164.

Mavko, G. and Nur, A., Wave attenuation in partially saturated rocks, Geophysics 44 (1979) 161-178.

Nur, A., Effects of stress on velocity anisotropy in rocks with cracks, J. Geophys. Res. 76 (1971) 2022-2034.

Nur, A., Viscous phase in rocks and the low-velocity zone, J. Geophys. Res. 76 (1971) 1270-1277.

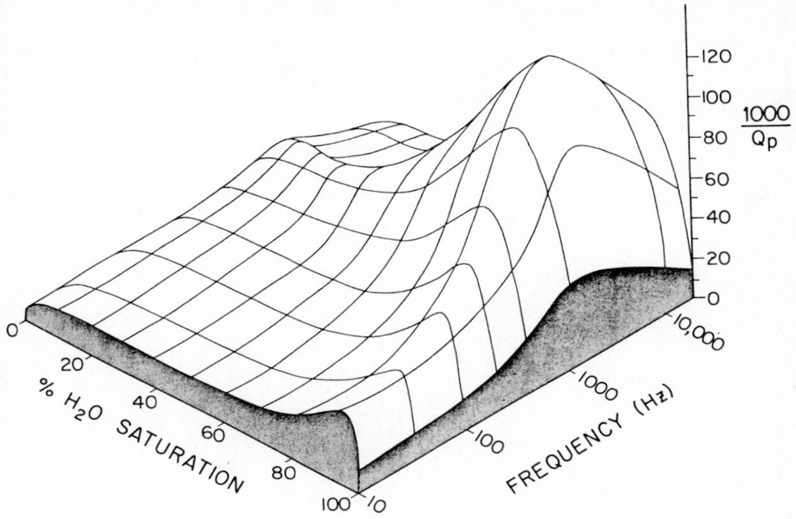

Figure 18
Schematic presentation of the dissipation factor Q^{-1} as a function of frequency and partial saturation.

Nur, A. and Simmons, G., Stress-induced velocity anisotropy in rock: an experimental study, J. Geophys. Res. 74 (1969) 6667.

Nur, A. and Simmons, G., The effect of viscosity of a fluid phase on velocity in low porosity rocks, Earth Planet. Sci. Lett. 7 (1970) 99-108.

O'Connell, R.J. and Budiansky, B., Viscoelastic properties of fluid-saturated cracked solids, J. Geophys. Res. 82, 36 (1977) 5719-5736.

Tittmann, B.R., Lunar rock seismic Q in 3000-5000 range achieved in laboratory, Phil. Trans. R. Soc. Lond. A 285 (1977) 475-479.

Walsh, J.B., Seismic wave attenuation in rock due to friction, J. Geophys. Res. 71 (1966) 2591-2599.

Walsh, J.B., New analysis of attenuation in partially melted rock, J. Geophys. Res. 74 (1969) 4333-4337.

Winkler, K. and Nur, A., Pore fluids and seismic attenuation in rocks, Geophys. Res. Lett. 6 (1979) 1-4.

Winkler, K., Nur, A. and Gladwin, M., Friction and seismic attenuation in rocks, Nature 277 (1979) 528-531.

Wyllie, M.R.J., Gardner, G.H.F. and Gregory, A.R., Studies of elastic wave propagation in porous media, Geophysics 27 (1962) 569-589.

Continuum Models of Discrete Systems 4
eds. O. Brulin and R.K.T. Hsieh

PHASE CHANGES IN CRYSTAL MECHANICS

Gareth P. Parry

School of Mathematics, University of Bath,
Bath, Avon, England.

Classical thermoelasticity denies the possibility of phase transitions which correspond to changes of symmetry in the constitutive law. By analysing a problem of crystal mechanics, we try to understand the extensions of thermoelasticity which are needed to deal with this phenomenon. The theory which emerges has many novel aspects, and it gives perspective to the conventional view of the situation. Notably, the change in symmetry is shown to induce a roughness of the thermoelastic potential, so confirming Landau's own suspicions regarding the general validity of his theory of phase transitions of the second kind.

Transitions from solid to liquid and changes in symmetry of elastic crystals are commonly grouped together under the label of 'phase transitions'. However, these two phenomena should be viewed as quite distinct, according to the classical theory of thermolelasticity, where it is presumed that the symmetry of a material does not change with deformation or temperature. In thermoelasticity, a solid can never become a fluid, for such a transition corresponds to a change in invariance of the free energy, so that the solid and fluid phases are effectively viewed as different materials, with different free energies. On the other hand, thermoelasticity can accommodate some changes of symmetry in elastic crystals, specifically those changes which correspond just to changes in the geometrical symmetry of equilibrium configurations of the crystal, not to changes of invariance in the free energy. Such changes are analogous to the changes of symmetry which are apparent when an isotropic material is referred to different reference configurations. In configurations gained via dilatation of the natural state, the solid is isotropic, when just one of the principal stretches of the deformation is different from the other two, the solid appears to be transversely isotropic, and when the principal stretches are all distinct, it seems to be orthorhombic. These changes of symmetry are in some sense 'apparent' changes because the symmetry of the constitutive law is the same in each 'phase' of the solid, they are to be distinguished from the 'real' solid-fluid transition.

Here we describe a prototype model, which is not quite thermoelasticity, which allows changes in invariance of the Helmholtz free energy by introducing an internal variable to describe the geometry of the material. For different choices of the internal variable, the free energy has different symmetries. One of the main tasks of the analysis is the derivation of the thermoelastic picture which obtains from this model on elimination of the internal variable.

The most straightforward setting for the problem appears to be crystal mechanics, for we have some understanding of crystal symmetry and some internal variables leap to the eye. Envisage the change which occurs when the lattice of 'centre atoms' in a body centred tetragonal (b.c.t.) lattice moves off-centre. Before the change, the atoms are situated at the points of a simple lattice, after the change the atoms may not generally be so located. This change in fact corresponds to a change in invariance of the free energy. Suitable internal variables would be the relative separation of the lattice of centre atoms and the lattice remaining, measured by the vector difference of any two points, one from each lattice, or the displacement of a typical 'centre atom'. The analysis deals only with continuous changes in the geometry of the crystal, in the physics jargon, with transitions 'of the second kind'. We shall imagine that the control variable producing change is the temperature θ and that the displacement of the lattice of centre atoms is incipiently nonzero at $\theta = \theta_o$. There is no difficulty in extending the analysis to cover changes induced by various types of loading.

The passage from this internal variable theory to thermoelasticity involves two distinct steps. Firstly it involves the elimination of the internal variables in terms of the other geometric parameters, which are the lattice vectors. If the change in structure is spontaneous in the sense that it does not occur in response to the 'switching-on' of some force conjugate to the internal variable, then we presume that the internal variable takes the value corresponding to globally stable configurations of the crystal, which as such minimises the free energy absolutely. Secondly, the elimination of the lattice vectors in favour of the macroscopic deformation gradient, or the Cauchy Green tensor, requires that some lattice vectors change smoothly through the change in structure. It is evident, even at this stage, that there is no reason why the basis vectors of the lattice of 'centre atoms' should not vary smoothly, so that the scheme is feasible. However, some choices of lattice vectors do not fit the requirements, so that care is needed.

The thermoelastic theory which emerges is not standard, and there are novel features in the analysis of symmetry and in the thermoelastic potential itself. Perhaps most strikingly, it transpires that the change in invariance of the free energy induces a roughness in the thermoelastic potential. Specifically, some second derivatives of the potential are discontinuous where there is such a change in structure.

Symmetry

We have to allow the lattice of centre atoms in a b.c.t. configuration of a crystal to move independently of the remaining lattice. So we allow geometric configurations of the crystal which correspond to a superposition of two simple lattices, in brief, to 2-lattices [1]. The two simple lattices which compose the 2-lattice are congruent and parallel and so may be brought into coincidence by uniform translation. Generally, four vectors are required to specify the points of a 2-lattice, they may be taken as any three basis vectors of one simple lattice, $\underset{\sim}{e}_a$ say, (a=1,2,3), and a vector $\underset{\sim}{p}$, the shift, representing the relative positions of any two atoms in distinct simple lattices.

In a tetragonal configuration of the simple lattices, the conventional choice of the $\underset{\sim}{e}_a$ would be the three orthogonal vectors which are the edges of the unit cell first introduced by Brillouin, which are such that

$$|\underset{\sim}{e}_1| \le |\underset{\sim}{e}_2| = |\underset{\sim}{e}_3| \quad , \tag{1}$$

say, for definiteness. In b.c.t. configurations of the 2-lattice, an obvious choice of the shift would be

$$\underset{\sim}{p} = \frac{1}{2}\sum_a \underset{\sim}{e}_a \, , \tag{2}$$

However, in such configurations, the whole arrangement of atoms in the 2-lattice may also be considered as a simple lattice, with lattice vectors $\underset{\sim}{\bar{e}}_a$ given by

$$\bar{e}_1 = p,\ \bar{e}_2 = e_2,\ \bar{e}_3 = e_3\ . \tag{3}$$

That is to say, the description of the arrangement is not unique. B.c.t. configurations may be described as simple lattices, or as 2-lattices. This complicates the analysis somewhat.

The points of a simple lattice do not uniquely determine the basis vectors (lattice vectors), for any choice

$$e^*_a = \gamma_a^{\ b} e_b\ ,\quad \bar{e}^*_a = \bar{\gamma}_a^{\ b} \bar{e}_b\ , \tag{4}$$

is also a set of basis vectors, for the appropriate simple lattice, provided that the $\gamma_a^{\ b}$, $\bar{\gamma}_a^{\ b}$ are integers, and that

$$\det \gamma = \pm 1\ ,\ \det \bar{\gamma} = \pm 1 \tag{5}$$

where γ and $\bar{\gamma}$ are the matrices with elements $\gamma_a^{\ b}, \bar{\gamma}_a^{\ b}$, and where det stands for the determinant. Conversely, all sets of basis vectors are of the form given by (4) and (5) [2]. Neither do the points of a 2-lattice determine the shift uniquely, for any vector

$$q = p + x\ , \tag{6}$$

with x any integer linear combination of the e_a, serves equally to represent the separation of the constituent simple lattices.

It is natural to suppose that the Helmholtz free energy per unit mass of the crystal may be represented, in b.c.t. configurations, in the form

$$\phi = \phi_1(\bar{e}_a, \theta)\ , \tag{7}$$

where the subscript 1 denotes that the crystal is represented as a simple lattice. Correspondingly, when the crystal is no longer a simple lattice, but a 2-lattice, we write

$$\phi = \phi_2(e_a, p, \theta)\ . \tag{8}$$

As remarked earlier, the b.c.t. configuration may be thought of either as a simple lattice or as a 2-lattice, so that

$$\phi_1(\bar{e}_a, \theta) = \phi_2(e_a, \tfrac{1}{2}\sum_a e_a, \theta)\ . \tag{9}$$

The free energy depends only on the current configuration of the atoms in the crystal, it takes no account of the history of deformation, so that it must be invariant with respect to any rearrangement of the indistinguishable atoms which make up the crystal. Thus

$$\phi_1(\bar{e}_a, \theta) = \phi_1(\bar{\gamma}_a^{\ b} \bar{e}_b, \theta)\ , \tag{10}$$

with $\bar{\gamma}_a^{\ b}$ as specified, and

$$\phi_2(e_a, p, \theta) = \phi_2(\gamma_a^{\ b} e_b, p+x, \theta)\ , \tag{11}$$

with $\gamma_a^{\ b}$ and x as specified.

We have also to account for the relation (9). Since the mapping

$$\bar{e}_a \rightarrow \bar{\gamma}_a^{\ b} \bar{e}_b\ , \tag{12}$$

which regenerates the b.c.t. lattice, induces the mapping

$$e_a \to (M\bar{\gamma}M^{-1})_a^{\ b} e_b \ , \tag{13}$$

of the basis vectors of the 2-lattice, with

$$M = (M_a^{\ b}) = \begin{Vmatrix} 2 & -1 & -1 \\ 0 & 1 & 0 \\ 0 & 0 & 1 \end{Vmatrix} \tag{14}$$

it follows that, in additon to (11), we have

$$\phi_2(e_a, \frac{1}{2}\sum_a e_a, \theta) = \phi_2(\tilde{\gamma}_a^{\ b} e_b, \frac{1}{2}\sum_a \tilde{\gamma}_a^{\ b} e_b, \theta) \tag{15}$$

with

$$\tilde{\gamma} = M\bar{\gamma}M^{-1} \ . \tag{16}$$

When the configuration is not body centred (b.c.), (9) does not hold and so neither does (15). Therefore there is a change of invariance of the Helmholtz free energy in the passage from a simple lattice to a 2-lattice, for (11) does not include (15), because a typical $\tilde{\gamma}$ has half integer elements [3] [4].

The analysis of the bifurcation involved in the change of invariance requires that we appreciate the various constraints on the material parameters. This necessitates a brief discussion of the crystallographic point groups.

The elements of the crystallographic point group, P say, corresponding to the simple lattice with basis vectors e_a is the set of orthogonal transformations Q such that

$$Qe_a = \gamma_a^{\ b} e_b \tag{17}$$

for some integers $\gamma_a^{\ b}$ satisfying (5). That is to say, the point group consists of the rotations and reflections which regenerate the lattice. It is clear that the point group depends on the lattice vectors. The point group $\bar{P}$ defined via the basis vectors $\bar{e}_a$, in the obvious way, generally includes P. This is perhaps clear, the 2-lattice may be 'more symmetrical' than the constituent simple lattices [3]. There are configurations where the constituent simple lattices are b.c.t., but the 2-lattice is face centred cubic.

By frame indifference, ϕ_2 may be written as

$$\phi_2(\eta, p_c, \theta) \tag{18}$$

with η the matrix of elements $e_a \cdot e_b$ and p_c the triplet with elements $p \cdot e_a$ (a=1,2,3) (the subscript c denotes the covariant components). To each element of the point group Q there corresponds an integer linear transformation satisfying (5), via (17). The set of such $\gamma_a^{\ b}$ forms the lattice group, L say, and it follows from (17) that

$$\gamma\eta\gamma' = \eta \ , \quad \gamma \varepsilon L \ , \tag{19}$$

with the dash denoting the transpose.

The transformation Q regenerates one of the constituent simple lattices, by virtue of (17). If we imagine that each vector in the 2-lattice is subject to this transformation, then it is clear that the whole 2-lattice is regenerated provided

$$Qp = p + x \tag{20}$$

via (6), with x as specified. Equation (11) translates to

$$\phi_2(\eta, \underset{\sim}{p}\cdot\underset{\sim}{e}_a, \theta) = \phi_2(\gamma\eta\gamma', (\underset{\sim}{p}+\underset{\sim}{x})\cdot\gamma_a{}^b\underset{\sim}{e}_b, \theta) . \tag{21}$$

With (18) and (20)

$$\underset{\sim}{p}\cdot\underset{\sim}{e}_a = (\underset{\sim}{p}+\underset{\sim}{x})\cdot\gamma_a{}^b\underset{\sim}{e}_b \quad , \gamma\varepsilon L \quad , \tag{22}$$

and it is evident that the derivatives of ϕ_2 are constrained at the fixed points corresponding to (19) and (22).

There are also constraints which arise through the 1-lattice description. They correspond to the common fixed points of the mappings

$$\eta \to \tilde{\gamma}\eta\tilde{\gamma}' \tag{23}$$

with $\tilde{\gamma}$ given by (16), being associated with the orthogonal transformations which regenerate the whole 2-lattice. These constraints are generally stronger than those corresponding to the 2-lattice description, since $\bar{P}$ includes P.

Elimination of the internal variable

It is convenient, in this section, to employ the variable

$$\underset{\sim}{\pi} = \underset{\sim}{p} - \frac{1}{2}\sum_a \underset{\sim}{e}_a \quad , \tag{24}$$

representing the displacement of a typical centre atom, in preference to the shift $\underset{\sim}{p}$. The use of $\underset{\sim}{\pi}$ simplifies the calculations, because the monatomicity of the crystal requires that

$$\phi_2(\underset{\sim}{e}_a, \underset{\sim}{p}, \theta) = \phi_2(\underset{\sim}{e}_a, -\underset{\sim}{p}, \theta) \tag{25}$$

so that, in a loose notation,

$$\begin{aligned}
\phi_2(\underset{\sim}{e}_a, \underset{\sim}{\pi}, \theta) &= \phi_2(\underset{\sim}{e}_a, \underset{\sim}{p}-\tfrac{1}{2}\sum_a \underset{\sim}{e}_a, \theta) = \phi_2(\underset{\sim}{e}_a, \underset{\sim}{p}+\tfrac{1}{2}\sum_a \underset{\sim}{e}_a, \theta) && \text{via (11)}\\
&= \phi_2(\underset{\sim}{e}_a, -\underset{\sim}{p}-\tfrac{1}{2}\sum_a \underset{\sim}{e}_a, \theta) && \text{via (25)}\\
&= \phi_2(\underset{\sim}{e}_a, -\underset{\sim}{p}+\tfrac{1}{2}\sum_a \underset{\sim}{e}_a, \theta) && \text{via (11)}\\
&= \phi_2(\underset{\sim}{e}_a, -\underset{\sim}{\pi}, \theta) . && (26)
\end{aligned}$$

Thus ϕ_2 is odd in $\underset{\sim}{\pi}$, so that in body-centred configurations, where $\underset{\sim}{\pi}$ vanishes, odd derivatives with respect to $\underset{\sim}{\pi}$ vanish.

Stable equilibrium configurations of the crystal minimise ϕ_2 absolutely. We choose to perform the minimisation first with respect to $\underset{\sim}{\pi}$, at fixed $\underset{\sim}{e}_a$ and θ, giving

$$\underset{\sim}{\pi} = \bar{\underset{\sim}{\pi}}(\underset{\sim}{e}_a, \theta) , \tag{27}$$

say, as the absolute minimiser such that

$$\phi_2(\underset{\sim}{e}_a, \underset{\sim}{\pi}, \theta) \geqslant \phi_2(\underset{\sim}{e}_a, \bar{\underset{\sim}{\pi}}, \theta) . \tag{28}$$

In practice, this minimum is determined in two steps. First of all we find the turning points of ϕ_2 with respect to $\underset{\sim}{\pi}$ via

$$\frac{\partial\phi_2}{\partial\underset{\sim}{\pi}}(\underset{\sim}{e}_a, \underset{\sim}{\pi}, \theta) = 0, \tag{29}$$

and index the relative minima so obtained by the superscript i, so that

$$\underset{\sim}{\pi} = \underset{\sim}{\pi}^{i}(\underset{\sim a}{e},\theta) \tag{30}$$

are solutions of (21). The absolute minimum value of ϕ_2, at given $\underset{\sim a}{e}, \theta$, which is $\phi_2(\underset{\sim a}{e}, \bar{\underset{\sim}{\pi}}(\underset{\sim a}{e},\theta),\theta)$ is then given by

$$\bar{\phi}_2(\underset{\sim a}{e},\theta) \stackrel{\text{def}}{=} \min_i \phi_2(\underset{\sim a}{e},\underset{\sim}{\pi}^{i}(\underset{\sim a}{e},\theta),\theta). \tag{31}$$

The procedure is straightforward when (29) has only one solution for $\underset{\sim}{\pi}(\underset{\sim a}{e},\theta)$ in the neighbourhood of the b.c.t. configuration. Via (18), this one solution must be $\underset{\sim}{\pi} = \underset{\sim}{0}$, when the crystal remains b.c. On the other hand, if

$$\det \frac{\partial^2\phi}{\partial \underset{\sim}{\pi}^2}(\underset{\sim a}{e},0,\theta_0) = 0 \quad , \tag{32}$$

then there is generally more than one solution of (29), and relative displacement of the constituent simple lattices is possible. We shall illustrate the procedure in the simplest possible case.

Firstly, to account for frame indifference, write ϕ_2 in the form

$$\phi_2(\eta,\underset{\sim c}{\pi},\theta) \tag{33}$$

with η the matrix with elements $\underset{\sim a}{e}\cdot\underset{\sim b}{e}$, and $\underset{\sim c}{\pi}$ denoting the triplet of covariant components of $\underset{\sim}{\pi}$ with respect to the basis $\underset{\sim a}{e}$, so that $\underset{\sim c}{\pi} = (\underset{\sim}{\pi}\cdot\underset{\sim a}{e})$. Expanding (29) in a Taylor series about $\underset{\sim c}{\pi} = 0$, making use of (26), we find that in one case there are locally exactly three solutions of (21), which are given to lowest order in increments of η by

$$\underset{\sim c}{\pi} = 0 \quad ,$$
$$\underset{\sim c}{\pi} = \pm\,(\{-N_1/4b_1\}^{\frac{1}{2}},0,0) \; , \tag{34}$$

where

$$N_1 \stackrel{\text{def}}{=} \sum_i \frac{\partial^2\phi_2}{\partial\pi_1{}^2\partial\eta_{ii}}\,\delta\eta_{ii} \quad , \quad b_1 \stackrel{\text{def}}{=} \frac{1}{4!}\frac{\partial^2\phi_2}{\partial\pi_1{}^4} \; . \tag{35}$$

Let us denote the nontrivial solutions by $\underset{\sim c}{\pi}^{+}$ and $\underset{\sim c}{\pi}^{-}$. The three solutions are relative minima provided that b_1 is positive [5] [6]. By inspection, the nontrivial solutions are real solutions if and only if N_1 is negative. Substituting these three solutions into the Taylor expansion of $\phi_2(\eta,\underset{\sim c}{\pi},\theta)$ we find that the two nonzero solutions coalesce to give just one 'sheet' (as they must via (18)) given by

$$\phi_2{}^1(\eta,\theta) \stackrel{\text{def}}{=} \frac{-N_1{}^2}{16b_1} + \phi_2(\eta,0,\theta) \tag{36}$$

valid if $N_1 < 0$, and the trivial solution gives

$$\phi_2{}^0(\eta,\theta) \stackrel{\text{def}}{=} \phi_2(\eta,0,\theta). \tag{37}$$

Clearly $\phi_2{}^1 < \phi_2{}^0$, if $N_1 < 0$, so that the absolute minimum of the free energy is given finally by

$$\bar{\phi}_2(\eta,\theta) = \begin{cases} \phi_2{}^0 \;, & N_1 > 0 \\ \phi_2{}^1 \;, & N_1 < 0 \;. \end{cases} \tag{38}$$

Provided that the equilibrium path of the crystal is not locally $N_1 = 0$, it follows from (36) and (38) that the derivatives $\frac{\partial \phi_2}{\partial \eta}$ are continuous on that path, but that there is a finite discontinuity of the second derivatives $\frac{\partial^2 \phi_2}{\partial \eta^2}$ precisely at the solutions of (32). (In [4], genericity arguments are invoked to discount the possibility that $N_1 = 0$).

Derivation of thermoelasticity

The behaviour of the lattice vectors must now be correlated with the macroscopic deformation. The Cauchy-Born hypothesis, that lattice vectors are embedded in the macroscopic deformation, clearly fails for some choices of lattice vectors. Specifically, the lattice vectors $\bar{e}_a$ defined via (3), which generate a b.c. lattice, no longer constitute a basis of the lattice when the centre atoms are displaced infinitesimally. On the other hand, the basis e_a of the constituent simple lattices varies smoothly through the change in structure, so that if the e_a were E_a in some reference configuration of the crystal,

$$e_a = FE_a \quad , \tag{39}$$

with F the matrix of macroscopic deformation gradients. With (39), η may be interpreted as the Cauchy Green tensor based on the triad reciprocal to E_a. The potential $\bar{\phi}_2(\eta,\theta)$ then assumes the role of the thermoelastic potential, and we have the result that the isothermal elastic moduli are discontinuous when there is a change of invariance of the free energy.

It is of interest to determine how the symmetry of the free energy translates into the symmetry of thermoelastic potential, and particularly to discover the consequences of the change in invariance of the free energy.

Via (11),

$$\phi_2(\eta,p_c,\theta) = \phi_2(\gamma\eta\gamma',p_c,\theta) \tag{40}$$

It follows that if $p_c^{\,i}(\eta,\theta)$ is a solution of

$$\frac{\partial \phi_2(\eta,p_c,\theta)}{\partial p_c} = 0 \ , \tag{41}$$

so then is $p_c^{\,i}(\gamma\eta\gamma',\theta)$ (the superscript i is again an index labelling one solution of (41))

Thus

$$\bar{\phi}_2(\gamma\eta\gamma',\theta) = \min_i \phi_2^{\,i}(\gamma\eta\gamma',p_c^{\,i}(\gamma\eta\gamma',\theta),\theta) = \min_i \phi_2^{\,i}(\gamma\eta\gamma',p_c^{\,i}(\eta,\theta),\theta)$$

$$= \min_i \phi_2^{\,i}(\eta,p_c(\eta,\theta),\theta) = \bar{\phi}_2(\eta,\theta) \quad , \tag{42}$$

via (40). In the sense of (40) and (42), the thermoelastic potential has the symmetries of the free energy which are independent of the internal variable.

However, this is not the whole story. The thermoelastic potential is the absolute minimum amongst the sheets

$$\phi_2^{\,i}(\eta,\theta) \stackrel{\text{def}}{=} \phi_2(\eta,p_c^{\,i}(\eta,\theta),\theta) \quad , \tag{43}$$

and these sheets generally correspond to functions which have symmetries different to that expressed by (42). For example, the sheet $\phi_2^{\,o}(\eta,\theta)$ corresponding to the trivial solution has the symmetries

$$\phi_2^{\,o}(\eta,\theta) = \phi_2(\tilde{\gamma}\eta\tilde{\gamma}',\theta) \tag{44}$$

via (15). We can be a little more explicit if we consider the particular case of the previous section. In that case there are three relative minima, $\underset{\sim}{\pi}_c = 0$ and $\underset{\sim}{\pi}_c = \underset{\sim}{\pi}_c^{\pm}$ say. There are two sheets, $\phi_2^{\,o}(\eta,\theta) = \phi_2(\eta,0,\theta)$ and $\phi_2^{\,1}(\eta,\theta)$ $= \phi_2(\eta,\underset{\sim}{\pi}_c^{\pm}(\eta,\theta),\theta)$. The symmetries of $\phi_2^{\,o}$ are given by (44), the symmetries of $\phi_2^{\,1}$ are derived by an argument like that in (42). Specifically, since $\underset{\sim}{\pi}_c^{+}$, say, is a solution of

$$\frac{\partial\phi_2}{\partial\underset{\sim}{\pi}_c}(\eta,\underset{\sim}{\pi}_c,\theta) = 0 , \tag{45}$$

so is $\underset{\sim}{\pi}_c^{+}(\gamma\eta\gamma',\theta)$. The analysis of the last section is only valid locally, say close to $\eta = \eta_o$. Therefore, provided that $\gamma\eta_o\gamma' = \eta_o$, we must have that $\pi_c^{+}(\gamma\eta\gamma',\theta)$ is one of $0,\underset{\sim}{\pi}_c^{\pm}(\eta,\theta)$. Since $\underset{\sim}{\pi}_c^{+}(\eta,\theta)$ is non zero, so is $\underset{\sim}{\pi}_c^{+}(\gamma\eta\gamma')$. It follows that $\underset{\sim}{\pi}_c^{+}(\gamma\eta\gamma',\theta)$ is one of $\underset{\sim}{\pi}_c^{\pm}(\eta,\theta)$ and finally that

$$\phi_2^{\,1}(\eta,\theta) = \phi_2^{\,1}(\gamma\eta\gamma',\theta) \quad , \quad \gamma\varepsilon L \tag{46}$$

It seems difficult to obtain a stronger statement than this with local analysis. However, (44) and (46) are sufficient to demonstrate that <u>the sheets have different symmetries</u>, and this is the sense in which the change of invariance of the free energy translates to thermoelasticity.

References

[1] Pitteri, M., Reconciliation of local and global symmetries of crystals, to appear.

[2] Ericksen, J.L., Nonlinear elasticity of diatomic crystals, Int. J. Solids Structures 6, 951-957 (1970).

[3] Ericksen, J.L., Changes in symmetry in elastic crystals, to appear.

[4] Parry, G.P., On phase transitions involving internal strain, to appear in Int. J. Solids Structures

[5] Landau, L.D., On the theory of phase transitions, in Collected Papers of L.D. Landau (ed. D. Ter Haar), 193-216. Pergamon Press and Gordon and Breach, London, 1965.

[6] Tisza, L., On the general theory of phase transitions, in Phase Transformations in Solids (ed. R. Smoluchowski, J.E. Mayer, W.A. Weyl), John Wiley & Sons, New York and Chapman & Hall Ltd. London, 1951.

Continuum Models of Discrete Systems 4
eds. O. Brulin and R.K.T. Hsieh

NON-LINEAR CONTINUUM MODELS OF DISCRETE CRYSTAL DEFECTS

C. Teodosiu

Department of Solid Mechanics
Institute for Physics and Technology of Materials
70701 Bucharest
Romania

The understanding of the plastic flow and fracture of crystalline materials rests ultimately on knowledge of the atomic arrangements around lattice imperfections. The present paper concerns the determination of these atomic arrangements by means of semidiscrete methods that make use of the lattice theory for the close proximity of crystal defects and of non-linear elasticity theory for the remaining of the crystal. The significance of the results available so far is discussed in some detail and further possible applications are outlined.

INTRODUCTION

A thorough understanding of the strength of crystalline materials rests ultimately on knowledge of the atomic arrangements around dislocations, point defects, grain and phase boundaries, since it is these imperfections that make possible the plastic flow and the nucleation and propagation of cracks at applied forces that are several orders of magnitude lower than those necessary to fracture a perfect crystal.

Unfortunately, the well-developed techniques of continuum mechanics break down near crystal defects, since they lead to infinite values of the stress and displacement fields, which are physically unacceptable. On the other hand, the fully atomic models based on lattice theory lead to correct predictions of the atomic arrangements around imperfections, but must be restricted to bounded atomic blocks containing a not very large number of atoms, in order to save computation time. A third possibility - which proved to be the most successfull one - is to apply a semidiscrete method, which makes use of the lattice theory for the close proximity of crystal defects and of the elasticity theory for the remaining of the crystal, each of these theories providing the boundary conditions necessary for the other.

Semidiscrete methods have been successfully applied during the last 25 years to study atomic configurations around dislocations, point defects, and cracks. The development of high-speed computers, coupled with advances in elaborating some realistic interatomic potentials has greatly improved the simulation of crystal defects by semidiscrete methods. In the last few years there was a growing interest for applying non-linear elasticity for the continuum part of the model, ever since the necessary analytical solutions have become available and combined atomistic and continuum calculations done for straight dislocations have shown that non-linear effects play an important role in the determination of the core configurations and the estimation of the overall dilatation produced by dislocations.

The present paper gives an account of recent progress obtained in coupling non-linear elasticity theory with semidiscrete methods. For the sake of simplicity, we shall confine ourselves to consider straight dislocations, although the methods described could be equally applied to calculate atomic configurations around dislocation loops and point defects or to study interaction problems.

RIGID AND FLEXIBLE BOUNDARIES

In order to simulate a single dislocation by a semidiscrete method it is customary to divide the crystal into two regions. Region I, which is the next neighbourhood of the dislocation line, is treated as a discrete lattice; atom positions are considered individually, with some interatomic potential being assumed to give the potential energy in terms of the relative atom positions. Region II, the remainder of the crystal, is considered as an elastic continuum. Clearly, any displacement field satisfying the equilibrium equations of the elasticity theory provides also the equilibrium atom positions when evaluated at the discrete lattice points. However, the only atoms which need to be explicitly considered in region II are those whose positions are required for the calculation of the forces exerted on the atoms of region I.

On coupling the atomistic and continuum parts of the model, suitably boundary conditions must be introduced on the separation surface Σ_0 between regions I and II. For a straight dislocation this internal boundary is generally chosen as a circular cylindrical surface of radius r_0 having the dislocation line as axis (Fig. 1), or as the lateral boundary of a rectangular parallelepiped having the dislocation line in the middle. In both cases, region I is limited to a repeat distance in the direction of the dislocation line, with periodic boundary conditions being introduced on the surface normal to the dislocation line, in order to simulate an infinitely deep crystallite.

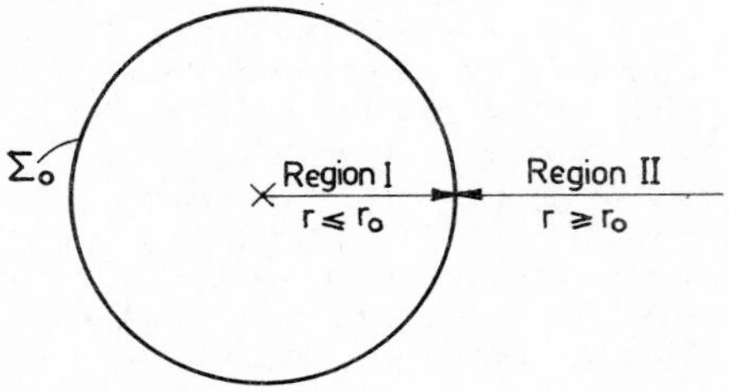

Figure 1
Concentric arrangement of regions for the rigid-boundary method and for Sinclair's flexible-boundary method (Flex-S).

Until recently, the most frequently used method for simulating crystal dislocations has been the semidiscrete method with rigid boundary, in short: the rigid-boundary method. This method consists in the calculation of the relaxed atom positions in region I, while the atoms situated on Σ_0 and in region II are kept fixed in their positions given by the linear elastic displacement field, as obtained by neglecting the core boundary conditions.

The rigid-boundary method has been successfully used to study the influence of an applied stress on the core configuration and on the dislocation mobility, the interaction between dislocations and point defects, the dislocation kinks and jogs, the dissociation of a full jog into half-jogs, and the dissociation of a dislocation into partials. Most of these studies have employed the anisotropic elasticity theory for a more exact evaluation of the atomic displacements in region II. In addition, region I has been gradually increased from 20 atomic rows in the pioneering work of Huntington, Dickey, and Thompson [1] to 2565 atomic rows in the work of Duesbery, Vitek, and Bowen [2].

In spite of this progress, it is apparent that the rigid-boundary method has some essential drawbacks. The most important of them is that the elastic solution used for region II does not take into account the boundary conditions on Σ_0, and hence cannot be improved during the calculation of the equilibrium atom positions in region I. Moreover, when using the linear elastic solution for region II, the rigid boundary introduces some artificial constraints upon region I. For example, the mean volume dilatation of region I must vanish, at variance with theoretical considerations and experimental evidence. Furthermore, the atom positions given by the linear elasticity theory, which is based on a harmonic interaction potential, abruptly change across Σ_0 into atom positions computed on the basis of anharmonic potentials, thus making impossible a smooth passage from the dislocation core to its surroundings. Finally, calculations with an increasing number of atoms in region I show a rather slow convergence, even for the atom positions in the very centre of the dislocation core, which should be less sensitive to the boundary conditions imposed on Σ_0.

Due to the shortcomings of the rigid-boundary method, several semidiscrete methods have been developed in the last decade, which allow the relaxation of Σ_0 and the modification of the displacement field used for region II together with the atom positions in region II. These methods, which are called flexible-boundary methods permit a considerable reduction of the number of atoms in region I, together with the improvement of the elastic solution used for region II, and release some of the strong constraints imposed by a rigid boundary. At the same time, by appreciably reducing the number of the atoms which need to be treated independently for a given level of accuracy, the flexible-boundary methods make possible the consideration of more complex dislocation configurations and interactions.

In the remaining part of this paper we shall focus our attention upon combining the semidiscrete methods with a non-linear elastic description of region I. We shall see that this approach considerably improves the smoothness of the discrete-continuum transition and allows a better description of some typical non-linear dislocation effects.

STATIC RELAXATION

Before examining the semidiscrete methods we will briefly review the method of static relaxation, which is the main procedure used to obtain the atom positions in region I by successive approximations.

Let $\underset{\sim}{X}$ denote the vector of the generalized co-ordinates of a system, for instance the Cartesian co-ordinates of the atoms in region I. Assume that the potential energy E of the system is known as a function of $\underset{\sim}{X}$. The generalized force $\underset{\sim}{F}$ associated to the vector $\underset{\sim}{X}$ is defined by

$$\underset{\sim}{F}(\underset{\sim}{X}) = - \frac{\partial E(\underset{\sim}{X})}{\partial \underset{\sim}{X}} . \qquad (1)$$

If $\underset{\sim}{X} = \underset{\sim}{X}^{(e)}$ is an equilibrium configuration of the system which corresponds to a minimum of the potential energy, then, obviously, $\underset{\sim}{F}(\underset{\sim}{X}^{(e)}) = \underset{\sim}{O}$. On the other hand, when the system is not in equilibrium, but is closed to the equilibrium configuration, we may write in the first approximation

$$\underset{\sim}{F} = \underset{\sim}{M}(\underset{\sim}{X} - \underset{\sim}{X}^{(e)}), \tag{2}$$

where

$$\underset{\sim}{M} = \left.\frac{\partial \underset{\sim}{F}(\underset{\sim}{X})}{\partial \underset{\sim}{X}}\right|_{\underset{\sim}{X}=\underset{\sim}{X}^{(e)}} . \tag{3}$$

Clearly, (2) may be rewritten as

$$\underset{\sim}{X}^{(e)} = \underset{\sim}{X} - \underset{\sim}{M}^{-1} \underset{\sim}{F}. \tag{4}$$

Since $\underset{\sim}{M}$ depends on $\underset{\sim}{X}^{(e)}$, the determination of $X^{(e)}$ from (4) is generally done by successive approximations. Let $\underset{\sim}{X} = \underset{\sim}{X}_o$ be an initial configuration of the system. From (3) it results $\underset{\sim}{M}_o = (\partial \underset{\sim}{F}/\partial \underset{\sim}{X})_{\underset{\sim}{X}=\underset{\sim}{X}_o}$ and (4) yields $\underset{\sim}{F}_o = \underset{\sim}{F}(\underset{\sim}{X}_o)$. Introducing this result into (4) gives the first approximation, $\underset{\sim}{X}_1 = \underset{\sim}{X}_o - \underset{\sim}{M}^{-1}\underset{\sim}{F}_o$. If required, this iteration step can be repeated until the magnitude of $\underset{\sim}{F}$ becomes sufficiently small. Of course, the numerical inversion of the matrices $\underset{\sim}{M}_o, \underset{\sim}{M}_1, \ldots$, is very time-consuming. Therefore, this method is adequate only when merely small adjustments of the atomic configuration are required. In such cases, although the linear relation (2) does not hold exactly and (4) gives only a first approximation, it is usually possible to obtain convergence by repeating the application of (4) with $\underset{\sim}{M}$ replaced by $\underset{\sim}{M}_o$, even though its components change slightly (Sinclair [3]).

Another variant of the static relaxation, which proves to be advantageous when large adjustments of the atomic configuration are required, consists in releasing each atom, independently, to a provisional equilibrium position, dictated by the vanishing of the resultant force exerted on this one atom by its neighbours. Then, this step is repeated until the whole atomic array is covered, and the relaxation of the whole array is repeated until the changes in co-ordinates of all atoms between successive iterations and/or the residual forces are within preset limits (see, e.g. Chang and Graham [4]).

SINCLAIR'S METHOD (FLEX-S)

Consider a straight dislocation lying in an infinite anisotropic elastic medium, and take the positive direction of the dislocation line as x_3- axis of a Cartesian frame $\{\underset{\sim}{e}_1, \underset{\sim}{e}_2, \underset{\sim}{e}_3\}$. Choose as boundary Σ_o of region I, as before, a circular cylindrical surface of radius r_o and axis x_3. Let us denote by Γ_o the intersection line of Σ_o with the x_1x_2-plane, and by Δ the region outside Γ_o within this plane. The displacement vector field is assumed as independent of x_3, single-valued and of class C^2 in the region obtained from Δ by removing its points situated on the negative x_1-axis, and having across the cut $x_2 = 0$, $x_1 \leqslant -r_o$ a jump given by

$$\underset{\sim}{u}(x_1, 0^+) - \underset{\sim}{u}(x_1, 0^-) = -\underset{\sim}{b} \text{ for } x_1 \in (-\infty, r_o] , \tag{5}$$

where $\underset{\sim}{b}$ is the true Burgers vector of the dislocation. In addition, the displacement gradient must vanish at infinity and satisfy some traction boundary conditions on Σ_o . It has been shown by Eshelby, Read, and Shockley [5], that the linear elastic displacement field produced by the dislocation may be expressed in the general anisotropic case as

$$\underset{\sim}{u}(x_1, x_2) = \underset{\sim}{u}_V(x_1, x_2) + 2\mathrm{Re} \sum_{\alpha=1}^{3} \underset{\sim}{A}_\alpha \sum_{m=1}^{\infty} a_{m\alpha}(x_1 + p_\alpha x_2)^{-m}, \tag{6}$$

where

$$\underset{\sim}{u}_V(x_1,x_2) = \frac{1}{\pi}\sum_{\alpha=1}^{3} \underset{\sim}{A}_\alpha D_\alpha \ln[(x_1+p_\alpha x_2)/r_o] + \underset{\sim}{u}^o \qquad (7)$$

is the so-called Volterra solution, which gives the principal part of the displacement field at large distances from the dislocation line, the complex vectors $\underset{\sim}{A}_\alpha$ and scalars D_α, p_α are completely determined by the elastic constants, the Burgers vector, and the orientation of the dislocation line, while the coefficients $a_{m\alpha}$ depend on the initially unknown tractions acting on Σ_o from the dislocation core. Finally, $\underset{\sim}{u}^o$ is an arbitrary translation.

Sinclair [4] approximates the series (6) by the finite sum

$$\underset{\sim}{u}(x_1,x_2) = \underset{\sim}{u}_V(x_1,x_2) + 2\mathrm{Re}\sum_{\alpha=1}^{3} \underset{\sim}{A}_\alpha \sum_{m=1}^{\infty} a_{m\alpha}(x_1+p_\alpha x_2)^{-m} \qquad (8)$$

and interprets $a_{m\alpha}$ as 3n adjustable parameters to be determined together with the 2N co-ordinates x_{1i}, x_{2i} of the N atomic rows situated inside Σ_o, i.e. within region I. Thus, Sinclair's method consists in looking for the minimum of the potential energy

$$E = E(x_{1i}, x_{2i}, a_{m\alpha}), \qquad (9)$$

the 2N + 3m parameters x_{1i}, x_{2i}, and $a_{m\alpha}$ playing the role of generalized co-ordinates of the system. Sinclair has given the explicit expression of the components of the force gradient $\underset{\sim}{M}$ intervening in the static relaxation for the case of a two-body (central) interatomic potential. In fact, he used both variants of the static relaxation explained above, namely the individual atom relaxation at the beginning of the iteration, when large adjustments of the configuration are required, and the simultaneous relaxation of the whole atomic array for the final steps of the calculation.

By applying the above procedure to the ⟨100⟩ edge dislocation in α iron, Sinclair [4] has shown that the introduction of a flexible boundary between regions I and II may lead to a substantial computer-time saving. In particular, the results concerning the bond length in the centre of the dislocation core obtained by Gehlen, Rosenfield, and Hahn [6] on a rigid-boundary model with 780 atoms could be recovered by Sinclair on a flexible-boundary model with only 100 atoms in region I, which obviously means a considerable reduction of the required computation.

Sinclair's method, which is sometimes called Flex-S, made use originally of linear elasticity theory. However, a similar approach, which consists in improving the elastic solution by using second-order elasticity and satisfying the boundary conditions on Σ_o, has been proposed by Seeger as early as 1968, a brief outline of this method being given by Teodosiu and Nicolae [7].

The principle of the method is straightforward. The stress vector (or, alternatively, the displacement vector) on Γ_o is taken as a Fourier series of the polar angle with initially undetermined coefficients. Since higher harmonics lead to terms in the elastic solution that vanish rapidly with increasing distance from the dislocation core, it is in general sufficient to consider only the first two or three harmonics. The non-linear elastic solution is found by an iterative procedure involving the solution of a linear elastic boundary-value problem at each step. Then, the total potential energy is minimized as a function of the boundary conditions and of the position vectors of the atoms located in region I.

Let us consider, for example, the case of an edge dislocation lying along a two-fold symetry axis in an infinite anisotropic elastic medium. If we attempt to find out the most significant correction to the Volterra solution, then we should retain terms up to the order $0(\rho^{-1})$ in the expression of the displacement field and $0(\rho^{-2})$ in that of the stress field as $\rho = (x_1^2 + x_2^2)^{1/2} \to \infty$. The second-order elastic solution corresponding to this approximation has been derived by Seeger, Teodosiu, and Petrasch [8], and subsequently, in a corrected Eulerian formulation, by Petrasch [9] and by Teodosiu and Soós [10]. The u_3 displacement component vanishes, while the complex displacement $U = u_1 + iu_2$ is given by

$$U(x_1, x_2) = U_V(x_1, x_2) + U_N(x_1, x_2) + \sum_{\alpha} (\delta_\alpha A_\alpha / z_\alpha + \varrho_\alpha \bar{A}_\alpha / \bar{z}_\alpha), \tag{10}$$

where $U_V(x_1, x_2)$ is the Volterra solution; $z_\alpha = z + \tau_\alpha \bar{z}$, $\alpha = 1, 2$ with $z = x_1 + ix_2$; δ_α, ϱ_α, τ_α are complex constants depending on the second-order elastic constants, the Burgers vector, and the dislocation orientation; $U_N(x_1, x_2)$ is the non-linear elastic correction, which is a linear combination of terms having the form z_α^{-1}, $\bar{z}_\alpha z_\alpha^{-2}$, $z_\alpha^{-1} \ln z_\beta$, $z_\alpha^{-1} \ln \bar{z}_\beta$ $(\alpha, \beta = 1, 2)$ and of their complex conjugates, and whose coefficients depend on the second-and third-order elastic constants, the Burgers vector, and the dislocation orientation. Finally, A_1 and A_2 are arbitrary complex parameters.

By comparing (10) with (8), it is obvious that Sinclair's method can be applied exactly as before, the only difference being that U_N introduces a known non-linear contribution to the displacement field. In particular, (9) becomes

$$E = E(x_{1i}, x_{2i}, A_1, A_2), \qquad i = 1, 2, \ldots, N, \tag{11}$$

and hence the potential energy of the system depends on the 2N co-ordinates of the atoms located in region I and on the two adjustable complex parameters A_1, A_2.

The above method has been applied by Granzer, Belzner, Bücher, Petrasch, and Teodosiu [11] to the simulation of the ⟨110⟩ edge dislocation in NaCl. In particular, these authors have found a non-vanishing mean dilatation of the order of one atom volume per plane, as was to be expected. A different attempt to including second-order elastic effects in the Flex-S method without considering, however, the boundary conditions on Σ_0, has been developed by Bullough and Sinclair [12], by using Willis' solution [13] for the non-linear anisotropic elastic field of a screw dislocation.

THE FLEXIBLE-BOUNDARY METHOD WITH OVERLAPPING REGIONS (FLEX-I)

This method has been proposed in 1972, independently, by Teodosiu and Nicolae [7] and by Gehlen, Hirth, Hoagland, and Kanninen [14]. The basic idea of the method is to make regions I and II overlap. For a straight dislocation we may take for instance, the interior of a circular cylinder of radius r_1 as region I and the exterior of a circular cylinder of radius $r_0 < r_1$ as region II, both cylinders having the dislocation line as axis (Fig. 2).

Denote by Σ_0 and Σ_1 the circular cylindrical surfaces of radius r_0, respectively r_1. One can now use the following scheme of successive approximations. First, the static relaxation is performed for region I, keeping fixed the atoms situated on and outside Σ_1 in their positions given by the Volterra solution. The next step is to calculate the linear elastic solution for region II, by taking into account this time the displacement or the traction boundary conditions on Σ_0 as derived from the first step, by using some interpolation procedure. Then, one performs again the static relaxation of the atoms in region I, but keeping fixed the atoms situated on and outside Σ_1 in their positions resulted from the second step, and so on. The linear elastic solution satisfying prescribed boundary conditions on Σ_0 has been derived by Teodosiu and Nicolae [7] for an edge dislocation lying along a two-fold axis of material symmetry, and by Teodosiu, Nicolae, and Paven [15] for an arbitrary straight dislocation lying in an elastic medium with general anisotropy.

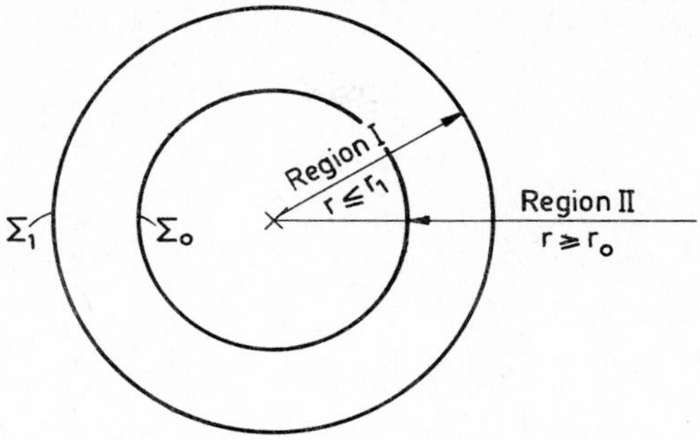

Figure 2
Concentric arrangement of regions for the flexible-boundary method with overlapping regions (Flex-I)

The method of the overlapping regions has been applied by Gehlen et al. [4], again to the ⟨100⟩ edge dislocation in α iron, by using, however, the linear isotropic elastic solution. An interesting remark made by these authors is that, when applied to a perfect lattice, the Flex-I method would yield small but non-zero atomic displacements. The explanation lies in the non-locality of the interaction potential, which prevents the forces exerted on region-II atoms from being replaced by a perfectly equivalent distribution of surface tractions acting on Σ_0. In order to remove this secondary effect, the tractions acting on Σ_0 must be calculated as differences between the actual interatomic forces and the forces acting across Σ_0 in the perfect lattice.

Clearly, the non-linear elastic solution presented above can be also used in conjunction with the Flex-I method. For instance, this can be done for an edge dislocation by starting from equation (10) and using the following iteration steps:

(i) Elastic computation of the initial atom positions in the whole array according to (10) with $A_1 = A_2 = 0$.

(ii) Static relaxation of the atoms in region I with atoms on and outside Σ_1 being kept fixed.

(iii) Determination of the adjustable complex parameters A_1, A_2 from the Fourier analysis of the atomic displacements on Σ_0.

(iv) Elastic recomputation of the atom positions in the whole array, this time according to (10).

(v) Repetition of steps (ii) - (iv) until the changes in the atom positions and in the values of the adjustable parameters between successive iterations lie within preset limits.

A procedure of this type has been employed by Petrasch and Belzner [16] to determine the core configuration of edge dislocations in NaCl and AgCl. In fact, they supplemented the non-linear elastic solution (10) by terms in z_α^{-2} and $\bar{z}_\alpha^{-2}$ with adjustable coefficients, which certainly leads to a better approximation of higher-order effects. Moreover, the use of three-body interatomic potentials, in addition to the two-body potentials corresponding to the Born-Mayer repulsion and to the Van der Waals attraction, allowed a much better fit to the elastic constants, which was beneficial, especially for AgCl. The results obtained after 13 cycles for NaCl and 9 cycles for AgCl show a significant improvement against the rigid-boundary methods as regards the continuity across Σ_0 of the residual forces exerted on the ions at the end of the iteration.

THE METHOD OF HOAGLAND, HIRTH, AND GEHLEN (FLEX-II)

This flexible-boundary method has been proposed by Hoagland [17] and developed by Hoagland, Hirth, and Gehlen [18]. Recently, Sinclair, Gehlen, Hoagland and Hirth [19] introduced several refinements of the method and extended it to allow the computation of the mean volume dilatation of a dislocated crystal. In what follows we give a brief description of the method in its latest form (Flex-II); for details the reader is referred to the original papers cited above.

In Flex-II three regions are explicitly considered around a straight dislocation. For illustration, these regions are represented in a concentric arrangement in Fig. 3, although their shape need not be circular. Like in the other flexible-boundary methods, the atoms in region I are relaxed individually, while the atoms in regions II, III, and in the remainder of the crystal are displaced collectively, according to linear elasticity theory. However, in Flex-II, the atoms of all three regions are supposed to interact via the same interatomic potential. Region II contains all atoms on which a force may be exerted by at least one region-I atom, while region III is that part of the remainder of the crystal whose atoms interact with region-II atoms. Clearly, the thickness of both regions II and III should equal the maximum range of the interatomic force law.

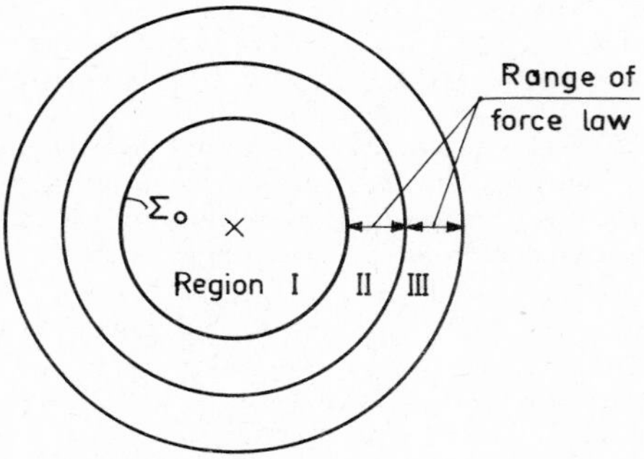

Figure 3
Concentric arrangement of regions for the flexible-boundary method of Hoagland, Hirth, and Gehlen (Flex-II)

The initial positions of the atoms throughout the crystal are usually chosen according to the Volterra solution. The first step is to relax the atoms of region I towards their equilibrium positions given by the minimum potential energy, keeping fixed the boundary Σ_0 between regions I and II. The second step is to calculate the forces acting on region-II atoms, and to determine the displacements produced by them in the whole crystal by using Green's tensor function of linear elasticity. These two steps are iterated until the residual force exerted on each atom in regions I and II is less than a preset limit.

Since the crystal outside region I is supposed to behave linearly elastic, no unequilibrated forces will develop in region III or beyond. Indeed, all atoms of these regions interact only with atoms that are displaced according to the equilibrium equations of linear elasticity. Moreover, the atom positions beyond region III need not be stored during the computation, for they can be calculated at the end of the iteration, by summing up the effects of the unequilibrated forces exerted on region-II atoms at all iteration steps.

Comparative studies of the Flex-II method with other flexible-boundary methods have shown that, in terms of computational efficiency for a given accuracy, Flex-II is superior. Consequently, the size of region I can be considerably reduced for the same level of accuracy. For instance, in the computation of Woo and Puls [20] region I has been reduced to a rectangle 4.5b x 6.5b, where b is the magnitude of the Burgers vector, while the boundary between regions II and III has been located only 7b away from the dislocation line. In exchange, the diminishing of the number of atoms in region I requires caution, since it implies the treatment by linear elasticity of regions relatively close to the dislocation line, where the strains are still of the order of a few percentages.

Sinclair, Gehlen, Hoagland, and Hirth [19] have recently used the Flex-II method for the calculation of the overall dilatation of a finite body due to a straight dislocation. In order to account for non-linear effects beyond region II, they have first calculated the distribution of quasi-body forces $\underset{\sim}{f}$ corresponding to the non-linear part of the second-order elastic constitutive equations in region III and beyond. Then, the supplementary dilatation produced by the moments of this force distribution have been added to the linear elastic dilatation. Clearly, this approach involves several severe approximations with respect to a fully non-linear analysis based on second-order elasticity. Even so, the analysis of Sinclair et al. has emphasized once more the importance of the non-linear elastic contribution to the overall dilatation produced by dislocations and has shown that, at least for the $\langle 100 \rangle$ edge dislocation in α iron and the crystal array investigated, this contribution is quantitatively comparable with that of the linear elastic displacements corresponding to the relaxation of the boundary between regions I and II.

An improvement of the non-linear version of Flex-II could be achieved by calculating the supplementary displacement field produced in the crystal by the quasi-body forces $\underset{\sim}{f}$ as a surface integral involving again Green's tensor function and extended to region III and beyond.

CONCLUSIONS

Non-linear elastic versions of the semidiscrete methods Flex-S and Flex-I require the rather complicated analytical solution of the non-linear boundary-value problem for the continuum part of the model. However, after determining the adjustable parameters occuring in the solution, this particularity becomes a major advantage, since the analytical expression of the elastic far-field includes both non-linear elastic and core effects. On the other hand, combining non-linear elasticity with Flex-II allows reducing the discrete part of the model to a relatively small number of atoms, by taking fully into account, however, the non-linear effects arising from the high strains at short and moderate distances from the dislocation core. This situation is most promising for the future handling of more complex problems, such as the defect-defect and boundary-defect interactions at small separation distances, as well as the simulation of dislocation loops and grain boundaries.

REFERENCES

[1] Huntington, H. B., Dickey, J. E., Thompson, R., Dislocation energies in NaCl, Phys. Rev. 100 (1955) 1117-1129.

[2] Duesbery, M. S., Vitek, V., Bowen, D. K., The effect of shear stress on the screw dislocation core structure in body-centred cubic lattices, Proc. R. Soc. London A332 (1973) 85-111.

[3] Sinclair, J. E., Improved atomistic model of a bcc dislocation core, J. Appl. Phys. 42 (1971) 5321-5329.

[4] Chang, R., Graham, L. T., Energy and atomic configuration of complete and dissociated dislocations, phys. stat. sol. 18 (1966) 99-103.

[5] Eshelby, J. D., Read, W. T., Shockley, W., Anisotropic elasticity with application to dislocation theory, Acta Met. 1 (1953), 251-259.

[6] Gehlen, P. C., Rosenfield, A. R., Hahn, G. T., Structure of the <100> edge dislocation in iron, J. Appl. Phys. 39 (1968) 5246-5254.

[7] Teodosiu, C., Nicolae, V., Edge dislocation in an infinite anisotropic elastic medium under consideration of the core conditions, Rev. Roum. Sci. Tech. -Méc. Appl. 17 (1972) 919-934.

[8] Seeger, A., Teodosiu, C., Petrasch, P., Second-order effects in the anisotropic elastic field of a straight edge dislocation, phys. stat. sol. (b) 67 (1975) 207-224.

[9] Petrasch, P., Doctor Thesis, University of Frankfurt am Main, 1978.

[10] Teodosiu, C., Soós, E., Non-linear effects in the elastic field of single dislocations, Rev. Roum. Sci. Tech. -Méc. Appl. 26 (1981) (in print).

[11] Granzer, F., Belzner, V., Bücher, M., Petrasch, P., Teodosiu, C., Atomistic calculations on edge dislocations in ionic crystals of rock salt structure, J. Physique (Colloque C9, suppl. no. 11-12) 34 (1973) C9 - 359-365.

[12] Bullough, R., Sinclair, J. F., A.E.R.E. Progress Report TP29 (1974).

[13] Willis, J. R., Second-order effects of dislocations in anisotropic crystals, Int. J. Engng. Sci. 5 (1967) 171-190.

[14] Gehlen, P. C., Hirth, J. P., Hoagland, R. G., Kanninen, M. F., A new representation of the strain field associated with the cube-edge dislocation in a model of α iron, J. Appl. Phys. 43 (1972) 3921-3933.

[15] Teodosiu, C., Nicolae, V., Paven, H., The influence of the core conditions on the linear anisotropic elastic field of a straight dislocation, phys. stat. sol. (a) 27 (1975) 191-204.

[16] Petrasch, P., Belzner, V., Elasto-atomistic coupling procedure to determine the core configuration of edge dislocations in ionic crystals, J. Physique (Colloque C7, suppl. no. 12) 37 (1976) C7 - 553-556.

[17] Hoagland, R. G., Ph. D. Thesis, Ohio State Univ. (1973).

[18] Hoagland, R. G., Hirth, J. P., Gehlen, P. C., Atomic simulation of the dislocation core structure and Peierls stress in alkhali halides, Phil. Mag. 34 (1976) 413-439.

[19] Sinclair, J. E., Gehlen, P. C., Hoagland, R. G., Hirth, J. P., Flexible boundary conditions and nonlinear geometric effects in atomic dislocation modeling, J. Appl. Phys. 49 (1978) 3890 - 3897.

[20] Woo, C. H., Puls, M. P., Atomistic breathing shell model calculations of dislocation core configurations in ionic crystals, Phil. Mag. 35 (1977) 727-756.

Part 7
LATTICE DEFECTS, GRANULAR MEDIA, NONLOCAL MEDIA

Continuum Models of Discrete Systems 4
eds. O. Brulin and R.K.T. Hsieh

BOUNDS ON NON-LOCAL EFFECTIVE CONSTITUTIVE LAWS FOR COMPOSITES

G. Diener and J. Weißbarth
Sektion Physik
Technische Universität Dresden
8027 Dresden, G.D.R.

Linear effective constitutive laws for static inhomogeneous mean fields in random media are considered. The random medium is described by local constitutive laws with stochastic material parameters. The effective parameters show dispersion, i.e. they depend on the "wave vector" $\underline{k}$ of the mean field. Upper and lower bounds for the permittivity $\varepsilon(k)$ and the Lamé's parameters $\lambda(k)$, $\mu(k)$ are derived. In particular, the Hashin-Shtrikman bounds are generalized to $k \neq 0$. Moreover, improved bounds for $\varepsilon(k)$ involving three-point correlations are given. For $k \to \infty$, the bounds converge to the exact values.

1. INTRODUCTION

In this paper we deal with linear constitutive laws in random media. In particular, our interest is focused on multicomponent mixtures. We start from a local constitutive law with randomly varying material parameters. As examples let us consider electrostatics and linear elasticity

$$\underline{D}(\underline{r}) = \varepsilon_{st}(\underline{r})\,\underline{E}(\underline{r}) \quad , \qquad \sigma(\underline{r}) = C_{st}(\underline{r})\,\varepsilon(\underline{r}). \tag{1.1}$$

Here, $\underline{E}$ and $\underline{D}$ mean the electric field and the displacement field, respectively, and ε_{st} is the dielectric permittivity. The Hooke's law contains the stress tensor σ, the strains ε and the stochastic Hooke's tensor C_{st}. The mean fields are related by so-called effective material parameters

$$\langle \underline{D} \rangle = \varepsilon \langle \underline{E} \rangle \ , \qquad \langle \sigma \rangle = C \langle \varepsilon \rangle. \tag{1.2}$$

If we define the mean fields as ensemble averages, they may depend on position. It is well-known that in this case the effective constitutive laws become non-local, i.g.

$$\langle \underline{D}(\underline{r}) \rangle = \int d^3\underline{r}' \ \varepsilon(\underline{r},\underline{r}') \langle \underline{E}(\underline{r}') \rangle . \tag{1.3}$$

Further, if statistical homogeneity is assumed, a Fourier transformation reduces eq. (1.3) to

$$\langle \underline{D}(\underline{k}) \rangle = \varepsilon(k) \langle \underline{E}(\underline{k}) \rangle . \tag{1.4}$$

Analogous equations may be written down for other physical phenomena. Thus, the effective material parameters depend on $\underline{k}$, they show dispersion. The case of homogeneous mean fields corresponds to $\underline{k} = 0$.

Attempts to calculate an effective material parameter from statistical information must be based on the corresponding stochastic field equation, i.g.

$$L_{st}\, u = -\frac{\partial}{\partial \underline{r}} \cdot \varepsilon_{st}(\underline{r}) \frac{\partial}{\partial \underline{r}}\, u = q \quad , \quad \underline{E} = -\frac{\partial u}{\partial \underline{r}} \quad , \tag{1.5}$$

where q is a non-random source term. Because of statistical homogeneity assumed above, we have to consider an infinite body. The sources may be chosen in such a way that the fields vanish at infinity.

The effective operator for the mean field is defined by

$$L \langle u \rangle = -\frac{\partial}{\partial \underline{r}} \cdot \int d^3\underline{r}'\ \varepsilon\,(\underline{r}-\underline{r}') \frac{\partial}{\partial \underline{r}'} \langle u(\underline{r}') \rangle = q(\underline{r}) \tag{1.6}$$

or, in Fourier representation

$$L(\underline{k}) \langle u(\underline{k}) \rangle = \underline{k} \cdot \varepsilon\,(\underline{k})\, \underline{k} \langle u(\underline{k}) \rangle = q(\underline{k}). \tag{1.7}$$

The effective operator is related to the stochastic operator by the formal equation

$$L = \langle L_{st}{}^{-1} \rangle^{-1} . \tag{1.8}$$

The non-local behaviour of effective constitutive laws has been investigated in [1,2] by means of a perturbational treatment. A self-consistent calculation for $\varepsilon(k)$ is given in [3] . This approach has been extended to wave propagation in [4] .

On the other hand, variational methods have been very successful in determining bounds on the effective parameters for homogeneous mean fields $k = 0$ (see i.g. [5-10]). Our present purpose is to extend these powerful procedures to $k \neq 0$ in order to derive bounds for $\varepsilon(k)$ as well as for $C(k)$ and to compare them to the results of the self-consistent approach.

2. VARIATIONAL PRINCIPLES

Equations (1.5-1.8) are equivalent to the following variational principle [9,10]

$$\int d^3\underline{r}\ \langle \tilde{u}\, L_{st}\, \tilde{u} \rangle \geq \int d^3\underline{r}\ \langle \tilde{u} \rangle\, L \langle \tilde{u} \rangle \tag{2.1}$$

where $\tilde{u}$ is an arbitrary trial function. The inequality also holds for inhomogeneous mean fields $\langle \tilde{u} \rangle$ provided that the sources q are non-random. In order to prove (2.1), we must assume L_{st} to be self-adjoint and positive definite. In the special case of electrostatics eq. (2.1) takes the form

$$\int d^3\underline{r}\ \langle \underline{\tilde{E}}\ \varepsilon_{st}\ \underline{\tilde{E}} \rangle \geq \int d^3\underline{r}\ \langle \underline{\tilde{E}} \rangle\, \varepsilon \langle \underline{\tilde{E}} \rangle \ , \ \underline{\tilde{E}} = -\frac{\partial \tilde{u}}{\partial \underline{r}} \tag{2.2}$$

where ε means an integraloperator according to (1.3).

If we use a trial function $\tilde{u}(\underline{r}) = A \langle \tilde{u} \rangle$ with A being an arbitrary operator except $\langle A \rangle = 1$, we obtain after a Fourier-transformation

$$\underline{k} \cdot \varepsilon_+(\underline{k})\, \underline{k} := \left\langle \left(\frac{\partial}{\partial \underline{r}} A\, e^{i\underline{k}\cdot\underline{r}} \right)^* \varepsilon_{st}(\underline{r}) \frac{\partial}{\partial \underline{r}} A\, e^{i\underline{k}\cdot\underline{r}} \right\rangle \geq \underline{k}\, \varepsilon(\underline{k})\, \underline{k}. \quad (2.3)$$

For any A, $\langle A \rangle = 1$ this relation provides us an upper bound ε_+ on the effective permittivity $\varepsilon(\underline{k})$. Lower bounds on $\varepsilon(\underline{k})$ may be obtained from an analogous variational principle, where $\underline{E}$ and ε_{st} are replaced by $\underline{D}$ and $1/\varepsilon_{st}$, respectively.

$$\int d^3\underline{r} \, \langle \tilde{\underline{D}}\, \varepsilon_{st}^{-1}\, \tilde{\underline{D}} \rangle \geq \int d^3\underline{r}\, \langle \tilde{\underline{D}} \rangle\, \varepsilon^{-1} \langle \tilde{\underline{D}} \rangle, \quad \frac{\partial \tilde{\underline{D}}}{\partial \underline{r}} = \frac{\partial \langle \tilde{\underline{D}} \rangle}{\partial \underline{r}}. \quad (2.4)$$

In the elastic case similar variational principles may be formulated. But, due to the tensorial character, their exploitation is somewhat more complicated.

We restrict our considerations to the case that both the local and the overall properties are isotropic. Then, ε_{st} and $\varepsilon(\underline{k})$ are scalars. Moreover, the latter depends only on the absolute value of $\underline{k}$.

Hooke's tensors C_{st} and C reduce to the two stochastic Lamé's parameters $\mu_{st}(\underline{r})$, $\lambda_{st}(\underline{r})$ and to the effective parameters $\mu(k)$, $\lambda(k)$, respectively. In fact, the mean elastic energy may be written, in the Fourier representation, as

$$\frac{1}{2} \int d^3\underline{k} \;\; \mathrm{Tr} \langle \tilde{\varepsilon}(\underline{k}) \rangle^* C(\underline{k}) \langle \varepsilon(\underline{k}) \rangle = \frac{1}{2} \int d^3\underline{k} \left\{ 2\mu(k) \mathrm{Tr} \langle \tilde{\varepsilon} \rangle^* \langle \tilde{\varepsilon} \rangle + \lambda(k) \left| \langle \mathrm{Tr}\, \tilde{\varepsilon} \rangle \right|^2 \right\} \quad (2.5)$$

(see [2]). Then, from variational principles analogous to (2.2-2.4), bounds may be derived for $\mu(k)$ (if a mean shear deformation $\underline{k} \perp \langle \tilde{\underline{u}} \rangle$, $\mathrm{Tr} \langle \tilde{\varepsilon} \rangle = 0$ is assumed) and for $2\mu(k) + \lambda(k)$ (longitudinal mean field $\underline{k} \parallel \langle \tilde{\underline{u}} \rangle$)

$$\mu_+(k) \geq \mu(k) \geq \mu_-(k) \;;\; (2\mu + \lambda)_+ \geq 2\mu + \lambda \geq (2\mu + \lambda)_- \quad (2.6)$$

3. GENERALIZED HASHIN-SHTRIKMAN BOUNDS FOR SCALAR PARAMETERS

The simplest trial functions we may insert in eqs. (2.2) to (2.4) are the mean fields $\tilde{\underline{E}} = \langle \tilde{\underline{E}} \rangle$ (A = 1) and $\tilde{\underline{D}} = \langle \tilde{\underline{D}} \rangle$. They yield the famous Voigt- and Reuss bounds:

$$\langle \varepsilon_{st} \rangle \geq \varepsilon(k) \geq \langle 1/\varepsilon_{st} \rangle^{-1}. \quad (3.1)$$

Consequently, these bounds hold not only for k = 0, but also for arbitrary $\underline{k}$. In the case k = 0, more restricting bounds have been found by Hashin and Shtrikman [5.6]. According to [9] these bounds may be derived from the variational principles given in Section 2 by the aid of a so-called local approximation. The starting point is an exact, but formal expression for $\underline{E}$

$$\underline{E} = \left\{ 1 + \Gamma P (1 - \Gamma \delta\varepsilon)^{-1} \langle (1 - \Gamma \delta\varepsilon)^{-1} \rangle^{-1} \right\} \langle \tilde{\underline{E}} \rangle \quad (3.2)$$

where Γ is a Green's operator for a homogeneous reference medium ε_o; $\delta\varepsilon = \varepsilon_{st} - \varepsilon_o$ are the fluctuations and P is a projection operator which only retains the fluctuating part of the following expression. In the "local" ansatz, the Γ in the denominators is replaced by a number γ

$$A = \left\{ 1 + \Gamma P(1 - \gamma\delta\varepsilon)^{-1} \langle (1 - \gamma\delta\varepsilon)^{-1} \rangle^{-1} \right\}. \tag{3.3}$$

Insertion into (2.3) yields, among others, a term containing the three-point correlation function. This term may be omitted, if ε_o is chosen as $\varepsilon_o = \max \varepsilon_{st}$. The remaining terms may be minimized by an appropriate choice of γ. Of course, the optimal γ depends on k. The minimization may be carried out explicitly for arbitrary binary systems and for multiphase cell materials defined in [8]. The result is

$$\varepsilon_+(k) = \left\langle \frac{\varepsilon_{st}}{\varepsilon_o + I_2(k)\delta\varepsilon} \right\rangle \Big/ \left\langle \frac{1}{\varepsilon_o + I_2(k)\delta\varepsilon} \right\rangle, \quad \varepsilon_o = \max \varepsilon_{st}, \quad \delta\varepsilon = \varepsilon_{st} - \varepsilon_o. \tag{3.4}$$

An analogous procedure for (2.4) yields the lower bound

$$\varepsilon_-(k) = \left\langle \frac{\varepsilon_{st}}{\hat{\varepsilon}_o + I_2(k)\delta\hat{\varepsilon}} \right\rangle \Big/ \left\langle \frac{1}{\hat{\varepsilon}_o + I_2(k)\delta\hat{\varepsilon}} \right\rangle, \quad \hat{\varepsilon}_o = \min \varepsilon_{st}, \quad \delta\hat{\varepsilon} = \varepsilon_{st} - \hat{\varepsilon}_o. \tag{3.5}$$

The function $I_2(k)$ only depends on the geometry of the mixture but not on the material properties. It contains a two-point correlation function which, essentially, provides us a correlation length we need as a scaling parameter for k.

In the limit k = 0, the function I_2 becomes independent of the geometry: $I_2(0) = 1/3$, and the bounds (3.4), (3.5) agree with the usual Hashin-Shtrikman bounds [5,11]. In the opposite limit $k \to \infty$, the function I_2 tends to 1, and both bounds then coincide with the exact result [2,3]

$$\varepsilon_+(k = \infty) = \varepsilon_-(\infty) = \varepsilon(\infty) = \langle 1/\varepsilon_{st} \rangle^{-1}. \tag{3.6}$$

To illustrate expressions (3.4), (3.5), they have been evaluated for a special binary cell mixture consisting of spherical grains (radius R). In figs. 1 results are plotted for some characteristic values of the permittivities ε_1, ε_2 and the volume fractions v_1, v_2 of both components (dash-dotted lines). The spread is relatively small and diminishes with increasing k. Notice that moderate heterogeneties $\varepsilon_2/\varepsilon_1 = 2$ and 4 are considered, for which perturbational methods are not applicable.

4. BOUNDS INVOLVING THREE-POINT CORRELATIONS

Improved bounds may be obtained, if more information about the geometry is taken into account. In a next step we may include three-point correlation functions of $\delta\varepsilon$. There are two ways to do this. The first one, commonly used is based on a truncated perturbation series, i.g.

$$\underset{\sim}{\tilde{E}} = (1 + \Gamma P \delta\varepsilon \Lambda) \langle \underset{\sim}{\tilde{E}} \rangle \tag{4.1}$$

where an additional open parameter Λ is introduced [7,8,10]. The method may also be applied to $k \neq 0$. Again, the optimal value for the parameter $\Lambda(k)$ can be calculated for binary systems and cell materials. In this way, we obtain bounds of the following form

$$\varepsilon_+^{(3)}(k) = \langle \varepsilon_{st} \rangle - \frac{\langle \varepsilon'^2 \rangle^2 (I_2(k))^2}{\langle \varepsilon'^2 \rangle \langle \varepsilon_{st} \rangle I_2(k) + \langle \varepsilon'^3 \rangle I_3(k)}$$

$$\left[\varepsilon_-^{(3)}(k)\right]^{-1} = \langle \eta_{st} \rangle - \frac{\langle \eta'^2 \rangle^2 (1 - I_2(k))^2}{\langle \eta'^2 \rangle \langle \eta_{st} \rangle [1 - I_2(k)] + \langle \eta'^3 \rangle [1 - 2I_2(k) + I_3(k)]} \tag{4.2}$$

$$\varepsilon' := \varepsilon_{st} - \langle \varepsilon_{st} \rangle, \qquad \eta_{st} := 1/\varepsilon_{st}, \qquad \eta' := \eta_{st} - \langle \eta_{st} \rangle.$$

The function $I_3(k)$ is a functional of the three-point correlation but again depends only on the geometrical structure of the phase mixture. It reflects some information about the cell shapes.

Another way to derive bounds including three-point correlations is to use the "local" approximation (3.3) and, contrarily to section 3, to retain the three-point correlation term in the variational principle (2.3). Then, the most convenient choices for ε_o are $\langle \varepsilon_{st} \rangle$ (upper bound) and $\langle 1/\varepsilon_{st} \rangle^{-1}$ (lower bound), respectively. A minimization with respect to the open parameter $\gamma(k)$ yields for binary mixtures as well as for multiphase cell materials

$$\varepsilon_+^{(\ell)}(k) = \langle \varepsilon_{st} \rangle + \frac{(I_2(k))^2}{I_3(k)} \left\langle \frac{\varepsilon'}{\langle \varepsilon_{st} \rangle I_2 + \varepsilon' I_3} \right\rangle \left\langle \frac{1}{\langle \varepsilon_{st} \rangle I_2 + \varepsilon' I_3} \right\rangle^{-1}$$

$$\left[\varepsilon_-^{(\ell)}(k)\right]^{-1} = \langle \eta_{st} \rangle + \frac{(1 - I_2)^2}{1 - 2I_2 + I_3} \left\langle \frac{\eta'}{\langle \eta_{st} \rangle (1 - I_2) + \eta'(1 - 2I_2 + I_3)} \right\rangle \tag{4.3}$$

$$\times \left\langle \frac{1}{\langle \eta_{st} \rangle (1 - I_2) + \eta'(1 - 2I_2 + I_3)} \right\rangle^{-1}.$$

For arbitrary binary mixtures these bounds are identical with those given in (4.2). They take the form

$$\varepsilon_+^{(\ell)}(k) = (v_1\varepsilon_1 + v_2\varepsilon_2) - \frac{(\varepsilon_2 - \varepsilon_1)^2 v_1 v_2 (I_2(k))^2}{(v_1\varepsilon_1 + v_2\varepsilon_2) I_2(k) + (\varepsilon_2 - \varepsilon_1)(v_1 - v_2) I_3(k)}$$

$$\left[\varepsilon_-^{(\ell)}(k)\right]^{-1} = (v_1\eta_1 + v_2\eta_2) - \frac{(\eta_2 - \eta_1)^2 v_1 v_2 (1 - I_2)^2}{(v_1\eta_1 + v_2\eta_2)(1 - I_2) + (\eta_2 - \eta_1)(v_1 - v_2)(1 - 2I_2 + I_3)} \tag{4.4}$$

where $\varepsilon_i = 1/\eta_i$ and v_i are the permittivity and the volume fraction of the i-th component, respectively.

In the limit $k \to \infty$ ($I_2, I_3 \to 1$) expressions (4.3) coincide with the exact value (3.6). In the limit $k \to 0$ a dependence on the geometry remains because of I_3. In the case of multiphase cell mixtures the bounds (4.3) turn out to lie always within the bounds given by (4.2), although the statistical information involved is the same. In particular, for $k \to \infty$ the bounds (4.3) always yield the exact values, whereas in (4.2) only the lower bound always agrees with the exact result.

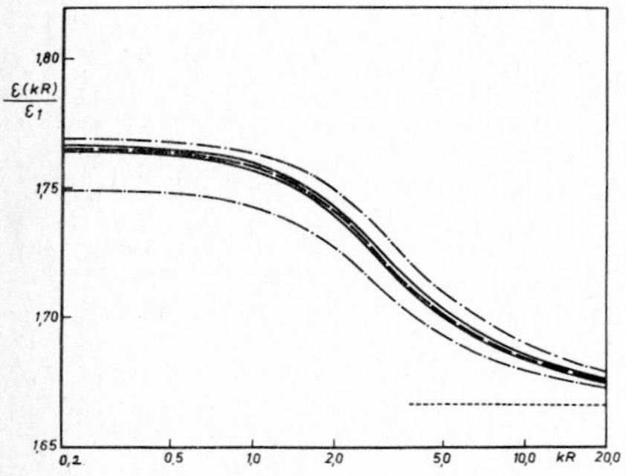

$v_1=0.2$, $v_2=0.8$, $\varepsilon_2/\varepsilon_1=2$.

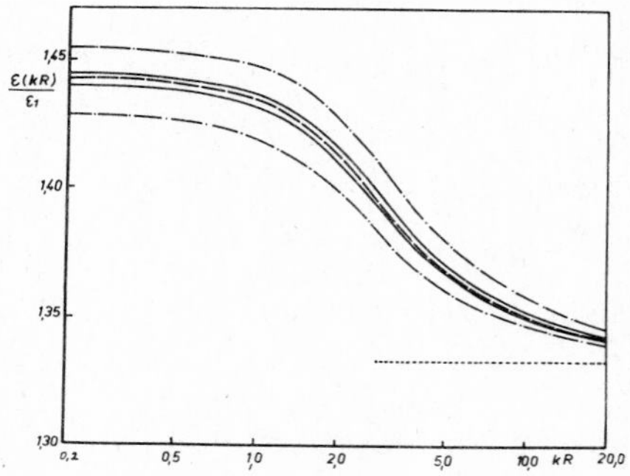

$v_1=v_2=0.5$, $\varepsilon_2/\varepsilon_1=2$.

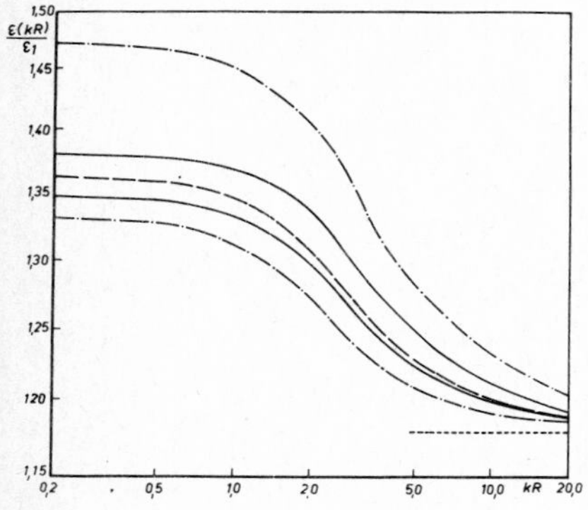

$v_1=0.8$, $v_2=0.2$, $\varepsilon_2/\varepsilon_1=4$.

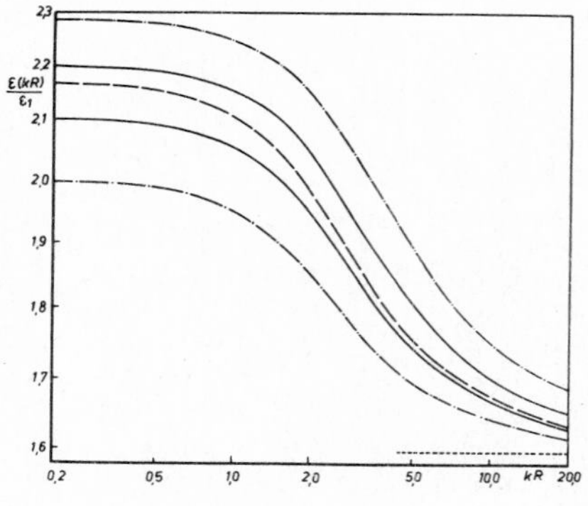

$v_1=v_2=0.5$, $\varepsilon_2/\varepsilon_1=4$.

Fig. 1. Effective permittivity for binary cell mixtures composed of nearly spherical grains with radius R.
-.-.-.- Generalized Hashin-Shtrikman bounds (3.4), (3.5).
——— Bounds including three-point correlation (4.3).
- - - - Self-consistent solution according [3]
...... Asymptotic value for $k = \infty$

Also for $k \to 0$ the bounds (4.3) improve the results (4.2) first derived in [7]. Hence, it is always reasonable to give preference to the bounds (4.3) instead of (4.2).

In comparison with the Hashin-Shtrikman bounds (3.4), (3.5), a considerable improvement is achieved by including three-point correlations according to (4.3). This fact is illustrated in Figs. 1 (solid lines). For a moderate heterogeneity $\varepsilon_2/\varepsilon_1 = 2$ of the two-component mixture, the upper and lower bounds nearly coincide. For comparison, the results of a self-consistent treatment [3] are also indicated (dashed lines). They lie between the bounds (4.4). This confirms the reliability of the self-consistent approach proposed in [3].

5. THE ELASTIC CASE

As quoted in section 2, bounds on the effective Lamé's parameters $\mu(k)$ and $2\mu(k) + \lambda(k)$ may be derived analogously to sections 3 and 4. However, if we use a "local" ansatz according to (3.3), the variational parameter γ must be replaced, in principle, by a tensor of fourth rank. Fortunately, for a shear deformation (bounds on μ) it reduces to a number and, for the longitudinal case, to a 2 x 2 matrix.

In order to construct upper bounds analogously to section 3, we have to choose a homogeneous reference material with the shear modulus $\mu_o = \max \mu_{st}$ and the bulk modulus $K_o = \max K_{st}$. To get a lower bound, we have to put $\hat{\mu}_o = \min \mu_{st}$, $\hat{K}_o = \min K_{st}$. The bounds obtained for μ in this manner are

$$\mu_+(k) = \left\langle \frac{\mu_{st}}{\mu_o + \alpha(k)\delta\mu} \right\rangle \Big/ \left\langle \frac{1}{\mu_o + \alpha(k)\delta\mu} \right\rangle \tag{5.1}$$

$$\mu_-(k) = \left\langle \frac{\mu_{st}}{\hat{\mu}_o + \hat{\alpha}(k)\delta\hat{\mu}} \right\rangle \Big/ \left\langle \frac{1}{\hat{\mu}_o + \hat{\alpha}(k)\delta\hat{\mu}} \right\rangle$$

where $\quad \overset{(\wedge)}{\alpha} = \frac{1}{2}\left(1 + I_2(k) - 4\,\dfrac{\overset{(\wedge)}{\mu}_o + 3\overset{(\wedge)}{K}_o}{4\overset{(\wedge)}{\mu}_o + 3\overset{(\wedge)}{K}_o}\,J_2(k)\right).$

Here, a new functional $J_2(k)$ of the two-point correlation occurs. It varies between the limiting values $J_2(0) = 2/15$ and $J_2(\infty) = 0$ according to the geometry of the mixture. The bounds for $2\mu(k) + \lambda(k)$ involve the same functions but exhibit very complicated forms. Therefore, they are not reproduced here.

For $k \to \infty$ the bounds yield again the exact values

$$k = \infty: \quad \mu_+ = \mu_- = \mu = \langle 1/\mu_{st} \rangle^{-1}$$
$$(2\mu+\lambda)_+ = (2\mu+\lambda)_- = 2\mu + \lambda = \langle (2\mu_{st} + \lambda_{st})^{-1} \rangle^{-1}. \tag{5.2}$$

For $k = 0$ we come back to the Hashin-Shtrikman bounds for μ and for $2\mu + \lambda = 4\mu/3 + K$ [6]. Therefore, eq. (5.1) may be considered as a generalization of the Hashin-Shtrikman bounds to $k \neq 0$.

Some numerical results for a binary cell mixture consisting of spherical grains are shown in Figs. 2. Moderate and strong

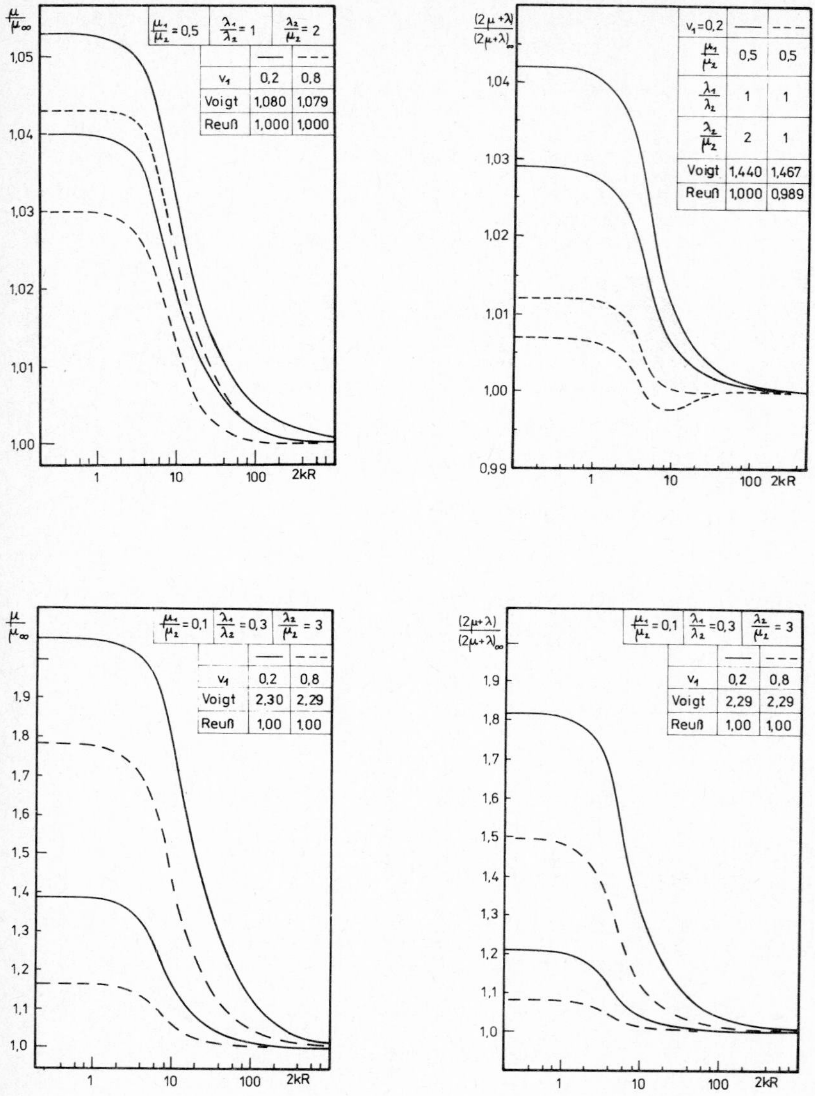

Fig. 2. Generalized Hashin-Shtrikman bounds for the effective Lamé's parameters of binary cell mixtures (spherical grains, radius R; μ_i, λ_i-Lamé's parameters of the components). The values of the Voigt- and Reuss bounds are also indicated.

heterogenities μ_2/μ_1 = 2 and 10, λ_2/λ_1 = 1,2,10 have been chosen. The dependence on k is not very strong. Of course the bound spread increases with increasing heterogeneity. The bounds converge for increasing k. All curves are normalized to the asymptotic values for k = ∞.

The results scetched in sections 3 to 5 are derived and discussed in detail in three consecutive papers submitted to J.Mech.Phys. Solids.

Of course, it would be desirable to extend the treatment to the dynamical case (wave propagation). Unfortunately, such an extension is not straightforward because the corresponding wave operators are not positive definite.

REFERENCES

[1] Beran, M.J. and McCoy, J.J., Mean field variation in random media, Q.Appl. Math. 28 (1970) 245.

[2] Beran, J.J. and McCoy, J.J., Mean field variations in a statistical sample of heterogeneous linearly elastic solids, Int.J.Solids Structures 6 (1970) 1035.

[3] Diener, G. and Käseberg, F., Effective linear response in strongly heterogeneous media - self-consistent approach, Int.J.Solids Structures 12 (1976) 173.

[4] Diener, G. and Budde, H.-H., Wave propagation in strongly heterogeneous media, Arch.Mech. 32 (1980) 59.

[5] Hashin, Z. and Shtrikman, S., A variational approach to the theory of effective magnetic permeability of multiphase materials, J.Appl.Phys. 33 (1962) 3125.

[6] Hashin, Z. and Shtrikman, S., A variational approach to the theory of the elastic behaviour of multiphase materials, J.Mech.Phys.Solids 11 (1963) 127.

[7] Beran, M.J., Use of the variational approach to determine bounds for the effective permittivity in random media, Nuovo Cimento 38 (1965) 771.

[8] Miller, M.N., Bounds for effective electrical, thermal, and magnetic properties of heterogeneous materials, J.Math.Phys. 10 (1969) 1988.

[9] Dederichs, P.H. and Zeller, R., Variational treatment of the elastic constants of disordered materials, Z.Physik 259 (1973) 103.

[10] Kröner, E. and Koch, H., Effective properties of disordered materials, Sol.Mech.Arch. 1 (1976) 183.

[11] Willis, J.R., Bounds and self-consistent estimates for the overall properties of anisotropic composites, J.Mech.Phys. Solids 25 (1977) 185.

Continuum Models of Discrete Systems 4
eds. O. Brulin and R.K.T. Hsieh

A POINT SOURCE OF HEAT IN A COMPOSITE MATERIAL

D. J. Jeffrey
Department of Applied Mathematics and Theoretical Physics,
University of Cambridge,
Silver St, Cambridge CB3 9EW

A point source of heat is placed in a composite material that consists of spheres of conductivity λ_2 embedded in a matrix of conductivity λ_1. The ensemble averages of the temperature field and the heat flux are calculated over all realisations of the composite. We consider two independent asymptotic limits, the limit of large distance from the source and the limit of small volume fraction of the particles. The results obtained show the first non-local effects of particle interactions.

INTRODUCTION

When this series of conferences started in 1975, those of us interested in particle interactions in particulate composites were busy calculating the effective properties of a suspension placed in a uniform field, be it temperature field, flow field or strain field. This type of calculation had already collected a variety of approaches and a long bibliography, in spite of being a special case of the more general problem of the response of a composite medium to a non-uniform applied field. One of the attractions of any special case is that it allows some particular difficulty in the general case to be bypassed; in this instance, the difficulty was the presence of an interaction between a large length scale L, which might be imposed, say, by outer boundaries or curvature of the applied field, and the small length scale a of the inhomogeneities in the medium. It turned out, however, that if one tried to calculate effective properties by summing interactions between particles, one discovered that the difficulty had not been completely bypassed (Jeffrey 1978, 1980), because non-convergent integrals appeared, and a grudging acknowledgement had to be made of the existence of some large length scale L. In the 1977 meeting, Goddard (1978) suggested that Kellogg's (1953, pp.206-211) approach of considering the flux from a point source might circumvent the divergence difficulties. The work reported here shows what happens when this program is undertaken, but that is not the justification of the present investigation, which is concerned not with looking back to effective-property calculations, but with moving on to cases in which the externally imposed field is non-uniform, that is, the length scale L is allowed to play some part in the response of the medium.

There are at least two simple ways in which to generate a non-uniform applied field. One way is to impose a periodic field, perhaps a wave-like disturbance - the periodicity then defines L; the other is to place a point source in the medium. In this latter case we can consider the field at large distances from the source and expand the field locally around some distant point in a Taylor series. The Taylor expansion sounds a simple idea to implement, but, as we shall see below, there are some snags that must be recognised if one is not to arouse dormant non-convergent integrals only just put to sleep in effective-property calculations (Jeffrey 1978).

Among the antecedents of the present paper, two publications can be singled out as bearing particularly on this one. Beran & McCoy (1970) considered the non-uniform temperature field in a composite with a general structure. They derived an equation for the mean field that was non-local - it was an integro-differential equation - and examined the kernel in the integral for the particular case in which the properties of the composite varied only weakly. The other paper is by Diener & Käseberg (1976) who derived a non-local equation for the mean field by using a periodic non-random source. The remarkable prediction they made was that the length scale in the kernel could be much longer than the 'obvious' length scale provided by the micro-structure, the length a. The point-source approach and the periodic-source approach are complementary, although we shall not demonstrate it here, being Fourier transforms of each other.

GOVERNING EQUATIONS

We consider a composite material cosisting of a matrix of conductivity λ_1 and embedded spherical particles, each with conductivity λ_2 and radius a. The volume fraction of the spheres will be c. We place a point source of heat of strength $4\pi Q$ in the medium and choose it as our origin. At any point $\underset{\sim}{x}$ the equations governing the heat flux $\underset{\sim}{q}$ and the temperature T are

$$\nabla \cdot \underset{\sim}{q} = -4\pi Q \delta(\underset{\sim}{x}) \quad \text{and} \quad \underset{\sim}{q} = \lambda(\underset{\sim}{x}) \nabla T \ .$$

Define a function $H(\underset{\sim}{x})$ that equals 1 inside a particle and 0 outside, then the conductivity can be written

$$\lambda(\underset{\sim}{x}) = \lambda_1 + (\lambda_2 - \lambda_1) H(\underset{\sim}{x}) \ .$$

Taking the ensemble averages of these equations, we obtain

$$\nabla \cdot \langle \underset{\sim}{q} \rangle = -4\pi Q \delta(\underset{\sim}{x}) \quad \text{and} \quad \langle \underset{\sim}{q} \rangle = \langle \lambda(\underset{\sim}{x}) \nabla T \rangle \ .$$

From these we see that $\langle \underset{\sim}{q} \rangle = -Q\underset{\sim}{x}/x^3$ and is unchanged from the non-random case. The other equation becomes

$$\langle \nabla T \rangle = \langle (1-H)\lambda_1^{-1}\underset{\sim}{q} \rangle + \langle H\lambda_2^{-1}\underset{\sim}{q} \rangle = \lambda_1^{-1}\langle \underset{\sim}{q} \rangle + (\lambda_2^{-1} - \lambda_1^{-1})\langle H\underset{\sim}{q} \rangle \ .$$

The calculation is thus reduced to calculating $\langle H\underset{\sim}{q} \rangle$, which is to say, the average of $\underset{\sim}{q}(\underset{\sim}{x})$ over those cases in which a particle overlaps $\underset{\sim}{x}$.

TRANSFORMATION OF HARMONICS

One of the main operations we shall use in the following calculations is the transformation of spherical harmonics from one origin to another. It is therefore convenient to gather the various results together in this section for later use. We consider the triangle ABC shown in figure 1.

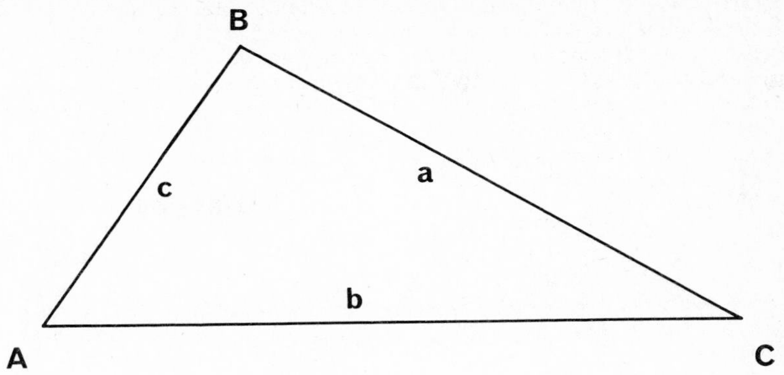

Figure 1. Transformation of harmonics from A to B.

Let A and B be two origins of spherical polar co-ordinates, each set of co-ordinates having the pole AB. They also share the same azimuthal angle ϕ. This choice of angles gives the maximum symmetry to the relations below, because they have the same form whether one goes from A to B or vice versa, but it means that not both co-ordinate systems are right-handed. Handedness is not important here (because there are no vector products), but in other applications it might be more convenient to use π-B instead of B, if a change in handedness is more inconvenient than lack of algebraic symmetry. Let harmonics at A be of the forms

$$b^{-n-1}Y_{mn}(\cos A,\phi) \quad \text{or} \quad b^{n}Y_{mn}(\cos A,\phi) \;,$$

where $Y_{mn}(\cos A,\phi)=P_n^m(\cos A)\exp(im\phi)$. Then (Morse & Feshbach 1953, pp. 1271-1272; Hobson 1931, §89)

(a) for all b and c,

$$a^n\, Y_{mn}(\cos B,\phi) = c^n \sum_{s=m}^{n} (-1)^{s+m} \binom{n+m}{n-s} \left(\frac{b}{c}\right)^s Y_{ms}(\cos A,\phi)$$

(b) for b>c,

$$a^{-n-1}\, Y_{mn}(\cos B,\phi) = (-1)^{n+m}\, c^{-n-1} \sum_{s=n}^{\infty} \binom{s-m}{n-m} \left(\frac{c}{b}\right)^{s+1} Y_{ms}(\cos A,\phi).$$

(c) for b<c,

$$a^{-n-1}\, Y_{mn}(\cos B,\phi) = c^{-n-1} \sum_{s=m}^{\infty} \binom{n+s}{s+m} \left(\frac{b}{c}\right)^{s} Y_{ms}(\cos A,\phi).$$

Corresponding formulae using the angle π-B are obtained from the relation $Y_{mn}(\cos(\pi-B),\phi)=(-1)^{n+m}Y_{mn}(\cos B,\phi)$.

POINT SOURCE AND ONE SPHERE

The basic building block for our calculations is the temperature field around a solitary sphere induced by a point source of heat. Stratton

(1941, §3.23) gives the solution to this problem when the source is outside the sphere and using the results of the last section, one can easily derive the solution for the source inside. From now on, we shall non-dimensionalize all lengths with respect to a and the source Q with respect to a and λ_1; thus the radius of a sphere is now 1. The geometry of the problem is shown in figure 2.

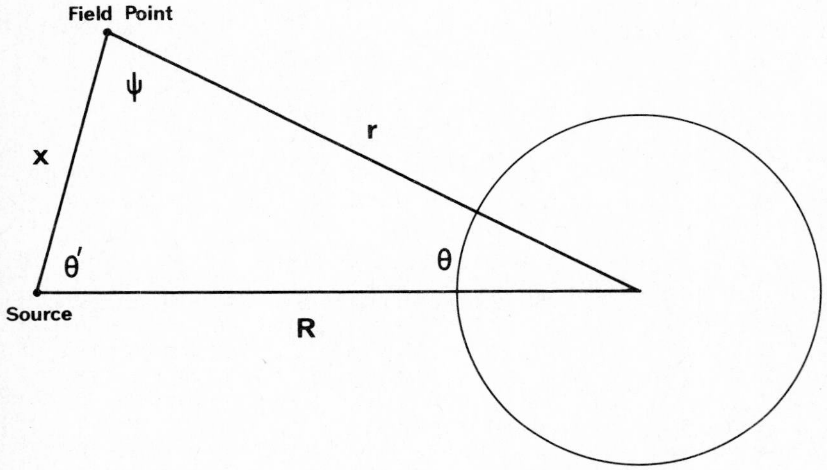

Figure 2. Co-ordinate systems for sphere and source.

The field point is a distance x from the source; the sphere centre is a distance R; and spherical polar co-ordinates are taken at the sphere centre as shown. There are two cases.

(a) Source outside sphere, i.e. R>1.

The temperature field outside the sphere (r>1) is

$$T_o = Q/x - Q\sum_{n=1}^{\infty} n\beta_n R^{-n-1} r^{-n-1} P_n(\cos\theta) .$$

The temperature inside is

$$T_i = Q\sum_{n=0}^{\infty} \gamma_n R^{-n-1} r^n P_n(\cos\theta).$$

Here $\beta_n=(\alpha-1)/(n\alpha+n+1)$, $\gamma_n=(2n+1)/(n\alpha+n+1)$, and $\alpha=\lambda_2/\lambda_1$.

(b) Source inside sphere, i.e. R<1.

The temperature outside the sphere is

$$T_o = Q\sum_{n=0}^{\infty} \gamma_n R^n r^{-n-1} P_n(\cos\theta).$$

The temperature inside is

$$T_i = Q/\alpha x + (Q/\alpha)\sum_{n=0}^{\infty} (n+1)\beta_n R^n r^n P_n(\cos\theta) .$$

MEAN TEMPERATURE TO O(c)

As is well known from effective-property calculations, we can obtain the mean of some quantity correct to O(c), where c is the volume fraction, by averaging that quantity in the presence of just one sphere. The surprising result that we derive in this section is that for all points $\underset{\sim}{x}$ with x>2, i.e. points such that any sphere overlapping $\underset{\sim}{x}$ (as it must do for the averaging) cannot overlap the source, the mean temperature field shows no effect of the non-uniformity of $\langle q\rangle$. We denote the temperature field at $\underset{\sim}{x}$ given that the sphere centre is at $\underset{\sim}{r}$ by $\nabla T(\underset{\sim}{x};\underset{\sim}{r})$. Then we require

$$\langle H\underset{\sim}{q}\rangle = \lambda_2\langle H\nabla T\rangle = \lambda_2\int_V \nabla T(\underset{\sim}{x};\underset{\sim}{r})\ p(\underset{\sim}{r})\ d\underset{\sim}{r}\ .$$

where the integration is over the volume V given by $|\underset{\sim}{r}-\underset{\sim}{x}|<1$ and $p(\underset{\sim}{r})$ is the probability density for the sphere being at $\underset{\sim}{r}$, which will be taken to be constant. Since by symmetry we expect $\langle H\nabla T\rangle$ to be in the $\underset{\sim}{x}$ direction, we wish to calculate (see figure 2)

$$\langle H\nabla T\rangle.\underset{\sim}{x}/x = p(\underset{\sim}{r})\int_0^1\int_{-1}^1(\cos\psi\,\partial T/\partial r + \sin\psi\ r^{-1}\partial T/\partial\theta)\ d(\cos\psi)2\pi r^2dr.$$

Using the solution for the temperature field and the transformation relations, we find that the integrand equals

$$2\pi Q\sum_{n=1}^{\infty}\ \sum_{s=n}^{\infty} n\gamma_n r^{-2}(-1)^n\binom{s}{n}\left(\frac{r}{x}\right)^{s+1}P_{s-1}(\cos\psi)\ ,$$

giving
$$\langle H\nabla T\rangle.\underset{\sim}{x}/x = -\tfrac{4}{3}\pi p(\underset{\sim}{r})Q\gamma_1 x^{-2}\ .$$

Thus
$$\langle\nabla T\rangle = \lambda_1^{-1}\langle q\rangle + (1-\alpha)\langle H\nabla T\rangle = (1-3\beta_1 c)\langle q\rangle,$$

and the composite behaves as a homogeneous material with an effective conductivity $\lambda_1(1+3\beta_1 c)$. This result is remarkable but negative in that there is no effect of curvature to this order, disappointing when one hopes to show something new. For this reason, we turn to two-sphere interactions.

MEAN TEMPERATURE TO $O(c^2,x^{-2})$

We now consider two spheres, one of which (call it sphere 1) overlaps a point $\underset{\sim}{x}$, where x>>1. We observe that, in the first approximation, when the spheres are close together, they see the field produced by the source as uniform, which leads us to expect a link with effective-property calculations. It is obvious that if we try to average simply over all positions of the non-overlapping sphere (sphere 2), we shall land ourselves with non-convergent integrals. Therefore, we must, as stated in the introduction, acknowledge the existence of the large length scale, and divide the interactions into those that must be treated exactly, by which we mean solving a problem which includes the source and both spheres, and those that can be approximated by the uniform-field solution.

We use the method of reflexions to extend the solution for a point source and one sphere to a point source and two spheres. Referring to figure 2, we take the sphere shown there to be the one not overlapping $\underset{\sim}{x}$. To order $R^{-2}r^{-2}$, the sphere responds to the source by acting as a dipole of strength $Q\beta_1$ (as can be seen by inspection of the exact solution). The dipole creates a temperature gradient in the neighbourhood of sphere

1 and the point $\underset{\sim}{x}$ equal to $\nabla(Q\beta_1 R^{-2}r^{-2}\cos\theta)$. Sphere 1 responds to the temperature gradient by gaining an internal uniform temperature gradient of size $3\beta_1\nabla(Q\beta_1 R^{-2}r^{-2}\cos\theta)$. When the two spheres are close together, this term is asymptotically equal to the one that causes the averaging integrals to be non-convergent. Thus our way is clear, we delete the offending r^{-3} interaction from the uniform-field approximation, and average instead the full non-local version of the same interaction.

The result of integrating the various terms, and including the result from the previous section, is

$$\langle H\nabla T\rangle = (-3\beta_1 c + 6\beta_1^2 c^2 + c^2\Sigma)Q\underset{\sim}{x}/x \ ,$$

where Σ stands for all the terms in the summation given in Jeffrey (1973, eqn 6.3). Substituting this into the expression for $\langle\nabla T\rangle$ and inverting, we obtain

$$\langle\underset{\sim}{q}\rangle = \lambda_1(1 + 3\beta_1 c + 3\beta_1^2 c^2 + c^2\Sigma)\ \langle\nabla T\rangle + O(x^{-3},c^2).$$

Thus the composite is again acting as a homogeneous medium, but this time only because the calculation was stopped at $O(x^{-3})$. The effective conductivity agrees with previous results, so we recover the previous work from the new point of view. In future work, the term $O(x^{-3},c^2)$ will be calculated for comparison with other, more general, theories of the response of a composite to a non-uniform applied field.

REFERENCES

[1] Beran, M.J. and McCoy, J.J., Mean field variation in random media, Quart. Appl. Math. 28 (1970) 245-258.

[2] Diener, G. and Käseberg, F., Effective linear response in strongly heterogeneous media - self-consistent approach, Int. J. Solids Structures, 12 (1976) 173-184.

[3] Goddard, J.D., Advances in the rheology of particulate dispersions, in: Provan, J.W. (ed.), Continuum Models of Discrete Systems (University of Waterloo Press, 1978).

[4] Hobson, E.W., The Theory of Spherical and Ellipsoidal Harmonics (Cambridge University Press, 1931).

[5] Jeffrey, D.J., Conduction through a random suspension of spheres, Proc. R. Soc. Lond. A335 (1973) 355-367.

[6] Jeffrey, D.J., The physical significance of non-convergent integrals in expressions for bulk quantities, in: Provan, J.W. (ed.), Continuum Models of Discrete Systems (University of Waterloo Press, 1978).

[7] Jeffrey, D.J., The averaged-equation approach to finding mean-field equations for a particulate random medium, in: Kröner, E. and Anthony, K.-H. (eds), Continuum Models of Discrete Systems (CMDS3) (University of Waterloo Press, 1980).

[8] Kellogg, O.D., Foundations of Potential Theory (Dover, New York, 1953).

[9] Morse, P.M. and Feshbach, H., Methods of Theoretical Physics (McGraw-Hill, New York, 1953).

[10] Stratton, J.A., Electromagnetic Theory (McGraw-Hill, New York, 1941).

Continuum Models of Discrete Systems 4
eds. O. Brulin and R.K.T. Hsieh

THE MEAN STRESS RESULTING FROM INTERPARTICLE COLLISIONS IN A RAPID GRANULAR SHEAR FLOW

J.T. Jenkins

Department of Theoretical and
Applied Mechanics
Cornell University
Ithaca, New York, U.S.A.

and

S.B. Savage

Department of Civil Engineering
and Applied Mechanics
McGill University
Montreal, Quebec, Canada

A continuum description of the stresses generated during the rapid shear flow of a granular material is developed by a consideration of interparticle collisions. A balance law for the average specific kinetic energy of the particle velocity fluctuations associated with the collisions is employed. This balance law, which is similar in form to that used in modeling fluid turbulence, relates the local rate of change of the fluctuation energy to its production by the mean flow, its dissipation into thermal energy, and the diffusive energy flux from neighboring points of the flow. It is indicated how these quantities, and the mean stress, may be calculated by an extension of the methods used by Savage and Jeffrey (J. Fluid Mech. 1981) to determine the collisional exchange of momentum in a granular shear flow.

INTRODUCTION

When a mass of granular material is rapidly sheared, interparticle collisions inevitably result. If the shear is maintained and the particles possess some degree of elasticity, then because of the collisions the particles develop a 'random' fluctuating velocity component in addition to the mean shearing motion. These random fluctuations are in some ways analogous to the molecular velocity fluctuations associated with the temperature of a dense fluid; but, there are important differences. In the molecular description of dense fluids, the mean shear usually causes only a slight perturbation to that average of the random velocity fluctuations which defines the thermal energy. In a granular flow, the velocity scales associated with the mean shear and the random fluctuations are more nearly equal. Because of dissipative mechanisms such as interparticle friction and inelasticity present in the granular flows, continued shearing motion is essential for the maintenance of the particle fluctuations. Without shear the fluctuations rapidly die out. Clearly, the balance law for the fluctuation kinetic energy must play an essential role in determining the behavior of granular shear flows; it is the main concern of the present paper.

We seek to describe rapid shearing motions of an idealized granular material consisting of identical hard spheres having a gaseous interstitial fluid. The inertial and viscous properties of the gas are assumed to be such that the momentum transfered from the gas to the grains is negligible compared to that transfered from grain to grain in a collision. By a description of the motion of such a material we mean the formulation of the initial-boundary value problem governing the behavior of the relevant continuum mechanical variables. Certainly the most relevant mechanical variable is the mean velocity. However, such a continuum formulation of what is essentially a discrete problem generally necessitates the identification of other continuum variables and the determination of how these and the mean velocity enter into the constitutive relations of the continuum theory.

Bagnold [1], in the first determination of this kind for granular materials focused his attention on the mean velocity and calculated the shear stress and normal stress necessary to maintain a homogeneous shearing flow by considering the transfer of momentum occurring in the collisions experienced by a particle moving with the mean velocity. Here, because both the frequency of collisions and the change of momentum parallel to and perpendicular to the flow are proportional to the shear rate, the magnitude of each component of the stress varies with the square of the shear rate. Bagnold's experiments [1] in his 'grain inertia' regime support this conclusion. McTigue [2,3] has recently elaborated upon Bagnold's calculation, and Savage [4] and Savage and Sayed [5] have presented additional experimental evidence confirming the main trends of his stress relations. However, Jenkins and Cowin [6] have pointed out that there exist shear flows for which the use of Bagnold's stress relations leads to predictions inconsistent with observations. For example, in the steady fully developed flow down a vertical channel the balance of momentum across the channel requires that the horizontal normal stress be constant. But with a symmetric velocity profile, Bagnold's relation predicts that the normal stress is zero at the channel centerline and increases towards the walls. Jenkins and Cowin attribute this failure of Bagnold's stress relations to the facts that they are purely local equations which depend upon mean shear rate and that the balance of fluctuation kinetic energy was ignored.[1]

Blinowski [7] and Ogawa [8] have recently provided continuum formulations of theories for flowing granular materials in which the velocity fluctuations are explicitly taken into account. Blinowski writes down local forms of the balance laws for mass, momentum, and energy and an equation of balance for the mean of that product of the velocity fluctuations that corresponds to the turbulent Reynold's stress. Ogawa, in addition to the standard local balances, proposes an energy balance for the specific kinetic energy associated with the velocity fluctuations. As in the kinetic theory of gases this energy is identified with a temperature. Such balance laws must be closed by the addition of constitutive relations for the fluxes and for the internal production of those quantities that are not conserved.

Here we shall adopt the additional balance law for specific kinetic energy proposed by Ogawa. We wish first to exploit the similarities between rapid shear flows of granular materials and the turbulent shear of an incompressible Newtonian fluid. In this we are guided by the direct modeling of the evolution of the turbulence energy as described in detail by Launder and Spalding [9]. From this we receive reinforcement for the utilization of the novel balance law and suggestions as to the detailed form of the constitutive relations that must be used in it. Secondly, we would like to exploit the analogy between a rapidly sheared fluctuating granular material and the hard sphere models used in the molecular theories of dense fluids (Croxton [10], Hansen and McDonald [11], Chapman and Cowling [12]) to calculated the transport and dissipative properties of the granular material. Ogawa, et al. [13], for example, using relative rough arguments of this type, have recently calculated the dependence of the mean stress and the dissipation of the fluctuation kinetic energy upon the density, strain rate, fluctuation kinetic energy, and parameters governing the imperfection of collisions. In a more careful and elaborate treatment which considers the details of the collision kinematics, Savage and Jeffrey [14] have numerically determined the dependence of the components of the stress upon the ratio of the square of the shear rate and the fluctuation kinetic energy for smooth, perfectly elastic grains. From calculations like these we can, in principle, determine the form of the entire constitutive theory. Of particular interest would be the form of the flux of fluctuation kinetic energy. This quantity we shall see is crucial to the theory if it is to describe the simplest shearing flows. As yet we have not performed the detailed calculations for such quantities, so here we merely employ the form suggested by turbulence modeling.

THEORY

In Ogawa's work [8] and in the equivalent specialization of Blinowski's results [7] described by Jenkins and Cowin [6], the balance laws for mass and linear momentum have, respectively, the standard forms

$$\dot{\rho} + \rho v_{i,i} = 0 \tag{1}$$

where ρ is the mean mass per unit volume, $\underline{v}$ is the mean velocity, and the superposed dot denotes the time derivation calculated with respect to the mean motion, and

$$\rho \dot{v}_i = T_{ik,k} + \rho f_i \tag{2}$$

where $\underline{T}$ is the stress tensor and $\underline{f}$ is the specific body force.

The specific kinetic energy k of the velocity fluctuations $\underline{v}'$ is half the trace of the second correlation tensor $\underline{K}$ of these fluctuations,

$$K_{ij} \equiv \langle v_i' v_j' \rangle \tag{3}$$

where the brackets denote an ensemble average. The additional balance law associated with the fluctuation kinetic energy k has the form

$$\rho \dot{k} = -q_{i,i} + T_{ij} D_{ij} - \gamma \tag{4}$$

where $\underline{q}$ is the flux of this energy, $\underline{D}$ is the symmetric part of the mean velocity gradients,

$$2D_{ij} = v_{i,j} + v_{j,i} \tag{5}$$

the term $tr(\underline{TD})$ is the rate of production of fluctuation kinetic energy by the mean motion, and γ is the rate of dissipation of fluctuation kinetic energy into heat.

Here we shall focus on the purely mechanical theory and not write down the local form of the energy equation describing the degradation of mean and fluctuating velocity into heat. Consequently, we ignore changes in the temperature of the individual grains resulting from the dissipation rate γ.

The balance laws (1), (2), and (4) must be completed by relating the stress, fluctuation energy flux, and the energy dissipation rate to, for example, the mean density, the mean velocity gradients, the fluctuation energy, and the spatial gradients of the fluctuation energy.

Using standard techniques in continuum mechanics we could write down general forms for these constitutive relations. We do not do this because we are not interested in the theory with greatest generality. We would prefer somehow to obtain the simplest theory capable of describing the important physical phenomena. However, if we had written down general constitutive relations we would have adopted forms of these that were invariant under rigid motions of the material. The corresponding energy equation for a turbulent fluid does possess this invariance [15], however Lumley [16] indicates why such objective balance laws and constitutive relations are, in general, not to be expected when modeling turbulence.

As part of a general exposition of models for turbulence, Launder and Spalding [9] discuss an equation for the fluctuation kinetic energy that is identical to equation (4). When the turbulent diffusion is supposed to be similar to molecular diffusion, the energy flux is taken to have the familiar form

$$q_i = -c_1 k_{,i} \quad , \tag{6}$$

where c_1 is a positive constant. When the energy dissipation is governed by the cascade of energy from larger to smaller eddies, the dissipation of energy is independent of the fluid viscosity and its form follows from dimensional arguments,

$$\gamma = c_2 \rho d^{-1} k^{3/2} \tag{7}$$

where c_2 is a constant, ρ is the density of the incompressible fluid, and d is the length-scale of the turbulence. Launder and Spalding also point out that in fluids it is often not sufficient to model the evolution of the energy fluctuations because the transport effects on the length-scale are as important. However, for rapidly sheared granular materials the length scale of the fluctuations can naturally be identified with the particle diameter; so, for identical particles, the length-scale is fixed. Because of this, we might anticipate that energy modeling could be more successful for granular materials than for turbulence. However, an additional complication in granular materials is the variation of the mean density, raising the question of how this variable density enters into the constitutive relations. In any event, the analog between turbulence and rapidly sheared granular material seems real and suggestive, and the kinetic energy of the velocity fluctuations in both theories seems to be as significant a kinematic and dynamic variable as the mean velocity.

This can be seen in other ways that are, perhaps, more convincing in that they involve the calculation of the constitutive relations for the stress, energy flux, and dissipation from considerations of momentum exchanged and energy transferred or lost in collisions. These methods exploit the analogy between the fluctuation kinetic energy of the granular material and the thermal energy of a gas of hard spheres.

In a calculation of this type Ogawa et al. [13], for example, determine the dissipation rate and the stress tensor for tacky, rough, imperfectly elastic spheres to be

$$\gamma = -\alpha_o \phi b^{-1} k^{3/2} \tag{8}$$

and

$$T_{ij} = \alpha_1 \phi k \delta_{ij} + \phi b k (\alpha_2 D_{ij} + \alpha_3 D_{\ell\ell} \delta_{ij}) \tag{9}$$

where b is the density dependent radius of the sphere on which collisions are assumed to take place, ϕ is a known function of the density, and the coefficients α are known functions of the coefficient of restitution, friction coefficient, and tackiness. Note that the form of the dissipation (8) determined by Ogawa, et al. is the same as that given in (7), with the appropriate identification of corresponding terms.

Ogawa et al. [13] do not calculate a form for the energy flux $\underline{q}$, nor is it clear from their rather idealized consideration of the collision process how they might do so. In this respect, their resulting continuum theory is seriously deficient. As Jenkins and Cowin [6] pointed out, if this flux is omitted, the theory is incapable of describing the gravity flow of a granular material in a vertical channel.

Savage and Jeffrey [14] have also recently proposed a method for calculating the stresses in a rapidly sheared granular material based upon the determination of the exchange of momentum in collisions. The relative position and velocity of two particles in a binary collision are determined by plausible probability density functions. For smooth, perfectly elastic spheres of radius a, their numerical calculations of the components of the stress tensor determined in this way may be fit by an equation of the form

$$T_{ij} = 16\rho a^2 g_o \left[\left(\frac{4}{35} - \frac{k}{8a^2 II} \right) II\delta_{ij} + \left(\frac{32}{35\pi} + \frac{3\sqrt{2}k^{1/2}}{10a\sqrt{\pi}|II|^{1/2}} \right) |II|^{1/2} D_{ij} - \right.$$
$$\left. - \frac{8}{35} D_{ik}D_{kj} \right] , \tag{10}$$

where g_o is a known function of ρ and $2II \equiv (\text{tr}\underline{D})^2 - \text{tr}(\underline{D}^2)$. Note that this approximate form of the stress calculated for collisions in which energy is conserved is much different than that calculated by Ogawa, et al. [13] for dissipative collisions. However, the difference in form is not due to the difference in the energetics of the collisions; but rather to the different approximations made in calculating the averages.

It is a relatively straight-forward matter to use the method of Savage and Jeffrey to treat rough, inelastic spheres and to calculate the stress and dissipation in this case. What is more important, it seems to be possible to extend this method in order to calculate the flux of fluctuation energy. When this is done, the entire set of constitutive relations will have been calculated from a detailed consideration of the collisions between rough, inelastic, spherical particles.

Until this calculation has been completed we will assume, for the purpose of discussing the features of the continuum theory that would result, that the energy flux is given by (6), the energy dissipation rate by (8), and the stress by, say, (10). Then the system of differential equations (1), (2) and (4) determine, in principle, the fields of mean density ρ, mean velocity $\underline{v}$, and fluctuation energy k, once the appropriate initial conditions and boundary conditions have been supplied.

For the mean velocity at rough boundaries, it appears plausible to adopt the standard no-slip condition; however, for smooth boundaries the appropriate form of a slip condition is not clear. When the flux of energy is proportional to the spatial gradient of energy, as we anticipate it to be in all cases of physical interest, the energy balance for the fluctuation is second order in the spatial derivatives of k, and a boundary condition for the fluctuation energy is required. When the boundary is smooth and perfectly elastic, there will be no flux of energy through the boundary. Likewise, we anticipate no energy flux through a free surface in a gravitational field. When the energy flux is given by (6), this is a familiar homogeneous boundary condition for the normal derivative of k. For inelastic and rough boundaries we anticipate that the flux of energy through the boundary will be a function of the incident energy. When the flux is proportional to the gradient of energy, this boundary condition is a generalization of Newton's law of cooling.

We have, then, the desired formulation of the initial-boundary value problem for rapidly sheared granular materials. Solutions of this problem should exhibit features observed in physical flows of these materials. We anticipate that the predictions of a theory with this structure will improve as the calculations of the constitutive relations from detailed consideration of the collisions are made more precise. However, we emphasize again that, unless the diffusion of fluctuation energy is included in such theories, the system of equations has solutions involving indeterminancies or physical impossibilities.

Finally, we should underline those features of the physical problem that we have tacitly ignored. We have neglected the kinetic energy associated with the rotations of individual particles, we have focused our attention on the isotropic part of the second correlation tensor of the velocity fluctuation tensor and said nothing about the evolution of the anisotropy of this measure of fluctuation velocity, and, perhaps the most serious omission, we have ignored any static forces of interaction between grains that may exist because of the possibility of enduring contacts being established, particularly in those regions of the flow where the shear rate is small and particle concentration is high.

ACKNOWLEDGEMENT

The authors are indebted to M. Sayed for helpful discussions. The work of S.B. Savage was supported by the Natural Sciences and Engineering Research Council Grant A3369.

FOOTNOTE

1. Another possible reason for the failure is that the flow is outside the grain inertia regime described by Bagnold's relations; sustained particle contacts may occur and quasi-static Coulomb stresses could be dominant in parts of the flow field.

REFERENCES

[1] Bagnold, R.A., Experiments on the gravity-free dispersion of large solid spheres in a Newtonian fluid under shear, Proc. Roy. Soc. A 225 (1954), 49-63.

[2] McTigue, D.F., A model for stresses in shear flow of granular materials, in: Cowin, S.C. and Satake, M. (eds.), Proceedings of the U.S.-Japan Seminar on Continuum Mechanics and Statistical Approaches in the Mechanics of Granular Materials (Gakujutsu Bunken Fukyu-Kai, Tokyo, 1978).

[3] McTigue, D.F., A Nonlinear Continuum Theory for Flowing Granular Materials, Ph.D. Thesis, Dept. of Geol., Stanford University (August 1979).

[4] Savage, S.B., Experiments on shear flows of cohesionless granular materials, in: Cowin, S.C. and Satake, M. (eds.) Proceedings of the U.S.-Japan Seminar on Continuum Mechanics and Statistical Approaches in the Mechanics of Granular Materials (Gakujutsu Bunken Fukyu-Kai, Tokyo, 1978).

[5] Savage, S.B. and Sayed, M., Experiments on dry cohesionless granular materials in an annular shear cell at high shear rates, EUROMECH 133 - Statics and Dynamics of Granular Materials, Oxford University, 14-17 July 1980.

[6] Jenkins, J.T. and Cowin, S.C., in: Cowin, S.C. (ed.) Mechanics Applied to the Transport of Bulk Materials, AMD 31 (Amer. Soc. Mech. Eng., New York, 1979).

[7] Blinowski, A., On the dynamic flow of granular media, Archives of Mechanics 30 (1978) 27-34.

[8] Ogawa, S., Multitemperature theory of granular materials in: Cowin, S.C and Satake, M. (eds.), Proceedings of the U.S.-Japan Seminar on Continuum Mechanics and Statistical Approaches in the Mechanics of Granular Materials (Gakujutsu Bunken Fukyu-kai, Tokyo, 1978).

[9] Launder, B.E. and Spalding, D.B., Lectures in Mathematical Models in Turbulence (Academic Press, London and New York, 1972).

[10] Croxton, C.A., Liquid State Physics (Cambridge University Press, Cambridge, 1974).

[11] Hansen, J.P. and McDonald, I.R., Theory of Simple Liquids (Academic Press, London and New York, 1976).

[12] Chapman, S., and Cowling, T.G., The Mathematical Theory of Non-Uniform Gases (Cambridge University Press, Cambridge, 1970).

[13] Ogawa, S., Umemura, A., Oshima, N., On the equations of fully fluidized granular materials, ZAMP 31 (1980) 483-493.

[14] Savage, S.B. and Jeffrey, D.J., The stress tensor in a granular flow at high shear rates, J. Fluid Mech. (in press).

[15] Speziale, C.G., Invariance of turbulent closure models, Phys. Fluids 22 (1979) 1033-1037.

[16] Lumley, J.L., Toward a turbulent constitutive relation, J. Fluid Mech. 41 (1970) 413-434.

Continuum Models of Discrete Systems 4
eds. O. Brulin and R.K.T. Hsieh

THE SOLUTION OF SOME LATTICE DEFECT PROBLEMS IN NONLOCAL ELASTICITY

I. Kovács and G. Vörös

Institute for General Physics
Eötvös Loránd University
Budapest
Hungary

A quantitative treatment of some lattice defect problems is given applying nonlocal elasticity. This theory makes it possible to solve a set of problems unsolvable in classical elasticity. The self-energy of a dilatation centre and its interaction with other defects such as another dilatation centre, spherical homogeneous volume defect and edge dislocation are discussed in detail. Finally the self-energy of a spherical homogeneous volume defect is calculated.

1. INTRODUCTION

Lattice defects are always present in crystalline solids. They change many properties of the crystal considerably. A quantitative treatment of their properties can be given by the continuum model of lattice defects /1-5/ . Previous calculations were carried out within the frame of linear classical elasticity which leaded, for example, to divergent interaction energies with decreasing distance between point defects and dislocations.

In a set of papers Eringen and Edelen /6/ and Eringen /7,8/ have developed a new theory of elasticity called nonlocal elasticity. This theory takes into account long range (nonlocal) interactions in the determination of the elastic stresses originating from a displacement field, U_ℓ, eliminating therefore the stress singularities which appear in classical elasticity. The theory was successfully applied to treat the stress field and self-energy of a screw /9/ and an edge /10/ dislocation.

Using the concept of elastic multipoles all the possible stationary lattice defects can be described in linear classical elasticity by surface distribution of elastic monopoles characterized by strength P_{jk} /11/. This theory has been generalized to nonlocal elasticity as well /12/.

The purpose of this paper is to calculate some quantitative properties of lattice defects applying nonlocal elasticity.

2. BASIC CONCEPTS

Our treatment will be restricted to linear, isotropic, elastic solids with classical elastic moduli tensor

$$C_{ik\ell m} = \lambda\delta_{ik}\delta_{\ell m} + \mu(\delta_{i\ell}\delta_{km} + \delta_{im}\delta_{k\ell}) . \quad (1)$$

where λ and μ are the Lame constants.

The displacement field, U_ℓ, of a monopole with strength P_{jk} placed at $\underline{r}'$ in an infinite medium is given by /11/

$$U_\ell(\underline{r}) = P_{jk}(\underline{r}')G_{\ell j,k'}(\underline{r}-\underline{r}') ,\qquad (2)$$

where the usual indical notation is used (k' means differentiation according to x'_k), and $G_{\ell j}(\underline{r}-\underline{r}')$ is the classical Green's function tensor satisfying /11/:

$$C_{ik\ell m}G_{\ell j,mk}(\underline{r}-\underline{r}') + \delta_{ij}\delta(\underline{r}-\underline{r}') = 0.\qquad (3)$$

Here $\delta(\underline{r}-\underline{r}')$ is the Dirac δ function and δ_{ij} is the Kronecker symbol.

Substituting the displacement (2) into the classical equilibrium condition given in the form

$$C_{ik\ell m}U_{\ell,mk} + f_i = 0,\qquad (4)$$

and using eq.(3) one can define a body force due to an elastic monopole in classical elasticity as

$$f_i(\underline{r}) = P_{ik}(\underline{r}')\ \delta_{,k}(\underline{r}-\underline{r}') .\qquad (5)$$

Using these results and specifying P_{ik} for lattice defects it is possible to derive all the important properties of them /11/.

To extend the theory for nonlocal elasticity let us first summarize its basic equations. According to Eringen /8,9/ the basic equations of linear isotropic, nonlocal elastic solids, for the static case and vanishing body forces are (with somewhat different denotations) as follows

$$t_{ik,k} = 0\qquad (6)$$

$$t_{ik} = \int\tau_{ik}(\underline{r}-\underline{r}')dV',\qquad (7)$$

$$\tau_{ik} = \alpha(\underline{r},\underline{r}')\ C_{ik\ell m}\ \varepsilon_{\ell m}(\underline{r}'),\qquad (8)$$

$$\varepsilon_{\ell m} = \tfrac{1}{2}(U_{\ell,m} + U_{m,\ell}).\qquad (9)$$

Here $\alpha(\underline{r}-\underline{r}')$ is a continuous function characterizing the nonlocal interactions,and

$$\int\alpha(\underline{r}-\underline{r}')dV' = 1.\qquad (10)$$

Clearly, if $\alpha=\delta(\underline{r}-\underline{r}')$ then eqs. (6) and (7) give the classical equilibrium condition. If $\alpha\neq\delta(\underline{r}-\underline{r}')$ then the above equations show that the stress, t_{ik}, at a point $\underline{r}$ is a function of strain at all points $\underline{r}'$ in the body having volume V.

Eringen has shown /9,10/ that eq. (6) is satisfied if and only if

$$\sigma_{ik,k} = 0,\quad \text{in } V,$$

where

$$\sigma_{ik} = C_{ik\ell m}\ \varepsilon_{\ell m}.$$

This statement is very important, because it means that the solution of eq.(6) can be given by using the classical solution of the same problem. In other words it means that the displacement field due to a certain problem remains the same in the nonlocal elasticity as in the classical one. A difference arises, however, in the stresses for the two cases because of the nonlocal interactions.

In the treatment of the properties of lattice defects the real body forces can be neglected. So an equilibrium condition of (6) must be solved as did by Eringen for the case of a screw and an edge dislocation /9,10/. However, in the general theory of static lattice defects we have introduced fictitious body forces in the way outlined above. This concept turned out to be very useful in the formulation of the different properties of the defects. To keep this concept for the nonlocal elasticity as well and to remain in accordance with the above statement of Eringen we have to solve the more general equilibrium condition of

$$t_{ik,k} + F_i = 0, \tag{11}$$

with the requirement that what body force, F_i, leads to the same displacement field as what is obtained from the classical equation

$$\sigma_{ik,k} + f_i = 0. \tag{12}$$

The solution of this equation is given by (for an infinite medium)

$$U_\ell(\underline{r}')= \int_V f_j(\underline{r}'') \, G_{\ell j}(\underline{r}'-\underline{r}'')dV'' , \tag{13}$$

where $G_{\ell j}$ satisfies eq. (3). Substituting this expression into eqs. (7)-(9) one obtains for the nonlocal stresses

$$t_{ik}(\underline{r}) = \iint C_{ik\ell m}\alpha(\underline{r}-\underline{r}') f_j(\underline{r}'') G_{\ell j,m}{}'(\underline{r}'-\underline{r}'')dV'dV'' . \tag{14}$$

From eq. (11) we get for the nonlocal body forces /12/

$$F_i(\underline{r}) = \int\alpha(\underline{r}-\underline{r}'') \, f_j(\underline{r}'') \, dV'' . \tag{15}$$

This expression gives the connection between body forces leading to the same displacement fields in local and nonlocal elastic medium.

Substituting the body force given by (5) into eq. (15) we get the body force due to an elastic monopole, in nonlocal elasticity as

$$F_j(\underline{r})= P_{jk}(\underline{r}')\alpha,_{k}{}'(\underline{r}-\underline{r}'), \tag{16}$$

where we used the properties of the Dirac δ function. The result shows the important fact that the effect of a monopole in nonlocal elasticity can be characterized by a nonsingular body force.

Let now an additional displacement field, U_i^A independent of the monopole, P_{jk}, also present in the nonlocal elastic medium. To produce this displacement field in the presence of the monopole additional work done against the forces F_j must also be supplied. This work done is due to the interaction energy, U, between the monopole and the displacement field U_i^A, which is given by /12/

$$U(\underline{r}') = - \int P_{jk}(\underline{r}') \, \alpha(\underline{r}''-\underline{r}') \, \varepsilon^A_{jk}(\underline{r}'')dV''. \tag{17}$$

This interaction energy depends not only on the position of the monopole in the strain field ε^A_{jk} as in the classical case, but because of the long range interactions it depends on all the strain values belong to the different points of the body. It seems, however, that if $\alpha=\delta(\underline{r}''-\underline{r}')$ then the classical result /11/

$$U(\underline{r}')=-P_{jk}\varepsilon^A_{jk} \, (\underline{r}') \tag{18}$$

is obtained.

The interaction force acting on the monopole can also be obtained from eq. (17). By definition

$$f_n=-U,_{n}{}' = \int P_{jk}(\underline{r}') \, \alpha(\underline{r}''-\underline{r}') \, \varepsilon^A_{jk,n}(\underline{r}'')dV . \tag{19}$$

3. THE PROPERTIES OF A DILATATION CENTRE

The displacement field of a dilatation centre placed at $\underline{r}'$ in an infinite medium is

$$U_\ell(\underline{r},\underline{r}') = \frac{c}{4\pi} \left(\frac{1}{R}\right),_{\ell} , \tag{20}$$

where $R=|\underline{r}-\underline{r}'|$ and c is the volume increment of the body containing the defect:

$$c = \int U_{\ell,\ell}(\underline{r},\underline{r}')dV . \tag{21}$$

The definition of a dilatation centre as a monopole can also given by its strength

$$P_{jk} = P\delta_{jk} , \tag{22}$$

where /12/:

$$P = -2\mu \frac{1-\nu}{1-2\nu} c . \tag{23}$$

The classical stress field can easily be obtained as

$$\sigma_{ik} = c\lambda\delta(\underline{r}-\underline{r}')\delta_{ik} - \mu\frac{c}{2\pi}\left(\frac{1}{R}\right)_{,ik} . \tag{24}$$

With the use of this expression from eq. (7) we get for the nonlocal stresses of a dilatation centre

$$t_{ik} = c\lambda\alpha(\underline{r}-\underline{r}')\delta_{ik} - \mu\frac{c}{2\pi}\int \alpha(\underline{r}-\underline{r}'')\left(\frac{1}{|\underline{r}''-\underline{r}'|}\right)_{,ik} dV''. \tag{25}$$

The hydrostatic stress field is

$$t_{\ell\ell} = 2\mu c \frac{1+\nu}{1-2\nu} \alpha(\underline{r}-\underline{r}') . \tag{26}$$

This result shows the important fact that, because of the nonlocal effects, the hydrostatic stress field of a dilatation centre differs from zero and remains finite everywhere in the material contrary to classical elasticity.

Instead of its strength, P_{ik}, a monopole can also be characterized by the tensor

$$a_{ik} = S_{ik\ell m} P_{\ell m} , \tag{27}$$

where

$$S_{ik\ell m} = -\frac{\nu}{2\mu(1+\nu)}\delta_{ik}\delta_{\ell m} + \frac{1}{4\mu}(\delta_{i\ell}\delta_{km} + \delta_{im}\delta_{k\ell})$$

is the stiffness tensor. Applying eqs. (22) and (23) we get for a dilatation centre

$$a_{ik} = -\frac{1-\nu}{1+\nu} c\, \delta_{ik} . \tag{28}$$

4. THE INTERACTION OF A DILATATION CENTRE WITH OTHER STRAIN FIELDS

Applying eq. (16) for a strain field $\varepsilon^A_{jk}(\underline{r}')$ and for a dilatation centre of strength $P_{jk}=C_{jk\ell m}\, a_{\ell m}$ placed at position $\underline{r}$, we obtain for the interaction energy

$$U(\underline{r}) = \frac{1-\nu}{1+\nu} c\, t^A_{\ell\ell}(\underline{r}) , \tag{29}$$

where we used eq. (28) and

$$t_{\ell\ell} = 2\mu\frac{1+\nu}{1-2\nu}\int_V \alpha(\underline{r}-\underline{r}')\, \varepsilon^A_{\ell\ell}(\underline{r}')dV' . \tag{30}$$

In the following the detailed analysis of a few special cases is given.

4.1. DILATATION CENTRE

Substituting expression (25) into eq. (29) we get for the interaction energy of two dilatation centres:

$$U(\underline{r},\underline{r}') = 2\mu \frac{1-\nu}{1-2\nu} cc^A\alpha(\underline{r}-\underline{r}') . \tag{31}$$

It can be seen that if $\alpha(\underline{r}-\underline{r})=\delta(\underline{r}-\underline{r}')$ (classical elasticity), then no interaction exists between dilatation centres.

In nonlocal elasticity for the function, α Eringen suggested the expression /9,10/

$$\alpha = \frac{k^3}{\pi^{3/2}a^3} e^{-k^2\frac{R^2}{a^2}} \tag{32}$$

where $R=|r-r'|$ is the distance between the defects, a is the lattice parameter and k constant which can be estimated for face centred cubic (fcc) metals in the following way. Considering the size effect /13/ of alloying atoms in fcc metals we have

$$c = \frac{a}{4}\varepsilon , \tag{33}$$

where $\varepsilon = \frac{V-V_o}{V_o}$, the relative volume difference between host an alloying atoms.

Substituting these last two expressions into eq. (30) we get (with $\nu \simeq \frac{1}{3}$)

$$U(R) = \mu \frac{(1-\nu)k^3a^3}{\pi^{3/2}8(1-2\nu)} \varepsilon\varepsilon^A e^{-k^2\frac{R^2}{a^2}} = 0{,}0449k^2\mu a^3 \varepsilon\varepsilon^A e^{-\frac{k^2R^2}{a^2}} . \tag{34}$$

It is clear that the self-energy of a dilatation centre can be obtained from this expression as

$$U_s = \frac{1}{2} U(0) \simeq 0{,}0225k^3\mu a^3\varepsilon^2 . \tag{35}$$

This result can be used to determine the value of k by adjusting the self-energy to the one obtained by Flinn and Marududin on the basis of a discrete atomic model /14/. According to their result $U_s \simeq 0{,}06\mu a^3\varepsilon^2$. Comparing this expression to eq. (35) we get $k \simeq 1{,}40 \simeq \sqrt{2}$. This value is somewhat larger as the one estimated by Eringen /9/. However, if we take the atomic radius in fcc metals as $r_o=a/\sqrt{2}$, then the nonlocal effect attenuates to 1,6% of its value at R=0 at a radial distance, $R \simeq 2r_o$ which is quite reasonable comparing it to the behaviour of the usual interatomic potentials.

We can now use eq. (34) for numerical estimations. Let us take as an example the aluminium with $a=4.10^{-8}$cm, $\mu=27.10^3$MPa. For this case

$$U(R) = 0{,}485k^3\varepsilon\varepsilon^A e^{-k^2\frac{R^2}{a^2}} = 1{,}33\varepsilon\varepsilon^A e^{-k^2\frac{R^2}{a^2}} \quad (\text{eV}) .$$

If we take for Si impurity atom $\varepsilon=-0{,}16$ and for Mg $\varepsilon^A=0{,}41$ /13/ we get for the interaction energy of these impurity atoms placed at the first neighbouring positions in the lattice $(R=a/\sqrt{2})$ U=0,024eV. This value shows that the elastic interaction can lead to formation of impurity atom pairs only at relatively low temperatures.

4.2. SPHERICAL HOMOGENEOUS VOLUME DEFECT

Consider a spherical homogeneous volume defect with stress free transformation /2/ of

$$\varepsilon^T_{jk} = \frac{1}{3}\varepsilon^T\delta_{jk}. \tag{36}$$

The strain field of such a defect is

$$\varepsilon^A_{jk} = \begin{cases} \varepsilon^c_{jk} - \varepsilon^T_{jk} ; & \underline{r}' \in V_h \\ \varepsilon^c_{jk} ; & \underline{r}' \notin V_h \end{cases} \tag{37}$$

where V_h is the volume of the defect and /2/

$$\varepsilon^{c}_{jk} = \frac{1+\nu}{12\,\pi(1-\nu)}\,\varepsilon^{T} \int_{V_h} \left(\frac{1}{|\underline{r}'-\underline{r}''|}\right)_{,jk} dV'' \, . \tag{38}$$

With the use of these expressions and eq. (29) we get for the interaction energy of a volume defect and a dilatation centre

$$U(\underline{r}) = 2\mu c\,\frac{1-\nu}{1-2\nu} \int_{V_h} \alpha(\underline{r}-\underline{r}')\left[\varepsilon^{c}_{\ell\ell}(\underline{r}')-\varepsilon^{T}_{\ell\ell}(\underline{r}')\right] dV' + \int_{V_m} \alpha(\underline{r}-\underline{r}')\,\varepsilon^{c}_{\ell\ell}(\underline{r}')\, dV' \, , \tag{39}$$

where the second integral must be taken over the volume of the body outside the defect, that is $V_m = V-V_h$, where V is the volume of the body. Using the properties of the Dirac δ function it is easy to see that

$$\varepsilon^{c}_{\ell\ell} = \begin{cases} -\dfrac{1+\nu}{3(1-\nu)}\,\varepsilon^{T} \; ; & \underline{r}' \in V_h \, , \\ 0 \; ; & \underline{r}' \notin V_h \, . \end{cases} \tag{40}$$

Substituting these expressions into eq. (39) we have

$$U(\underline{r}) = -\frac{4}{3}\,\mu c \varepsilon^{T}\,\frac{2-\nu}{1-2\nu} \int_{V_h} \alpha(r-r')\,dV' \, . \tag{41}$$

It can be seen again, that if $\alpha(\underline{r}-\underline{r}') = \delta(\underline{r}-\underline{r}')$, $(\underline{r}\notin V_h)$ then $U(\underline{r})=0$, that is no interaction exists between a spherical volume defect and a dilatation centre in position outside the defect. If $\underline{r}\in V_h$, which is due to the case of a point defect within the inclusion, from eq. (39) we obtain for the classical case

$$U_o = -\frac{4}{3}\,\mu c \varepsilon^{T}\,\frac{2-\nu}{1-2\nu} \, . \tag{42}$$

In a real material the inclusions are always foreign phases (zones, precipitates, etc.). The description of their properties as homogeneous volume defects means the neglection of the modulus effect /2/. In first approximation this can be done, however in the present case the difference in the properties of the inclusion and the matrix may cause a change in c within and outside the inclusion for the same alloying atom.

Let the centre of the volume defect with radius R_o in the origin of the coordinate system. Introducing the variable $x=R/(a/k)$, where R is the distance of the dilatatation centre from the origin, eq. (41) can be written after integration with respect to the angle variables in the form

$$U(R) = -U_o\,\frac{e^{-x^2}}{\sqrt{\pi}\,x} \int_0^{x_o} \left(e^{-2xx'} - e^{2xx'}\right) e^{-x'^2}\, x'dx' \, , \tag{43}$$

where $x_o=R_o/(a/k)$, and U_o is given by eq. (42). This integral can be calculated only numerically, and the result is shown in Fig. 1. The curves are due to the case when c is the same within and outside the volume defect. Since $a/k \approx a/\sqrt{2}$, therefore the unit of the abscissa is approximately the atomic distance in fcc metals. The circles along the curves show the interaction energies belonging to the three, nearest positions of the dilatation centre along the surface outside of the volume defect. It can be seen that there is an appreciable deviation from the classical case, when U=0 outside and $U=U_o$ inside the volume defect. For numerical estimation let us consider again fcc metals with $\nu\approx\frac{1}{3}$. In this case $U_o \approx -18\varepsilon\varepsilon^{T}$(eV). With $\varepsilon=\varepsilon^{T}=0{,}1$ we get $U_o=-0{,}18$ eV. On the basis of Fig. 1. it can be seen that outside the volume defect the interaction energy can be neglected around room temperature, but a significant binding energy appears along the surface of the inclusion.

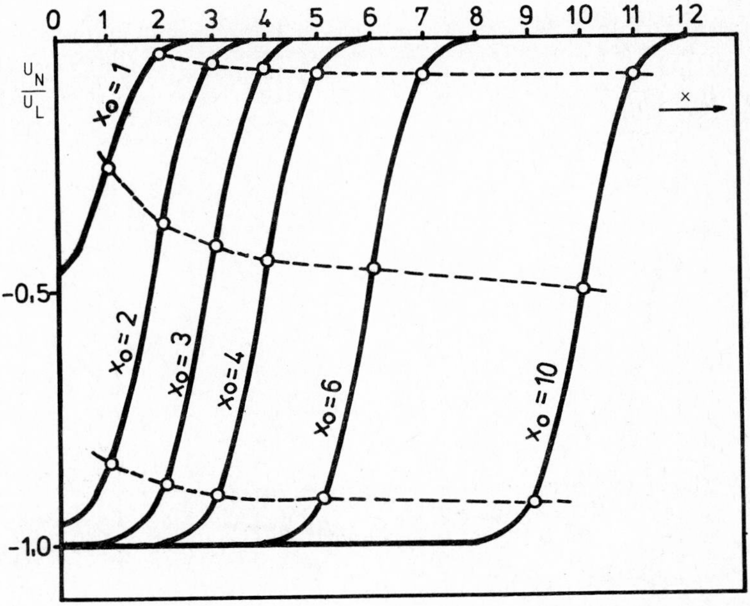

Figure 1

4.3. EDGE DISLOCATION

The nonlocal elastic stress field of an edge dislocation has been calculated by Eringen /10/. Using his results we get for the hydrostatic stress field (for infinite medium)

$$t_{\ell\ell} = \mu b \frac{1+\nu}{\pi(1-\nu)} \frac{k}{a\rho} \left(1 - e^{-\rho^2}\right) \sin\theta \, , \tag{44}$$

where $\rho=r/(a/k)$, r, θ are the polar coordinates of the defect /10/, b is the Burgers vector. Applying eq. (29) the interaction energy is

$$U(r,\theta) = U_{oe} \frac{1-e^{-\rho^2}}{\rho} \sin\theta \quad , \tag{45}$$

where

$$U_{oe} = \frac{\mu bkc}{\pi a} \, . \tag{46}$$

The most important feature of this result is that contrary to the classical result the interaction energy remains finite if $r \to 0$. With $a \to 0$ the classical Cottrell--Bilby result is obtained /1/.

Eq. (45) can be used to define the force acting on the defect along the direction r as

$$f = -\frac{\partial U}{\partial r} = -f_o \frac{(1+2\rho^2)e^{-\rho^2}-1}{\rho^2} \sin\theta , \qquad (47)$$

where $f_o = kU_o/a$.

Fig.2. shows the interaction energy and force as a function of distance between the dislocation and the point defect. It can be seen that near the dislocation a bound state appears. Its position is at $\rho \simeq 1,121$. The maximum attracting force is exerted on the defect at the position $\rho \simeq 1,793$. The curves on the figure belong to the cases $c<0$, $\sin\theta>0$ (compressed zone) and $c>0$, $\sin\theta<0$ (dilatation zone).

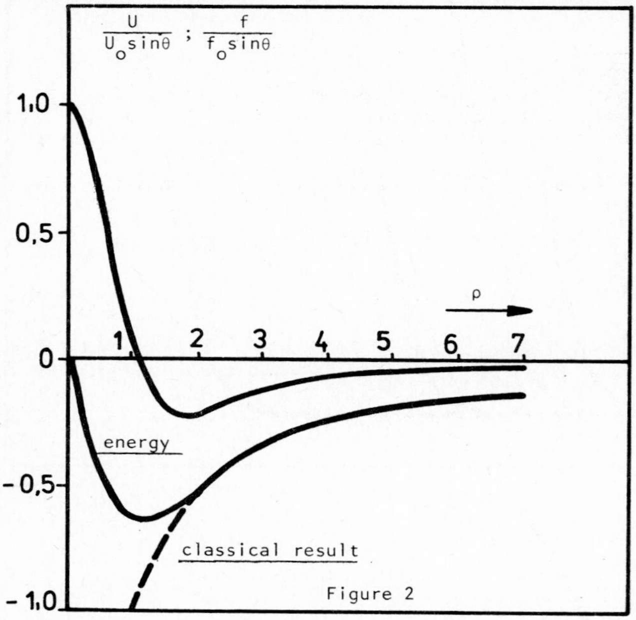

Figure 2

As an example let us apply our results to face centred cubic metals. In this case $b=a/\sqrt{2}$, by which we obtain $U_{oe}=0,078\mu a^3\varepsilon$(eV). The maximum binding energy ($\sin\theta=1$):

$$U_b^{max} = 0,638U_{oe} = 0,05\ \mu a^3\varepsilon. \qquad (48)$$

For Al $U_b^{max}=0,54\varepsilon$ eV, which gives for Mg and Si impurities 0,22 eV and 0,09 eV respectively.

5. THE SELF-ENERGY OF A HOMOGENEOUS SPHERICAL VOLUME DEFECT

Considering the homogeneous volume defect as space distribution of monopoles with a volume density of $\sigma^T_{jk}=C_{jk\ell m}\varepsilon^T_{\ell m}$ /11/, we can obtain its self-energy from eq. (17) in the form

$$U_s = -\frac{1}{2}\int_{V_h} \alpha(\underline{r}-\underline{r}')\ \varepsilon_{jk}(r)\ dV\ dV' , \qquad (49)$$

where ε_{jk} is given by eq. (37). Using the properties of ε_{jk} the self-energy can expressed as

$$U_s = \frac{2}{9}\mu\frac{1}{1-\nu}\nu\,(\varepsilon^T)^2 \int_{V_h}\int_{V_h} \alpha(\underline{r}-\underline{r}')\,dV\,dV' \ . \tag{50}$$

If $\alpha=\delta(\underline{r}-\underline{r}')$, then the integral is equal to the volume of the defect, V_h and we obtain therefore the classical result, U_o. In nonlocal elasticity the integral can be calculated again only numerically. Eq. (50) can be transformed into the form:

$$\frac{U_s}{U_o} = \frac{3}{\sqrt{\pi}x_o^3}\int_0^{x_o}\int_0^{x_o} ky\ \ e^{-(x-y)^2} - e^{-(x\ y)^2}\ \ dxdy\ , \tag{51}$$

where $x_o=R_o/(\frac{k}{a})$, and R_o is the radius of the defect.

The result of the numerical calculation is shown in Fig.3. It can be seen that for large size defects the self-energy tends to the classical value, but significant deviation appears in the case of small size (with radius of a few atomic distance) defects.

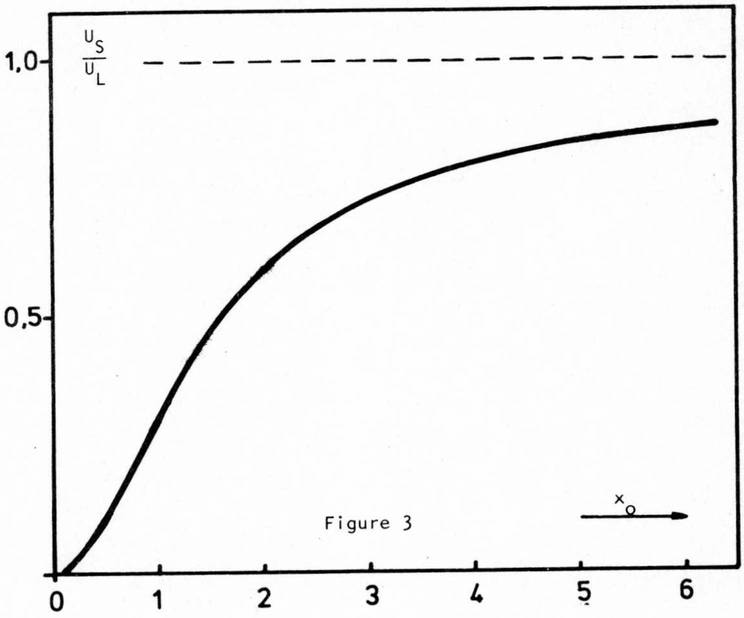

Figure 3

6. CONCLUSION

The quantitative properties of static lattice defects can be treated in general in nonlocal elasticity. The results contain nonlocal stresses which have no singularities. This property makes it possible to solve a set of problems unsolvable in classical elasticity, such as the interaction and self-energy of dilatation centres, etc.

REFERENCES

/1/ A.H. Cottrell and B.A. Bilby, Proc. Phys. Soc. A62. (1949) 49

/2/ J.D. Eshelby, Progr. in Solid Mech., Vol.11., North-Holland, Amsterdam, (1961) p. 89

/3/ R. Bullough and R.C. Newman, Phil. Mag. 7. (1962) 529

/4/ N.F. Fiore and C.L. Bauer, Progr. Mat. Sci. 13. (1968) 85

/5/ I. Kovács and L. Zsoldos, Dislocations and Plastic Deformation, Pergamon Press, London, 1973.

/6/ A.C. Eringen and D.G. Edelen, Int. J. Engng. Sci. 10. (1972) 233

/7/ A.C. Eringen, Int. J. Engng. Sci. 10. (1972) 425

/8/ A.C. Eringen, Continuum Physics, Vol.IV, Part 11. Academic Press, New York, 1976.

/9/ A.C. Eringen, J. Phys. D: Appl. Phys. 10. (1977) 671

/10/ A.C. Eringen, Int. J. Engng. Sci. 15. (1977) 177

/11/ I. Kovács, Physica, 94B. (1978) 177

/12/ I. Kovács and G. Vörös, Physica, 96B. (1979) 111

/13/ H.W. King, J. Mat. Sci. 1. (1966) 79

/14/ P.A. Flinn and A.A. Maradudin, Ann. Phys. 18. (1962) 81

Continuum Models of Discrete Systems 4
eds. O. Brulin and R.K.T. Hsieh

BEHAVIOUR OF POLYCRYSTALLINE METALS UNDER COMPLEX LOADING CONDITIONS: TESTING, MODELLING AND COMPUTATION

J. Kratochvíl*, Y. Ohashi**, M. Šatra***, M. Tokuda****

* Institute of Physics, Czechoslovak Academy of Sciences, Prague, Czechoslovakia
** Department of Mechanical Engineering, Nagoya University, Nagoya, Japan
*** National Institute for Machine Design, Prague, Czechoslovakia
**** Department of Mechanical Engineering, Mie University, Tsu, Japan

The results of the complex tension-torsion experiments on polycrystalline metals are interpreted in terms of the simplified model of a polycrystal of Lin´s type. A satisfactory agreement between the experiments and the theory is achieved for two sets of strain-controlled deformation processes. The suggested model of the polycrystal is implemented into the finite element method of solution of elasto-plastic boundary-value problems. A method of the solution of a modified plane strain problem is studied as an example.

INTRODUCTION

The main factor which limits the reliability of elasto-plastic analysis is the lack of analytical models of plastic behaviour comparable in accuracy with the current numerical methods of solution of boundary-value problems. The principal difficulty is to get an adequate mathematical description of the strong influence of the deformation history upon the plastic material response in complex loading conditions.

In this paper we summarize the basic features of the plastic response of polycrystalline metals as observed in complex tension-torsion experiments [1-6] and outline a relatively simple mathematical model of this phenomenon [7-8]. We discuss an example of a modified plane strain problem to ilustrate how to built the suggested model of the polycrystal into the finite element method of solution of elasto-plastic stress analysis problems.

PLASTIC RESPONSE IN COMPLEX TENSION-TORSION TESTS

Typical plastic behaviour of polycrystalline metals under complex loading conditions is demonstrated in Figs. 1, 2a, 3a, 4, 5a, 6a (similar results have been obtained also at stress or plastic strain controlled conditions [9-11]). The two types of strain trajectories shown in Figs. 1 and 4 are achieved by combined extension and twist of a thin-walled tubular specimens. The trajectories are represented in the vector space (e_{11}, $2e_{12}/\sqrt{3}$) of deviatoric strain $\underset{\sim}{e} = \underset{\sim}{\varepsilon} - \mathrm{tr}\underset{\sim}{\varepsilon}/3$; $\underset{\sim}{\varepsilon}$ is the strain tensor and e_{11}, e_{12} are the normal and shear components of deviatoric strain $\underset{\sim}{e}$, respectively. The trajectories in Fig. 1 come from extension of the tube up to the strain L_o followed by simultaneous extension and twist such that a sudden change of the strain vector direction arises at L_o. The angle at the "corner" at L_o is denoted by Θ_c. The strain trajectories in Fig. 4 consist of two perpendicular straight segments connected by arcs of various radii R.

The stress response of elasto-plastic polycrystalline materials along the described strain trajectories is represented by the deviatoric stress vector $\vec{s} = (3s_{11}/2, \sqrt{3}s_{12})$; s_{11} and s_{12} are the normal and shear components of the deviatoric stress tensor, respectively. The arrows in Figs. 1 and 4 indicate the deviatoric stress vector at the current points of the strain trajectories.

The torsion - tension experiments show that the direction of the deviatoric stress vector and the direction of the deviatoric strain increment (represented by the direction of the tangent to the strain trajectory) coincide only, if the curvature along the strain trajectory is zero or small, as in the case of the initial straight segments of the trajectories in Figs. 1 and 4 or a trajectory of the type Fig.4 with large R. But when the curvature is large, e.g. the corner at L_o in Fig. 1 or small R trajectories in Fig. 4, the stress direction <u>delays</u> from that of the strain increment. The delay occures just after the large curvature part of the strain trajectory and gradually disappears with an increase of succesive deformation of zero or low curvature. The delay effect is shown schematically in Figs. 1 and 4.

The measured angle Θ between the stress and strain increment directions, called the delay angle, are shown in Figs. 2a and 5a as the functions of the length of strain trajectory $L=\int [(de_{11})^2 + (2de_{12}/\sqrt{3})^2]$ (the parts of the curves along the first straight segments of the trajectories, where $\Theta = 0°$, are omitted). The curves were obtained for polycrystalline brass at the constant strain rate 3×10^{-6} sec^{-1} and are described in detail in [4,6] . Similar behaviour was found on polycrystalline mild steel and aluminium alloys [3,2,12] . In Fig. 2a there are shown the delay angles Θ for two straight segment trajectories with various "corner" angles Θ_c as shown in Fig. 1. At each curve the angle Θ is equal Θ_c at the "corner" and gradually disappears with successive plastic deformation along the second straight segment. In Fig. 5a, where the delay angle curves along the trajectories in Fig. 4 are shown, we observe an analogical behaviour. That means the material "remembers" the direction of anisotropy induced by plastic deformation just temporarily, it exhibits the directional memory of a <u>fading</u> type.

The changing strain direction influences the magnitude of the deviatoric stress vector too. Figs. 3a and 6a show the diagrams[1] of the relative magnitude of the stress vector s/s_o vs. the length of the strain trajectory L, $s = [(3s_{11}/2)^2 + 3(s_{12})^2]^{1/2}$ and s_o is the value of s at $L = L_o$. The stress magnitude decreases just after the change of the strain direction suddenly, but increases again successively along the straight segments of the trajectories. Finally the diagrams become parallel or practically coincide (the curve $\Theta = 180°$ becomes parallel out of Fig. 3a, approximately at $L-L_o = 1.5$ %).

Generally the delay angle and stress magnitude curves in Figs. 2a, 3a, 5a and 6a depend on the prestrain L_o. But as long as the prestrain exceedes a certain limit value which is a characteristic of the material (for brass the limit strain is approximately 1 % [5] , for mild steel 2 % [2]), the curves are practically independent of the prestrain [2,5] , the effect becomes <u>saturated</u>.

MODEL OF PLASTIC BEHAVIOUR OF A POLYCRYSTAL

To keep computing time of elasto-plastic stress analysis within tole-

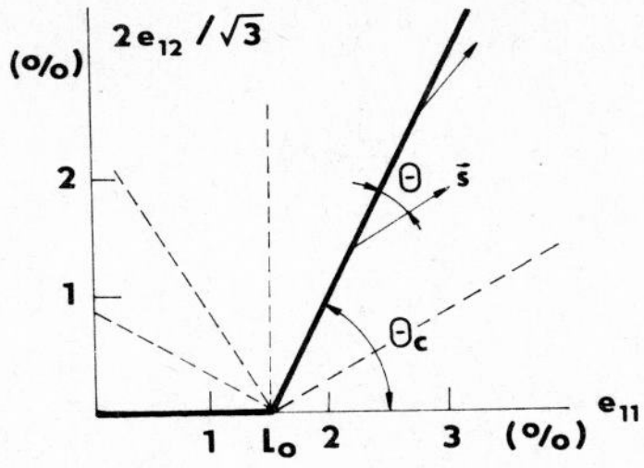

Fig. 1

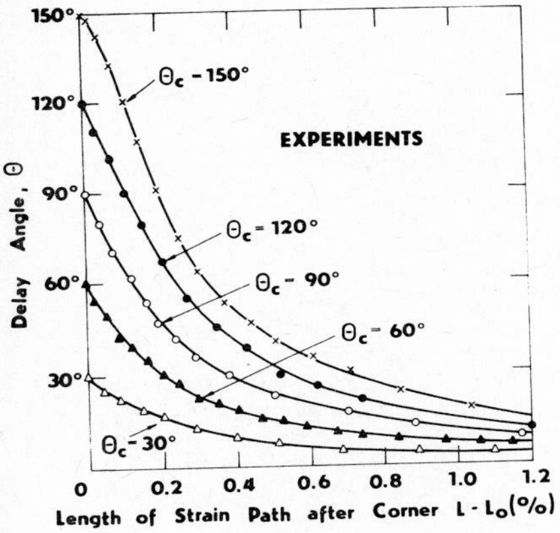

Fig. 2a

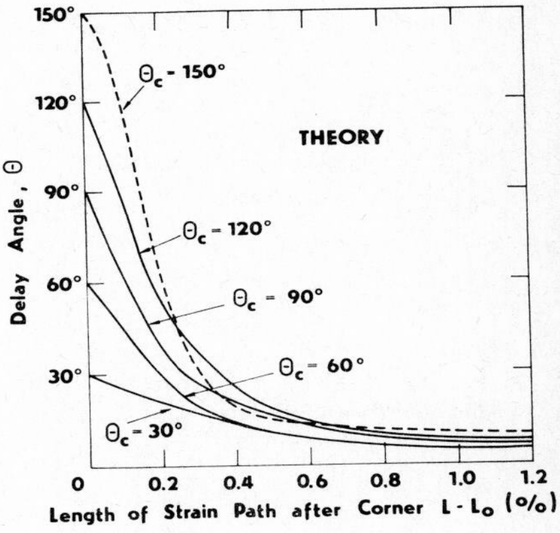

Fig. 2b

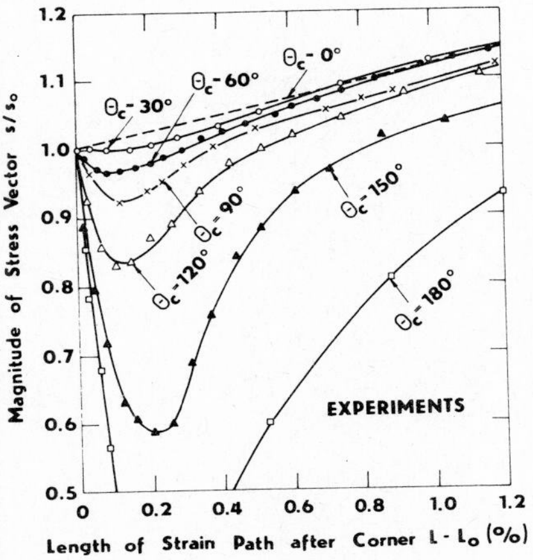

Fig. 3a

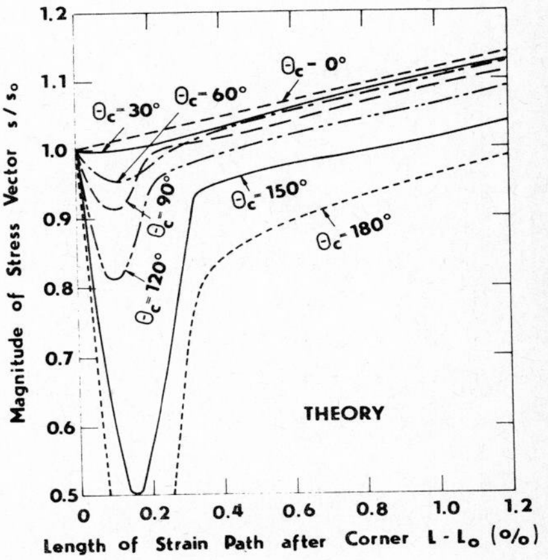

Fig. 3b

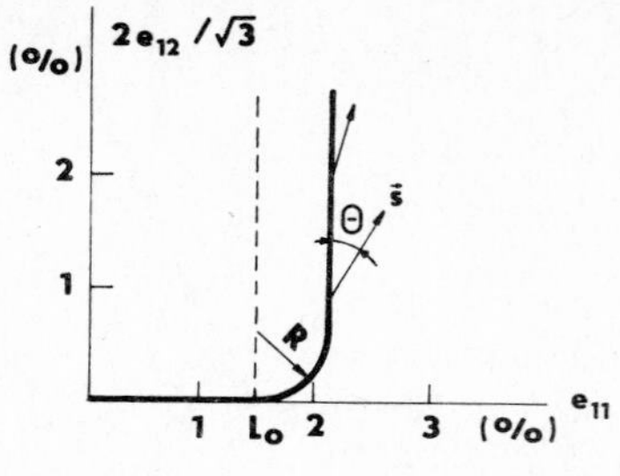

Fig. 4

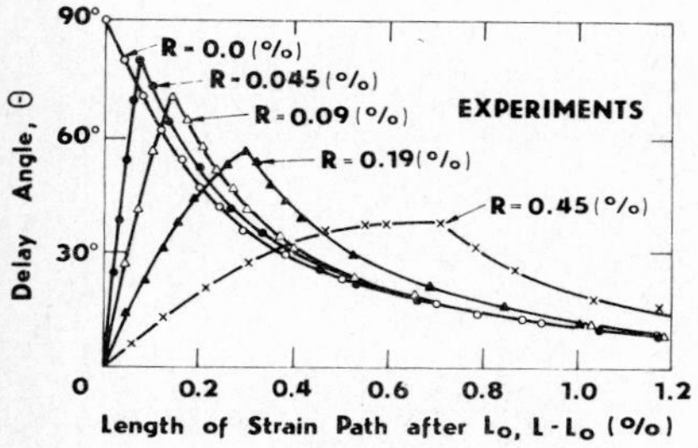

Fig. 5a

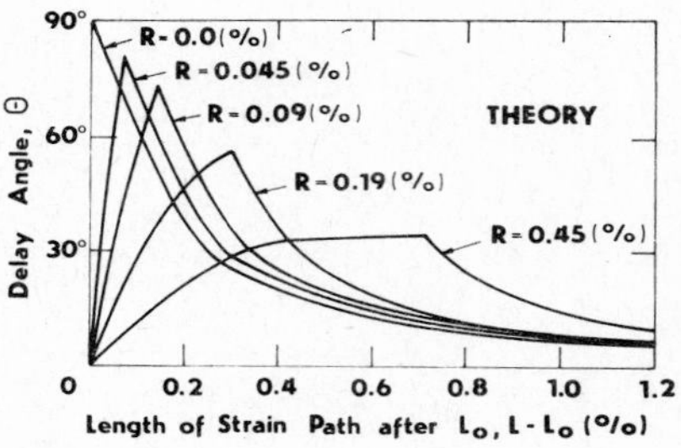

Fig. 5b

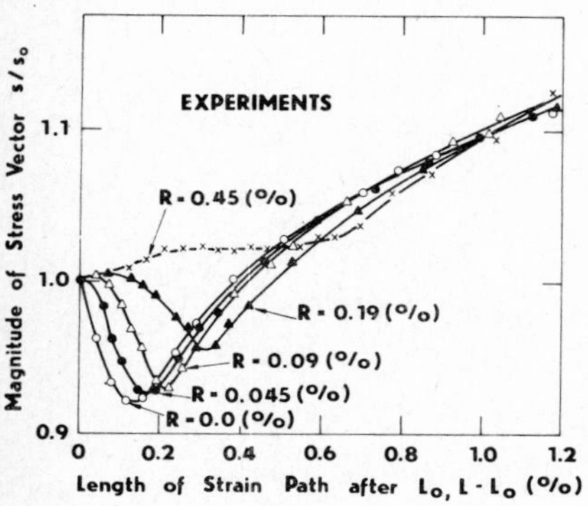

Fig. 6a

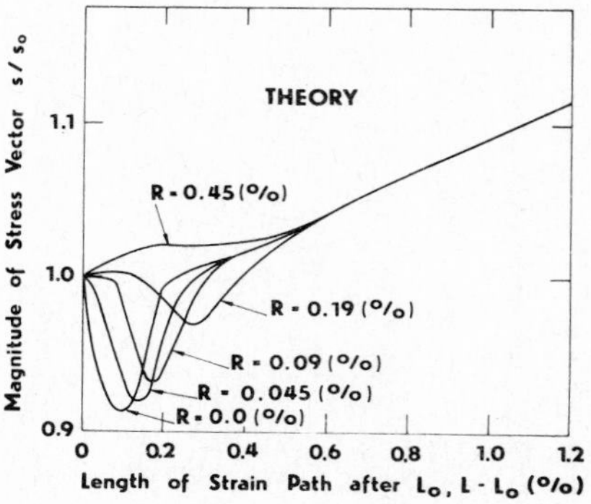

Fig. 6b

rable limits the mathematical model of the memory of plastic materials has to be as simple as possible. On the other hand it is desirable to represent the observed memory effects with an adequate accuracy. As a compromise we chose the model of polycrystal plasticity of Lin´s type [7-8] . The model is both relatively simple and simulates fairly well the essential features of the plastic memory as revealed by the tension-torsion experiments (see Figs. 2, 3, 5, 6).

The mechanism of plastic deformation of polycrystal can be sketched as follows. An isotropic polycrystal in an ideal sense contains a great number of single-crystal grains of all orientations. In each grain several slip systems, consisting of a slip plane and slip direction, can carry plastic deformation. The number and orientations of the slip systems in a grain are determined by the crystallographic structure of the solid. A slip system starts to yield, if the shear stress resolved on that slip plane and slip direction reaches a critical value (the Schmid law). The critical resolved shear stress is generally a function of deformation history of the single-crystal grain in which the slip system operates.

In a single-crystal grain the slip systems interact with each other causing the mutual work-hardening by changing their critical values of the resolved shear stress. Both active and inactive (latent) systems may be hardened. Moreover, there may occur a Bauschinger effect, i.e. the slip system that is exact opposite to the active one may harden less than the active system.

Plastic deformation in a polycrystal starts to develop in the grains favourably oriented with respect to the external stress applied to the polycrystal. The differences in plastic deformation among grains cause the internal stress. Such build up of a directional deformation inhomogenity and the accompaning internal stress reflexes the deformation history experienced by the material. The internal stress together with external load control the slip process in the grains and in this way determine the current plastic response of the polycrystal. The interference of the internal stress on the deformation process causes that the directions of the plastic strain increment and the external stress need not coincide. Due to this history dependent anisotropy the delay effect in the polycrystal may arise.

The developed internal stress makes further plastic deformation in the favourably oriented grains more difficult and facilitates deformation in the other grains. Keeping the direction of the deformation increment fixed we can reach a point when inhomogenity of plastic deformation ceases to increase, the directional memory becomes saturated.

If the direction of loading is changed, the grains favourably oriented with respect to the new direction start to yield and the internal stress developed due to previous deformation process is gradually re-oriented. The record of the deformation history preserved in the polycrystal is step by step overriden by the new deformation experience. The mechanism of the directional memory has a fading property.

Sketched polycrystal plasticity is in Lin´s approach [7] strongly simplified. We use the following approximations:

(i) The critical resolved shear stress in any slip system (both active and latent) in a single-crystal grain increases proportionally to the algebraic sum of plastic shears on all slip systems in the grain (the Taylor hardening law [13]). To incorporate the Bauschinger effect we assume additionally that the critical resolved shear stress of the system opposite to the active one is proportionally reduced with respect to the plastic shear of the active slip system. No strain rate effect or grain boundary influence is considered.

(ii) The complex structure of the slip systems of the grains is approximated by a single effective slip plane per grain with three pairs of effective slip directions. The directions are taken to be $\pi/3$ apart (the number of the slip directions were selected by trial [8] : for two or four pairs of effective slip directions there was less satisfactory agreement between the experimental and calculated curves that it is in Figs. 2, 3, 5 and 6).

(iii) The deformation processes in the grains in the polycrystal are coupled together by the assumption that the total strain of all grains is the same and is equal to strain of the polycrystal[2].

The simplified model of the polycrystal is thus fully specified by the elastic constants and three constants of the plastic response, namely, the initial value of the critical resolved shear stress τ_0, the coefficient K of the linearized Taylor hardening, and the coefficient of the Bauschinger effect K_B.

The Lin´s model is particularly suitable for evaluation of the strain controlled deformation processes. Due to the assumption (iii) the grains of different orientations can be considered separately. Strain of a grain is equal to the strain of the polycrystal and corresponding stress in the grain is determined using the Hooke law for the elastic part of strain and the Schmid law for the plastic part. The stress in the polycrystal is then obtained as the average of the stresses in the grains over all orientations. Using this procedure the theoretical curves for brass as shown in Figs. 2b, 3b, 5b and 6b have been obtained ($\tau_0 = 140/\sqrt{3}$ MPa, K = 500 MPa, K_B = 500 MPa). For such simplified model the agreement between theory and experiment in Figs. 2, 3, 5 and 6 is surprisingly good.

The main disadvantage of Lin´s model is that the internal stress responsible for directional memory is overestimated. The simplified assumption (iii) leads to the violation of the conditions of equilibrium at the grain boundaries which results in a higher stress level. However, the more realistic self-consistent model of Kröner [15] leads to the same results[3] as shown in Figs. 2b, 3b, 5b and 6b. Therefore the second reason for the overestimation of the internal stresses analysed in [16] seems to be more important. In Lin´s and Kröner´s models the interaction among grains that causes the internal stress remain always purely elastic, while in reality this interaction is somewhat softer as it becomes plastic as well. Therefore the level of the internal stress is lower than predicted by these models (a way of "softening" of Kröner´s model for a unidirectional proportional loading has been suggested in [16]).

The artifically high level of the internal stress as predicted by Lin´s model (may be together with the inadequate representation of the number and orientation of the slip systems and the Bauschinger effect in the grains) probably causes that the experimental and theoretical curves in Figs. 2, 3, 5 and 6 do not agree in detail and the curves $\Theta = 150°$ in Fig. 2a and Fig. 2b are too far from each other.

METHOD OF SOLUTION OF A ELASTO-PLASTIC BOUNDARY-VALUE PROBLEM

The model of the polycrystal sketched in the previous section has been formulated and tested for the strain trajectories represented by the plane curves. Therefore the model may be used directly in elasto-plastic boundary-value problems where only such type of strain trajectories occurs. As an example we outline here the way of implementation of the model into the finite element method of solution of a modified elasto-plastic plane strain problem.

We consider a plane strain problem where strain tensor $\underset{\sim}{\varepsilon}$ satisfies the conditions $\varepsilon_{13} = \varepsilon_{23} = 0$, $\varepsilon_{33} = (\varepsilon_{11} + \varepsilon_{22})/2$. The components of the corresponding strain tensor deviator $\underset{\sim}{e} = \underset{\sim}{\varepsilon} - (1/3)\mathrm{tr}\underset{\sim}{\varepsilon}$ are $e_{11} = (\varepsilon_{11} - \varepsilon_{22})/2$, $e_{11} = -e_{22}$, $e_{12} = \varepsilon_{12}$, $e_{13} = e_{23} = e_{33} = 0$. There is $\mathrm{tr}\, \underset{\sim}{e} = 0$ expressing approximately (i.e., if the elastic part of strain is neglected) that plastic deformation preserves volume. The nature of the deviatoric strain is the same as in tension-torsion experiments. The strain trajectories are fully determined by the components e_{11}, e_{12} and are represented by plane curves in the vector space (e_{11}, $2e_{12}/\sqrt{3}$) similarly as in Figs. 1 and 4.

To solve our problem by the finite element method the incremental initial stress approach seems to be propriate [17] . The solution uses the iterative schema that consists of applying load increments to the body and carrying out iterations to restore the equilibrium throughout the body. The iteration process [18]

$$(1) \qquad \underset{\sim}{K} \Delta \underset{\sim}{\delta}^{(i)} = \Delta \underset{\sim}{R}^{(i-1)},$$

gives in i-iteration cycle the improved displacement vector $\underset{\sim}{\delta}^{(i)} = \underset{\sim}{\delta}^{(i-1)} + \Delta \underset{\sim}{\delta}^{(i)}$ of the nodal points of a finite element mesh. In the iteration procedure the stiffness matrix $\underset{\sim}{K}$ is not updated and is taken to be equal to the elastic stiffness matrix of the considered modified plane strain problem. The unbalanced nodal forces $\Delta \underset{\sim}{R}^{(i-1)}$ are given by the relation

$$(2) \qquad \Delta \underset{\sim}{R}^{(i-1)} = \underset{\sim}{R} - \sum_{e=1}^{N} \underset{\sim}{a}_e^T \left(\int_{V_e} \underset{\sim}{B}_e^T \underset{\sim}{\sigma}_e^{(i-1)} dV_e \right),$$

where $\underset{\sim}{R}$ represents the applied external forces and the terms in the sum represent the internal nodal forces due to prevailing stresses $\underset{\sim}{\sigma}_e^{(i-1)}$ in the finite elements; N is the total number of the elements, $\underset{\sim}{a}_e^T$ is the connectivity matrix and V_e is the volume of the element e. The matrix $\underset{\sim}{B}_e$ connects the displacements $\underset{\sim}{\delta}_e$ in the nodal points of the element e and the strain $\underset{\sim}{\varepsilon}_e$

$$(3) \qquad \underset{\sim}{\varepsilon}_e = \underset{\sim}{B}_e \underset{\sim}{\delta}_e .$$

In each iteration cycle (1) the increment of the nodal displacement vector $\Delta \underset{\sim}{\delta}^{(i)}$ is determined and by (3) the increments of strain $\Delta \underset{\sim}{\varepsilon}^{(i)}$ and the corresponding components of deviatoric strain $\Delta e_{11}^{(i)}$, $\Delta e_{12}^{(i)}$ are evaluated in each finite element. To get the stress $\Delta \underset{\sim}{\sigma}_e^{(i)}$ appearing in (2) we use the suggested model of the polycrystal. The increment of deviatoric strain $\Delta \underset{\sim}{e}^{(i)}$ causes in the grains of the polycrystal either only the elastic response or the resolved shear stress in one or two slip systems of some grains reaches the current critical value (in the present model in a given grain the maximum of two out of six available slip systems can operate simultaneously). The increments of plastic shears $\Delta\gamma_n^{(i)}$, $n = 1, \ldots, 6$ in the active slip systems in a grain are governed by the system of equations [19,8]

$$(4) \qquad 2G \left[\mathrm{tr} \left(\Delta \underset{\sim}{e}^{(i)} \underset{\sim}{\beta}_n \right) - \sum_{m=1}^{6} \Delta\gamma_m^{(i)} \mathrm{tr} \left(\underset{\sim}{\beta}_m \underset{\sim}{\beta}_n \right) \right] = \Delta\tau_n^{(i)},$$

where G is the shear modulus, $\underset{\sim}{\beta}_n$ the orientation tensor

$$\beta_n = (m_n \otimes n_n + n_n \otimes m_n)/2 ;$$

n_n is the normal vector to the slip plane and m_n unit slip direction of slip system n), $\tau_n^{(i)}$ is the critical resolved shear stress in the slip system n at i-iteration cycle determined by the linearised Taylor hardening and modified by the Bauschinger effect.

From the solution of (4) we get plastic strain increment in the grain $\Delta p_G^{(i)} = \Sigma \Delta\gamma_n^{(i)} \beta_n$, n = 1,...,6, and then the deviatoric stress increment in the grain $\Delta s_G^{(i)} = 2G (\Delta e^{(i)} - \Delta p_G^{(i)})$. The deviatoric stress increment $\Delta s_e^{(i)}$ is gained by averaging the $\Delta s_G^{(i)}$ over all grains of the polycrystal representing the finite element e. The stress $\sigma_e^{(i)}$ is given by $\sigma_e^{(i)} = s_e^{(i)} + (E/3 (1 - 2\nu)) \mathrm{tr}\,\varepsilon^{(i)}$, where ν is the Poisson´s ratio. Having the new value of $\sigma_e^{(i)}$ the next iteration cycle can be performed. The iteration continues until the norm

$$\left\| \frac{\Delta \delta^{(i)T} \Delta \delta^{(i)}}{\delta^{(i)T} \delta^{(i)}} \right\| \leq \varepsilon$$

is sufficiently close to zero.

[1] In diagrams in Figs. 3a and 6a the effect of the third invariant of deviatoric stress is eliminated approximately with the use of a factor of modification proposed in paper [2] . After such modification the torsional and tensional stress-strain diagrams coincide. The same simplification is used in the model of a polycrystal sketched in the next section.

[2] In this respect our approach is similar to the so called sublayer model of plasticity suggested e.g. by Besseling [14] . Modelling of plastic behaviour of the grains is, however, in both approaches different.

[3] This was confirmed by our direct calculation as well as supported by theoretical arguments in [16] .

REFERENCES

[1] Lensky, V. S., Plasticity, in: Lee, E. H. and Symonds, P. S. (eds.) Proc. Second Symp. on Naval Structural Mechanics (Pergamon, Oxford, 1960).
[2] Ohashi, Y. and Tokuda, M., Precise Measurement of Plastic Behaviour of Mild Steel Tubular Specimens Subjected to Combined Torsion and Axial Force, J. Mech. Phys. Solids 24 (1973) 241-261.
[3] Ohashi, Y., Tokuda, M. and Mizuno, S., Plastic Behaviours of Mild Steel for the strain Trajectory of Two Straight Branches, Trans. JSME 40 (1974) 680-687.
[4] Ohashi, Y., Tokuda, M., Ito, S. and Miyake, T., Experimental Investigation on History Dependence of Plastic Behaviour of Brass under Combined Loading, Trans. JSME 45 (1978) 1575-1584.

English translation in: Bulletin of JSME 23 (1980) 1305-1312.
[5] Ohashi, Y., Tokuda, M. and Tanaka, Y., Precise Experimental Results on Plastic Behaviour of Brass Under Complex Loading, Bull. l´Acad. Polonaise Sci. 26 (1978) 261-272.
[6] Ohashi, Y., Kurita, Y., Suzuki, T., Tokuda, M., Effect of curvature of the Strain Trajectory on the Plastic Behaviour of Brass, J. Mech. Phys. Solids 29 (1981) 69-86.
[7] Lin, T. H., Analysis of Elastic and Plastic Strains of a Face-Centred Cubic Crystal, J. Mech. Phys. Solids 5 (1975) 143-149.
[8] Tokuda, M., Kratochvíl, J. and Ohashi, Y., Mechanism of Induced Plastic Anisotropy in Polycrystalline Metals, Trans. JSME (Submitted) (Pre-prints of the JSME, No. 814-1 (1981) 1-7).
[9] Ohashi, Y., Kawashima, K. and Yokochi, T., Anisotropy due to Plastic Deformation of Initially Isotropic Mild Steel and its Analytical Formulation, J. Mech. Phys. Solids 23 (1975) 277-294.
[10] Shiratori, E., Ikegami, K., Kaneko, K., Murakami, T. and Yoshida, S., Plastic Deformation Behaviour in Combined Loading along Stress Path, Trans. JSME 43 (1977) 1231.
[11] Shiratori, E., Ikegami, K. and Kaneko, K., Stress and Plastic Strain Increment after Corners on Strain Paths, J. Mech. Phys. Solids 23 (1975) 325-334.
[12] Ohashi, Y. and Kawashima, K., Plastic Deformation of Aluminium Alloy under Abruptly-Changing Loading of Strain Paths, J. Mech. Phys. Solids 25 (1977) 409-421.
[13] Taylor, G. I., Plastic Deformation of Metals, J. Inst. Metals 62 (1938) 307-324.
[14] Besseling, J.F., A Theory of Elastic, Plastic and Creep Deformations of an Initially Isotropic Materials Showing Anisotropic Strain-Hardening, Creep, Recovery and Secondary Creep, Trans. ASME J. Appl. Mech. 25 (1958) 529-536.
[15] Kröner, E., Zur plastischen Verformung des Vielkristalls, Acta Metall. 9 (1961) 155-161.
[16] Berveiller, M. and Zaoui, A., Extension of the Self-Consistent Scheme to Plastically-Flowing Polycrystals, J. Mech. Phys. Solids 26 (1979) 325-344.
[17] Zienkiewicz, O. C., The Finite Element Method in Engineering Science (McGraw-Hill, London, 1971).
[18] Aamondt, B. and Bergan, P. G., Propagation of Elliptical Surface Cracks and Nonlinear Fracture Mechanics by Finite Element Method, Proc. 5th. Conf. on Dimensioning and Strength Calculations and 6th Congress on Material Testing, (Budapest, 1974) 31-42.
[19] Hutchinson, J. W., Plastic Stress-Strain Relations of F.C.C. Polycrystalline Metals Hardening According to Taylor´s Rule, J. Mech. Phys. Solids 12 (1964) 11-24.

Part 8
HYDRODYNAMICS, SUSPENSIONS, MULTIPHASE MATERIALS

Continuum Models of Discrete Systems 4
eds. O. Brulin and R.K.T. Hsieh

CLOSURE PROCEDURES IN THE CLASSICAL STATISTICAL THEORY OF DENSE FLUIDS

G.H.A. Cole
Physics Department
University of Hull
Hull, N. Humberside
ENGLAND

The observed macroscopic properties of dense fluids, whether in thermodynamic equilibrium or not, are related to the corresponding molecular properties by means of a hierarchy of probability distribution functions in configuration and momentum spaces. The constituent molecules are in continuous interaction one with another and the calculation of the macroscopic properties must involve some decoupling (closure) procedure to allow results to be obtained. The difficulties raised by such calculations will be reviewed and unsolved problems will be highlighted.

§1 MACROSCOPIC AND MICROSCOPIC MATTER

The response of a macroscopic fluid to the action of forces of various kinds is accounted for by appealing to the conservation principles of mass, momentum and energy (see eg. Landau and Lifschitz (1958)). Dissipative effects are included by introducing a set of transport coefficients, such as the coefficient of dynamical (or shear) viscosity, of dilatational (or bulk) viscosity, thermal conductivity, electrical conduction, and so on. For an electrically conducting fluid permeated by a magnetic field, the equations of the theory are supplemented by the electromagnetic equations of Maxwell.

Entities such as pressure and temperature, fully defined only for conditions of strict thermodynamic equilibrium, are included in the description by making the assumption of local thermodynamic equilibrium (see eg. Cole (1967), Gyarmati (1970), Zubarev (1974)). The immediate neighbourhood of every point in the fluid is supposed to be in thermodynamic equilibrium within a prescribed accuracy of measurement even though the equilibrium conditions appropriate to one point will differ from those for a neighbouring point in the fluid. The difference will be smaller the nearer together the two points become. Special arguments are necessary if the fluid contains shock discontinuities or if the fluid speed locally is too high (turbulence). This description of macroscopic conditions is entirely self contained and does not require reference to the molecular composition of the fluid.

The macroscopic behaviour of the fluid is, in fact, determined by the properties of the constituent molecules and particularly by the collective interactions between them. The link between the macroscopic and microscopic worlds is made by relating the conservation statements for mass, momentum and energy at the macroscopic level to those applying to the microscopic description (Eisenschitz (1955), Zubarev (1974), Cole (1978)). By these means formulae are derived relating macroscopic quantities to the properties of the molecules. Although the procedure is well defined the associated severe mathematical difficulties

for fluids have led to a wide variety of methods for carrying the programme through. The theory is most developed for a fluid in full thermodynamic equilibrium although mathematical difficulties have prevented its exploitation except in the simplest cases. For non-equilibrium the situation is far from satisfactory and matters of principle still remain, especially for other than simple fluids. Theories of restricted validity have been available for a long time and the theory of the Brownian motion (see eg. Chandrasekhar (1943)) and the theory of the dilute non-uniform gas (see eg. Chapman and Cowling (1957)) still have an important influence on the development of theories of dense fluids.
The essential feature of these theories is the way in which the element of irreversibility is introduced. For the Brownian motion this involves a friction constant introduced so that the deceleration of a suspended particle of semi-colloidal size can be treated differently from the direct acceleration due to the net effect of many impacts with the smaller molecules of the suspending liquid. For the dilute gas, with impulsive binary molecular interactions, the element of irreversibility is introduced through the Boltzmann hypothesis of molecular chaos. For a simple dense gas or liquid system irreversibility is assumed to arise from the particular nature of the extended interactions between a representative molecule and its immediate neighbours. The correlation of the force on the molecule over a time interval is presumed to vanish if the time interval is large on the scale of a single representative interaction between a pair of molecules and yet is small on a macroscopic time scale. This hypothesis concerning the correlation of the force was first introduced by Kirkwood (1946) and plays the same role for liquids that molecular chaos plays for a dilute gas.

An important contribution over the last decade has been the development of computational attacks on the problem through molecular dynamics and Monte Carlo methods (see eg. Wood (1974)) but the matters of principle still remain to be resolved. It is opportune here to consider again some aspects of these general difficulties. We shall restrict ourselves to simple insulating fluids where the molecules are non-dipolar particles without internal degrees of freedom.

§2 SPACIAL ORDERING OF MOLECULES

The fluid state is distinguished by the particular short-ranged arrangement of the constituent molecules in space. The molecules are neither free (except for infrequence pair collisions) as in a dilute gas nor highly ordered over appreciable volumes of space as in a solid crystal. Instead the molecules show a mean spatial order from any one chosen as centre over a distance of a few molecular diameters after which order is essentially absent. This short ranged order is expressed in terms of the probabilities for the occurrence of pairs, triplets, quadruplets, and so on, of molecules about any molecule chosen as a representative centre in the fluid. The details of these probabilities in particular cases are determined by the form of the forces between the molecules and for non-uniform conditions also by the fluxes of momentum or energy within the fluid. In this way the macroscopic fluid parameters are related directly to the distribution of groups of molecules in space and to the details of the inter-molecular forces (Cole (1967), Zubarev (1974), Ravché and Mountain (1978)). Of particular importance in these formulae for simple fluids are the distributions for pairs and triplets of molecules in space.

§3 DISTRIBUTION FUNCTIONS

The fluid is represented by N identical particles, without internal degrees of freedom, each of mass m contained in a volume V, subject to the condition $N, V \to \infty$ but with $N/V = n$ finite. The interaction force between the particles is assumed derivable from the potential $\Psi(\underline{r}^N)$ where $\underline{r}^N = (\underline{r}_1, \underline{r}_2 \ldots \underline{r}_N)$ and $\underline{r}_i$ is the location of the i-th particle at time t. For simple fluids Ψ is the superposition of the simultaneous contributions from pair, triplet,..... interactions and we introduce the separate potentials

$$\Psi(\underline{r}^N) = \frac{1}{2}\sum_{i=1}^{N}\sum_{j>i}^{N}\psi(i,j) + \sum_{i=1}^{N}\sum_{j=1}^{N}\sum_{\ell=1}^{N}\psi(i,j,\ell) + \dots \tag{1}$$

where $\psi(i,j)$ is the interaction potential between the pair of particles, $\psi(i,j,\ell)$ that between the three particles (i,j,ℓ) and so on.

The discrete structure of the system of particles is replaced by a description involving a hierarchy of continuous probabilities each relating to the phase of a specific grouping of particles. For the total system, the probability that, at time t, a particle will be found in the spacial volume $d\underline{r}_1$ about $\underline{r}_1$ with momentum in the range $d\underline{p}_1$ about $\underline{p}_1$ and corresponding simultaneous specifications for the remaining particles is written in the obvious notation

$$N!\,D^{(N)}(\Gamma^{(N)},t)\,d\Gamma^{(N)} \tag{2}$$

where $\Gamma^{(N)} = (\underline{p}^N, \underline{r}^N)$ and $d\Gamma^{(N)} = (d\underline{p}^N, d\underline{r}^N)$. $D^{(N)}$ is called the N-th order phase distribution. Analogous distributions are defined for the lower order distributions $D^{(s)}$ where $s < N$ obtained from (2) by integrating away the details of the phase of $(N-s)$ unwanted particles.

If all the information concerning the momenta is removed in this way we are left with the N-th order configuration function $n^{(N)}(\underline{r}^N, t)$. The further elimination of $(N - s)$ configurational details provides the s-th order configurational distribution $n^{(s)}(\underline{r}^s, t)$. For a dilute gas, where the motion of single particles determines the macroscopic behaviour, we set $s = 1$. For a dense fluid we need to know $n^{(2)}(\underline{r}_{12}, t)$. For simple molecules, $\underline{r}_{12} = |\underline{r}_1 - \underline{r}_2| = r_{12}$ but $n^{(2)}$ may still depend on the relative orientation of particles reflecting the macroscopic non-uniform distortion

§4 THE LIOUVILLE EQUATION AND ITS FORMAL SOLUTION

Conservation of probability in 6N-dimensional phase is expressed by the Liouville equation

$$i\,\frac{\partial D^{(N)}}{\partial t} = \mathcal{L}^{(N)} D^{(N)} \tag{3}$$

where $\mathcal{L}^{(N)}$ is the Hermitian Liouville operator for N particles

$$\mathcal{L}^{(N)} = -i\sum_{j=1}^{N}\left(\frac{\underline{p}_j}{m}\cdot\frac{\partial}{\partial \underline{r}_j} + \underline{F}_j\cdot\frac{\partial}{\partial \underline{p}_j}\right) \tag{3a}$$

and $\underline{F}_j$ is the force on the particle j: in the absence of external forces this is the resultant of the interaction with neighbouring particles.

The formal solution of (3) is

$$D^{(N)}(\Gamma^{(N)},t) = \exp\left(-i\mathcal{L}^{(N)}t\right) D^{(N)}(0)$$

where the time propagator operator exp $(-i\mathcal{L}^{(N)}t)$ is defined by the conventional expansion. An alternative expression in the resolvant formalism (Foster and Cole (1971a), Prigogine (1964))is

$$D^{(N)}(\Gamma^{(N)},t) = -\frac{1}{2\pi i}\oint_c dz\,(\mathcal{L}^{(N)} - z)^{-1}\exp(-izt)\,D^{(N)}(0) \tag{4}$$

where z is the complex number and $(\mathcal{L}^{(N)} - z)^{-1}$ is the resolvant. The contour integral may be taken around any contour enclosing the real axis and a closed path in the lower half plane.

§ 5 ACCOUNTING FOR THE INTERACTIONS

The Hamiltonian for the N-particles is written

$$H(\Gamma^{(N)}) = \sum_{j=1}^{N} \left(\frac{p_j^2}{2m} + \lambda \sum_{l>j}^{N} \psi(j,l) \right)$$

where λ is a parameter which measures the intensity of the interaction energy between pairs of particles, and the interaction energy between larger groupings of particles is neglected. The Liouville operator is separated into a kinetic part $\mathcal{L}_0^{(N)}$ and a potential part $\delta\mathcal{L}^{(N)}$ by

$$\mathcal{L}^{(N)} = \mathcal{L}_0^{(N)} + \lambda\, \delta\mathcal{L}^{(N)} \tag{5}$$

where

$$\mathcal{L}_0^{(N)} = -i \sum_{j=1}^{N} \frac{\underline{p}_j}{m} \cdot \frac{\partial}{\partial \underline{r}_j} \tag{5a}$$

$$\delta\mathcal{L}^{(N)} = \sum_{l>j=1}^{N} \delta L_{jl} = i \sum_{l>j=1}^{N} \frac{\partial \psi(j,l)}{\partial \underline{r}_j} \cdot \left(\frac{\partial}{\partial \underline{p}_j} - \frac{\partial}{\partial \underline{p}_l} \right). \tag{5b}$$

Reduced distributions follow from (3) and (5) by integrating over the phase of the subgroup of particles not of interest. Introducing the renormalised Bogoliubov distribution $f^{(s)}$ by

$$f^{(s)} = V^s \int d\Gamma^{(N-s)} D^{(N)}$$

we obtain the equation for $f^{(s)}$ in the form

$$\frac{\partial f^{(s)}}{\partial t} + \sum_{j=1}^{N} \frac{\underline{p}_j}{m} \cdot \frac{\partial f^{(s)}}{\partial \underline{r}_j} = I \tag{6}$$

where

$$I = -i\lambda\, \delta\mathcal{L}^{(s)} - i\lambda n \int d\Gamma^{(s+1)} \sum_{j=1}^{s} \delta L_{j,s+1} f^{(s+1)}. \tag{6a}$$

This non-linear equation involves the two distributions $f^{(s)}$ and $f^{(s+1)}$ and forms a hierarchy of coupled equations for the calculation of $f^{(s)}$. It is precisely the set derived independently by Bogoliubov, (1946), Born and Green (1946) Kirkwood (1946) and Yvon (1935) but written now in terms of the Liouville operator rather than the Hamiltonian. The hierarchy is closed if the distribution $f^{(s+1)}$ is related to the lower order distributions. The various theories of dense fluids are in essence different ways of achieving such a closure in practice. The associated formulae are very complicated and need not be set down here but have been considered in detail by Foster and Cole (1971a,b) to which the reader is referred. The equations are non-Markovian (and so show a memory) in their general form. For application to dense fluids, where we set s = 1 or 2 in (6), concern is with calculating the average fluxes of mass, momentum and energy within the fluid and the details of the non-Markovian form can be suppressed (Foster and Cole (1971b)).

The inclusion of all the possible interaction contributions for all strengths (all λ) is the central problem still unsolved. Dominant interactions have, however, been accounted for and kinetic equation derived which can be reduced to forms already known (Foster and Cole (1971b)), provided the relaxation time for the system is much larger than the collision time. We list these cases now.

(i) Dilute, weakly coupled systems Here we suppose both λ and n are small, and collisions are instantaneous (Markov form). Setting $s = 1$ in (6) we obtain the Landau irreversibility equation describing two-body correlations. If the second particle has an equilibrium phase we can describe the case where a test particle is injected into an equilibrium system composed of particles of the same mass. This is the microscopic form of the more usual Brownian motion. The interaction term in (6a) is denoted by P and reduces to the sum of a mean and mean square increment of momentum during a relaxation time interval $\tau_r \gg \tau_c$ where τ_c is the interaction time. A parameter, $f^{(1)}$, is involved (proportional to $\lambda^2(3mk_BT)^{-1}$ where k_B is the Boltzmann constant and T the equilibrium temperature) which can be identified as the ordinary friction constant of the Brownian motion if it is independent of the momentum. Then (6) with $s = 1$ reduces to the Fokker-Planck equation (Eisenschitz (1958)).

(ii) Dilute, hard sphere gas Although n is small (dilute gas), λ is now large. The impulsive nature of the interactions is associated with terms of order $\lambda^q(\lambda^2 n)$ where $0 \leq q \leq \infty$ in (6a) and I $\equiv$ J in this case. J now reduces to the two-body scattering operator and the full equation becomes the Boltzmann transport equation for $f^{(1)}$.

(iii) Weakly coupled system: long range interaction The range of interaction is longer than scale of particle separation and the description of the time evolution of the system by purely binary encounters is invalid λ is small but λn is not small and terms of order $\lambda(\lambda n)^q$ are retained. The various terms depend linearly on t so the order of λ can be only one greater than that on n. In this case we write $I \equiv M$. The analysis of (6a) also allows the additional term $-\frac{\partial}{\partial p_1}\cdot\left(\underline{F}_1^* \frac{f^{(1)}}{m}\right)$ where $\underline{F}_1^*$ is the mean force on particle 1 due to the remaining particles, to be isolated in addition. This term can be called the Vlasov term (Allen and Cole (1968a)), denoted by $-M^*$.

(iv) Combination of terms Improved kinetic equations result if the various forms of I are combined appropriately. The early extension of the Kirkwood (1946) ideas was made by Rice and Allnatt (1961) who introduced an interparticle pair potential being the sum of a hard sphere repulsion at small separation distance with a soft attraction at large distances. The extension to a more general potential form (for example, the (12-6) potential) was made by Allen and Cole (1968b) using the interaction (resolvant) formalism of Prigogine (1962) (Explicitly, the cycle, ring and chain diagrams were included as dominant contributions, the chain diagrams becoming important at small separation distances). While the Rice-Allnatt theory is based primarily on the identification I = P + J, the Allen-Cole theory (1968b,c,) makes the identification I = J + M + M*. Using the (12-6) potential (with, in conventional notation ϵ/k_B= 121K and σ = 3.418 x 10^{-10}m for Argon) the thermal conductivity was calculated for the range of number concentrations $10^{19} \leq n \leq 10^{22}$ where it was shown that the calculated thermal conductivity has the correct order of magnitude at the various concentrations and the correct temperature dependence in each case. The analysis was taken further by Allen (1970, 1971).

We can notice that the equations are formally applicable to plasmas if λ is replaced by e^2, where e is the electronic charge, and the collisionless plasma results from

(v) Other Derived Theories The linear form of (6) for inhomogenous

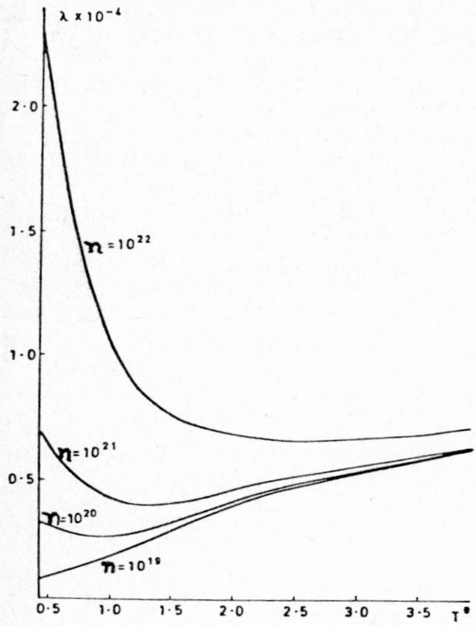

Fig. 1 Calculated dependence of thermal conductivity (in units 10^{-4} cal. $cm^{-1}s^{-1}K^{-1}$) on reduced temperature T^* $(= kT/\epsilon)$ for (12-6) potential, for four number densities, n, for Argon

systems provides new possibilities. This new equation has a "vacuum" component and a correlation component (Foster and Cole 1971b) and is in fact the fundamental kinetic equation of Balescu (1963) derived using generalised projection operators. The vacuum component of the equation reduces to the equation of Severne (1965) for the inhomogneous case and to the equation of Prigogine and Resibois (1961) for homogeneous systems, when in each case the conversion is made to wave vector notation.

Developing these equations in expansions in the ratios of the collision to hydrodynamic times (μ) and the relaxation to hydrodynamic times (ξ) leads to very general kinetic equations involving correlations linear in the gradients (Foster and Cole (1971b)). The equivalent non-linear equations reduce to the equations of Allen and Cole (1968b,c) for $\xi > \mu$ and alternatively to the equation of Prigogine, Nicolis and Misguich (1965) for $\xi < \mu$.

§ 6 INTERACTION REPRESENTATION

An alternative approach to accounting for interactions can be based on the resolvant formalism (4). In this case using (5) we have $\delta\mathcal{L}^{(N)} = (\mathcal{L}^{(N)} - z) - (\mathcal{L}_0^{(N)} - z)$ and

$$(\mathcal{L}^{(N)} - z)^{-1} = (\mathcal{L}_0^{(N)} - z)^{-1}\left[1 - (\mathcal{L}_0^{(N)} - z)^{-1}(\lambda\,\delta\mathcal{L}^{(N)})(\mathcal{L}^{(N)} - z)^{-1}\right],$$

the interaction effects being obtained successively by iteration, providing an expansion in powers of λ . The approach has an especial utility when the theory is developed in wave vector space as was shown by Prigogine (1962) and his school. The various interaction contributions are then isolated in terms of a diagram technique. The diagrams most investigated are those associated with the hard binary encounter and the extended (Fokker-Planck) interaction.

§7 REDUCTION TO CONFIGUARATION

For simple fluids not far from equilibrium a simplification is possible and particularly so if further the momentum components reach equilibrium faster than the spatial co-ordinates. In this case the equations can be reduced to those to determine the correction of the corresponding equilibrium configuaration and the full equation in phase is reduced to an equation in configuration space. For the Fokker-Planck contributions this amounts to the use of the Smokichowski (diffusion) equation (Eisenschitz (1958)), Chandrasekhar (1943), Cole (1967), under appropriate boundary conditions. It is in this reduced form that many of the numerical checks of the theory have been made (see eg. Rice and Gray (1965)). In particular the pair distribution is written

$$n^{(2)}(1,2) = n_0^{(2)}(1,2) + n_1^{(2)}$$

where $n_0^{(2)}$ is the equilibrium distribution and $n_1^{(2)}$ is the distortion due to the small departure from equilibrium. The theory is then applied to the calculation of the small distortion term $n^{(1)}$.

§8 EQUILIBRIUM STRUCTURE

Particular importance is attached to a knowledge of the fluid equilibrium structure both from the view-point of the calculation of the non-uniform structure and from that of the specification of thermodynamic variables. This problem is considerably more amenable to discussion than that considered so far because the N-th order distribution is known in this case through the canonical or grand canonical forms. The non-linear equation still pertains and the closure problem remains in the interaction term, I. In particular it is necessary to have a knowledge of $f^{(s+1)}$ in terms of $f^{(s)}$. For pair interactions this means $s = 2$ but if triplet forces must also be retained in the potential (1), $s = 3$ will be involved as well.

The short ranged spatial order for a simple fluid is isotropic about any point as centre so $\underline{r}_{12} = |\underline{r}_1 - \underline{r}_2| = r_{12}$. It is convenient to define the radial distribution $g^{(2)}(1,2)$ by

$$n_0^{(2)}(1,2) = n^2 g^{(2)}(1,2)$$

which is accessible to experimental measurement. Impenetrability of particles and the short range spatial ordering are accounted for by the conditions:

$$g^{(2)}(1,2) \to 0 \quad \text{as } r_{12} \to 0 \; ; \; g^{(2)}(1,2) \to 1 \quad \text{as } r_{12} \to \infty .$$

Higher order distributions are defined in a corresponding way and the hierarchy of functions $g^{(s)}(1,2,..s)$ $(s \geq 1)$ which results are related directly (according to the arguments of §3) by

$$g^{(s)}(1,2,\dots s) = \frac{n}{V} \int g^{(s+1)}(1,2,\dots, s, s+1)\, d\underline{r}_{s+1} . \tag{7}$$

Again a closure decoupling approximation must be appealed to if $g^{(1)}$ or $g^{(2)}$ are to be calculated.

The inaccuracies introduced by the approximation can possibly be reduced by using equations of the form (6) which relate to the effects of changes of the relative particle separations (Born and Green (1946)) or deviations of the particle separation about the mean (Cole (1958)).

§9 FORM OF THE SUPERPOSITION (CLOSURE) APPROXIMATION

Equilibrium conditons suppress the particular characteristics of the various time scales which can be used for non-uniform conditions. Appeal now can be made to the mean interaction potential energy $\Phi_s(1,2,\dots s)$ for a subgroup of s $(< N)$

particles defined by

$$\Phi_s(1,2,\dots s) = -k_B T \log g^{(s)}(1,2,\dots s). \tag{8}$$

Closure can be achieved if the total interaction for s particles is expressed in terms of the corresponding functions for (s - 1) particles. We assume

$$\Phi_s(1,2,\dots s) = \sum \Phi_{s-1}(1,2,\dots(s-1)) + W_s(1,2,\dots s)$$

where the summation is over all groupings of (s - 1) particles and W_s is an indirect correlation energy. From (8) we have for $g^{(3)}$ and $g^{(4)}$

$$g^{(3)}(1,2,3) = \left[\prod_{j>i=1}^{3} g^{(2)}(i,j)\right] S_3(1,2,3) \tag{9a}$$

$$\left[\prod_{j>i=1}^{4} g^{(2)}(i,j)\right] g^{(4)}(1,2,3,4) = \left[\prod_{k>j>i=1}^{4} g^{(3)}(i,j,k)\right] S_4(1,2,3,4) \tag{9b}$$

with

$$W_s(1,2,\dots s) = -k_B T \log S_s(1,2,\dots s) \tag{9c}$$

The indirect correlations S_s must be assigned. Approximations to this end were earlier proposed by Kirkwood (1935) ($S_3 = 1$) and by Fischer (1964) ($S_4 = 1$) . These approximations are inadequate as they stand and must be improved.

§10 APPROACH OF FUNCTIONAL DIFFERENTIATION

A group of N interacting particles are supposed in the external field of a pair of particles (denoted by land 2) and the behaviour of the system as a whole is investigated as one particle is removed to infinity. Introduce the functions x_1, x_2 and x_{12} by

$$x_i(j) = \psi(i,j) \text{ with } i = 1, 2 \;; \quad x_{12}(j) = x_1(j) + x_2(j).$$

Select a function $F(3|x_{12})$ involving the three particles (1,2,3) regarded as a functional of the second function $G(4|x_1)$ involving the particles 1 and 4. Make a functional Taylor expansion for F in terms of G about the point (particle 1 "switched off") : then

$$F(3|x_{12}) = F(3|x_2) + \sum_{s=1}^{\infty} \frac{1}{s!} \iint \prod_{i=4}^{s+3} \left[G(i|x_1) - G(i|0)\right] \times \left[\frac{\delta^s F(3|x_{12})}{\prod_{i=4}^{s+3} \delta G(i|x_1)}\right]_{x_1=0} d\underline{r}_4 \dots d\underline{r}_{N+s}. \tag{10}$$

The functions F and G are open to choice (see eg. Verlet (1964)). In order to achieve a closure for the triplet function (Moreton and Cole (1968)) set

$$F = n^{(1)}(3|x_{12})\left[\exp\left(\frac{x_1(3)}{k_B T}\right) - 1\right]$$
$$G = n^{(1)}(4|x_1)\left[\exp\left(\frac{x_1(4)}{k_B T}\right) - 1\right]. \tag{11}$$

For the closure of the quadruplet function we set instead

$$F = n^{(1)}(4|x_{23})\left[\exp\left(\frac{x_1(4)}{k_B T}\right) - 1\right]$$
$$G = \frac{n^{(1)}(5|x_{12})\, n^{(1)}(5|x_{13})}{n^{(1)}(5|x_1)}\left[\exp\left(\frac{x_1(5)}{k_B T}\right) - 1\right]. \tag{12}$$

§11 THE CLOSURE APPROXIMATIONS

Inserting (11) into (10) gives the following results:

(i) to first functional derivative; $S_3^{(1)} = 1$ (Kirkwood approximation)

(ii) to second functional derivative;

$$S_3^{(2)} = S_3^{(1)} + n\int d\underline{r}_4\, f(1,4)\, y(1,4)\, \Xi_3(2,3,4) \qquad (13)$$

where

$$f(1,4) = \exp\left(-\frac{\psi(1,4)}{k_B T}\right) - 1 \qquad \text{(Mayer function)}$$

$$y(1,4) = g^{(2)}(1,4)\exp\left(\frac{\psi(1,4)}{k_B T}\right)$$

$$\Xi_3 = g^{(3)}(1,2,3)\left[g^{(2)}(2,3)\right]^{-1} - g^{(2)}(2,4) - g^{(2)}(3,4) + 1:$$

(iii) to triplet functional derivative

$$S_3^{(3)} = S_3^{(2)} + n^2\iint d\underline{r}_4\, d\underline{r}_5\, f(1,4)f(1,5)\, y(1,4)\, y(1,5)\, \Xi_4(2,3,4,5) \qquad (14)$$

where

$$\Xi_4 = \frac{1}{2}\frac{g^{(4)}(2,3,4,5)}{g^{(2)}(2,3)} - \frac{g^{(3)}(2,3,4)}{g^{(2)}(2,3)}\sum_{j=2}^{4} g^{(2)}(j,5) - \frac{1}{2}\sum_{j=2}^{3} g^{(3)}(j,4,5)$$
$$+ g^{(2)}(2,4)\left[\sum_{j=2}^{4} g^{(2)}(j,5) - 2\right] + \left[g^{(2)}(3,4) - 2\right]\sum_{j=3}^{4} g^{(2)}(j,5) - 2.$$

Further terms can be added involving higher powers of the density.

The same procedure follows for the quadruplet distribution by inserting (12) into (10). The results obtained are

(iv) to first functional derivative; $S_4^{(1)} = 1$ (Fischer approximation)

(v) to second functional derivative;

$$S_4^{(2)} = S_4^{(1)} + n\int d\underline{r}_3\, f(1,5)\, \Xi_5(1,2,3,4,5)\exp\left(\frac{\psi(1,5)}{k_B T}\right) \qquad (15)$$

with

$$\Xi_5 = \frac{g^{(3)}(1,2,5)\, g^{(3)}(1,3,5)}{g^{(2)}(1,2)\, g^{(2)}(1,3)\, g^{(2)}(1,5)\, g^{(2)}(2,5)\, g^{(2)}(3,5)}\; \textcircled{H}_4$$

and

$$\textcircled{H}_4 = \frac{g^{(4)}(2,3,4,5)}{g^{(3)}(2,3,4)} - \sum_{k>j=2}^{4}\frac{g^{(3)}(j,k,5)}{g^{(2)}(j,k)} + \sum_{j=2}^{4} g^{(2)}(j,5) - 1.$$

Again the addition of higher order terms will introduce higher powers of the number density n.

It should be pointed out that these expressions do not amount to a simple expansion of S_s in powers of n because the distributions $g^{(s)}$ are themselves dependent of the density

§12 A NUMERICAL TEST

One immediate test of the theory, although at lower than liquid densities, is the calculation of the virial coefficients B_j, with j=2,3,4,5...., in the expansion of the pressure in a power series in the inverse of the volume. We use the equation for the change of $g^{(2)}$ with n given by Schofield (1966)

$$\frac{\partial \log g^{(2)}(r)}{\partial n} = \frac{\int\left[\frac{g^{(3)}(1,2,3)}{g^{(2)}(1,2)} - 1\right]d\underline{r}_3 - 8\pi\int\left[g^{(2)}(r) - 1\right]r^2 dr}{1 + 4\pi n\int\left[g^{(2)}(r) - 1\right]r^2 dr}.$$

Two potential forms are involved, viz hard spheres with the pair potential

$$\psi(r) = \infty \quad (r \leq 1) \quad \text{and} \quad \psi(r) = 0 \quad (r > 1)$$

and the Lennard-Jones (12-6) form

$$\psi(r) = 4\epsilon\left[\left(\frac{\sigma}{r}\right)^{12} - \left(\frac{\sigma}{r}\right)^{6}\right]$$

with $\epsilon/k_B = 121$ K and $\sigma = 3.418 \times 10^{-10}$ m.

The results are collected in Table 1 (hard spheres) and Table 2 for a ((12-6) potential). The effects of the additional terms in the closure expressions is clear.

$B_2 = b$; $B_3 = 0.6250b^2$	
$B_4(K) = 0.1431\ b^3$	(Kirkwood: $S_3^{(1)} = 1$);
$B_4(S_3^{(1)}) = 0.2869\ b^3$	(known exact value: eqn. (13));
$B_5(K) = -0.0592\ b^4$	(Kirkwood : $S_3^{(1)} = 1$)
$B_5(S_3^{(1)}) = 0.0641\ b^4$	(exact B_4: eqn. (13)
$B_5(S_3^{(2)}) = 0.1104\ b^4$	(exact B_4: eqn. (14) and $S_4^{(1)} = 1$)

$B_4(S_3^{(1)})$ and $B_5(S_3^{(2)})$ are exact and further coefficients can be added to the list (as usual $b = 2\pi/3$).

Table 1: Second, Third, Fourth and Fifth virial coefficients for a hardsphere gas for the different closure formulae.

T^* $\left(= \frac{kT}{\epsilon}\right)$	$B_4(T^*)$		$B_5(T^*)$		
	$(S_3^{(1)} = 1)$	exact B_4	$S_3^{(1)} = 1$	$S_3^{(2)}$	Exact B_5
0.75	61.99	-18.84	221.23	-254.03	-185.85
1.00	11.11	-0.28	17.522	-16.641	-2.860
1.25	2.03	0.33	4.195	-3.986	+0.043
1.60	-0.31	0.19	1.585	-1.201	-0.041
3.00	-0.18	0.12	0.311	-0.026	+0.058
20.00	+0.09	0.08	0.026	+0.019	0.021

TABLE 2

$B_4(T^*)$ and $B_5(T^*)$ as function of T^* using eqn. (13)

§ 13 SUMMARY

The problem of acconting for the simultaneous interactions between many particles is very severe and remains unsolved in its general form. Recent work in the field has concentrated on the computer simulations of physical systems, taking advantage of the important advances in computation techniques which have occurred over the last decade. Useful and important results have been obtained even though only relatively small groupings of particles can be treated by modern techniques, but has taken some of the emphasis away from the fundamental developments considered in this paper. Great interest could centre around the numerical evaluation of the various integrals associated with the different theoretical developments considered here and additional insight gained of the physical processes involved. For instance, a through going numerical analysis of the effects of the different combined terms of § 5 (iv) would be highly informative and could well have practical value. For equilibrium, a test of the closure approximation of §11 at liquid densities could be most instructive. The fundamental study of fluid behaviour using the general approach considered here, and developments of it, can still repay further study and could well provide the background to a deeper understanding of the properties of fluids at high density.

REFERENCES

(1) Allen, P.M., 1970, Mol.Phys., 18, 349; 1971, Physica, 52, 237

(2) Allen, P. and Cole, G.H.A., 1968a, Mol.Phys., 15, 549; 1968b, ibid., 14, 413; 1968c, ibid., 15, 557

(3) Balescu, R., 1963, Statistical Mechanics of Charged Particles, New York, Interscience.

(4) Bogoliubov, N.N., 1946, Problems of a Dynamical Theory in Statistical Physics in Studies in Statistical Mechanics, Vol. (Ed.J.de Boer and G.E. Uhlenbeck), Amsterdam, North-Holland.

(5) Born, M. and Green, H.S., 1949, The kinetic Theory of Liquids, Cambridge University Press.

(6) Chandrasekhar, S., 1943, Rev.Mod.Phys., 15, 1.

(7) Chapman, S. and Cowling, T.G., 1957, The Mathematical Theory of Non-Uniform Gases, Cambridge University Press.

(8a) Cole, G.H.A., 1958, J.Chem.Phys., 28, 912

(8b) Cole, G.H.A., 1967, Introduction to the Statistical Theory of Dense Classical Fluids, Oxford: Pergamon Press.

(8c) Cole, G.H.A., 1978. Chapter 1 of Progress in Liquid Physics (Ed.C.A.Croxton) New York, Wiley.

(8d) Cole, G.H.A. and Moreton, A., 1967, Mol.Phys., 13, 501

(9) Eisenschitz, R.K., 1955, Phys. Rev. 99, 1059.

(10) Eisenschitz, R.K., 1956, Statistical Theory of Irreversible Processes, Oxford University Press.

(11) Foster, M.J. and Cole, G.H.A., 1971a, Mol.Phys., 20, 417: 1971b, ibid., 21, 385.

(12) Fischer, I.Z., 1964, Statistical Theory of Liquids, Chicago, University Press.

(13) Gyarmati, I., 1970, Non-Equilibrium Thermodynamics, Berlin, Springer-Verlag.

(14) Kirkwood, J.G., 1935, J.Chem.Phys., 3, 300

(15) Kirkwood, J.G., 1946, J.Chem.Phys., 14, 180.

(16) Landau, L.D. and Lifshitz, E.M., 1959, Fluid Mechanics, Oxford: Pergamon Press.

(17) Prigogine, I., 1962, Non-Equilibrium Statistical Mechanics, New York, Interscience.

(18) Prigogine, I. and Resibois, P., 1961, Physics, 27, 629.

(19) Prigogine, I., Nicolis, G. and Misguich, J., 1965, J.Chem.Phys., 43, 4516

(20) Ravechè, H.J. and Mountain, R.D., 1978, Chapter 12 of Progress in Liquid Physics, (Ed. C.A. Croxton), New York, Wiley.

(21) Rice, S.A. and Allnatt, A.R., 1961, J.Chem.Phys., 34, 2144

(22) Rice, S.A. and Gray, P., 1965, Statistical Theory of Liquids, New York, Interscience.

(23) Severne, G., 1965, Physica, 31, 877

(24) Schofield, P., 1966, Proc.Phys.Soc., 88, 149.

(25) Verlet, L., 1964, Physica, 30, 95.

(26) Wood, W.W., 1974, In Fundamental Problems in Statistical Mechanics, III, (Ed. E.G.D. Cohen), Amsterdam, North-Holland.

(27) Yvon, J. 1935, Actualities Scientifiques et Industrielles, p203, Paris, Hermann and Cie.

(28) Zubarev, D.N., 1974 Non-Equilibrium Statistical Mechanics, New York, Consultants Bureau.

Continuum Models of Discrete Systems 4
eds. O. Brulin and R.K.T. Hsieh

CLUSTER EXPANSION FOR THE DIELECTRIC CONSTANT OF A SUSPENSION OF SPHERES

B.U. Felderhof

Institut für Theoretische Physik A
RWTH Aachen
5100 Aachen
West-Germany

We discuss the cluster expansion for the susceptibility kernel and the effective dielectric constant of a suspension of polarizable spheres. The cluster integrals involve ensemble averages of many-body scattering kernels. We have proven the absolute convergence of the cluster integrals. The Clausius-Mossotti formula emerges naturally as a first approximation.

We consider a collection of polarizable spheres suspended in a uniform dielectric background with dielectric constant ε_1 . Inside the spheres the dielectric constant is characterized by

$$\varepsilon(\underset{\sim}{r}) = \varepsilon(|\underset{\sim}{r} - \underset{\sim}{R}|) \quad \text{for} \quad |\underset{\sim}{r} - \underset{\sim}{R}| < a, \tag{1}$$

where $\underset{\sim}{R}$ is the center of one of the spheres and a its radius. In a particular configuration of N spheres the dielectric constant of the suspension is given by

$$\begin{aligned} \varepsilon(1,\ldots,N) &= \varepsilon(|\underset{\sim}{r} - \underset{\sim}{R}_j|) \quad \text{for } |\underset{\sim}{r} - \underset{\sim}{R}_j| < a,\ j=1,..,N, \\ &= \varepsilon_1 \text{ otherwise.} \end{aligned} \tag{2}$$

We assume that the spheres do not overlap. The electric field satisfies Maxwell's equations of electrostatics

$$\nabla\cdot\varepsilon\ (1,\ldots,N)\ \underset{\sim}{E}\ (1,\ldots,N) = 4\pi\rho_0,\ \nabla\times\ \underset{\sim}{E}(1,\ldots N) = 0, \tag{3}$$

where $\rho_0(r)$ is a given charge density. The induced polarization is defined by

$$\underset{\sim}{P}(1,\ldots,N) = \frac{1}{4\pi}\left[\varepsilon(1,\ldots,N)-\varepsilon_1\right]\underset{\sim}{E}(1,\ldots,N). \tag{4}$$

The field equations can in principle be solved. We express the electric field $\underset{\sim}{E}$ (1,...,N) in terms of the imposed field $\underset{\sim}{E}_0(r)$ generated by $\rho_0(\underset{\sim}{r})$ in the absence of spheres. The fields are related by a linear integral kernel

$$\underset{\sim}{E}(1,\ldots,N) = \underset{\approx}{K}(1,\ldots,N)\cdot \underset{\sim}{E}_0. \tag{5}$$

Similarly the induced polarization is expressed by an N-particle polarizability kernel

$$\underset{\sim}{P}(1,\ldots,N) = \underset{\approx}{A}(1,\ldots,N)\ \cdot \underset{\sim}{E}_0. \tag{6}$$

The kernels $\underset{\approx}{K}$ and $\underset{\approx}{A}$ are related by (4), but also

$$\underset{\approx}{K}(1,\ldots,N) = \underset{\approx}{I} + \underset{\approx}{G}_0\ \underset{\approx}{A}(1,\ldots,N)\ , \tag{7}$$

where $\underset{\approx}{I}$ is the identity operator and $\underset{\approx}{G}_0$ is the Green function for the uniform medium with dielectric constant ε_1.

We assume that the system is disordered and that it makes sense to average over an ensemble of configurations described by a probability distribution $P(\underset{\sim}{R}_1, \ldots, \underset{\sim}{R}_N)$. The average field equations are

$$\nabla \cdot \varepsilon_1 \langle E \rangle + 4\pi \langle \underset{\sim}{P} \rangle = 4\pi\rho_0 , \quad \nabla \times \langle E \rangle = 0 . \tag{8}$$

Averaging the integral relations (5) and (6) and eliminating $\underset{\sim}{E}_0(\underset{\sim}{r})$ one obtains a linear relationship between $\langle \underset{\sim}{P}(\underset{\sim}{r}) \rangle$ and $\langle \underset{\sim}{E}(\underset{\sim}{r}) \rangle$ of the form

$$\langle \underset{\sim}{P}(\underset{\sim}{r}) \rangle = \int \underset{\approx}{X}(\underset{\sim}{r}, \underset{\sim}{r}') \cdot \langle \underset{\sim}{E}(\underset{\sim}{r}') \rangle \, d\underset{\sim}{r}' , \tag{9}$$

where $\underset{\approx}{X}(\underset{\sim}{r}, \underset{\sim}{r}')$ is a generalized susceptibility kernel. One can derive a cluster expansion for this kernel by making a cluster decomposition of the operators $\underset{\approx}{A}(1, \ldots, N)$ and $\underset{\approx}{K}(1, \ldots, N)$. The cluster decomposition is defined by

$$\underset{\approx}{A}\{s\} = \sum_{\{r\} \subset \{s\}} \underset{\approx}{B}\{r\} , \qquad \underset{\approx}{K}\{s\} = \sum_{\{r\} \subset \{s\}} \underset{\approx}{L}\{r\} , \tag{10}$$

where $\{s\}$ is some set of s labels and the sum is over all subsets $\{r\}$ of $\{s\}$. We include the empty set, putting $\underset{\approx}{A}(\emptyset) = \underset{\approx}{B}(\emptyset) = 0$ and $\underset{\approx}{K}(\emptyset) = \underset{\approx}{L}(\emptyset) = \underset{\approx}{I}$. The relations (10) can be inverted to

$$\underset{\approx}{B}\{s\} = \sum_{\{r\} \subset \{s\}} (-1)^{s-r} \underset{\approx}{A}\{r\} , \quad \underset{\approx}{L}\{s\} = \sum_{\{r\} \subset \{s\}} (-1)^{s-r} \underset{\approx}{K}\{r\} . \tag{11}$$

The cluster expansion of the generalized susceptibility kernel has the form [1]

$$\underset{\approx}{X} = \sum_{l=1}^{\infty} \frac{1}{(l-1)!} \underset{\approx}{X}_l , \tag{12}$$

where $\underset{\approx}{X}_l$ involves integration over l particle positions. Explicitly $\underset{\approx}{X}_l$ is given by

$$\underset{\approx}{X}_l = \sum_{\substack{\text{part.} \\ \Sigma_j s_j = l}} (-1)^{k-1} \overline{\underset{\approx}{B}_1\{s_1\}} \; \overline{\underset{\approx}{L}\{s_2\}} \ldots \overline{\underset{\approx}{L}\{s_k\}} , \tag{13}$$

where the sum is over all partitions of l objects into blocks with the condition that label 1 occurs in the first block. The overhead bar indicates integration over an s_j-particle density. In the first factor we have used symmetry to single out the polarization of particle 1 and have put $\overline{\underset{\sim}{B}\{s\}} = s \, \overline{\underset{\sim}{B}_1\{s\}}$. The 2- and 3-particle terms read explitly

$$\underset{\approx}{X}_2 = \int d\underset{\sim}{R}_1 d\underset{\sim}{R}_2 \left[n(1,2)\underset{\approx}{B}_1(1,2) - n(1)\, n(2)\underset{\approx}{B}_1(1)\underset{\approx}{L}(2) \right] ,$$

$$\underset{\approx}{X}_3 = \int d\underset{\sim}{R}_1 d\underset{\sim}{R}_2 d\underset{\sim}{R}_3 \Big[n(1,2,3)\underset{\approx}{B}_1(1,2,3) - n(1)n(2,3)\underset{\approx}{B}_1(1)\underset{\approx}{L}(2,3)$$

$$- 2n(1,2)n(3)\, \underset{\approx}{B}_1(1,2)\underset{\approx}{L}(3) + 2n(1)n(2)n(3)\underset{\approx}{B}_1(1)\underset{\approx}{L}(2)\underset{\approx}{L}(3) \Big] , \tag{14}$$

where in $\underset{\approx}{X}_3$ we have used the symmetry of the integrand. The above expansion is equivalent to one derived by Finkel'berg [2] some years ago, but he did not pursue the analysis any further.

As evident from (13) the l-particle cluster integral involves an l-particle cluster polarizability kernel which is integrated over the l-particle density, corrected by sums of products of polarizability kernels of lower order which are integrated over corresponding factorized particle densities. The latter factorization allows overlap configurations to occur. Since in such configurations the particles are at very close range they may be expected to give a major contribution to the integral.

In order to derive a corresponding cluster expansion for the effective dielectric constant it is convenient to write the susceptibility kernel $\underset{\approx}{X}(\underset{\sim}{r},\underset{\sim}{r}')$ as an integral over a single particle polarizability kernel

$$\underset{\approx}{X}(\underset{\sim}{r},\underset{\sim}{r}') = \varepsilon_1 \int d\underset{\sim}{R}_1 n(\underset{\sim}{R}_1)\ \underset{\approx}{a}(\underset{\sim}{r},\underset{\sim}{r}';\underset{\sim}{R}_1)\ . \tag{15}$$

In a spatially uniform region the kernel $\underset{\sim}{a}$ will be of the form $\underset{\approx}{a}(\underset{\sim}{r},\underset{\sim}{r}';R_1) = \underset{\approx}{a}(\underset{\sim}{r}-R_1\ ,\ \underset{\sim}{r}'-R_1)$, and will be of relatively short range. For slowly varying fields one can define an effective dielectric tensor

$$\underset{\approx}{\varepsilon}^* = \varepsilon_1\left[\underset{\approx}{1} + 4\pi n\underset{\approx}{\alpha}^*\right], \tag{16}$$

where the polarizability tensor $\underset{\approx}{\alpha}$ is given by

$$\underset{\approx}{\alpha}^* = \langle O|\ \underset{\approx}{a}|\ O\rangle = \int \underset{\approx}{a}(\underset{\sim}{r},\underset{\sim}{r}')d\underset{\sim}{r}d\underset{\sim}{r}'\ . \tag{17}$$

It can be expressed as a sum of cluster integrals

$$\underset{\approx}{\alpha}^* = \sum_{l=1}^{\infty} \frac{1}{(l-1)!}\ \underset{\approx}{\alpha}_l^* \tag{18}$$

with

$$\underset{\approx}{\alpha}_l^* = \varepsilon_1^{-1} \sum_{\substack{\text{o.part.}\\ \sum_j s_j = l}} (-1)^{k-1} \frac{(l-1)!}{(s_1-1)!\prod_{j=2}^{k} s_j!} < O|\ \overline{\underset{\approx}{B}_1\{s_1\}}\ \overline{\underset{\approx}{L}\{s_2\}}\ldots\overline{\underset{\approx}{L}\{s_k\}}|O\rangle, \tag{19}$$

where the sum is over all ordered partitions of l labels.

In order to make further progress it is necessary to make a judicious estimate. As is well known, the Clausius-Mossotti formula

$$\frac{\varepsilon_{CM} - \varepsilon_1}{\varepsilon_{CM} + 2\varepsilon_1} = \frac{4\pi}{3}\ n\alpha\ , \tag{20}$$

where α is the dipole polarizability of a sphere, gives an excellent approximation to the dielectric constant of a dense polarizable fluid. For the case that the permittivity inside a sphere is a constant $\varepsilon_2 > \varepsilon_1$ the value ε_{CM} coincides with the lower bound of Hashin and Shtrikman [3]. In the present scheme ε_{CM} emerges as an overlap contribution to the integrals. In the l-body cluster integral we define the overlap volume as that part of configuration space where the spheres $1,2,\ldots,l$ overlap sequentially. In the overlap volume any distribution function $n\{s_j\}$ in (19) for which $\{s_j\}$ contains more than one particle, vanishes identically. Hence only the full partition $(1|2|\ldots|l)$ contributes in the overlap volume. The corresponding contribution to α_l^* is given by

$$\underset{\approx}{\alpha}_l^{*,ov} = \varepsilon^{-1}(-1)^{l-1}(l-1)!\ \int_{ov} d\underset{\sim}{R}_2\ldots d\underset{\sim}{R}_l n(2)\ldots n(l) \langle O|\underset{\approx}{B}_1(1)\underset{\approx}{L}(2)\ldots\underset{\approx}{L}(l)|O\rangle, \tag{21}$$

where the integrations are restricted by the overlap condition $|R_j - R_{j-1}| < 2a$ $(j = 2,\ldots,l)$. Assuming that the density is uniform in a sphere of radius at least $2(l-1)a$ one can perform the integrals sequentially, starting with $\underset{\sim}{R}_1$. One finds

$$\underset{\approx}{\alpha}_l^{*,ov} = (l-1)!\ \alpha\left(\frac{4\pi}{3}\ n\alpha\right)^{l-1} \underset{\approx}{1} \tag{22}$$

so that

$$\underset{\approx}{\alpha}^{*,ov} \equiv \alpha_{CM}\ \underset{\approx}{1} = \alpha\left[1 - \frac{4\pi}{3}n\alpha\right]^{-1}\ \underset{\approx}{1}\ , \tag{23}$$

which with (16) leads to the Clausius-Mossotti formula(20).

The remaining part of configuration space yields corrections to the Clausius-Mossotti formula. From the explicit expression (14) it is fairly evident that the 2-body cluster integral is absolutely convergent. We have proven absolute convergence for the cluster integrals of all orders [1] . As a consequence, in a limiting procedure in which the system grows beyond bounds in a sequence of congruent shapes one finds a well-defined effective dielectric constant, independent of the sample shape. The 2-body integral in (14) can be evaluated with great numerical accuracy [1] . The higher order terms cannot be evaluated accurately, but one can make a fair guess at the major contribution. Here the correction terms of Kirkwood [4] and Yvon [5] serve as a starting-point.

REFERENCES

[1] Cohen, E.G.D., Felderhof, B.U., and Ford, G.W., to be published.
[2] Finkel'berg, V.M., Sov. Phys. JETP 19 (1964) 494.
[3] Hashin, Z., and Shtrikman, S., J. Appl. Phys. 33 (1962) 3125.
[4] Kirkwood, J.G., J. Chem. Phys. 4 (1936) 592.
[5] Yvon, J., Recherches sur la théorie cinétique des liquides (Hermann et Cie, Paris, 1937).

Continuum Models of Discrete Systems 4
eds. O. Brulin and R.K.T. Hsieh

ELECTROVISCOUS EFFECTS

E.J. Hinch

Department of Applied Mathematics and Theoretical Physics,
University of Cambridge,
England.

The rheology of a suspension of small particles can change dramatically if the particles become electrically charged. In this review I try to explain the various mechanisms which lead to this so called electroviscous effect. One difficulty of the subject is that there are a number of effects governed by six or more non-dimensional groups, and it is only recently that the significance of some of these groups has been understood. The review does not pretend to be a complete historical account of the subject, nor a careful comparison of experiments and theory.

THE ELECTRICAL DOUBLE LAYER

When a solid surface is placed in contact with a liquid, it usually becomes electrically charged to a potential of typically 0.1V. There are several sources of the charge:- in a metal the electron cloud can spill over the edge of the solid, in a crystaline material the ions on the surface are free to lift up a little from their lattice sites, and ions of an impurity can be embedded in the surface. It is also possible that from an electrolyte solution certain ions can be firmly adsorbed onto the surface by Van der Waals and other selective molecular forces. Thus the solid surface carries a net charge. The important measure of the charge for all the electrokinetic phenomena is the potential just outside the surface at the level where the continuum fluid velocity appears to drop to zero, i.e. on the effective 'no slip' plane. This potential is called the zeta potential, ζ .

Now with the exception of liquids of inert molecules, the liquid will contain mobile ions of different charges. In particular an electrolyte solution will contain a large number of dissociated ions. The ions in the liquid with opposite charges to that of the solid surface (the counter-ions) are attracted towards the surface, while ions of the same sign are repelled. Considerable thermal or Brownian motions prevent the ions moving far towards or away from the surface. Thus few ions come out of the liquid and the result is a diffuse cloud of charge near to the surface with a net charge opposite to that of the surface. This diffuse cloud along with the charged surface constitute the electrical double layer.

The basic theory of the diffuse cloud of charge dates back to Gouy in 1908. In thermodynamic equilibrium the number density of the ith species of ion varies in space according to the Boltzmann distribution

$$n^i = n_0^i e^{-\frac{ez^i\phi}{kT}}$$

in which n_0^i is the number density far from the surface, e the charge of an electron, z^i the valency of the ith species, $\phi(\underline{x})$ the electrostatic potential with $\phi \to 0$ far from the surface where $n^i \to n_0^i$, and kT is the Boltzmann temperature. If the liquid is electrically neutral away from the surfaces, i.e. not too many ions are adsorbed onto the surfaces, then $\Sigma_i n_0^i e z^i = 0$. Now for small surface potentials ($\zeta < 25$mV i.e. $e\zeta/kT < 1$) the number density of the ions differs little from its value far from the surface and we may linearize the exponential in the Boltzmann distribution for

$$n^i = n_0^i \left(1 - \frac{e z^i \phi}{kT} + \cdots\right)$$

Summing over the species, and assuming that the liquid is electrically neutral, we have the total charge density that is proportional to the potential

$$\rho = \Sigma_i n^i e z^i = -\left(\Sigma_i n_0^i z^{i2} e^2/kT\right)\phi$$

Substituting this into Poisson's equation for the potential

$$\nabla^2 \phi = -\rho/\epsilon$$

in which ϵ is the dielectric constant of the liquid, we obtain the governing equation

$$\nabla^2 \phi = \kappa^2 \phi$$

where $\kappa^2 = \Sigma_i n_0^i z^{i2} e^2/\epsilon kT$. The solution of this equation for a sphere of radius a is

$$\phi = \zeta \frac{a}{r} e^{-\kappa(r-a)}$$

This shows that the diffuse charge cloud has a thickness κ^{-1}, the Debye-Hückel length which occurs in the theory of solvents. For the very dilute 10^{-5} Molar solution of NaCl salt in water, $\kappa^{-1} = 0.1\,\mu m$. The charge cloud is thick if the electrolyte is dilute (n_0^i small) and if the dielectric constant is large ($\epsilon \sim 80$ for water, cf. $\epsilon \sim 3$ hydrocarbon solvents).

There are several limitations in the simple Gouy theory of the electrical double layer. The restriction that the potentials are small ($\zeta < 25$mV) may be removed by not linearizing the exponential. The resulting nonlinear equation governing the potential must then be solved numerically. The other limitations reflect the fact that ions do not quite move according to the macroscopic electric field. The electrostatic field seen by an ion differs from the macroscopic field if there are many ions present. The ions can be affected by Van der Waals and other molecular forces. Finally the high dielectric constant for water is not constant in the layer of molecules near the surface if the layer is highly ordered or highly charged.

ELECTROKINETIC PHENOMENA

In the various electrokinetic phenomena the diffuse charge cloud is distorted by an applied electric field or an applied flow. The equations governing the distorted charge cloud are first the equations of electrostatics

$$\nabla^2 \phi = -\rho/\epsilon \qquad \text{with} \qquad \rho = \Sigma_i n^i e z^i$$

In all the phenomena the fluid motion is slow and on a short length scale (low Reynolds number) and so is governed by the Stokes equations, including an electrostatic body force

$$\nabla \cdot u = 0 \qquad \text{and} \qquad 0 = -\nabla p + \mu \nabla^2 u - \rho \nabla \phi$$

Finally the ion densities n^i are governed by a conservation equation

$$\frac{\partial n^i}{\partial t} + \nabla \cdot (n^i v^i) = 0$$

and a dynamical law for the velocity v^i of the ith species

$$v^i = u + \omega^i (-ez^i \nabla \varphi - kT \nabla \log n^i)$$

which says that the ions drift relative to the fluid due to the electrostatic and entropic forces acting on them, with ω^i the mobility of the ith species of ion. The boundary conditions associated with the above field equations are straightforward for the fluid flow, i.e. no slip relative to the possibly moving rigid surface. The electrostatic and the ion boundary conditions are more contentious. In some circumstances the charge on the surface is fixed, e.g. when the surface charge is due to embedded impurities. In other circumstances something like the surface potential ζ remains fixed, when for example the surface charge is due to the adsorption of ions which are relatively freely exchanged with the liquid. In the second situation it is necessary for the chemical reaction rate to be much faster than any other time scale of interest. Fortunately some of the electrokinetic phenomena turn out to be insensitive to whether the boundary condition is constant charge or constant potential or something in between.

Before proceeding to a more complete review of the electroviscous effects, two other electrokinetic phenomena will be briefly reviewed, because these other phenomena provide an independent way of measuring the controlling surface potential ζ , as well as being of interest in their own right.

In the phenomenon of electro-osmosis an electric field E is applied along a pipe and this induces a fluid motion along the pipe. The details of the solution of the governing equations are sufficiently simple for them to be presented here. In most experiments the charge cloud is very thin compared with the diameter of the pipe. Within these thin charge clouds, the electric charge density is unchanged by the applied field, i.e.

$$\rho = -\epsilon \zeta \kappa^2 e^{-\kappa y}$$

where y is the distance normal to the wall. The equation for the viscous flow along the pipe then becomes

$$\mu \frac{d^2 u}{d y^2} + \rho E = 0$$

So outside the thin wall layers

$$u \to \frac{\epsilon \zeta}{\mu} E$$

Because this result depends on few parameters, and is independent of κ , it provides a convenient way of measuring the potential of the surface ζ . For $\zeta = 25\,mV$, an applied field $E = 1\,V/cm$ would induce a velocity of $2\,\mu m/s$. Such small velocities require a microscope to measure them in a pipe, although in a porous rock they can lead to large and more easily measured pressure gradients.

In electrophoresis an applied electric field E causes a suspended particle to move with a velocity V . The theory for this phenomenon was developed by Smoluchowski 1921, Henry 1931, Booth 1950a, Wiersema, Loeb & Overbeek 1966, D.A. Saville 1977 and O'Brien & White 1978. In the case of a thin double layer compared with the size of the particle we have from the electroosmosis result above

$$v = \frac{\epsilon \zeta}{\mu} E$$

While the electric field and hence the velocity of the fluid just outside the double layer will vary over the surface of the particle, both the electric field and the flow are governed by the same potential problem (if the dielectric constant of the liquid is much greater than that of the particle) and so the variations cancel. Hence the result for thin double layers is independent of the shape, as well as the size, of the particle. For thick double layers, i.e. very small particles, the electrophoretic velocity v is obtained by balancing the Stokes viscous drag $6\pi\mu a v$ (for a spherical particle of radius a) with the electric force on a point particle QE, where the total charge Q is related to the surface potential by $Q = 4\pi\epsilon a \zeta$ (for a sphere). Thus for a spherical particle

$$v = \frac{2}{3} \frac{\epsilon \zeta}{\mu} E$$

At intermediate thicknesses of the double layer it is necessary to calculate the small distortion of the charge cloud caused by the flow, this distortion tending to retard the particle.

The electroviscous effects are a number of different effects which cause a suspension of charged particles to have a different rheology to the same suspension but with uncharged particles. Typically the presence of charges on the particles increases the viscosity and leads to some elasticity. The so called primary electroviscous effect concerns the distortion of the charge cloud around an individual particle caused by the shearing of the suspension. The forces pulling the distorted charge cloud back to equilibrium transmit a force through a distance and constitute an extra stress in the suspension. In the so called secondary electroviscous effect one ignores the distortion of the charge cloud and instead focuses on the interactions between particles as they pass one another in the shearing suspension. Finally in the tertiary electroviscous effect one studies the influence of the charges on the deformation of a particle, for example a charged flexible polymer molecule.

The analysis of the electroviscous effects proceeds as in the study of any other suspension. First one examines the detailed motion of the particles, including the electrical double layers. Then one calculates the bulk stress by averaging over a large volume the stress found on the particle scale, remembering to include the electric Maxwell stress along with the viscous, Brownian and other stresses.

THE PRIMARY ELECTROVISCOUS EFFECT

This is the effect due to the distortion of the charge cloud around an individual particle. In practice it is a small effect, with no significant application, although it can be measured, just. The early theories were given by Smoluchowski 1916 for thin double layers and by Booth 1950b. Booth's theory was tested by experiments by Bull 1940, Dobry 1953 and Stone-Masui & Watillon 1968. Recent theoretical contributions have been made by Russel 1978a, Lever 1979 and Sherwood 1980.

Booth's theory rests on a number of restrictions. He studied only rigid spherical particles. The suspension was dilute so that interactions between the particles could be ignored. The ion equation was linearized for small potentials ($e\zeta / kT < 1$) by replacing n^i by n_0^i in the term multiplying the electric field. The shearing flow was assumed to be weak so that the charge cloud was only slightly distorted. This assumption requires a certain ion Peclet number to be small. When it is small, the n^i multiplying the flow u in the ion equation may be replaced by its values in the undistorted charge cloud, and also the quadratic body force $-\rho\nabla\varphi$ may be similarly linearized around the undistorted charge cloud. The final assumption in Booth's basic theory is that the electrical forces are small compared with the viscous forces, so that to a first approximation the shearing flow past the particle is that calculated for an uncharged particle. This final assumption requires a certain electric Hartman number to be small. Instead of reproducing Booth's lengthy calculation, below I present just a dimensional analysis which estimates the magnitude of the effect in the two cases of thick and thin double layers. Afterwards the recent developments which lift some of Booth's restrictions will be reviewed.

When the electrical double layer is thick compared with the size of the particle, the particle behaves like a point charge $Q = 4\pi\epsilon a\zeta$, at least in the first approximation for the primary electroviscous effect. Then within the charge cloud, i.e. on the length scale κ^{-1}, the potential is $O(Q\kappa/\epsilon)$ so that the electric field is $O(Q\kappa^2/\epsilon)$. To calculate the magnitude of the distortion of the charge cloud, we balance the Brownian forces on an ion with the fluid forces. In a shearing flow with a shear rate γ, the velocities across the charge cloud are $O(\gamma\kappa^{-1})$, and so dividing by the ion mobility we have the fluid forces on the ions are $O(w^{-1}\gamma\kappa^{-1})$. If the distortion of the charge cloud changes the number density of the ions by Δn from n, then the Brownian forces opposing the distortion are $O(kT\kappa\,\Delta n/n)$. Balancing the forces we obtain

$$\frac{\Delta n}{n} = \frac{\gamma}{\kappa^2 wkT}$$

The nondimensional group on the right hand side is known as the Peclet number. It will be small, and so Booth's theory is applicable, for all but the highest shear rates, e.g. for the dilute 10^{-5} Molar solution of NaCl in water at room temperature, which has $\kappa^{-1} = 0.1\,\mu m$, we require $\gamma < 10^5\,s^{-1}$. With our estimate above for the distortion of the charge cloud, we can now proceed to estimate the increase in the bulk stress. The volume fraction of the fluid affected is $N\kappa^{-3}$, where N is the number density of the particles in the suspension. The magnitude of the Maxwell stresses is ϵE^2 so the magnitude of the deviatoric Maxwell stresses caused by the distortion of the charge cloud is $\epsilon E^2\,\Delta n/n$. Bringing together these estimates, we find that the extra bulk stress in the suspension due to the distortion of the charge clouds is

$$\mu\gamma \,.\, N\kappa^{-3} \,.\, \frac{\kappa^2 Q^2}{\mu\epsilon wkT} \,.\, \frac{1}{240\pi}$$

where the numerical factor of $1/240\pi$ comes from Booth's full analysis. For this case of thick double layers, the assumption that the particles do not interact requires the charge clouds not to overlap, i.e. $N\kappa^{-3}$ be small. The assumption that the viscous forces dominate the electric forces requires the group $\kappa^2 Q^2/\mu\epsilon wkT$ to be small. For a surface potential of $25\,mV$ and a $0.05\,\mu m$ sphere with a $0.1\,\mu m$ thick charge cloud in water (need $e\zeta/kT < 1$ and $a < \kappa^{-1}$), the surface charge is $100e$ and this nondimensional group is then 40, which is not small, but see later. While the charge cloud contribution to the bulk stress is small, being the product of two small non-dimensional groups, this small stress can exceed the Einstein term for rigid uncharged spheres $\mu\gamma\,.\,Na^3\,.\,20\pi/3$ if the particles are very much smaller than the charge cloud. Finally we can observe that, because the particle behaves as a point charge, the thick double layer result is not restricted to spherical particles.

The estimate of the primary electroviscous effect for thin double layers follows the basic arguments for the case of thick double layers with a couple of minor modifications. First the cloud charge is distorted by the velocity normal to the surface. As this grows quadratically from the surface like $\gamma(r-a)^2/a$ (a consequence of the no slip boundary condition and incompressibility), within the double layer of thickness κ^{-1} the distorting velocity is $O(\gamma\kappa^{-2}a^{-1})$. Then balancing the Brownian forces on an ion with the fluid forces using this estimate of the velocities, we obtain an estimate for the distortion of the charge cloud

$$\frac{\Delta n}{n} = \frac{\gamma}{\kappa^2 w k T} \cdot \frac{1}{a\kappa}$$

The second modification required for thin double layers is that the volume fraction of the suspension within the layers is $Na^2\kappa^{-1}$. Bringing together these modified estimates, and using the surface potential ζ which is more natural for thin double layers than the total charge Q, we find that the extra bulk stress in the suspension due to the distortion of the charge clouds is

$$\mu\gamma \,.\, Na^3 \,.\, \frac{\epsilon\zeta^2}{a\kappa\,\mu w k T} \,.\, \frac{1}{a\kappa} \,.\, 20\pi$$

where again the numerical factor of 20π comes from Booth's full analysis. For this case of thin double layers, the assumption that the particles do not interact is that Na^3 is small, while the assumption that the viscous forces dominate the electric forces is that $\epsilon\zeta^2/a\kappa\mu w k T$ is small. [For a surface potential of $25\,mV$ and a $2\,\mu m$ sphere with a $0{\cdot}1\,\mu m$ double layer in water, this last non-dimensional group is $0{\cdot}02$.] Because the charge cloud contribution to the bulk stress is the product of three small quantities, it is a very small effect and difficult to detect in experiments. Finally it should be noted that the thin double layer analysis does not apply until the double layer is quite thin, $\kappa^{-1} < \frac{1}{20}a$.

All of the simplifying assumptions made by Booth have now been examined. Russel (1978a) and Lever (1979) have analysed the nonlinear distortion of the charge cloud which occurs when the Peclet number $P = \gamma/\kappa^2 w k T$ is not small. Russel used a boundary layer theory for thin double layers which is applicable while $P < a\kappa$ $(\gg 1)$. As all types of linear shearing flows could be studied, Russel was able to derive a constitutive equation for the suspension which was of an Oldroyd type incorporating only Jaumann derivatives. Thus the primary electro-viscous effect shows a shear thinning at $P = 1$ but no change from a Newtonian viscosity in pure extensional flows up to $P = a\kappa$. This strange behaviour reflects the first effect of the stronger shears is not to distort the charge cloud differently but instead to rotate the spherical particle with the vorticity. The distortion of the charge cloud does then change as it responds to the oscillating normal component of the velocity seen by the rotating particle. A similar behaviour is found in a suspension of liquid drops when their viscosity is much greater than that of the suspending liquid.

Lever's 1979 study of genuinely large distortions of the charge cloud was for thick double layers. In this case the particle becomes a point charge in an undisturbed linear shearing flow and this can be easily analysed using Fourier transforms. Lever found that the so called second order fluid theory for the first departures from Newtonian behaviour had a particularly small range of applicability $P < 0{\cdot}1$. At higher flow strengths, $P > 10$, the double layer contribution to the bulk stress decreases like $N\epsilon^{-1}Q^2\kappa^2(wkT/\gamma)^{1/2}$. In these strong flows the charge cloud is substantially stripped off the particle and spread over a long distance downstream from the particle. This much diluted charge cloud induces weak electric fields compared with that from the point charge particle, although these weak fields are essential to produce non-isotropic components in the bulk stress. Lever finds that the main contribution to the bulk stress comes from a small region near the particles of size $\ell = (wkT/\gamma)^{\frac{1}{2}}$; up to this region the fast flow can bring in neutral electrolyte solution before the electric field is strong enough to repell the co-ions and attract the counter-ions. Inside this small region the electric field from the point charge is $O(\varepsilon^{-1} Q \ell^{-2})$ and from the charge cloud is $O(\varepsilon^{-1} Q \kappa^2)$, and so the deviatoric contribution to the bulk stress from this small region is $O(N \varepsilon^{-1} Q^2 \kappa^2 \ell)$.

The effect of the electrical forces on the flow, which was ignored in Booth's basic analysis, has been explored by Sherwood (1980). As long as the shear flow is weak so that the distortion of the charge cloud is small, this problem requires the solution of some ordinary differential equations which couple the flow and the perturbation to the electric field. Solving these equations Sherwood found that there was a 4% decrease from Booth's theory when the non-dimensional group $\epsilon \zeta^2 / a\kappa \mu \omega kT$ exceeds 0.04 for the thin double layer with $a = 500\,\kappa^{-1}$, and that there was a 3% decrease from Booth's theory when the relevant non-dimensional group $\kappa^2 Q^2 / \epsilon \mu \omega kT$ exceeds 800 for the thick double layer $a = \frac{1}{2}\kappa^{-1}$. Thus the effect of the electrical forces on the flow is negligible in practice even when the appropriate non-dimensional group for thick double layers often exceeds unity (see estimates earlier). Sherwood went on to consider what would happen if the electrical forces were much larger than those found in nature. He found that close to the particle the fluid would be held at rest by the strong electric forces which would be created by any movement of the charge cloud. These strong electric forces however decay exponentially away from the particle, and so the viscous forces balance them at some radius R given by $R^2 e^{-2\kappa(R-a)} = a^2 \epsilon \zeta^2 / \mu \omega kT$. As the electrical forces become negligible a few κ^{-1} outside this R, the particle behaves like a rigid sphere with an effective radius R and contributes to the bulk stress according to the Einstein result $\mu\gamma \,.\, NR^3 \,.\, 20\pi/3$.

In a second study in the same paper Sherwood (1980) examined the modification of Booth's theory needed for high surface potentials. Although the static double layer now satisfies a nonlinear differential equation, the perturbation caused by the shearing flow is linear at low shear rates. For thick double layers Sherwood found that the charge cloud contribution to the bulk stress increases monotonically with the surface potential ζ, at low ζ growing quadratically according to Booth's theory and at high ζ tending to a limiting value. The first deviations from Booth's theory occur not when $e\zeta/kT = 1$ but at a higher value corresponding to the net counter-ion charge within a radius of the particle being comparable with the charge on the particle, i.e. $\frac{4\pi}{3}(2a)^3 e n_0 e^{\frac{e\zeta}{kT}} \gtrsim 4\pi a \epsilon \zeta$ or $a^2\kappa^2 e^p \gtrsim p$ with $p = e\zeta/kT$. Thus at $a\kappa = 0{\cdot}05$ Booth's theory is 5% too high only at $e\zeta/kT = 6$. At very high surface potentials, so many counter-ions are attracted to the particle that the charge on the particle is virtually neutralized within a radius of the particle by the cloud of counter-ions, the neutralizing ceasing as the potential drops to the critical value given above for the first deviations from Booth's theory. The distant κ^{-1} region outside the particle, from which the leading contribution from the bulk stress comes, thus sees a particle with this critical potential and so the bulk stress tends to a limiting value at high potentials.

For thin double layers Sherwood found a more complicated behaviour : the electrical contribution to the bulk stress first increased to a maximum and then dropped to a limiting value at high surface potentials. The first deviations from Booth's theory occur at $e\zeta/kT = 1$ for thin double layers, with a 5% underestimate at $e\zeta/kT = 2$ for $a\kappa = 10^3$. Beyond this surface potential the co-ions are effectively repelled from the surface to the point where the potential drops to $\varphi = kT/e$. With the co-ions excluded and the electric fields increased to $\kappa kT/e \,.\, e^{\frac{1}{2}\frac{e\zeta}{kT}}$ in a thin region $\kappa^{-1} e^{\frac{1}{2}e\zeta/kT}$, the counter-ions contribute twice Booth's value to the bulk stress, and so the bulk stress starts to increase more rapidly than Booth's theory. At the higher potential of $\zeta = 2kT/e \,.\, \log a\kappa$, perturbations to the counter-ions become inhibited by tangential electric fields, and so their contribution to the bulk stress drops. Thus at very high potentials we are left with just the contribution from the co-ions which see the surface neutralized by the co-ions down to an effective potential of $4kT/e$.

The only analysis of the primary electroviscous effect for a non-spherical particle is a paper by Sherwood (1981) on rods, motivated by an experiment by Domard (1976) who reported an intrinsic viscosity of completely ionized poly-α-L-glutamic acid three times that of uncharged rigid rods of the same length. Sherwood found that while the charge cloud contribution to the bulk stress increases with the length ℓ of the rod due to the cloud being in higher flows and transmitting the force over a longer distance (net effect proportional to ℓ^3 as in the contribution for an uncharged rod), such an increase had to be offset against a decrease in the charge on the rod per unit length if the total charge was held fixed (decrease proportional to ℓ^{-2}). In explaining Domard's observations, Sherwood was forced to a numerical calculation because the charge cloud was as thick as the length of the rod. In such circumstances there is an important interaction between the cloud from different parts of the particle which is not included in the two simple models of point charges widely spaced along the rod and of a long slender cloud charge.

Finally in the paper about rods, Sherwood makes an important remark concerning the testing of the theory of all electroviscous effects using the surface potential ζ measured by electrophoresis. Sherwood observed that the electroviscous effects depend on ζ^2 while electrophoresis depends on ζ . If the surface potential is not constant over the surface of the particle, then the appropriate average of ζ^2 for an electroviscous effect might be considerably different from the square of the appropriate average of ζ measured by electrophoresis.

THE SECONDARY ELECTROVISCOUS EFFECT

This is the effect due to the interactions between the particles. It is a large effect, the one responsible for the dramatic changes in rheology when suspended particles become charged. An early theory of the effect by Chan, Blanchford & Goring (1966) has now been superseded by the work of Russel (1978b). Experiments have been performed by Chan et al. and by Stone-Masui & Watillon (1968).

While the secondary electroviscous effect is large, it is not the most important colloidal phenomenon involving the interaction between the particles. Attractive Van der Waals forces can cause small suspended particles to aggregate. The first interest of a colloidal scientist is whether the electrostatic repulsion from the double layer can stop this aggregation, known as stabilizing the suspension. If the suspension does aggregate, a shearing flow can increase the aggregation rate by bringing together particles faster than Brownian motion, while at very high flow rates the shearing can tear apart aggregates, see van de Ven & Mason (1977) and Zeichner & Schowalter (1977). The following discussion of the secondary electroviscous effect presupposes that the particles are not aggregating, at least at any significant rate.

To consider the interactions between particles in a suspension we need the electrostatic repulsion law for two particles with double layers, and this has not yet been calculated for all possible conditions. (There are some interesting problems here at small separations with large surface charges, and with the time needed for some surfaces to come to electrochemical equilibrium as the particles change their separation). We shall use a simple repulsion law which applies to two spheres with low surface potentials $e\zeta/kT < 1$ not too close together $r - 2a > \mathrm{Min}(a, \kappa^{-1})$,

$$F = \frac{\epsilon \zeta^2 a^2}{r^2} (1 + \kappa r)\, e^{-\kappa(r-2a)}$$

where r is the distance between the centres of the spheres. Fortunately many results are not sensitive to the details of the force law. The magnitude of this repulsive force for a $0.5\mu m$ sphere in water with a surface potential $\zeta = 25\,mV$ at a separation of two double layer thicknesses $\kappa^{-1} = 0.1\mu m$ is $1.5 .10^{-13}$ Newtons. This should be compared with the Van der Waals attraction $Aa/12(r-2a)^2$ of 10^{-14} Newtons, the viscous drag $6\pi\mu a^2\gamma$ of 5.10^{-15} Newtons at a shear rate of $\gamma = 1s^{-1}$, and a Brownian force kT/a of 8.10^{-15} Newtons. Thus in practical situations all these forces are likely to be important. For simplicity however we consider a sequence of idealization in which one or more of the forces are suppressed.

One additional simplification is needed to make theoretical progress; we must assume that the suspension is dilute. Thus we study just the first effects of the interactions between particles, which mostly involve pairs of particles interacting, and which mostly are $O(c^2)$ in the bulk stress where $c = 4\pi Na^3/3$ is the volume fraction of the particles in the suspension. While the assumption of diluteness requires c to be small, the dramatic rheology will be revealed in abnormally large coefficients of c^2 .

We start at the high shear rate limit in which hydrodynamic interactions dominate. In the absence of electrostatic and Brownian forces, most pairs of particles move past one another according to pure hydrodynamics on open trajectories which separate indefinitely at large future and past times. In some flows with sufficient vorticity like simple shear, however, a small proportion of the pairs of particles orbit one another indefinitely, see Goldsmith and Mason (1964) and Lin, Lee & Sather (1970). Weak electrostatic repulsions disturb the open trajectories little, but have a major effect on the closed trajectories, totally depleting them after sufficient time. Thus using the work of Batchelor and Green (]972), we find that at high shear rates the suspension is characterized by an effective viscosity $\mu(1+2.5c+7.6c^2+\dots)$ in pure straining flows which only have open trajectories, and by an effective viscosity $\mu(1+2.5c+6c^2+\dots)$ in simple shear once the closed trajectories have all been depleted. Finally it should be noted that particles which start on a collision course do come very close during the hydrodynamic interaction, within $10^{-3}a$, and at these separations otherwise small Van der Waals attractions might become significant.

At lower shear rates the particles take longer to pass one another on the open trajectories and thus become deflected by the electrical repulsion, the lower the shear rate the more distant the particles affected. It is useful to estimate the range r_* of the significant electrical interaction by balancing a typical viscous force with the electrical force, i.e. r_* is given by

$$6\pi\mu a\,\gamma\, r_* \quad = \quad \frac{\epsilon\zeta^2 a^2}{r_*^2}\,(1+\kappa r_*)\,e^{-\kappa(r_*-2a)}$$

Inside $r = r_*$ the electrical repulsion dominates, while outside the hydrodynamic forces win. Thus a pair of particles approach one another on the open trajectories unaffected by the repulsion until r drops to r_* . Then the pair roll over one another keeping roughly to $r = r_*$, and once past they part deflected onto new, more distant open trajectories.

The range of interaction r_* , as noted above, increases as the shear rate is decreased. At sufficiently high shear rates our formula gives $r_* < 2a$. Under these conditions hydrodynamic forces dominate at all separations and we are in the high shear rate limiting regime. In weaker flows r_* is larger and once $r_* > 3a$ the complicated details of the hydrodynamic interaction and the full electrostatic law become unimportant. In the case of thick double layers there are two possible shear thinning regimes, a bare Coulombic regime when $3a < r_* < \kappa^{-1}$ which has

$$r_* \quad = \quad (\epsilon\zeta^2 a/6\pi\mu\gamma)^{1/3}$$

and a shielded regime when $3a < \kappa^{-1} < r_*$ which has

$$r_* = \kappa^{-1} \log [x/(\log x + 2a\kappa)^2] + 2a \quad \text{with} \quad x = \epsilon \zeta^2 a \kappa^3 / 6\pi\mu\gamma$$

In the shielded regime the pairs of particles roll past one another deviating little from $r = r_*$, because the electrical repulsion drops more abruptly through the exponential in this case than it decays algebraically in the Coulombic case. When the double layers are thin, there is only the shielded shear thinning regime once $r_* > 3a$.

To estimate the bulk stress in the shear thinning regimes first consider just those pairs of particles at the separation r exerting a force F on one another. The stress transmitted across a small macroscopic area δA by these pairs, $\sigma \cdot \delta A$, will be the force F multiplied by the number of pairs crossing δA, i.e. the number of particles in the volume $r \cdot \delta A$. Averaging over the different separations which occur with a probability density $p(r)$, we obtain the contribution to the bulk stress from the interaction of these effectively point particles

$$\sigma = N \int F\, r\, p(r)\, d^3 r$$

Using the interaction distance r_* for the typical separation we find the estimate

$$N \,.\, 6\pi\mu a \gamma r_* \,.\, r_* \,.\, N r_*^3$$

which corresponds to an effective viscosity of the suspension

$$\mu \left[1 + 2.5c + k c^2 \left(\frac{r_*}{a} \right)^5 \ldots \right]$$

with the constant k coming from the detailed calculations. [Russel (1978b) found that in the Coulombic regime $k = 2.87$ in simple shear and $k = 2.79$ in axisymmetric straining motion.] Had the factor in the effective viscosity been $(r_*/a)^6$ instead of the fifth power, it would have meant that the particles were effectively acting like rigid spheres of radius r_*. This is the source of the dramatic rheology.

We now turn to the low shear rate limit. As the shear rate decreases the interaction distance r_* increases, apparently without bound, and the effective viscosity increases. At sufficiently large separations, however, the Brownian motions of the particles become significant. They enter when the thermal energy kT exceeds the electrical interaction energy $\epsilon \zeta^2 a^2 e^{-\kappa(r-2a)}/r$, i.e. when increases to

$$r_{*\,max} = \kappa^{-1} \log [y/(\log y + 2\kappa a)] + 2a \quad \text{with} \quad y = \epsilon \zeta^2 a^2 \kappa / kT$$

The appropriate constant k to use with this value of r_* in the effective viscosity was calculated to be $3/40$ by Russel (1978b). He compared this prediction for the low shear rate viscosity with the measurements of Stone-Masui & Watillon (1968). Large values of $\frac{3}{40}(r_*/a)^5$ were measured, like 10^3, and were predicted within 20% (after the original data had been corrected for the ion concentration being higher than in the pure electrolyte through ions released from the surface of the particles). Finally it should be noted that there are indications that the first departures from the Newtonian low shear rate limit involve stresses proportional to $\gamma^{3/2}$ and $\gamma^2 \log \gamma$. If this is correct, the suspension does not have a second-order fluid approximation.

A second, alternative limit to the unbounded increase in the interaction distance r_* is set by the typical particle separation $N^{-1/3} = 1.6\, a\, c^{-1/3}$. When r_* rises to this bound, the particles no longer interact just in pairs but instead each particle interacts simultaneously with many others, leading to contributions to the effective viscosity $O(c^{1/3})$. As the particles behave like point charges in these multiparticle interactions if $c \ll 1$, the regime is suitable for numerical simulation. Some preliminary investigations have examined how initially randomly placed particles would move in the absence of a shearing flow. These investigations found that on a time scale $c^{1/3}\epsilon\zeta^2/6\pi\mu a$, the particles first all move to be at least $0.8\, a\, c^{-1/3}$ apart, then move into unfilled random gaps, and finally settle down into a locally ordered crystal with a fair number of crystal faults. The growth of a crystal structure has frequently been observed in dilute suspensions of charged polystyrene latices. It is amusing to note that, if a perfect such crystal were sheared along a crystal face, it should alternately store and release electrostatic potential energy with no net dissipation, thus producing a material with a yield stress $O(c^{3/4}\,\epsilon\zeta^2 a^{-2})$ and a zero viscosity. A non-zero viscosity would result from the particles not being precisely in their lattice sites due to Brownian motions, or from some crystal faults. Some important experiments which are relevant here are those by Hoffmann (1972 & 1974). He found in more concentrated suspensions that as the shear rate is increased the viscosity increases discontinuously from a low value as the suspension jumps to a disordered state from an ordered state (detected by X-ray diffraction). Some recent work by Clark (1981) has found the same behaviour, with a progression of ordered states, in some very dilute suspensions.

In practical situations the charge clouds are not significantly distorted by the low shear rates involved in the secondary electroviscous effects. Lever 1978 however has calculated what would happen if two particles with significantly distorted thick charge clouds were to collide, taking into account the changing distortion during the collision. He finds that, as the particles repell one another, they slip relative to the fluid and their clouds, and the resulting eccentricity with their clouds inhibits further slip, i.e. the repulsive forces are less effective. He also finds that under special conditions the particles can attract one another through the quadrupole nature of the far fields.

It is a pleasure to have an opportunity to thank W.B. Russel, D.A. Lever and J.D. Sherwood who have contributed so much to my understanding of this subject.

REFERENCES

Batchelor, G.K. & Green, J.T. 1972 J. Fluid Mech. 56, 401.
Booth, F. 1950a Proc. Roy. Soc. A203, 514.
Booth, F. 1950b Proc. Roy. Soc. A203, 533.
Bull, H.B. 1940 Trans. Faraday Soc. 36, 80.
Chan, F.S., Blanchford, J. & Goring, D.A.I. 1966 J. Colloid Interface Sci. 22, 378
Clark, N.A. 1981 Proc. IUTAM/IUPAC Symp. in Canberra on 'Interaction of particles in colloidal dispersions'.
Dobry, A. 1953 J. Chim. Phys. 50, 507.
Goldsmith, H.L. & Mason, S.G. 1964 Proc. Roy. Soc. A282, 569.
Gouy, G. 1908 Ann. Chim. Phys. 8, 294.
Henry, D.C. 1931 Proc. Roy. Soc. A133, 106.
Hoffman, R.L. 1972 Trans. Soc. Rheol. 16, 155.
Hoffman, R.L. 1974 J. Colloid Interface Sci. 46, 491.
Lever, D.A. 1978 Ph.D. thesis, Cambridge.
Lever, D.A. 1979 J. Fluid Mech. 92, 421.
Lin, C.J., Lee, K.J. & Sather, N.F. 1970 J. Fluid Mech. 43, 35.
O'Brien, R.W. & White, L.R. 1978 J.C.S. Faraday II 74, 1607.
Russel, W.R. 1978a J. Fluid Mech. 85, 673.
Russel, W.R. 1978b J. Fluid Mech. 85, 209.
Saville, D.A. 1977 Ann. Rev. Fluid Mech. 9, 321.
Sherwood, J.D. 1980 J. Fluid Mech. 101, 609.
Sherwood, J.D. 1981 to appear in J. Fluid Mech.
Smoluchowski, M. 1916 Kolloidzschr. 18, 190.
Smoluchowski, M. 1921 Handbuch der Electrizitat und des Magnetismus II, 366.
Stone-Masui, J. & Watillon, A. 1968 J. Colloid Interface Sci. 28, 187.
van de Ven, T.G.M. & Mason, S.G. 1977 Colloid & Polymer Sci. 255, 468 and 794.
Wiersema, P.H., Loeb, A.L. & Overbeek, J.Th.G. 1966 J. Colloid Interface Sci. 22, 79.
Zeichner, G.R. & Schowalter, W.R. 1977 A.I.Ch.E.J. 23, 243.

Continuum Models of Discrete Systems 4
eds. O. Brulin and R.K.T. Hsieh

AN ATTEMPT TO TREAT THE VISCOSITY AS A TRANSPORT PROPERTY OF TWO PHASE MATERIALS

W.-D. Sältzer, B. Schulz

Kernforschungszentrum Karlsruhe
Institut für Material- und Festkörperforschung
Postfach 3640
D-7500 Karlsruhe
Federal Republic of Germany

INTRODUCTION

The analysis of severe but hypothetical nuclear accidents have created recently enhanced interest for physical properties under special conditions. For instance a possible granulation of the nuclear fuel into particles, which are dispersed in a liquid metal,leads to the formation of a so-called particle bed in a core-catcher. Reliable data of their physical properties are required to calculate the heat removal from these particle beds using thermohydraulic codes. This question can be traced back to the problem of the viscosity of undiluted suspensions, where the fluid (the liquid metal) is a Newtonian fluid.

In the following, the discussion of known equations for the effective viscosity of these suspensions will lead to the conclusion that these have no function enabling one to estimate the viscosity of such suspensions at such high concentration with the desired reliability.
The consideration of the shape of suspended particles and hence their orientation, in particular, justifies the attempt to treat the viscosity as an analogue to other transport properties, for which these parameters and their influence on the respective effective physical properties are known.

LITERATURE SURVEY

Numerous equations for the effective viscosity η of suspensions can be found in the literature. These may be divided into purely empirical ones and into theoretical equations. The former group is characterized by two main features:

1. They are adapted to the experimental data as best as possible (e.g. /1/).
2. They are made to fit experimental data using theoretical relationships with empirical constants (e.g. /2/).

In the latter group we find only equations based on physical considerations. They may once again be subdivided into two groups:

- Those in which the final equation is based on pure hydrodynamics, i.e. continuum models.
- Those taking account of the influence of the interaction between individual particles.

All equations of the latter group are listed in tab.1 to 3. They give the effective relative viscosity η_r (ratio between the viscosity of the suspension η and that of the fluid η_0) as a function of the concentration of the dispersed phase and of other geometrical parameters. The equations are only valid making certain assumptions.

In table 1 one can find the functions of spherical particles only in a dilute suspension. Equation (1) derived by Einstein /3/ is physically exact, but valid only under the following assumptions:

- the size of the suspended particles is large compared to that of the molecules of the fluid and small compared to the dimensions of the whole aggregate.

- the fluid can be taken as incompressible.
- the densities of the fluid and the solid are equal.
- the second phase is distributed homogeneously in the fluid.
- a possible interaction of the single particles is neglegable.
- the flow is steady beyond the near viscinity of the particles.

No.	Equation	Author	Remarks
1	$\eta_r = 1 + 2.5\ c$	Einstein 1910	–
2	$\eta_r = 1 + \nu_1 c$	Guth/Simha 1936	$\nu_1 = 2.5 + 5/32 \cdot 2.5\ a/d$ a = sphere diameter d = diameter of the fluid

Tab.1: Relative viscosity $\eta/\eta_o = \eta_r$ as a function of concentration; dilute suspensions, spherical particles.

Equation (2) /4/ tries to overcome the restriction to infinite extended dimensions by introducing the parameters of the geometry of the measuring system and thus increasing the concentration dependence.
Table 2 summarizes equations which describe the extension of the Einstein equation to higher concentrations.
Here Brinkmann /5/ (equ.2 in tab.2) used the well known method of Bruggemann /6/ resulting in the exponential function of the effective viscosity versus concentration. Roscoe /7/ (equ.6 in tab.2) used the same type of equation but differentiated between suspensions of spheres of equal size and of spheres with an ideal size distribution. The first type can only reach a maximum concentration of 74 vol.% (1/0.74 ≡ 1.35), which also means the maximum viscosity possible, whereas for an ideal size distribution the equation is identical to Brinkmann's

No.	Equation	Author	Remarks
1	$\eta_r = 1 + 2.5\ c + \frac{109}{14} c^2 + \ldots$	Guth/Simha 1936	
2	$\eta_r = (1 - c)^{-2.5}$	Brinkmann 1952	
3	$\eta_r = 1 + 4.5\ c$	Hatschek 1910	
4	$\eta_r = (1 - 2.5\ c + 2.5 \cdot 0.62\ c^2)^{-1}$	de Bruijn 1943	
5	$\eta_r = 1 + 2.5\ c + 7.35\ c^2$	Vand 1948	
6	$\eta_r = (1 - m\ c)^{-2.5}$	Roscoe 1952	m = 1 size distribution m = 1.35 equal size, c_{max} = 0,74 = 1/1.35
7	$\eta_r = 1 + 2.5\ \lambda c + 6.25\ \lambda^2 c^2 + \ldots$	Simha 1952	$\lambda = \frac{4\ (1 - \frac{a^x}{a})^7)}{4(1 + (\frac{a^x}{a})^{10} - 25\ (\frac{a^x}{a})^3\ (1 + (\frac{a^x}{a})^4) + 42\ (\frac{a^x}{a})^5}$
8	$\eta_r = 1 + 5.5 \cdot \psi \cdot c\ \ldots$	Happel 1957	$\psi = \frac{1 + 5.5 \cdot c\ [4\ (\sqrt[3]{c})^7 - \frac{84}{11} (\sqrt[3]{c})^2 + 10]}{10\ [1 - (\sqrt[3]{c})^{10}] - 25\ c[1 - (\sqrt[3]{c})^7]}$

Tab.2: Relative viscosity $\eta/\eta_o = \eta_r$ as a function of concentration of spherical particles, no restriction for concentration, a^x/a-correction for interaction of spheres (radius - a) in a radius $r = a^x$.

Simha /8/ and Happel /9/ (equ.7 and 8 in tab.2) tried to take into account the interaction between the dispersed particles in deriving new constants, which increase further the concentration dependence of the relative viscosity. The constants have the disadvantage that they cannot be experimentally determined. Quite another method - a kinematic one - was published by Vand /11/ (equ.5 in tab.2).He firstly calculated the change in the shear tension of the fluid due to the presence of a single particle, then the change in shear tension caused by the presence of other particles,and lastly the case of collisions between indivicual particles.

All equations except No.3 and 8 in table 2 so far discussed are consistent in as much as they can be reduced to the Einstein relationship at small concentrations:

$$\eta_r = 1 + 2.5\ c \tag{1}$$

Equation 7 and 8 (in tab.2) try to take into account more realistic features, which characterize even dilute suspensions. The thus introduced parameters failed as they can not be determined in real systems with dispersed spheres.

Fig.1 shows the experimental data of the relative viscosity of suspensions with spheres. The main feature is that they differ up to several orders of magnitude for higher concentrations (The details of the curves in fig. 1 will be discussed below). This means that none of the equations can be used to describe all measured data on the relative viscosity of real suspensions of spheres, because there is only one single variable parameter in the equations - the concentration of the dispersed phase -. Some authors derive equations for dilute dispersions with particles having the shape of spheroids (tab.3).
Jeffrey /24/ (equ.1 in tab.3) assumed that the suspended particles have translation and rotational velocities which are identical to the respective values of the fluid. His equation includes a general constant which can only be calculated assuming the configuration of the suspension which causes a minimum or a maximum dissipation of frictional energy.
Peterlin /25/ (equ.2 in tab.3) avoided this general constant by introducing a time average for the movement of the spheroids. He restricted the validity of his equation and calculated the viscosity at the orientation angle $\theta = 0$ (the symmetry axis of the spheroids is perpendicular to the plane constructed by the direction of the velocity of flow and the shear tension).
Kuhn et.al. /26/ took two additive factors into account:

1. The Brown'ian motion (valid for ultrafine particles $< 1\ \mu m$).
2. A dependence of the relative viscosity of the velocity gradient.

The general form of their equation is:

$$\eta_r = 1 + c\,[(\tfrac{4}{15}\cdot A_1 + \tfrac{1}{3}\cdot A_2 + \tfrac{2}{3}\cdot A_3)\cdot f(\alpha^2_{rot}, A_1 . A_2, A_3) + B(1 - \tfrac{4}{9}\cdot \alpha^2_{rot}) \\ (1 + \frac{601}{1120}\cdot f'(\alpha^2_{rot}, A_1, A_2, A_3))] \tag{2}$$

The quantities A_n are complicated functions of the axial ratio of the spheroid and given in /27/ as well as B.

$$\alpha_{rot} = \frac{q}{4\,D_{rot}} \tag{2a}$$

q = velocity gradient; D_{rot} = rotation diffusion constant, according to /28/ (the influence of the Brown'ian movements).

Equation 3 and 4 (tab.3) are valid for prolates $a/b < 15$, and $a/b > 15$ for small velocity gradients ($q \to 0$) and allowing for the Brownian movement. All 4 equations in table 3 are of the form $\eta_r = 1 + \nu \cdot c$. This means that there is in principle no difficulty in applying Brinkmann's method to overcome the barrier of low concentrations. However, this seems most doubtful, because the authors take into account mechanisms in the movement of the particles, which are strongly influenced by an increase in the concentration. In this context the results published by Eirich, Margaretha and Bunzel /29/ are of great interest. They work with suspensions at concentrations of 0.0156 ÷ 4.0 vol.%, with rods of length/diameter ratio of 5:1 to 140:1, using two types of viscosimeters and making photographs of the suspensions. They found the orientation to be dependant on the type of the viscosimeter as well as a preferred orientation, which was connected to the axial ratio of the rods and a relative steady characteristic motion of the single particles. The authors measured a major dependence of the relative viscosity versus concentration, much too pronounced to apply one of the equations known at that time (1936). Table 4 gives the comparison of some of their data in Kuhn's equation 4

No.	Equation	Author	Remarks
1	$\eta_r = 1 + \nu_1 c$	Jeffrey 1923	$\nu_1 = f$ (axial ratio)
2	$\eta_r = 1 + \nu_2 c$	Peterlin 1939	$\nu_2 = f$ (axial ratio, shear tension)
3	$\eta_r = 1 + \nu_3 c$	Kuhn et.al. 1951	$\nu_3 = 2.5 + 0.4075\ (a/b-1)^{1.508}$ a/b-axial ratio of ellipsoid $a/b < 15$
4	$\eta_r = 1 + \nu_4 c$	Kuhn et.al.	$\nu_4 = 1.6 + [(a/b)^2/5]\cdot[(3\ln(2\,a/b) - 4.5)^{-1} + (\ln(2\,a/b) - 0.5)^{-1}]$ a/b-axial ratio of ellipsoid $a/b > 15$

Tab.3: Relative viscosity $\eta/\eta_o = \eta_r$ as a function of concentration, ellipsoidal particles, dilute dispersions.

in table 3 (1951):

c (vol.%) a/b	η_{rm} (Eirich)	a/b	η_{rc} (Kuhn)
0.5	1.031	17:1	1.151
1.0	1.076	"	1.302
2.0	1.170	"	1.604
0.125	1.035	50:1	1.221
0.25	1.075	"	1.442
0.5	1.173	"	1.885
1.0	1.285	"	2.770
2.0	1.900	"	4.541

Table 4: Measured relative viscosity η_{rm} compared to Kuhn's equation η_{rc} for prolates with high (> 15:1) axial ratios.

It is clearly seen, that the calculated values are much higher (max. factor 2.5) than the measured data. The reason might be that Eirich observed a tendency of the majority of the particles having high axial ratios to have a stable orientation with respect to the direction of flow. This fact cannot be considered in Kuhn's equation.
As a consequence it seems, as if the known equations for the relative viscosity of suspensions cannot be suitably extended to higher concentrations and irregular shaped particles.

THE VISCOSITY AS A TRANSPORT PROPERTY

If one defines transport properties by equations of the following type:

$$\dot{q} = -\lambda \cdot \text{grad } T \tag{3}$$

$\dot{q}$ - a flow density; $T = T(x,y,z)$ - a three dimensional field; λ - effective transport property;

then there are general relationships for the effective properties of two phase soli materials /30,/31/, which describe the dependence of the unlimited concentration, the shape and the orientation of the particles. They are confirmed by numerous experimental data /32,33/. In this model the particles are substituted by spheroids and it was shown /34,35/, that the axial ratio and the orientation in a given direction can be determined from plane sections of the microstructure. The influence of the shape on the effective transport property is described by the de-electrification factor F tabulated elsewhere /30/. If one considers the definition of viscosity:

$$\tau_{xy} = \eta \cdot \frac{dv}{dz} \qquad \tau_{xy} \text{ - shear tension in x,y plane} \quad dv/dz \text{ - velocity gradient in z-direction} \tag{4}$$

and analogue equations for solid materials:

$\sigma = E \cdot \varepsilon$ ε - strain due to normal stress; E - Youngs modulus (4a)

$\tau = G \cdot \phi$ ϕ - deformation angle due to shear stress; G - shear modulus (4b)

$\sigma = \eta \cdot \dot{\varepsilon}$ $\dot{\varepsilon}$ - creep rate due to normal stress - their similarity follows. (4c)

Furthermore it is possible to transform equation (4) for laminar flow and special geometrical conditions in such a way, that a direct analogy with equation (3) is shown:
For a laminar flow in a circular tube (radius a) with a flow velocity $\vec{v} = (0,0,w)$, a pressure $p = (0,0,p_z)$, the following is true:

$$\partial p/\partial z = \eta \cdot \mathrm{div} \cdot \mathrm{grad} \cdot w, \quad \text{which leads to} \tag{5}$$

$$\dot{Q}_S = -\frac{a^2}{8\eta} \cdot dp/dz = -\frac{a^2}{8\eta}\,\mathrm{grad}_z p, \quad \text{with } \dot{Q}_S \text{ - flow density.} \tag{6}$$

Besides this it is known, that the laws governing the potential flow are identical to those for the electrostatic field.

All these factors justify a treatment of the viscosity as a transport or field property.
Setting η for the conductivity and the ratio $\eta_{fluid}/\eta_{solid\ phase} \to 0$, one gets from /30,31/:

$$\eta_r = \eta/\eta_{fluid} \cdot (1-c)^{-\left(\frac{1-\cos^2\alpha}{F} + \frac{\cos^2\alpha}{1-2F}\right)} \tag{7}$$

In this equation F is the shape factor of the dispersed spheroids of axial ratio a/b, α the angle between the symmetry axis of the spheroid and the direction of the velocity gradient; $\cos^2\alpha = 1/3$ represents random orientation. This equation has the same mathematical form as Roscoes and Brinkmann's. It reduces for spheres to

$$\eta_r = (1-c)^{-3} \tag{7a}$$

It is important to point out, that using equation (7) for the calculation of the viscosity includes the assumption that the dispersed particles in a fluid can be considered in their behaviour by using time and spatial averages. Thus a time and spatial dependance of geometrical parameters is excluded. Furthermore, one should mention that the validity of equation (7) does not automatically include the acceptance of the numerical values for F as defined above for field properties, although this is done in the following comparison with experimental data. The theoretical and experimental data were firstly compared investigating spheres suspended in fluids (fig.1). From solid two phase materials it is known that when mixing particles of two phases a dispersion of single particles of phase one in a matrix of phase two is formed at low concentration (fig.2a). Increasing the concentration of phase one increases the probability of preferred contacts between particles of this phase, which leads to the formation of linear and branched chains and of agglomerates (fig.3). Thus a change of the shape of the particles takes place. The structure changes as shown in fig.2b and fig.2c, phase one becomes at last the matrix of phase two (fig.2c). It is now possible to determine the mean number of contacts and the mean number of descendants $\bar{k}$/per particle in a linear or branched chain /36,37/. The value $\bar{k}+1$ is taken as the axial ratio a/b of spheres building up chains (first approximation). The results measured in two different systems are shown in fig.4. The mean curve gives the possibility to introduce a nonspherical shape to the experimental data in fig.1. The second abscissia gives the axial ratio of chains built up of single spheres and derived from fig.4. If this is valid, the curve for $\cos^2\alpha = 1/3$ should give a mean description of the concentration dependance, whilst for $\cos^2\alpha = 0$ and 1 all cases must be covered. And indeed fig.1 shows clearly that this is true.
A survey on data worked out on systems with non-spherical particles is shown in fig.5, where the experimental data are enveloped by curves according to exponents m = 18 and 2.0 in equation (7). For a direct comparison only data on systems with particles of lower axial ratio are considered. The reason is the above mentioned Eirich observation: that for high axial ratios the orientation increasingly

changes to a preferred direction parallel to that of flow. However, the exact value of the orientation is not known. Therefore, only data on systems with particles of axial ratio 5:1 /29/, 1.83:1 /39/ and own data for oblates a/b = 0.279 /41/ are compared to equation (7) taking $\cos^2\alpha = 1/3$. As shown in fig.6 the agreement is reasonable, taking into account the lack of information on the microstructure of such systems.

SUMMARY

Finally it can be concluded that a direct transformation of equations developed for field properties of two phase solid materials to the viscosity of suspensions gives good agreement with experimental results. It is necessary for the future: to ensure the validity of the numerical data of the shape factor F and to work out methods for a description of the microstructure of suspensions at high concentration of the solid phase. This is most important for a proof of the main assumption of the transformation, i.e. to describe the microstructure by time and spatial averages.

LITERATURE

/1/ D.G. Thomas: J. Coll. Sci. 20 (1965), p.267

/2/ R. Rutgers: Rheol. Acta, Bd.2, H.3 (1962), p.202

/3/ A. Einstein: Ann. d. Phys. 19 (1906), p.289

/4/ E. Guth, R. Simha: Koll. Z. 74 (1936), p.266

/5/ H.C. Brinkman: J. Chem. Phys. 20, No.4 (1952), p.571

/6/ D.A.G. Bruggemann: Ann. Phys. 24 (1935), p.636

/7/ R. Roscoe: Brit. J. Appl. Phys. 3 (1952), p.267

/8/ R. Simha: J. Appl. Phys. Vol. 23, No.9 (1952), p.1020

/9/ J. Happel: J. Appl. Phys. Vol.28, 11 (1957), p.1288

/10/ E. Hatschek: Koll. Z. 8 (1910), p.301

/11/ V. Vand: J. Phys. Coll. Chem. 52 (1948), p.277

/12/ H. de Bruijn: Recueil 61 (1942), p.863

/13/ F. Eirich et al.: Koll. Z. 74 (1936), p.276

/14/ H. Eilers: Koll. Z. 96/97 (1941), p.313

/15/ J.V. Robinson: Phys. Coll. Chem. 53 (1949), p.1042

/16/ S.G. Ward. R.L. Whitmore: Brit. J. Appl. Phys. 1 (1950), p.286

/17/ J.V. Robinson: J. Phys. Coll. Chem. 55 (1951), p.455

/18/ P.S. Williams: J.Appl. Phys. 3 (1953), p.120

/19/ P.J. Ridgen: Road Res. Techn. Pap. 28 (1954)

/20/ H. Sweeny, R.D. Geckler: J. Appl. Phys. 25 (1954), 1135

/21/ G.H. Higginbotham et al.: Brit. J. Appl. Phys. 9 (1958), p.372

/22/ L. Nicodemo et al.: Chem. Eng. J. 29 (1974), p.729

/23/ S. Van Kao et al.: University of Washington, (1974)

/24/ G.B. Jeffrey: Roy. Soc. of Lond. A 102 (1923), p.161

/25/ A. Peterlin: J. Phys. 111 (1939), p.232

/26/ W. Kuhn et al.: Erg. exakt. Nautrw. Bd.25 (1951), p.1

/27/ R. Eisenschitz: Z. Phys. Chem. Abt. A 163 (1933), p.133

/28/ R. Gans: Ann. d. Phys. 86 (1928), p.628

/29/ F. Eirich et al.: Koll. Z. 75 (1936), p.20
/30/ B. Schulz: KfK 1988 (1974)
/31/ G. Ondracek: KfK 2688 (1978)
/32/ G. Ondracek: in print
/33/ B. Schulz: High Temp.,High Press. (1981), in print
/34/ G. Ondracek, B. Schulz: Prakt. Met. 10 (1973), p.16, p.67
/35/ R. Pejsa: Diss. 1981, Univ. Karlsruhe (F.R.G.)
/36/ J. Gurland: 4. Plansee-Seminar, Reutte/Tirol (1962), p.507
/37/ B. Schulz: in KfK-ext. 74/4, p.127
/38/ W. Harison: J. Soc. Col. Bradford 27 (1911), p.84
/39/ W.C.R. Powell: J. Chem. Soc. 55 (1914), p.1
/40/ L. Nicodemo, L. Nicolais: Chem. Eng. J. 8 (1974), p. 155
/41/ B. Schulz: Proc. ANS/ASME Meeting Nucl. Reactor Safety, Saragota USA, 1980

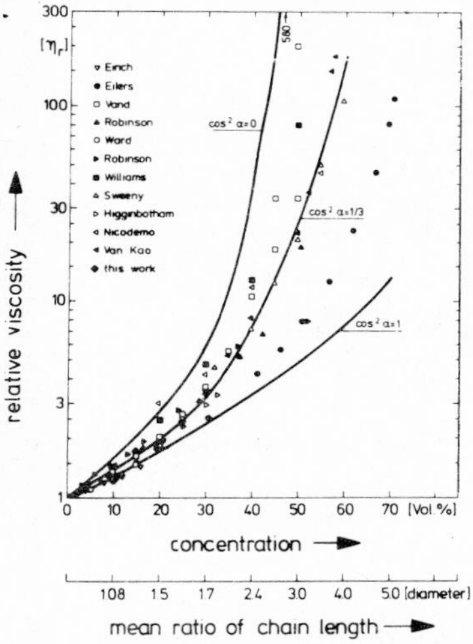

Fig. 1: Relative viscosity versus concentration, spheres.

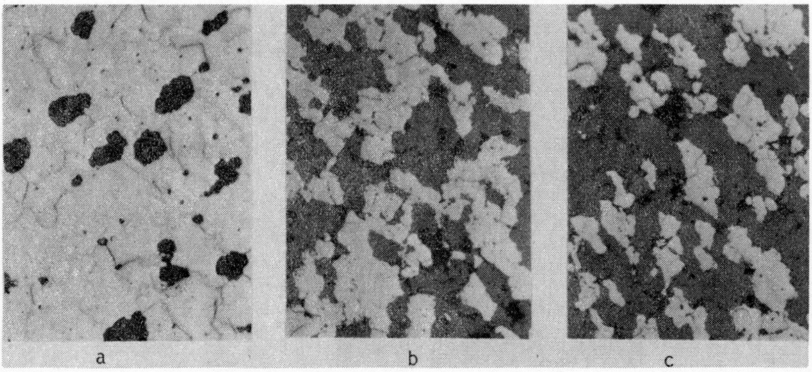

Fig.2: Microstructure of UO_2-Mo: a - Mo-matrix (10 vol.% UO_2), b - chains of UO_2 and Mo (50 vol.%), c - UO_2-matrix (70 vol.% UO_2).

Fig. 3 : Possible configuration of contacts of particles.

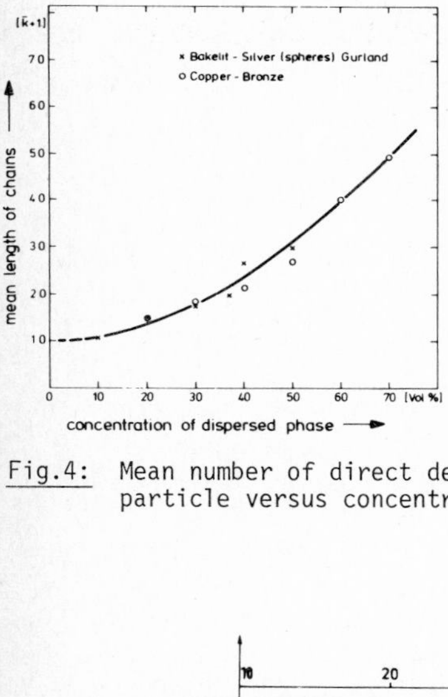

Fig.4: Mean number of direct descandants/ particle versus concentration.

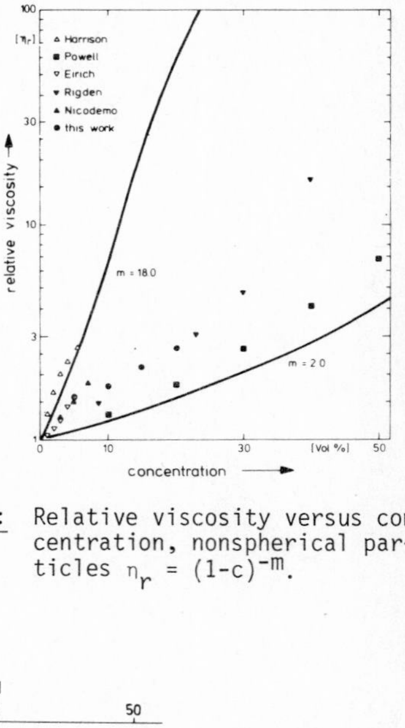

Fig.5: Relative viscosity versus concentration, nonspherical particles $\eta_r = (1-c)^{-m}$.

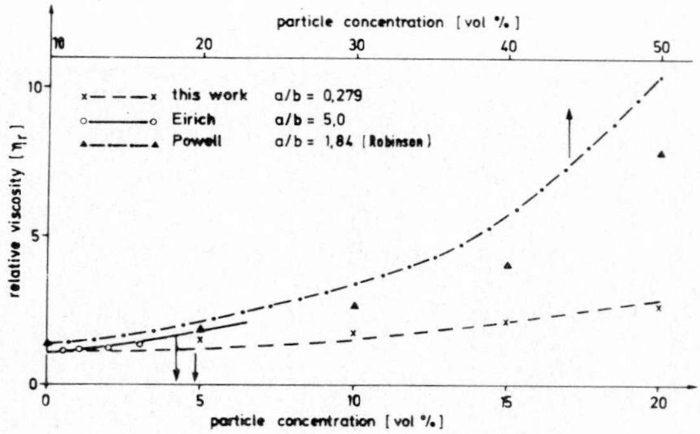

Fig.6: Relative viscosity versus concentration, comparison with calculated values, spheroids.

Part 9
COMPOSITE MATERIALS

Continuum Models of Discrete Systems 4
eds. O. Brulin and R.K.T. Hsieh

A CONTINUUM INTERPRETATION OF LONGITUDINAL THERMOELASTIC WAVE PROPAGATION IN A LAYERED COMPOSITE

Pencho As. Marinov

Institute of Mechanics and Biomechanics
Bulgarian Academy of Sciences
P.O.Box 373, 1090 Sofia
Bulgaria

An approximate continuum-discrete model of coupled thermoelastic layered composite on the basis of recently developed polarization theory is presented by considering the propagation of lingitudinal time-harmonic waves. The equation of thermal conductivity is obtained at a finite wave speed for the thermal wave front. This one and the equations of motion are investigated together. Finally, a 6-th degree dispersion equation in complex form is obtained.

INTRODUCTION

In recent years many approximate models in various degrees of exactness have been proposed for the description of the dynamic behaviour of elastic layered composite. It is worth mentioning effective modulus theory [1,2], effective stiffness theory [3,4], diffusing continuum theory [5], theory of interacting continua [6] and the mixture theory [7]. However the approximate continuum model a description of the dynamic behaviour of coupled thermoelastic layered composite has not been developed yet. Recently the author has proposed the polarization theory for coupled thermoelastic two component materials in which are included the mean continuum with motion x_i, polarization vector $\bar{u}_i$ and internal forces p_i [8]. Thermoelastic potential and local entropy (local Clausius-Duhem inequality) have been utilised too. Very recently Mengy [9] has proposed an uncoupled thermoelastic theory of layered composite including the theory of thermal stresses.

In this work an approximate continuum-discrete model of coupled thermoelectric layered composite is developed by considering the propagation of longitudinal time-harmonic waves in a semi-infinite space. The equation of thermal conductivity is derived from the thermoelastic potential and modified Fouries's law by retaining one and the same degree of exactness in comparison with the propagation of waves in elastic layered composite [10]. A system of differential equations of motion and thermal conductivity is investigated by considering the propagation of longitudinal time-harmonic waves. Finally, a dispersion equation to the 6-th degree in complex form is obtained which determines the dispersion of mechanical and thermal waves.

1. KINEMATICS, FIELD EQUATIONS AND NON-ISOTHERMAL CONSTITUTIVE EQUATIONS

We consider a continuum which is macroscopically homogeneous and isotropic but is made of two components which occupy the material volume V bounded by the surface S. From mathematical point of view such a continuum may be considered as being the assembly of two materials coincident at an initial reference configuration V. We assume that the two material interact with each other by an internal force p_i during the deformation. Let us consider at time t_o a material point P in the underformed body. After the deformation, at time t, the material point P occupies position $p^{(\ell)}$ ($\ell = 1,2$). We denote the coordinates of P and $p^{(\ell)}$ by a system of

Cartesian coordinates x_j (j = 1,2,3) and $x_i^{(\ell)}$ (i = 1,2,3), respectively, see Figure 1, [8].

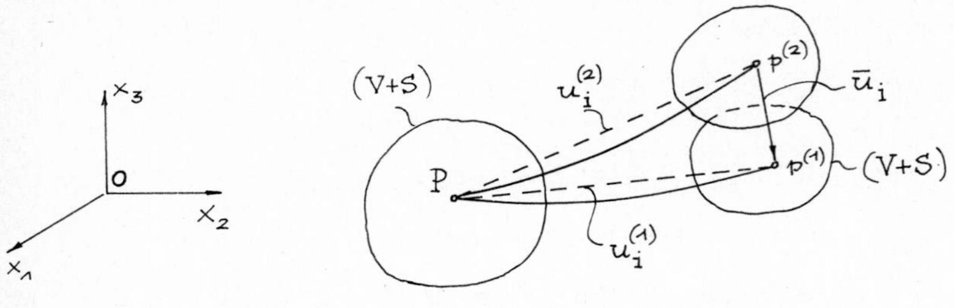

Figure 1

The individual motion of each constituent is given by

$$(1.1) \qquad x_i^{(\ell)} = x_i^{(\ell)}(x_j,t), \qquad \ell = 1,2$$

Displacement vectors of each constituent in a small deformation are

$$(1.2) \qquad u_i^{(\ell)} = x_i^{(\ell)}(x_j,t), \qquad \ell = 1,2$$

The assemption of macroscopical homogeneity and isotropy offers the possibility of approximately replacing the actual continuum by a fictitious continuum which is formed by a mean continuum with motion x_i and polarization vector $\bar{u}_j$. Then the motion of two component material may be described by the motion of centre of mass of constituents and the motion of polarization vector $\bar{u}_i$, i.e.

$$(1.3) \qquad x_i = \sum_{\ell} c^{(\ell)} x_i^{(\ell)}(x_j,t), \qquad \ell = 1,2$$

$$(1.4) \qquad \bar{u}_i = x_i^{(1)}(x_j,t) - x_i^{(2)}(x_j,t),$$

where:

$c^{(\ell)} = \rho^{(\ell)}/\rho$, $\rho = \sum_{\ell} \rho^{(\ell)}$, $\rho^{(\ell)}$ and ρ are partial and total mass densities.

From relations (1.3) and (1.4) yields

$$(1.5) \qquad u_i^{(1)} = u_i + c^{(2)}\bar{u}_i, \qquad u_i^{(2)} = u_i - c^{(1)}\bar{u}_i$$

Next we assume that particles which coincide before the deformation are interacting with the central internal forces p_i, i.e.

$$(1.6) \qquad \sum_{\ell} \rho^{(\ell)} p_i^{(\ell)} = 0, \qquad \sum_{\ell} \rho^{(\ell)} e_{ijk} \bar{u}_j p_k^{(\ell)} = 0, \qquad \ell = 1,2$$

The strain and polarization measures are

$$(1.7) \qquad u_{i,j}^{(\ell)} = x_{i,j}^{(\ell)} \quad \text{and } \bar{u}_i, \text{ where: } x_{i,j}^{(\ell)} = \partial x_i^{(\ell)} / \partial x_j$$

The field equations are :

- Balance of momentum

$$(1.8) \qquad \rho\dot{v}_i = t_{ji,j} + \rho F_i$$

- Balance of moment of momentum

(1.9) $$\rho c^{(1)} c^{(2)} e_{ijk}\bar{u}_j\ddot{\bar{u}}_k = e_{ijk}x_{j,p}t_{pk} + M_{pi,p} + \rho L_i$$

- Conservation of energy

(1.10) $$\rho\dot{E} = t_{ji}\dot{x}_{i,j} + p_i\dot{\bar{u}}_i + g_{ji}\dot{\bar{u}}_{i,j} - q_{i,i}$$

In these equations

$v_i = \dot{x}_i = \dot{u}_i$ - velocity vector of centre of mass,

(1.11) $t_{ji} = \sum_{\ell} t_{ji}^{(\ell)}$ - total stress tensor (true tensor, $t_{ij} \neq t_{ji}$),

$t_{ji}^{(\ell)}$ - partial stress tensor (true tensor, $t_{ij}^{(\ell)} \neq t_{ji}^{(\ell)}$),

(1.12) $F_i = \sum_{\ell} c^{(\ell)} F_i^{(\ell)}$ - total body force,

$F_i^{(\ell)}$ - partial body force,

(1.13) $M_{ji} = e_{jlk}\bar{u}_l(c^{(2)}t_{ik}^{(1)} - c^{(1)}t_{ik}^{(2)}) = e_{jlk}\bar{u}_l g_{ik}$ - total mechanical tensor which results from the polarization vector $\bar{u}_i$ (pseudo tensor, $M_{ji} \neq M_{ij}$),

(1.14) $g_{ji} = c^{(2)}t_{ji}^{(1)} - c^{(1)}t_{ji}^{(2)}$ - total couple stress tensor (pseudo tensor, $g_{ji} \neq g_{ij}$),

(1.15) $L_i = c^{(1)}c^{(2)}e_{ijk}\bar{u}_j(F_k^{(1)} - F_k^{(2)}) = e_{ijk}\bar{u}_j R_k$ - total mechanical force which results from the polarization vector $\bar{u}_i$ (pseudo vector),

$\dot{E}$ - time rate of change of internal energy per unit volume,

$p_i = \rho^{(1)}p_i^{(1)} = -\rho^{(2)}p_i^{(2)}$ - specific internal force,

$q_i = \sum_{\ell} q_i^{(\ell)}$ - total heat flux vector. Heat sources are absent.

Using the local Clausius-Duhem inequality the following non-isothermal constitutive equations are obtained [8]

(1.16) $$t_{ji} = \rho\partial W/\partial u_{i,j}, \quad p_i = \rho\partial W/\partial\bar{u}_i, \quad g_{ji} = \rho\partial W/\partial\bar{u}_{i,j},$$

$$\eta = -\partial W/\partial\theta \qquad \partial W/\partial\theta_{,i} = 0, \qquad q_i\theta_{,i} \geq 0,$$

where: W - total strain energy density,

$u_{i,j}$ and $\bar{u}_{i,j}$ - gradients of infinitesimal displacement vector u_i and polarization vector $\bar{u}_i$, respectivety,

θ - local non-equilibrium temperature,

$\eta = \sum_{\ell} c^{(\ell)}\eta^{(\ell)}$ - total entropy production.

It is assumed that $\theta^{(1)} = \theta^{(2)} = \theta$.

2. KINEMATICS, SMOOTHING OPERATION AND STRAIN ENERGY OF THERMOELASTIC COMPOSITE

An thermoelastic layered composite consisting of large number of alternating plane parallel layers of two homogeneous isotropic materials is considered. The densities, elastic constants, thicknesses, coefficients of linear thermal expansion, specific heats per unit volume at constant deformation of layers are denoted by $\rho^{(\ell)}$, $\lambda^{(\ell)}$, $\mu^{(\ell)}$, d_ℓ, $\lambda_t^{(\ell)}$ and $c_v^{(\ell)}$, $\ell = 1,2$, Figure 2.

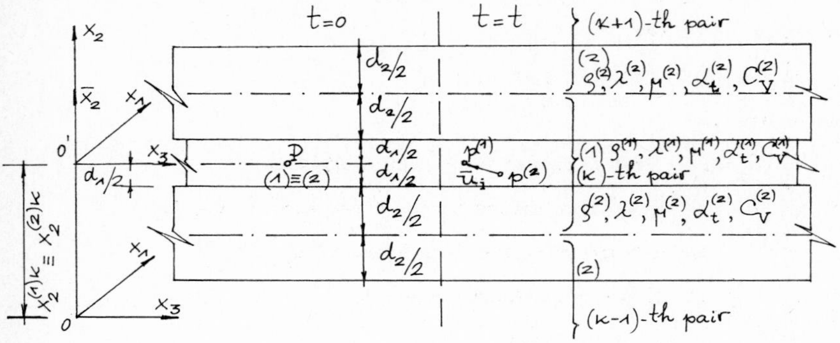

Figure 2

The kinematics and smoothing operation are the same, as well as at the elastic composite [10].

The continuity of displacement at the interface $\bar{x}_2 = \pm(d_1+d_2)/2$ yields to the following relation:

$$(2.1) \qquad \partial_2 u_i(x_1,x_2,x_3,t) = c^{(1)}\partial_2 \bar{u}_i(x_1,x_2,x_3,t)$$

The total strain energy density for k-th pair in non-isothermal case is

$$(2.2) \qquad W^k = W^k_{el} + W^k_\theta = \sum_\ell \left(1/2\,\lambda^{(\ell)} e^{(\ell)k}_{ii} e^{(\ell)k}_{jj} + \mu^{(\ell)} e^{(\ell)k}_{ij} e^{(\ell)k}_{ij} - \gamma^{(\ell)} e^{(\ell)k}_{ii}\theta^{(\ell)} - (\theta^\ell)^2 c_v^{(\ell)}/2T_o\right) = \sum_\ell \left(1/2\,\lambda^{(\ell)} e^{(\ell)k}_{ii} e^{(\ell)k}_{jj} + \mu^{(\ell)} e^{(\ell)k}_{ij} e^{(\ell)k}_{ij}\right) - \sum_\ell \left(\theta\gamma^{(\ell)} e^{(\ell)k}_{ii} + \theta^2 c_v^{(\ell)}/2T_o\right), \quad \ell = 1,2$$

where: $\theta^{(1)} = \theta^{(2)} = \theta$ and $\gamma^{(\ell)} = (3\lambda^{(\ell)} + 2\mu^{(\ell)})\alpha_t^{(\ell)}$.

The strain energy density W^k_{el} has been derived in [10]. We can obtain the strain energy density W^k_θ using the smoothing operation and replacing the components of strain tensor $e^{(\ell)}_{ii}$ from [10] into relation (2.2)

$$(2.3) \qquad W^k_\theta = \sum_\ell W^{(\ell)k}_\theta \cong 1/(d_1+d_2)\int (W^{(1)}_\theta + W^{(2)}_\theta)\,dx_2 = 1/(d_1+d_2)\Big[-(3\lambda^{(1)}+2\mu^{(1)})\alpha_t^{(1)}\theta\int_o^{d_1}(e^{(1)}_{11}+e^{(1)}_{22})\,dx_2 - (3\lambda^{(2)}+2\mu^{(2)})\alpha_\theta^{(2)}\theta\int_o^{d_2}(e^{(2)}_{11}+e^{(2)}_{22})\,dx_2 - \theta^2 c_v^{(1)}/2T_o\int_o^{d_1}dx_2 - \theta^2 c_v^{(2)}/2T_o\int_o^{d_2}dx_2\Big].$$

After some manipulation yields

$$(2.4) \qquad W^k_\theta = -(\gamma_1 c_1+\gamma_2 c_2)\theta(\partial_1 u_1+\partial_2 u_2) - (\gamma_1-\gamma_2)c_1 c_2\theta(\partial_1\bar{u}_1+\partial_2\bar{u}_2) - \theta^2 c_v/2T_o,$$

where $c_v = \sum_\ell c^{(\ell)} c_v^{(\ell)}$; the following designations to simplification are accepted: $c_1 = c^{(1)}$, $\rho_1 = \rho^{(1)}$, $\lambda_1 = \lambda^{(1)}$, $\mu_1 = \mu^{(1)}$, $\gamma^{(1)} = \gamma_1$,

$c_2 = c^{(2)}$, $\rho_2 = \rho^{(2)}$, $\lambda_2 = \lambda^{(2)}$, $\mu_2 = \mu^{(2)}$, $\gamma_2 = \gamma^{(2)}$.

3. EQUATIONS OF MOTION AND CONSTITUTIVE RELATIONS

The equations of motion in direction of layering ($n_1 = 1$, $n_2 = n_3 = 0$) yield

$$\rho_1 \ddot{u}_1^{(1)} = t_{11,1}^{(1)} + p_1 , \qquad \rho_2 \ddot{u}_1^{(2)} = t_{11,1}^{(2)} - p_1 , \tag{3.1}$$

$$\rho_1 \ddot{u}_2^{(1)} = t_{12,1}^{(1)} + p_2 , \qquad \rho_2 \ddot{u}_2^{(2)} = t_{12,1}^{(2)} - p_2$$

Using the relations (1.11), (1.14), (1.16), boundary conditions (2.1) and total strain energy density (2.2) the following constitutive relationa yield

$$t_{11}^{(1)} = (A_1c_1+A_2)c_1\partial_1u_1^{(1)} + (A_1c_1+A_2)c_2\partial_1u_1^{(2)} \tag{3.2}$$

$$+ (A_3c_1+A_4)c_1u_2^{(1)} + (A_3c_1+A_4)c_2u_2^{(2)} - (A_5c_1+A_6)\theta ,$$

$$t_{11}^{(2)} = (A_1c_2-A_2)c_1\partial_1u_1^{(1)} + (A_1c_2-A_2)c_2\partial_1u_1^{(2)}$$

$$+ (A_3c_2-A_4)c_1u_2^{(1)} + (A_3c_2-A_4)c_2u_2^{(2)} - (A_5c_2-A_6)\theta ,$$

$$t_{12}^{(1)} = (B_1c_1+B_2)c_1\partial_1u_2^{(1)} + (B_1c_1+B_2)c_2\partial_1u_2^{(2)}$$

$$+ (B_3c_1+\chi_1)c_1u_1^{(1)} + (B_3c_1+\chi_1)c_2u_1^{(2)} ,$$

$$t_{12}^{(2)} = (B_1c_2-B_2)c_1\partial_1u_2^{(1)} + (B_1c_2-B_2)c_2\partial_1u_2^{(2)}$$

$$+ (B_3c_2-\chi_1)c_1u_1^{(1)} + (B_3c_2-\chi_1)c_2u_1^{(2)} ,$$

$$p_1 = \chi_1c_1\partial_1u_2^{(1)} + \chi_1c_2\partial_1u_2^{(2)} , \qquad p_2 = A_4c_1\partial_1u_1^{(1)} + A_4c_2\partial_1u_1^{(2)} ,$$

where: $A_1 = 2(a_1d_1c_1 + a_2d_2c_2)$, $A_2 = 2c_1c_2\ (d_1a_1 - d_2a_2)$, $A_3 = b_1c_1 + b_2c_2$,

$A_4 = c_1c_2(b_1 - b_2)$, $A_5 = \gamma_1c_1 + \gamma_2c_2$, $A_6 = c_1c_2(\gamma_1 - \gamma_2)$,

$B_1 = k_1d_1c_1 + k_2d_2c_2$, $B_2 = c_1c_2(k_1d_1 - k_2d_2)$, $B_3 = k_1c_1 + k_2c_2$,

$\chi_1 = c_1c_2(k_1 - k_2)$, $a_1 = (\lambda_1/2 + \mu_1)/d_1$, $a_2 = (\lambda_2/2 + \mu_2)/d_2$,

$b_1 = \lambda_1/d_1$, $b_2 = \lambda_2/d_2$, $k_1 = \mu_1/d_1$, $k_2 = \mu_2/d_2$.

In the constitutive relations (3.2) the terms containing $\bar{u}_i$ and $\partial_1\bar{u}_i$ are neglected. This is the main assumption on the present theory.

Taking gradients of relations (3.2) and substituting ones into Eqs. (3.1) a system of four differential equations with five unknown

$u_1^{(1)}$, $u_1^{(2)}$, $u_2^{(1)}$, $u_2^{(2)}$

and the temperature θ is obtained. This system together with the equation of thermal conductivity constitute a complete system of the coupled thermoelastic layered composite.

4. EQUATION OF THERMAL CONDUCTIVITY

The equation of thermal conductivity is obtained on the basis of thermodynamics of irreversible processes [12]. Moreover, the modified Fouries's las is accepted,

$$(4.1)\qquad q_1(1+\tau\partial/\partial t) = -\lambda^o\theta_{,i}\ ,\qquad \vartheta = \theta - T_o\ ,$$

where: q_i - total heat flux, τ - retardation time for the heat flux,
θ - absolute temperature, T_o - temperature of the reference configuration,
λ^o - mean heat conduction coefficient, $\lambda^o = \sum_\ell c^{(\ell)}\lambda^{(\ell)o}$.

Combining the equation of entropy balance $\theta\dot{\eta} = -q_{i,i}$ and the modified Fouries's law, we arrive at the nonlinear equation of themal conductivity

$$(4.2)\qquad (1+\tau\partial/\partial t)(\theta\dot{\eta}) = \lambda^o\theta_{,ii}$$

The rate of entropy production is obtained by Eqs. $(1.16)_4$ and (2.4)

$$(4.3)\qquad \eta = -W_\theta/\partial\theta = A_5(c_1\partial_1 u_1^{(1)} + c_2\partial_1 u_1^{(2)}) + \theta C_v/T_o$$

In the constitutive relation (4.3) the terms containing $\bar{u}_i$ and $\partial_1\bar{u}_i$ are neglected too.

Substituting the constitutive relation (4.3) into equation (4.2) and linearizing one by means of the assumption that the temperature field differs only slightly from a prescribed uniform temperature T_o, i.e. $|\vartheta/T_o| \ll 1$, we arrive at the final form of equation of thermal conductivity

$$(4.4)\qquad \theta_{,ii} - (1+\tau\partial/\partial t)\dot{\theta}/\chi - A_5 c_1\frac{T_o}{\lambda^o}\dot{u}^{(1)}_{1,1} - A_5 c_2\frac{T_o}{\lambda^o}\dot{u}^{(2)}_{1,1} = 0$$

where: $\chi = \lambda^o/C_v$.

5. PROPAGATION OF LONGITUDINAL TIME-HARMONIC THERMOELASTIC WAVES

The displacement equations of motion (3.1) and equation of thermal conductivity (4.4) may be used to study the problem. We consider solutions of the form

$$(5.1)\qquad (u_1^{(1)},u_1^{(2)},u_2^{(1)},u_2^{(2)},\Theta) = (U_1^{(1)},U_1^{(2)},U_2^{(1)},U_2^{(2)},\Theta)e^{ik(n_1x_1+ct)}$$

where: $U_i^{(\ell)}$ are constant amplitudes, k - wave number, c - phase velocity and $n_1 = 1$.

The dispersion equation is obtained by requiring that the determinant of the coefficients vanishes. After some manipulations yields

$$(5.2)\qquad \begin{vmatrix} (\rho_2c^2 - A_1c_2)k & A_3c_2i & -A_5i \\ B_3c_2i & (\rho_2c^2 - B_1c_2)k & 0 \\ \frac{A_5T_oc_2}{\lambda^o}kc(1+\tau kci) & 0 & -[k+c(i-\tau kc)/\chi] \end{vmatrix} = 0$$

In nondimentional form the dispersion Eq. (5.2) can be written as

$$(5.3)\qquad \bar{\rho}\bar{\tau}\xi^4\bar{c}^6 - i\bar{\rho}\frac{1}{\bar{\chi}}\xi^3\bar{c}^5 - \left\{ \bar{\rho}\xi + \left[(\bar{A}_1+\bar{B}_1)\xi + \bar{M}\bar{\chi} \right]\bar{\tau} \right\} \xi^3\bar{c}^4$$

$$+ i\left[(\bar{A}_1+\bar{B}_1)\frac{1}{\bar{\chi}}\xi + \bar{M} \right]\xi^2\bar{c}^3 + \left[(\bar{A}_1+\bar{B}_1)\xi^3 + \frac{1}{\bar{\rho}}(\bar{A}_1\bar{B}_1\xi^3 + \bar{M}\bar{B}_1\bar{\chi}\xi^2 + \bar{A}_3\bar{B}_3)\bar{\tau} \right| \xi\bar{c}^2$$

$$- \frac{i}{\bar{\rho}}\left| \bar{M}\bar{B}_1\xi^2 + (\bar{A}_1\bar{B}_1\xi^3 + \bar{A}_3\bar{B}_3)/\bar{\chi} \right|\bar{c} - \frac{1}{\bar{\rho}}(\bar{A}_1\bar{B}_1\xi^3 + \bar{A}_3\bar{B}_3)\xi = 0 ,$$

where: $\xi = kd_1$, $\bar{c} = c/(\mu_2/\rho_2)^{1/2}$, $\bar{\rho} = \rho/\rho_2$, $\bar{A}_1 = A_1/\mu_2 = (a_1d_1c_1 + a_2d_2c_2)/\mu_2$,

$\bar{B}_1 = B_1/\mu_2 = (k_1d_1c_1 + k_2d_2c_2)/\mu_2$, $\bar{\chi} = \lambda_o/C_v d_1(\mu_2/\rho_2)^{1/2}$, $\bar{\tau} = \tau C_v\mu_2/\lambda_o\rho_2$,

$\bar{\tau}\bar{\chi} = (\mu_2/\rho_2)^{1/2}/d_1$, $\bar{M} = A_5^2 T_o\, d_1/\lambda_o(\mu_2\rho_2)^{1/2} = [(3\lambda_1+2\mu_1)c_1\lambda_t^{(1)} +$

$+ (3\lambda_2+2\mu_2)\lambda_t^{(2)}c_2]^2/\lambda_o(\mu_2\rho_2)^{1/2}$,

$\bar{A}_3 = A_3d_1/\mu_2 = (b_1c_1+b_2c_2)d_1/\mu_2$, $\bar{B}_3 = B_3d_1/\mu_2 = (k_1c_1+k_2c_2)d_1/\mu_2$

The equation (5.3) determines the dispersions of longitudinal mechanical and thermal waves. It is obvious that the thermal waves propagate over the longitudinal direction only.

In the case $\theta = 0$, $C_v = 0$, $\alpha_t^{(\ell)} = 0$, i.e. $\bar{\chi} = \infty$ and $\gamma_1 = \gamma_2 = A_5 = \bar{M} = 0$ the dispertion equation (5.2) transforms to

$$(5.4)\qquad \begin{vmatrix} (\rho_2c^2 - A_1c_2)k & A_3c_2 i \\ B_3c_2 i & (\rho_2c^2 - B_1c_2)k \end{vmatrix} = 0$$

which determines the dispersion of longitudinal elastic waves. The equation (5.4) is the same which has been obtained previously, equation (20) [10].

Finally we get

$$(5.5)\qquad \bar{c}^6 - i\frac{\rho_2}{\tau\mu_2}\frac{1}{\xi}\bar{c}^5 - \left[\frac{1}{\tau} + \frac{\rho_2}{\rho}\left(R_1 + \frac{I_1}{\xi}\right)\right]\bar{c}^4 + i\frac{\rho_2^2}{\rho}\frac{1}{\tau\mu_2}\left(R_1 + \frac{I_1}{\xi}\right)\frac{1}{\xi}\bar{c}^3 +$$

$$+ \frac{\rho_2}{\rho}\left\{ I_2 + \frac{\rho_2}{\rho}\left[R_2 + (c_1\gamma + c_2)\frac{I_1}{\xi} + \frac{R_3}{\xi^2}\right]\right\}\bar{c}^2 - i\frac{\rho_2^3}{\rho^2}\frac{1}{\tau\mu_2}\left[\frac{R_2}{\xi} + (c_1\gamma + c_2)\frac{I_1}{\xi^2} +\right.$$

$$\left. + \frac{R_3}{\xi^4}\right]\bar{c} - \frac{\rho_2^{7/2}}{\rho^2}\frac{\chi}{\tau\mu_2^{3/2}d_1}\left(R_2 + \frac{R_3}{\xi^2}\right) = 0, \text{ where}$$

$$R_1 = c_1\frac{3-4\nu_1}{1-2\nu_1}\gamma + c_2\frac{3-4\nu_2}{1-2\nu_2}, \quad \chi = \frac{\lambda^o}{C_V} = \frac{c_1\lambda_1^o + c_2\lambda_2^o}{c_1C_V^{(1)} + c_2C_V^{(2)}},$$

$$R_2 = 2\left(c_1\frac{1-\nu_1}{1-2\nu_1}\gamma + c_2\frac{1-\nu_2}{1-2\nu_2}\right)\left(c_1\gamma + c_2\right), \quad \gamma = {}^{\mu_1}/_{\mu_2}$$

$$R_3 = 2\left(c_1 \frac{\nu_1}{1-2\nu_1}\gamma + c_2 \frac{\nu_2}{1-2\nu_2}\alpha\right)\left(c_1\gamma + c_2\alpha\right) \quad , \alpha = {}^{d_1}/_{d_2}$$

$$I_1 = \frac{2\left[c_1\mu_1(1+\nu_1)\alpha_1^{\,t} + c_2\mu_2(1+\nu_2)\alpha_2^{\,t}\right]^2 T_o}{\left(c_1C_V^{(1)} + c_2C_V^{(2)}\right)\mu_2} ,$$

$$I_2 = \left(c_1 \frac{3-4\nu_1}{1-2\nu_1}\gamma + c_2 \frac{3-4\nu_2}{1-2\nu_2}\right)\rho_2^{3/2}\frac{\chi}{\tau\mu_2^{3/2}d_1} ,$$

ν_1, ν_2 - Poisson's ratio.

REFERENCES

1 Postma, G.W., Wave propagation in a stratified medium, Geophysics, 20 (1955) 780

2 Rytov, S.M., Acoustical propagation of a thinly laminated medium, Sviet.Phys.Acoustics, 2 (1956) 68

3 Sun, C.T., Achenbach, J.D. and Herr,amm, G., Continuum theory for a laminated medium, J.Appl.Mech., 35 (1968) 467

4 Achenbach, J.D. and Herrmann, G., Wave motion in solids with lamellar structuring, Dynamics of Structured Solids, The American Society of Mechanical Engineers, (1968) 23

5 Bedford, A. and Stern, M., A multicontinuum theory for composite elastic materials, Acta Mechanica, 14 (1972) 85

6 Hegemier, G.A. and Bache, T.C., A general continuum theory with microstructure for wave propagation in elastic laminated composites, J.Appl.Mech., 41 (1974) 101

7 McNiven, H.D. and Mengy, Y., A mixture theory for elastic laminated composites, Int.J.Solids Structures, 15 (1979) 281

8 Marinov, P.A., A microcontinuum mechanics approach to the characterization of two-composite linear elastic material, Teorijsla i primenjena mehanika, 3 Beograd (1977) 43

9 Mengi, Y., Birlik, G. and McNiven, H.D., A new approach for developing dynamic theories for structural elements, Part 2: Application to thermoelastic layered composites, Int.J.Solids Structures, 16 (1980) 1169

10 Marinov, P.A. and Valeva, V., A continuum interpretation of longidutinal wave propogation in an elastic two-layered composite, Theoretical and Applied Mechanics, No.1, Year IX, Sofia (1978) 36 (In Bulgarian)

11 Nowacki, W., Coupled fields in mechanics of solids, in: Koiter, W.T. (eds), Theoretical and Applied Mechanics (North-Holland Publishing Company) (1976) 171

12 Biot, M.A., Thermoelasticity and irreversible thermodynamics, J.Appl.Phys., 27 (1956) 240

Continuum Models of Discrete Systems 4
eds. O. Brulin and R.K.T. Hsieh

"ONE-PARTICLE" APPROXIMATIONS IN MECHANICS OF COMPOSITE MATERIALS

Konstantin Z. Markov

Faculty of Mathematics and Mechanics
University of Sofia
P.O.Box 373, Sofia 1090
BULGARIA

In this paper the effective elastic moduli and yield criterion for composite materials of matrix type are investigated. We outline the theories, whose predictions for the effective properties of such materials are based upon the solution of the elastic inhomogeneity problem for a single filler particle.

INTRODUCTION

We examine a two-phase composite material of matrix type - elastic or elastic-plastic matrix, containing elastic filler particles of one and the same shape (W). A fourth-rank tensor A_w is first of all introduced, which converts a homogeneous strain field applied to an unbounded elastic body at infinity into the average strain field within a single elastic inhomogeneity (W). We call "one-particle" approximation any theory, whose predictions for the effective properties of the composite material can be expressed through the tensor A_w.

We first outline four basic one-particle approximations in elasticity of composite materials, namely, a) The theory, linear with respect to the filler fraction. b) The self-consistent theory. c) The differential scheme. d) The theory of the effective field. A special attention is paid to the latter theory, because it appears, in particular, that the known results of Hashin [1] and Hill [2] can be obtained there without any spherical or cylindrical "composite elements".

We next propose a one-particle approximation that enables to specify approximately the effective yield locus for the composite material. The theory is illustrated by considering in more details the particulate and fiber-strengthened materials, as well as the perforated sheets, under the assumption that the matrix follows the Mises yield condition.

ONE-PARTICLE APPROXIMATIONS IN THE ELASTICITY OF COMPOSITES

Consider an unbounded elastic body (matrix) which contains a single particle (inhomogeneity), occupying the volume (W). We denote by L_1 and L_2 the tensors of the elastic moduli for the matrix and the inhomogeneity, respectively. Let a homogeneous strain field be applied to the matrix at infinity: T_e^{∞} = const, and $T_e^w = \frac{1}{w}\int_w T_e(\underline{x})dv$

be the average strain field within the inhomogeneity (W), which appears as a result of the strain T_e^∞ given. Then

$$(1) \qquad T_e^W = A_W(L_1, L_2) : T_e^\infty$$

where A_W is a certain fourth-rank tensor, the colon means the contraction with respect to two pairs of indices. The explicit form of A_W for ellipsoidal (W) could be extracted from the Eshelby results [3], providing that the matrix and the inhomogeneity are isotropic.

Suppose both constituents of the composite material in question to be elastic. In the sequel, all the quantities connected with the matrix are supplied wuth the subscript "m", and those connected with the filler - with the subscript "f"; in particular, L_m and L_f stand for the tensors of the elastic moduli for the matrix and filler, respectively.

Let (V) be a representative volume for the composite material. Then $V = V_m \cup V_f$, where (V_k) is the volume occupied by the constituent "k", k = m or f. Under the natural assumption of continuity for the displacement field within the composite, the "mixture rules"

$$(2) \qquad \langle T_e \rangle = c_m \langle T_e \rangle_m + c_f \langle T_e \rangle_f , \quad \langle T_\sigma \rangle = c_m \langle T_\sigma \rangle_m + c_f \langle T_\sigma \rangle_f$$

obviously hold, with T_σ standing for the stress tensor; brackets $\langle\ \rangle$ denote the averaging over (V), and $\langle\ \rangle_k$ - over the volumes (V_k), $c_k = V_k/V$ are the volume fractions of the constituents, k = m or f, so that $c_m + c_f = 1$.

According to the Hooke law

$$(3) \qquad \langle T_\sigma \rangle = L^* : \langle T_e \rangle, \quad \langle T_\sigma \rangle_k = L_k : \langle T_e \rangle_k \quad , \; k = m \text{ or } f .$$

Here L^* is the unknown tensor of the effective elastic moduli for the composite. Keeping in mind (3), we exclude $\langle T_e \rangle_m$ from (2) and get

$$(4) \qquad L^* : \langle T_e \rangle = L_m : \langle T_e \rangle + c_f [L] : \langle T_e \rangle_f \quad , \; [L] = L_f - L_m .$$

As the constituents are linear-elastic

$$(5) \qquad \langle T_e \rangle_f = A : \langle T_e \rangle$$

with a certain fourth-rank tensor A. Inserting (5) into (4), we get the basic relation

$$(6) \qquad L^* = L_m + c_f [L] : A$$

which reduces the problem of determining the overall elastic moduli tensor L^* to specifying the tensor A which is introduced in (5).

To specify the tensor A it is necessary, however, that the strain field within the representative volume should be evaluated. As the representative volume contains a great many filler particles, such an evaluation represents a hopelessly difficult problem. That is why approximate methods to specify the tensor A are needed. The simplest method of such a type consists in replacing the host of filler particles by a single one, embedded into a matrix allotted, for example with certain special elastic properties. Thus, the problem of specifying the tensor A reduces to the evaluation of the tensor A_w which pertains to the case of a single inhomogeneity - cf.(1).

DEFINITION. A theory of the composite material, whose predictions for the effective elastic moduli can be expressed through the tensor A_w, is called one-particle approximation.

We outline here four basic one-particle approximations in the elasticity of composite materials.

a) The Theory, Linear with Respect to the Filler Fraction.

We first consider dilute filler concentation $c_f \ll 1$. In this case it is possible to neglect the mutual influence of the filler particles and imagine that each such a particle is immersed into the unbounded matrix, whose strain at infinity is $\langle T_e \rangle$. Thus we can insert the tensor $A_w(L_m, L_f)$ as tensor A in (6) and reach the relation

$$(7) \qquad L^* = L_m + c_f[L] : A_w(L_m, L_f), \qquad c_f \ll 1 .$$

For isotropic constituents and spherical particle (W) the relations (7) turn into the known formulae for the effective bulk and shear moduli of a particulate composite material (cf., e.g.,[4]).

b) The Self-Consistent Theory.

When the filler fraction is not dilute, it appears possible to account, at least, to a certan extent, for the filler interactions making use of the self-consistent idea (cf., e.g., [5]). The gist of this idea lies in the assumption that each filler particle is embedded in a homogeneous material with unknown effective elastic properties. This assumption allows to insert the tensor $A_w(L^*, L_f)$ as tensor A in (6) and get in result the system

$$(8) \qquad L^* = L_m + c_f[L] : A_w(L^*, L_f)$$

from which we should determine the unknown components of the tensor L^* of the effective elastic moduli for the composite material.

For isotropic constituents and spherical particle (W) the system (8) turns into the "self-consistent" system for a particulate composite, first derived by Skorokhod [6] and later by Hill(1965) and Budiansky(1965).

c) The Differential Scheme.

The differential scheme leads to a differential equation for the tensor-valued function $L^* = L^*(c_f)$ on the base of the following consideration (cf.[7]). Let us increase the given filler fraction c_f with the infinitesimal amount dc_f. According to (7), the change of the effective tensor will be

$$dL^* = L^*(c_f + dc_f) - L^*(c_f) = (L_f - L^*(c_f)) : A_w(L^*(c_f), L_f)\, dr_f$$

$dr_f = dc_f/(1 - c_f)$, which yields the basic equation of the differential scheme, namely,

$$(9)\qquad \frac{dL^*}{dc_f} = \frac{1}{1-c_f}(L_f - L^*):A_w(L^*, L_f)\ ; \quad L^*(0) = L_m\ .$$

For isotropic constituents, the system (9) was closely investigated by McLaughlin [7] in two cases: the first, for (W) being a sphere which corresponds to a particulate composite and, the second, for (W) being a circular cylinder which corresponds to a unidirectional fiber-strengthened material.

d) The Theory of the Effective Field.

In a recent work [8], Kanaun proposed a method to calculate the overall elastic moduli for the composite material, which could be called the method of the effective field. As a matter of fact the same method was tacitly used by Levin,[9]. In those works [8],[9] the realization of the effective field idea was, however, rather cumbersome as a result, in our opinion, of usage of the complicated system of integral equations for the strain field in the composite. Here we propose, following [10], an elementary realization of this idea.

The idea of the effective field consists in assuming that each filler particle lies in a certain homogeneous strain field T_e^* (the effective field), which does not coincide with the macrostress $\langle T_e\rangle$ due to the influence of the rest of the filler particles. We therefore can imagine each filler particle to be embedded into the unbounded matrix whose strain tensor at infinity is T_e^*, so that

$$(10)\qquad \langle T_e\rangle_m = T_e^*\ , \quad \langle T_e\rangle_f = A_w(L_m, L_f):T_e^*$$

- cf. (1). Let us introduce the fourth-rank tensor B such that

$$(11)\qquad T_e^* = B:\langle T_e\rangle,$$

i.e., B transforms the macrostrain tensor into the effective strain tensor. Then, taking into account $(10)_2$ and (11), we get the tensor A to be

$$(12)\qquad A = A_w(L_m, L_f):B$$

- cf. (5). In order to specify B we should insert both relations (10) into $(2)_1$

$$\langle T_e\rangle = c_m T_e^* + c_f\langle T_e\rangle_f = (c_m\mathbb{1} + c_f A_w(L_m, L_f)):T_e^*$$

which, compared to (11), yields

$$(13)\qquad B = (c_m\mathbb{1} + c_f A_w(L_m, L_f))^{-1}\ .$$

Here $\mathbb{1}$ is the "unit" fourth-rank tensor with the components $\mathbb{1}_{pqrs} = \frac{1}{2}(\delta_{pr}\delta_{qs} + \delta_{ps}\delta_{qr})$. Keeping in mind (6), (12) and (13), we find the expression

$$(14)\qquad L^* = L_m + c_f[L]:A_w(L_m, L_f):(c_m 1 + c_f A_w(L_m, L_f))^{-1}$$

for the tensor of the effective moduli of the composite material, within the frame of the proposed realization of the idea of the effective field.

We shall only mention in short some possible applications and particular cases of the expression (14).

First of all (14) allows to specify, using the Eshelby results [3], the overall elastic moduli for a dispersion of identically oriented particles of spheroidal shape (W) [11]. In particular, if (W) be a sphere, the results for the bulk and shear moduli of the dispersion coincide with those obtained by Levin [9] ; moreover, the formula for the effective bulk modulus is the same as that found by Hashin [1] for the "spherical composite element". If (W) be a circular cylinder, the Levin results [9], concerning the effective moduli for a unidirectional fiber-reinforced material, are again obtained (these results coincide with those of Hashin & Rosen [12] and Hill [2] for the moduli of a "cylindrical composite element"). If (W) degenerate into a disc, the results for the effective moduli coincide with those obtained by Lifshitz & Rozenzveig [13] for laminated materials.

ONE-PARTICLE APPROXIMATION IN PLASTICITY OF COMPOSITE MATERIALS

Let us suppose now that the matrix of the composite under consideration is an elastic-plastic material, which obeys the Huber-Mises yield criterion, and the filler is again elastic. Generally speaking these suppositions correspond to the case of a plastic metal that contains brittle particles. Due to the filler specific shape and alignment, the composite material possesses a certain macroscopical symmetry, which we describe by means of an orthogonal subgroup G. Our aim consists in specifying the yield condition for the composite through the properties of its constituents, their volume fractions and macrosymmetry G.

Here we propose, following [14], an approximate method for prediction of the effective yield criterion of the composite. The method presents a one-particle approximation in the sense of the introduced above Definition, for it employs the same tensor A_W which pertains to the case of a single filler particle embedded in the matrix.

We first note that besides (2), the similar "mixture rule"

$$(15) \qquad \langle F \rangle = c_m \langle F \rangle_m + c_f \langle F \rangle_f \quad ; \quad F = \tfrac{1}{2} T_6 : T_e$$

obviously hold for the density F of the elastic energy of deformation.

We adopt now two basic simplifications. The first lies in the assumption that the relation (2) remains valid when the matrix yields plastically. The second consists in the supposition that the strain fields within the volumes (V_k), occupied by the constituent "K", are homogeneous, $T_e(\underline{x}) = T_e^k$, $\underline{x} \in V_k$, k = m or f.

The two assumptions just adopted enable to evaluate the average elastic energy stored in the filler through the macrostress tensor

$$(16) \qquad \langle F \rangle_f = \tfrac{1}{2} \langle T_6 : T_e \rangle = \tfrac{1}{2} \langle T_6 \rangle : \langle T_e \rangle = \tfrac{1}{2} T_6^f : L_f^{-1} : T_6^f = F_f(\langle T_6 \rangle) \ ,$$

For simplicity we choose the constituents isotropic. Then the quadratic function F_f in (16) should possess the macrosymmetry G of the composite, because $T_\sigma^f = C_w:\langle T_\sigma \rangle$, $C_w = L_f:A_w:L_f^{-1}$, and therefore C_w has the same symmetry as the tensor A_w, i.e. the symmetry G.

Suppose next that the whole volume (V_m), occupied by the matrix, turns plastic simultaneously. Then, at the fully plastic state of the matrix, we have

$$(17) \qquad \langle F \rangle_m = \frac{1}{6\mu_m}\sigma_o^2 + \frac{1}{18\,k_m}(\mathrm{tr}T_\sigma^m)^2 ,$$

as by supposition the Huber-Mises yield condition holds in (V_m). In (17) σ_o denotes the tensile yield stress for the matrix material, whose shear and bulk moduli are μ_m and k_m, respectively. The mean triaxial stress $\mathrm{tr}T_\sigma^m$ can be expressed through the macrostress $\langle T_\sigma \rangle$, provided $(2)_2$ and (1) are taken into account (we suppose for simplicity that $\langle T_e \rangle = T_e^\infty$, which holds for dilute filler concentration). Eventually

$$(18) \qquad \langle F \rangle_m = \frac{1}{6\mu_m}\sigma_o^2 + F_m(\langle T_\sigma \rangle) ,$$

where F_m is a certain scalar-valued quadratic function of $\langle T_\sigma \rangle$ which, like the function F_f entering (16), has the symmetry G of the composite material.

Finally, for the average elastic energy of the composite we have

$$(19) \qquad \langle F \rangle = \tfrac{1}{2}\langle T_\sigma \rangle : H^* : \langle T_\sigma \rangle = F^*(\langle T_\sigma \rangle) .$$

Here $H^* = L^{*-1}$ is the tensor of the effective stiffness moduli for the composite, so that F^* should be again a scalar-valued quadratic function of the macrostress tensor with the symmetry G.

To obtain the required yield criterion for the composite material, we should only insert (16), (18) and (19) into (15)

$$(20) \qquad F^*(\langle T_\sigma \rangle) = c_m\left(\frac{1}{6\mu_m}\sigma_o^2 + F_m(\langle T_\sigma \rangle)\right) + c_f F_f(\langle T_\sigma \rangle)$$

Since the functions F^*, F_m and F_f are quadratic scalar-valued G-invariants for the tensor $\langle T_\sigma \rangle$, the criterion (20) can be rewritten in the following final form

$$(21) \quad P_1 J_1^2 + P_2 J_1 J_2 + \dots + P_n J_r^2 + P_{n+1} I_1' + \dots + P_{n+s} I_s' = Q^2 \sigma_o^2 ,$$

where J_1 to J_r are the linear G-invariants for the symmetric second rank tensor $\langle T_\sigma \rangle$ and I_1' to I_s' are the quadratic ones, $n = \frac{1}{2}r(r+1)$. The scalar-valued coefficients P_1 to P_{n+s} and Q depend on the filler fraction c_f and on the dimensionless elastic characteristics of the constituents, i.e., on their Poisson ratios ν_m and ν_f and the ratio μ_f/μ_m. The explicit calculation of these coefficients could be fulfilled in accordance with the scheme just described, providing that a certain theory predicting the effective elastic moduli for the composite is chosen. Here we shall only mention some final results, concerning the particulate and fiber-reinforced composites.

a) Yield Criterion for Particulate Composite Materials

In this case the symmetry group G is the full orthogonal group, so that (21) should look as follows

$$(22) \qquad I_2' + P(c_f, \nu_m, \nu_f, \mu_f/\mu_m) J_1^2 = Q^2(c_f, \nu_m, \nu_f, \mu_f/\mu_m) \sigma_o^2 ,$$

where $I_2' = \frac{3}{2} T_d : T_d$, $T_d = T_\sigma - \frac{1}{3} J_1 I$ is the macrostress deviator, $J_1 = \mathrm{tr} T_\sigma$ (in what follows $T_\sigma = \langle T_\sigma \rangle$ for brevity; I is the unit second-rank tensor).

It is noteworthy that the yield criterion of the form (22) was proposed by Green [15] for a porous plastic solid.

Calculations, performed in accordance with the described scheme, yield that for dilute filler concentration

$$(23) \qquad P = p^2(\nu_m, \nu_f, \mu_f/\mu_m) c_f , \quad Q = 1 + q(\nu_m, \nu_f, \mu_f/\mu_m) c_f \ ; \ c_f \ll 1 .$$

For a porous solid for which $\mu_f/\mu_m = 0$, it appears that

$$(24) \qquad p = 0.5 , \quad q \approx -1.5 .$$

As a matter of fact this result was found by Skorokhod & Tuchinskii [16], from whom we borrowed the idea to employ the mixture rule (15) for the elastic energy of deformation.

For a rigid filler, when $\mu_f/\mu_m = \infty$, we obtain

$$(25) \qquad p = \frac{1 - 2\nu_m}{1 + \nu_m} , \qquad q \approx 0.5 .$$

b) Yield Criterion for Unidirectional Fiber-reinforced Materials

The scalar-valued functions entering (20) are transversely-isotropic in this case, so that the yield criterion (21) looks as follows

$$(26) \qquad I_2' + P_1 J_1^2 + P_2 J_1 J_2 + P_3 J_2^2 + P_4 I_3' = Q^2 \sigma_o^2 ,$$

where the invariants $J_2 = \sigma_{33}$, and $I_3' = d_{3k} d_{3k}$ are introduced which, together with J_1 and I_2 (cf. (22)), form a full system of linear and quadratic transversely-isotropic invariants for the macro stress tensor. The invariants are written in a Cartesian system x_1, x_2, x_3, for which the axis x_3 coincides with the fiber direction.

It is to note that the yield criterion of the form (26) was proposed in [17] for transversely-isotropic solids, without specifying the coefficients P_1 to P_4 and Q.

Calculations, performed in accordance with the described scheme, yield that the coefficients entering (26) for dilute filler (fiber, in this case) concentration can be written in the form

(27) $$P_k = p_k(\nu_m, \nu_f, \mu_f/\mu_m)c_f , \quad Q = 1 + q(\nu_m, \nu_f, \mu_f/\mu_m)c_f ,$$

$c_f \ll 1$, $k = 1$ to 4. (See [14] for more details.)

In particular, it is possible to extract from (27) the yield criterion for an in-plane loaded perforated sheet to be

(28) $$\sigma_{11}^2 + \sigma_{22}^2 + 3\sigma_{12}^2 - \sigma_{11}\sigma_{22} + \tilde{P}(\sigma_{11} + \sigma_{22})^2 = \tilde{Q}^2\sigma_o^2 ,$$
$$\tilde{P} = \frac{\nu_m}{1 + \nu_m} c_f , \quad \tilde{Q} = 1 - \frac{5 - \nu_m}{2(1 + \nu_m)} c_f \quad ; \quad c_f \ll 1 .$$

The yield criterion (28) shows good agreement with the experimental results [18] for $c_f \le 0.10$. A more detailed analysis of an anisotropic yield criterion for perforated sheets with square penetration pattern was performed in [19] making use of the above method.

REFERENCES:

[1] Hashin, Z., The elastic moduli of heterogeneous materials, Trans. ASME, ser. E, J.Appl.Mech. 29(1962) No 1.
[2] Hill, R., Theory of mechanical properties of fibre-strengthened materials 1. Elastic behaviour, J.Mech.Phys.Sol. 12 (1964) 199.
[3] Eshelby, J., Elastic inclusions and inhomogeneities, in:Sneddon, I. and Hill, R (eds), Progress in Solid Mechanics 2 (Interscience, N.Y., 1968).
[4] Krivoglaz, M. and Cherevko, A., On the elastic moduli of a solid mixture(in Russian), Fizika metallov i metalloved. 8(1959) 161.
[5] Kroner, E., Berechnung der elastischen Konstanten des Vielkristalls aus Konstantes des Einkristall, Z.Physik B151(1958) No 4.
[6] Skorokhod, Calculation of elastic moduli of solid mixture (in Russian), Poroshkovaja metallurgija 1961 No 1 50-55.
[7] McLaughlin, R., A study of the differential scheme for composite materials, Int.J.Engng Sci. 15(1977) 237 - 244.
[8] Kanaun, S., On the self-consistent field approximation for elastic composite medium(in Russian), JPMTF, 1977 No 2, 160 - 169.
[9] Levin, V., On the determination of elastic and thermoelastic moduli of composite materials(in Russian), MTT, 1976 No 6 137-45
[10] Markov, K., On the method of effective field in mechanics of composite materials, Proceedings Fourth Bulgarian Congress on Mechanics(Varna, September 1981), vol.1(to appear).
[11] Markov, K., Mechanical behaviour of micrononhomogeneous solids, Dissertation, Sofia University (1981).
[12] Hashin, Z. and Rosen, R., The elastic moduli of fiber-reinforced materials, Trans.ASME, ser. E, J.Appl.Mech., 31(1964) No 2.
[13] Lifshitz, I. and Rozenzveig, L., On the theory of elastic properties of polycrystals(in Russian), JETF, 16(1946) No 11 967.
[14] Markov, K., Yield surface for two-phase composite materials, C.R.bulg.Acad.sci., 34(1981) No 5(to appear).
[15] Green, R., A plasticity theory for porous solids, Int.J.Mech. Sci., 14(1972) 215 - 224.
[16] Skorokhod, V. and Tuchinskii, L., Yield condition for porous solids(in Russian), Poroshkovaja metallurgija, 1978 No 11 83 - 87.
[17] Boehler, J. and Sawczuk, A., Applications of representation theorems to describe yielding of ..., Mech.Res.Comm. 3(1976) 277-283.
[18] Litewka, A., Experimental study of the effective yield surface of perforated materials, Nucl.Engng Sci.57(1980) 417 - 425.
[19] Markov, K., An anisotropic yield criterion for perforated plates Res Mechanica Letters, 1(1981) (to appear).

Continuum Models of Discrete Systems 4
eds. O. Brulin and R.K.T. Hsieh

PULSE PROPAGATION IN COMPOSITES VIEWED AS INTERPENETRATING SOLID CONTINUA

M.F. McCarthy

National University of Ireland
University College
Galway, Ireland

H.F. Tiersten

Rensselaer Polytechnic Institute
Troy, New York 12181, USA

Modulated simple wave theory is used to study the propagation of one-dimensional, finite amplitude, high frequency pulses in composites which are modelled as interpenetrating solid continua. The exact equations which govern the propagation of pulses and which are generalizations of the equations which determine the manner in which acceleration waves propagate are derived. The implications of small amplitude finite rate theory are examined in detail and the influence of the structure of the composite on pulse propagation is assessed.

INTRODUCTION

In this paper, the continuum theory of composite materials modelled as interpenetrating solid continua developed in [1] is specialized and one-dimensional motions of solid composites with two identifiable elastic constituents are studied. Here it is shown how our earlier results on acceleration waves [2,3] may be generalized by applying concepts of modulated simple wave theory [4-9] to the study of finite amplitude, one-dimensional pulses in two-component composites. In particular, we seek to determine the manner in which the structure of the composite influences the propagation of high frequency pulses.

In Section 2 the equations which govern the one-dimensional motions of composites with two identifiable components are reviewed. It is shown in Section 3 that volumetric interaction effects between the constituents of the composite are negligible if the "wavelength" associated with any dynamical disturbance is short compared to a length scale defined by the interactions between the constituents, and then dynamical disturbances propagate as simple waves in the composite. The general theory of modulated simple waves is developed in Section 4 and its application to the study of "fast" pulses of arbitrary amplitude is outlined. In Section 5 these results are applied to the study of high frequency small amplitude pulses and it is shown how many of the results of the theory of acceleration waves [2,3] may be generalized to cover the propagation of high frequency small amplitude pulses. It is shown that the behavior of small amplitude high frequency pulses in composites in which the volumetric body force contains a dissipative term which depends on the relative velocity of the constituents is qualitatively the same as the behavior of such pulses in single component viscoelastic media.

Pulse propagation in two particular types of composite is treated in Section 6. First, in composites in which the stress in a particular continuum depends only on the state of deformation of that continuum it is shown that two types of pulse may propagate with speeds of propagation which are the same as those with which pulses would propagate in either of the constituent continua. It is shown that the motions induced by these pulses occur predominantly in one of the constituent continua and only as second-order effects in the other. The manner in which the amplitudes of such pulses vary as they propagate is determined and the possibility of shock formation is discussed. Finally, the propagation of pulses in chopped fiber composites is discussed and it is shown that only one type of pulse may propagate in a composite of this type. The motion induced by such a pulse is found to be predominantly associated with the matrix continuum and any motion induced in the fiber continuum is shown to be a second-order effect.

BASIC EQUATIONS FOR ONE-DIMENSIONAL MOTIONS

The macroscopic model which we consider consists of two distinct interpenetrating continua which initially occupy the same region of space and, hence, the location of the identifiable components of the composite may be specified by a single reference coordinate X. The motion of the center of mass of the combined continuum is described by the mapping $y = y\ (X,t)$ which gives the position at time t of the two particles of the constituent continua which are simultaneously located at X at $t = 0$. The position at time t of the particle of the ith continuum which was at X at $t = 0$ is denoted by $y^{(i)}\ (X,t) = y\ (X,t) + w^{(i)}(X,t)$ and as in [1] it is assumed that in any finite motion the fields $w^{(i)}$ are infinitesimal. The deformation gradients at the point X are defined by

$$F = F(X,t) = \partial_X y(X,t), \quad \overline{F}^{(i)} = \overline{F}^{(i)}(X,t) = \partial_X y^{(i)}(X,t) = F + F^{(i)} \tag{1}$$

where $F^{(i)} = \partial_X w^{(i)}(X,t)$ is the relative deformation gradient of the particle of the ith continuum which is located at X at $t = 0$. The mass densities of the constituent continua in the current and reference configurations are denoted by $\rho^{(i)}$ and $\rho_o^{(i)}$, respectively, so that the total densities in these configurations are $\rho = \rho^{(1)} + \rho^{(2)}$ and $\rho_o = \rho_o^{(1)} + \rho_o^{(2)}$. An elementary calculation [1] shows that

$$rw^{(1)} + w^{(2)} = 0 \tag{2}$$

where $r = \rho_o^{(1)}/\rho_o^{(2)}$ will be assumed constant. In what follows we consistently write w instead of $w^{(1)}$.

In the absence of extrinsic body forces, the equations of balance of the composite are

$$\rho_o\ddot{y} = \partial_X K, \quad \rho_o r\ddot{w} = \partial_X D + \mathfrak{F} \tag{3}$$

where

$$K = \tau^{(1)} + \tau^{(2)}, \quad D = \tau^{(1)} - r\tau^{(2)}, \quad \mathfrak{F} = (1+r)\,{}^{L}F^{12} . \tag{4}$$

In Eqs. (4) D and K represent the stress and relative stress of the composite, respectively, $\tau^{(1)}$ and $\tau^{(2)}$ are the stresses for each of the interpenetrating continua, while ${}^{L}F^{12}$ is the force exerted by continuum 2 on continuum 1 and a superposed dot denotes material time differentiation. The response of the composite is assumed to be the same as that studied earlier [2,3] so that

$$K = \partial_F \Psi, \quad D = \partial_{F^{(1)}}\Psi, \quad \tilde{\mathfrak{F}} = -\partial_w \Psi + \tilde{g}\,\dot{w} \tag{5}$$

where $\Psi = \rho_o \Sigma(F, F^{(1)}, w)$, Σ being the strain energy per unit mass in the reference configuration, while $g = g(F, F^{(1)}, w; \dot{w})$ is an even function of $\dot{w}$ which must be strictly negative.

The ensuing analysis may be written in more compact form by introducing a new notation through the definitions

$$u = \dot{y}, \quad v = \dot{w}, \quad \lambda = F - 1, \quad \gamma = F^{(1)} . \tag{6}$$

The motion of the composite is now governed by the system of equations

$$\dot{\underset{\sim}{u}} + \underset{\sim}{A}\partial_X \underset{\sim}{u} = \underset{\sim}{B} , \tag{7}$$

where

$$\underset{\sim}{A} = \begin{bmatrix} 0 & -\alpha_1 & 0 & -\alpha_2 \\ -1 & 0 & 0 & 0 \\ 0 & -r^{-1}\alpha_2 & 0 & -r^{-1}\beta_2 \\ 0 & 0 & -1 & 0 \end{bmatrix}, \quad \underset{\sim}{u} = \begin{bmatrix} u \\ \lambda \\ V \\ \gamma \end{bmatrix}, \quad \underset{\sim}{B} = \begin{bmatrix} \alpha_3\gamma \\ 0 \\ \beta_3\gamma + \mathfrak{F} \\ 0 \end{bmatrix} \tag{8}$$

and

$$\alpha_1 = \partial_\lambda^2\Sigma, \quad \alpha_2 = \partial_\lambda\partial_\gamma\Sigma, \quad \alpha_3 = \partial_\lambda\partial_w\Sigma, \quad \beta_2 = \partial_\gamma^2\Sigma,$$

$$\beta_3 = \partial_\gamma\partial_w\Sigma, \quad \Sigma = \Sigma(\lambda,\gamma,w), \quad \mathfrak{F} = -\partial_w\Sigma + gv, \quad g = \tilde{g}/\rho_o. \tag{9}$$

PROPAGATION OF SIMPLE WAVES

The relative influence of the difference stress D and the volumetric body force $\mathfrak{F}$ depends on the characteristic length $L_0 = (\beta_2/\mathfrak{F})^{1/2}$. The limiting situation which arises when L_0 is large compared to the wavelengths associated with dynamic disturbances corresponds to $\mathfrak{F} \to 0$. In this limit Σ is independent of w, $g = 0$ so that $\underset{\sim}{B} = \underset{\sim}{0}$ and the elements of the matrix $\underset{\sim}{A}$ are functions of λ and γ only. The system of equations (7) now admits simple wave solutions $\underset{\sim}{U} = \underset{\sim}{U}(\alpha)$, [10], in which the speed of propagation c of the wavelet α = constant is a root of the equation

$$c^4 - (\bar{c}_1^2 + \bar{c}_2^2)c^2 + (c_1^2 c_2^2 - \delta) = 0, \tag{10}$$

where

$$\bar{c}_1^2 = \bar{\alpha}_1, \quad \bar{c}_2^2 = \bar{\beta}_2/r, \quad \delta = \bar{\alpha}_2^2/r \tag{11}$$

and where a superposed bar is meant to indicate that a quantity is independent of w. The roots of (10) are

$$\bar{c}_\pm^2 = \frac{1}{2}\left\{(\bar{c}_1^2 + \bar{c}_2^2) \pm \sqrt{(\bar{c}_1^2 - \bar{c}_2^2)^2 + 4\bar{\delta}}\right\} \tag{12}$$

and, once c has been chosen, the components of $\underset{\sim}{u}(\alpha)$ are related by the formulae

$$u'(\alpha) = -c(\alpha)\lambda'(\alpha), \quad V'(\alpha) = -c(\alpha)\bar{H}(\alpha)\lambda'(\alpha), \quad \gamma'(\alpha) = \bar{H}(\alpha)\lambda'(\alpha), \tag{13}$$

where

$$\bar{H} = (c^2 - \bar{c}_1^2)/\bar{\alpha}_2 = \bar{\alpha}_2/r(c^2 - \bar{c}_2^2) \tag{14}$$

and here the notation $f'(\alpha) = df(\alpha)/d\alpha$ has been used.

Suppose that λ is prescribed as a function of t at some point X_1. Then $\lambda(X_1,t) = f(t/\tau_p)$, where f will be taken to represent a pulse if $f(y) = 0$ whenever $y < 0$ or $y > 1$. If the label $\alpha(X,t)$ of each wavelet is chosen so that $\alpha(X_1,t) = t/\tau_p$, then $\lambda(X,t) = f(\alpha)$ where

$$X - X_1 = c(\alpha)(t - \tau_p\alpha). \tag{15}$$

The functions $u(\alpha)$, $V(\alpha)$ and $\lambda(\alpha)$ now follow when Eqs. (13) are integrated and suitable constants of integration are chosen. Differentiation of (15) with respect to t yields the following expression for the acceleration

$$a(X,t) = u'(\alpha)\dot{\alpha} = u'(\alpha)\{\tau_p + (X - X_1)\tilde{G}(\alpha)u'(\alpha)/2c^4(\alpha)\}^{-1} \tag{16}$$

where

$$2(1 + r\bar{H}^2)\tilde{G} = \bar{\alpha}_{11} + 3\bar{H}\bar{\alpha}_{12} + 3\bar{H}^2\bar{\alpha}_{22} + \bar{H}^3\bar{\beta}_{22} \tag{17}$$

with

$$\bar{\alpha}_{11} = \partial_\lambda^3\bar{\Sigma}, \quad \bar{\alpha}_{12} = \partial_\lambda^2\partial_\gamma\bar{\Sigma}, \quad \bar{\alpha}_{22} = \partial_\lambda\partial_\gamma^2\bar{\Sigma}, \quad \bar{\beta}_{22} = \partial_\gamma^3\bar{\Sigma}. \tag{18}$$

It should be noted that (16) is valid throughout the pulse and not just at its front as is the case with Formula (6.14) of [2]. Just as in [2], the denominator of (16) may vanish and a shock will form at $X = X_s$, where

$$X_s - X_1 = \min_{0<\alpha<1} \{-2c^4(\alpha)\tau_p/\tilde{G}(\alpha)u'(\alpha)\}. \tag{19}$$

MODULATED SIMPLE WAVES

Our primary objective is to describe the propagation of a pulse which is propagating into the region $X > 0$ of a composite which is initially in equilibrium, in terms of the variation in λ as the pulse passes the point $X = 0$. Thus, we seek the solution of (7) subject to the initial conditions $u = v = 0$ for $t = 0$, $X > 0$ and the boundary condition

$$\lambda(0,t) = f(t/\tau_p), \quad 0 \le t \le \tau_p, \tag{20}$$

where $f(\xi) = 0$ whenever $\xi < 0$ or $\xi > 1$. If the speed of propagation of the acceleration wave generated at the head of the pulse is c, then the length of the pulse is $L_p = c\tau_p$. We have shown elsewhere [11], that the attenuation length L_a of the composite for such a pulse is given by the expression

$$L_a = 2(1 + rH^2)c/rH^2g, \quad H^2 = (c^2 - c_1^2)/(c^2 - c_2^2)r \tag{21}$$

and here $c_1^2 = \alpha_1$, $c_2^2 = \beta_2/r$.

In the limit when $\omega = L_a/L_p \gg 1$, Eq. (20) describes a short duration or high frequency pulse and it is appropriate to describe the propagation of such a pulse by modulated simple wave theory [4-9]. We have seen earlier that when $g = 0$ a pulse propagates as a simple wave. In modulated simple wave theory the simple wave is modulated by the dissipative mechanisms present in the composite and solutions of this type may be conveniently described in terms of the characteristic variable $\alpha(X,t)$ and X. If $t = T(X,\alpha)$ denotes the arrival time of the wavelet α at X and if we write $g(X,t) = G(X,\alpha)$, then

$$\dot{g} = \ell^{-1}\partial_\alpha G, \quad \partial_X g = \partial_X G - s\ell^{-1}\partial_\alpha G \tag{22}$$

where the incremental arrival time $\ell = \ell(X,\alpha)$ and the slowness of the wavelet $s = s(X,\alpha) = c^{-1}$ are given by

$$\ell = \partial_\alpha T, \quad s = \partial_X T, \tag{23}$$

respectively, so that

$$\partial_X \ell = \partial_\alpha s. \tag{24}$$

If α is chosen so that $\alpha = t/\tau_p$ at $X = 0$, then Eqs. (20), (22) and (23) together imply that

$$\Lambda(0,\alpha) = f(\alpha), \quad \ell(0,\alpha) = \tau_p, \qquad 0 < \alpha < 1. \tag{25}$$

Equations $(6)_4$ and (7) now assume the forms

$$\partial_\alpha W = s\ell^{-1}(\partial_X W - \Gamma) = \ell V \tag{26}$$

and

$$(\underset{\sim}{I} - s\underset{\sim}{A})\partial_{\alpha}\underset{\sim}{U} = \ell(\underset{\sim}{B} - \underset{\sim}{A}\partial_{X}\underset{\sim}{U}) \ . \tag{27}$$

The motion of the composite is now described by Eqs. (24), (26) and (27) for any choice of $s = s(X, \alpha)$. When the composite is in equilibrium, the right-hand sides of Eqs. (26), (27) vanish identically so that the equations (27) furnish us with a set of homogeneous linear equations for $\partial_{\alpha}U$ at the wavelet $\alpha = 0$ and, in view of Eq. (26), $\partial_{\alpha}W = 0$ at this wavelet. The homogeneous linear equations which follow from (27) have the same form as those which apply throughout the simple wave described in Section 3. Clearly, any changes in the right-hand sides of (26), (27) at any point X are only induced by the pulse as it passes this point. For this reason the modelling of the pulse by a simple wave at points very close to the front of the pulse should yield a reasonably accurate description of what is actually happening.

Let us choose $s = c_{+}^{-1}$, where c_{+} is the characteristic speed of Eqs. (7) corresponding to a fast wave moving in the direction of the X-axis. The matrix $(\underset{\sim}{I} - s\underset{\sim}{A})$ is now singular so that $\underset{\sim}{U}$ must satisfy the compatibility condition

$$\underset{\sim}{\mathcal{L}} \cdot \{\underset{\sim}{B} - \underset{\sim}{A}\partial_{X}\underset{\sim}{U}\} = 0 \tag{28}$$

where $\underset{\sim}{\mathcal{L}} = [1, -s^{-1}, rH, -rs^{-1}H]$ is the left-hand eigenvector of $\underset{\sim}{A}$ corresponding to the eigenvalue $s = c_{+}^{-1}$. Equation (28) is the nonlinear transport equation and it is to be noted that it contains no derivatives with respect to α.

The motion of the composite is now described by Eq. (26), any three of Eqs. (27) and the transport equation (28). Once these equations are solved, $y(X, \alpha)$ follows on integration of $\partial_{\alpha}y = \ell U$ and $t = T(X, \alpha)$ may then be obtained by integrating (23). The functions $y(X, t)$ and $w(X, t)$ may then be calculated.

The equations which govern the motion may be solved iteratively in the manner described in [4,9]. To a first approximation, conditions in the pulse can be determined by neglecting the right-hand sides of Eqs. (26), (27) and we then find that

$$\partial_{\alpha}\underset{\sim}{U} = \partial_{\alpha}\Lambda\underset{\sim}{\mathcal{R}}, \quad \partial_{\alpha}W = 0, \tag{29}$$

where $\underset{\sim}{\mathcal{R}} = [-c_{+}, 1, -Hc_{+}, H]$. Equations (29) yield $\partial_{\alpha}U$, $\partial_{\alpha}V$ and $\partial_{\alpha}\Gamma$ as linear functions of $\partial_{\alpha}\Lambda$ and, when Eqs. (29) are integrated at constant X subject to appropriate initial conditions, the structure of the wave is determined by relations of the form

$$\underset{\sim}{U} = \underset{\sim}{\hat{M}}(\Lambda, X) = \{\hat{U}(\Lambda, X), \Lambda(X, \alpha), \hat{V}(\Lambda, X), \hat{\Gamma}(\Lambda, X)\}, \quad w = W_{o}(X) \tag{30}$$

where

$$\partial_{\Lambda}\underset{\sim}{\hat{M}} = \underset{\sim}{\mathcal{R}}(\Lambda, \alpha) \ . \tag{31}$$

Formulae (30) determine U, Γ and V in terms of Λ at the point X. Substitution from (30) into (28) yields a first-order ordinary differential equation which determines the variation of $\Lambda(X, \alpha)$ with X at each wavelet α = constant.

The implications of the foregoing results for pulses of finite amplitude are examined in detail in [11].

HIGH FREQUENCY SMALL AMPLITUDE PULSES

Our objective here is to extend our earlier results on acceleration waves [2,3] to high frequency small amplitude pulses. When dimensionless variables are introduced through the definitions

$$x = L_a x^*, \quad y = L_a y^*, \quad w = L_a w^*, \quad t = L_a c_+^{-1} t^*, \quad \Sigma = c_+^2 \Sigma^*, \quad \mathfrak{F} = \rho_o c_+^2 L_a^{-1} \mathfrak{F}^* \tag{32}$$

and the asterisk is dropped in the resulting dimensionless forms of Eqs. (24), (26) and (27), we find that the motion of the composite is still governed by these equations which must be solved subject to the initial conditions $U = V = 0$ and the boundary condition

$$\lambda(0,t) = f(\omega t) \,. \tag{33}$$

We write $f(z) = \omega\varepsilon g(z)$ where $\varepsilon > 0$ and $\max|g(z)| = 1$, $z \in [0,1]$. We assume that $\omega \gg 1$ while $a = \omega^2 \varepsilon$ is finite so that we are dealing with small amplitude high frequency finite acceleration pulses for which

$$\lambda(0,t) = \frac{a}{\omega} g(\omega t) \,, \qquad 0 \le \omega t \le 1 \,. \tag{34}$$

We now assume that the composite is in equilibrium in its natural uniform state ahead of the pulse and seek solutions which formally satisfy Eqs. (24), (26) and (27) in the limit as $\omega \to \infty$. We content ourselves with solutions of the form

$$\Lambda = \frac{1}{\omega} \tilde{\Lambda}(X,\alpha), \quad U = \frac{1}{\omega} \tilde{U}(X,\alpha), \quad \Gamma = \frac{1}{\omega} \tilde{\Gamma}(X,\alpha), \quad V = \frac{1}{\omega} \tilde{V}(X,\alpha) \,,$$

$$w = \frac{1}{\omega^2} \tilde{W}(X,\alpha), \quad \ell = \frac{1}{\omega} \tilde{\ell}(X,\alpha), \quad c = c_o + \frac{1}{\omega} \tilde{c}(X,\alpha), \quad s = c_o^{-1} + \frac{1}{\omega} \tilde{s}(X,\alpha) \,. \tag{35}$$

where c_o is the speed of propagation of a fast acceleration wave. It follows from Eqs. (29) and (35) that

$$\tilde{U} = -c_o \tilde{\Lambda}, \quad \tilde{V} = -c_o H_o \tilde{\Lambda}, \quad \tilde{\Gamma} = H_o \tilde{\Lambda} \,, \tag{36}$$

where H_o is obtained from $(21)_2$ by setting $c = c_o$. Substitution from (35) and (36) into the transport equation (28) leads to an ordinary linear differential equation for $\tilde{\Lambda}(X,\alpha)$ which has the solution

$$\tilde{\Lambda}(X,\alpha) = a g(\alpha) \exp(-\theta X) \tag{37}$$

where

$$\theta = -g_o H_o^2 / 2c_o (1 + r H_o^2) \,. \tag{38}$$

Since $\theta > 0$, we see that the value of $\tilde{\Lambda}(X,\alpha)$ at each wavelet α = constant decays as the wavelet traverses the material. Thus, the amplitude of the pulse behaves qualitatively in the same manner as does the amplitude of a high frequency small amplitude pulse propagating in a viscoelastic material [7].

Since $s = c^{-1}$, we have

$$\tilde{s} = -\tilde{c}/c_o^2 = \tilde{\zeta}_o \tilde{\Lambda} \,, \quad \tilde{\zeta}_o = -\tilde{G}_o / c_o^3 \,, \tag{39}$$

where $\tilde{G}_o$ is the value of $\tilde{G}$ evaluated in the natural state of the composite. Equations (24), (37) and (39) together yield the formula

$$\tilde{\ell}(X,\alpha) = 1 + a\tilde{\zeta}_o g'(\alpha) [1 - \exp(-\theta X] / \theta \tag{40}$$

and when (40) is used, integration of (23) gives

$$\omega(t - X/c_o) = \alpha + a\tilde{\zeta}_o g(\alpha) [1 - \exp(-\theta X)] / \theta \,. \tag{41}$$

When (23) and (40) are used, the acceleration at the wavelet follows as

$$a(X,\alpha) = -c_o a g'(\alpha)\exp(-\theta X)\{1+\tilde{\zeta}_o a g'(\alpha)[1-\exp(-\theta X)]/\theta\}^{-1} \tag{42}$$

and it is to be emphasized that this formula holds throughout the pulse. The acceleration will become unbounded and a shock will form on the wavelet α_s at the point X_s where

$$X_s = -\theta^{-1} \ell n\{1+\theta/a\tilde{\zeta}_o g'(\alpha_s)\} \tag{43}$$

and the time at which the shock forms follows from (41) on setting $X = X_s$, $\alpha = \alpha_s$. If $\tilde{\zeta}_o < 0$, then the shock which forms nearest to $X = 0$ forms on the wavelet on which $g'(\alpha)$ is a maximum. A detailed analysis of shock formation and the propagation of weak shocks is given elsewhere [11].

Since $\partial_\alpha Y = \ell U$, Eqs. (35) and (36) imply that $\tilde{W} = H_o \tilde{y}$ where $y(X,\alpha) = \omega^{-2}\tilde{y}(X,\alpha)$. If Eqs. (26) and (36) are now combined, it follows that

$$\tilde{W}(X,\alpha) = -c_o H_o a \exp(-\theta X)\left\{\int_o^\alpha g(\xi)d\xi + a[1-\exp(-\theta X)]g^2(\alpha)/2\theta\right\}. \tag{44}$$

PULSE PROPAGATION IN SOME PARTICULAR COMPOSITES

First, let us consider the propagation of a pulse in a composite which is described by the constitutive equations

$$\Psi = \lambda_1 \Psi^{(1)}(\bar{F}^{(1)}) + \lambda_2 \Psi^{(2)}(\bar{F}^{(2)}) + \lambda_1\lambda_2\Psi^{(12)}(w), \quad \tau^{(i)} = \lambda_i T_i,$$

$$T_i = \partial_{\bar{F}^{(i)}}\Psi^{(i)}(\bar{F}^{(i)}), \quad \mathfrak{F} = -\partial_w \Psi^{(12)}(w) + g\dot{w}, \tag{45}$$

where λ_i is the volume fraction of the composite occupied by the ith constituent and g is assumed to be a negative constant. The motion of the composite is governed by Eqs. (7) with

$$\underset{\sim}{A} = \begin{bmatrix} 0 & -\bar{E}_1 & 0 & 0 \\ -1 & 0 & 0 & 0 \\ 0 & 0 & 0 & -\bar{E}_2 \\ 0 & 0 & -1 & 0 \end{bmatrix} \quad \underset{\sim}{u} = \begin{bmatrix} u^{(1)} \\ \lambda^{(1)} \\ u^{(2)} \\ \lambda^{(2)} \end{bmatrix} \quad \underset{\sim}{B} = \begin{bmatrix} \overset{*}{\mathfrak{F}}^{(1)} \\ 0 \\ -\overset{*}{\mathfrak{F}}^{(2)} \\ 0 \end{bmatrix} \tag{46}$$

where

$$u^{(i)} = \dot{y}^{(i)}(X,t), \quad \lambda^{(i)} = \bar{F}^{(i)} - 1, \quad \overset{*}{\mathfrak{F}}^{(i)} = [\lambda_i \bar{\rho}_o^{(i)}(1+r)]^{-1}\mathfrak{F},$$
$$\bar{\rho}_o^{(i)}\bar{E}_i = \partial_{\bar{F}^{(i)}} T_i. \tag{47}$$

An elementary analysis shows that two pulses may propagate with speeds $c_i = \bar{E}_i^{1/2}$ and these are the speeds with which simple waves would propagate in a material composed entirely of the ith constituent. We confine our attention to small amplitude high frequency pulses which propagate into a region which is initially at rest in its natural equilibrium state. We consider the pulse which propagates with speed c_1 and seek solutions of Eqs. (27) of the form

$$\Lambda^{(1)} = \frac{1}{\omega}\tilde{\Lambda}^{(1)}(X,\alpha)\,,\quad U^{(1)} = \frac{1}{\omega}\tilde{U}^{(1)}(X,\alpha)\,,\quad \Lambda^{(2)} = \frac{1}{\omega^2}\tilde{\Lambda}^{(2)}(X,\alpha)\,,$$

$$U^{(2)} = \frac{1}{\omega^2}\tilde{U}^{(2)}(X,\alpha)\,,\quad s = 1/c_1^{(0)} + \frac{1}{\omega}\tilde{s}(X,\alpha)\,,\quad \ell = \frac{1}{\omega}\tilde{\ell}(X,\alpha)\,, \tag{48}$$

where $c_1^{(0)}$ is the speed of propagation of the corresponding acceleration waves in the composite. An elementary calculation shows that in this pulse $\tilde{U}^{(1)} = -c_1^{(0)}\tilde{\Lambda}^{(1)}$ while the transport equation (28) reduces to an ordinary linear equation whose solution is

$$\tilde{\Lambda}^{(1)}(X,\alpha) = a_1 f_1(\alpha)\exp(-\xi_1 X)\,,\quad \xi_1 = -g\{2\bar{\rho}_o^{(1)}c_1^{(0)}(1+r)\}^{-1}\,, \tag{49}$$

where $\tilde{\Lambda}^{(1)}(0,\alpha) = a_1 f_1(\alpha)$. Just as in Section 5, we have

$$\tilde{s}_1 = \tilde{\zeta}_1\tilde{\Lambda}^{(1)}\,,\quad \tilde{\zeta}_1 = -\tilde{E}_1/2\rho_o^{(1)}c_1^{(0)^3}\,,\quad \tilde{E}_1 = \partial^2_{\bar{F}^{(1)}}T_1\big|_{\bar{F}^{(1)}=1} \tag{50}$$

and the analog of formula (40) follows as

$$\tilde{\ell}_1(X,\alpha) = 1 + a_1\tilde{\zeta}_1 f_1'(\alpha)\{1-\exp(-\xi_1 X)\}/\xi_1\,. \tag{51}$$

When (49) is used in Eq. $(7)_3$ we find that

$$\tilde{U}^{(2)} = -c_1^{(0)}\tilde{\Lambda}^{(2)} = -a_1 c_1^{(0)} g\,\exp(-\xi_1 X)\int_0^\alpha \ell_1(X,\xi)f_1(\xi)d\xi\,\cdot$$
$$\cdot\,\{\lambda_2\bar{\rho}_o^{(2)}(1+r)^2[c_2^{(0)^2}/c_1^{(0)^2} - 1]\}^{-1}\,. \tag{52}$$

The acceleration $\dot{u}^{(1)}$ is now given by a formula similar to (42) and, as usual, there is the possibility that shocks may form. Both $u^{(1)}$ and $\lambda^{(1)}$ will be discontinuous across such shocks but $u^{(2)}$ and $\lambda^{(2)}$ will not suffer jumps. It is clear from (48) and (52) that any motion induced in continuum 2 by the passage of the pulse will be a second-order effect. Similar results hold for pulses which propagate with slowness s_2 in composites described by Eqs. (45).

Finally, we examine the propagation of high frequency pulses in chopped fiber composites. Let continuum 2 represent the chopped fiber continuum so that $T^{(2)} = 0$ and

$$\Psi = \Psi(\bar{F}^{(1)},w)\,,\quad K = D = \partial_{\bar{F}^{(1)}}\Psi(\bar{F}^{(1)},w)\,,\quad \mathfrak{F} = -\partial_w\Psi(\bar{F}^{(1)},w) + g\dot{w}\,. \tag{53}$$

The equations of motion (3) now reduce to

$$\partial_X K + L_F^{12} = \rho_o^{(1)}\ddot{y}^{(1)}\,,\quad -L_F^{12} = \rho_o^{(2)}\ddot{y}^{(2)}\,. \tag{54}$$

As usual, we put $u^{(1)} = \dot{y}^{(1)}$, $\lambda^{(1)} = \bar{F}^{(1)} - 1$, $u^{(2)} = \dot{y}^{(2)}$, $\lambda^{(2)} = \bar{F}^{(2)} - 1$ and, in terms of the variables (X,α), the motion of the composite is now governed by the equations

$$\partial_\alpha U^{(1)} + s\zeta_1\partial_\alpha\Lambda^{(1)} = \ell\{H^{(1)} + \zeta_1\partial_X\Lambda^{(1)} + \zeta_2(\Lambda^{(1)} - \Lambda^{(2)})\}\,,$$
$$s\partial_\alpha U^{(1)} + \partial_\alpha\Lambda^{(1)} = \ell\partial_X U^{(1)} \tag{55}$$
$$\partial_\alpha U^{(2)} = -\ell H^{(2)}$$

where $H^{(i)} = {}^{L}F^{12}/\rho_o^{(i)}$, $\rho_o^{(1)}\zeta_1 = \partial_{\lambda^{(1)}}K(\lambda^{(1)},w)$, $\rho_o^{(1)}(1+r)\zeta_2 = \partial_w K(\lambda^{(1)},w)$. It follows from Eqs. (55) that only pulses whose wavelets propagate with speed $c_1 = \zeta_1^{1/2}$ exist in chopped fiber composites. The nonlinear transport equation for such pulses is

$$\zeta_1 \partial_X \Lambda^{(1)} + H^{(1)} + \zeta_2(\Lambda^{(1)} - \Lambda^{(2)}) - c_1 \partial_X U^{(1)} = 0 . \tag{56}$$

We now examine the behavior of small amplitude high frequency pulses which propagate into chopped fiber composites which are initially in a uniform state and seek solutions of (55) of the form

$$\Lambda^{(1)} = \frac{1}{\omega}\tilde{\Lambda}^{(1)}(X,\alpha), \quad U^{(1)} = \frac{1}{\omega}\tilde{U}^{(1)}(X,\alpha), \quad U^{(2)} = \frac{1}{\omega^2}\tilde{U}^{(2)}(X,\alpha),$$
$$\Lambda^{(2)} = \frac{1}{\omega^2}\tilde{\Lambda}^{(2)}(X,\alpha), \tag{57}$$
$$s = 1/c_1^{(0)} + \frac{1}{\omega}\tilde{s}(X,\alpha), \quad \ell(X,\alpha) = \frac{1}{\omega}\tilde{\ell}(X,\alpha).$$

A standard calculation now shows that $\tilde{U}^{(1)} = -c_1^{(0)}\tilde{\Lambda}^{(1)}$, while

$$\tilde{\Lambda}^{(1)}(X,\alpha) = f(\alpha)\exp(-\epsilon X), \quad \epsilon = g/2(1+r)c_1^{(0)} . \tag{58}$$

Formulae for $\tilde{s}$ and $\tilde{\ell}$ follow easily as

$$\tilde{s} = \tilde{\mu}_o\tilde{\Lambda}^{(1)}, \quad \tilde{\mu} = -\tilde{E}/2\rho_o^{(1)}c_1^{(0)^3}, \quad \tilde{E} = \partial_{\overline{F}^{(1)}}K(\overline{F}^{(1)},w) \tag{59}$$

and

$$\tilde{\ell} = 1 + \tilde{\mu}_o f'(\alpha)\{1-\exp(-\epsilon X)\}/\epsilon , \tag{60}$$

respectively. Of course, the acceleration $\dot{u}^{(1)}$ may become unbounded so that in certain circumstances shock waves may form. The velocity $u^{(1)}$ and the acceleration $\dot{u}^{(2)}$ will be discontinuous across such shocks. Finally, it follows from Eq. $(55)_3$ that

$$\tilde{U}^{(2)}(X,\alpha) = -\overline{\delta}\exp(-\epsilon X)\int_0^{\alpha}\tilde{\ell}(X,\xi)f(\xi)d\xi , \tag{61}$$

where

$$\overline{\delta} = -\{\partial_{\lambda^{(1)}}\partial_w\Psi(\lambda^{(1)},w)|_{\lambda^{(1)}=0} + (1+r)^{-1}gc_1^{(0)}\} . \tag{62}$$

We close by observing that in any chopped fiber composite, the motion of the fiber continuum induced by a pulse is a second-order effect.

ACKNOWLEDGEMENTS

This work was supported in part by the U.S. National Science Foundation under Grant No. ENG 7827637.

REFERENCES

1. Tiersten, H.F. and Jahanmir, M., A theory of composites modelled as interpenetrating solid continua, Arch. Rational Mech. Anal. 65 (1977), 153.
2. McCarthy, M.F. and Tiersten, H.F., One-dimensional acceleration waves in composite materials modelled as interpenetrating solid continua, (to appear).

3. McCarthy, M.F. and Tiersten, H.F., Wave propagation in composite media viewed as interpenetrating continua, in: Kroner, E. and Anthony, K.K. (eds), Continuum Models of Discrete Systems (University of Waterloo Press, 1980).
4. Varley, E. and Cumberbatch, E., Large amplitude waves in stratified media, J. Fluid Mech. 43 (1970), 513.
5. Seymour, B.R. and Mortell, M.P., Nonlinear geometrical acoustics, in: Nemat-Nasser, S., Mechanics Today, Vol.2 (Pergamon, Oxford, 1975).
6. Parker, D.F. and Seymour, B.R., Finite amplitude one-dimensional pulses in an inhomogeneous granular material, Arch. Rational Mech. Anal. 72 (1980), 265.
7. Varley, E. and Rogers, T.G., The propagation of high frequency finite acceleration pulses and shocks in viscoelastic materials, Proc. Roy. Soc. A296 (1967), 498.
8. Seymour, B.R. and Varley, E., High frequency, periodic disturbances in dissipative systems. 1. Small amplitude finite rate theory, Proc. Roy. Soc. A314 (1970).
9. Parker, D.F., Propagation of a rapid pulse through a relaxing gas, Phys. Fluids 15 (1972), 256.
10. Varley, E., Simple waves in general elastic materials, Arch. Rational Mech. Anal. 20 (1965), 309.
11. McCarthy, M.F., One-dimensional pulse propagation in composite materials modelled as interpenetrating solid continua (to appear).

Continuum Models of Discrete Systems 4
eds. O. Brulin and R.K.T. Hsieh

VARIATIONAL ESTIMATES FOR ATTENUATION AND DISPERSION IN ELASTIC COMPOSITES

D.R.S. Talbot

School of Mathematics
Bath University
Bath BA2 7AY
England

The propagation of plane waves in an n-phase medium is studied using a stochastic variational principle to be described by J.R. Willis. A long-wavelength dispersion relation and perturbations to it are derived which do not rely on isotropy of the phases or correlation functions. Both attenuation and dispersion are predicted; explicit formulae are given for a composite consisting of spheres embedded in a matrix when both phases and the distribution of inclusions are isotropic. The results obtained are compared with published experimental data.

1. INTRODUCTION

A stochastic variational principle derived by Willis [1], and described elsewhere in these proceedings, has been used by Talbot and Willis [2] to study wave propagation in randomly inhomogeneous elastic media. The purpose of this paper is to outline the derivation of the main results of [2] and show how they can be applied to a composite consisting of a matrix containing spherical inclusions. In the past, wave propagation in composites of this type has been discussed using a "multiple scattering" formalism. The total field is written as the sum of fields scattered from each inhomogeneity and a hierarchy of equations is generated by taking expectation values of the resultant equations conditional on one or more scatterers being fixed. The hierarchy is closed by making some approximating assumption, such as the quasicrystalline approximation of Lax [3], which is strictly valid only at low concentration. This method has been used, for example, by Datta [4] for ellipsoidal inclusions. The disadvantage of this method is that it is not easily applied to composites such as polycrystals. A more recent approach is that of Willis [5,6] who studied wave propagation in a matrix containing aligned spheroidal inclusions by formulating integral equations for "polarizations". By taking conditional averages of these equations, Willis generated a hierarchy which he closed using the quasicrystalline approximation. The integral equations can be applied to more general n-phase materials. However the physical significance of the quasicrystalline approximation is not obvious when applied to the equations for composites such as polycrystals, in contrast to its application to media of the matrix plus inclusion type. Another route followed to deal with this latter type of composite is that of Devaney [7], who used a quantum mechanical formalism. Devaney formally solved the Lippman-Schwinger equation for the Green tensor of the composite using a transition operator and again used the quasicrystalline approximation. In fact, the transition operator when applied to the Green tensor for a homogeneous comparison material produces a source term corresponding to the polarizations used by Willis. Also, the comparison material Devaney used to generate his equation was quite general and he proposed to estimate its properties self-consistently by requiring it to have the overall properties predicted for the composite.

All the studies mentioned above use a hierarchy of averaged equations. In contrast the approach adopted here and in [2] is to use the variational principle relating to configuration-dependent trial polarization fields derived in [1]. It was demonstrated in [1] that the principle does, in fact, generate the hierarchy obtained by averaging the integral equations, exactly. However, if the space of configuration-dependent trial fields is suitably restricted an optimal formulation is possible. The sense in which the equations obtained are optimal will be discussed later but they are similar to those derived in [5,6,7]. The formulation is applicable to composites having any number of constituent phases, which need not be isotropic, and arbitrary two point correlations.

In the following sections the derivation of a dispersion relation valid at low frequencies and real and imaginary corrections to it is described. Finally, the results obtained are applied to a particular composite.

For convenience of notation suffixes will be suppressed throughout.

2. FORMULATION

An arbitrarily inhomogeneous elastic body is assumed to occupy a region Ω and to have tensor of moduli L and density ρ. Stress and momentum polarizations τ, π are defined relative to a comparison material having tensor of moduli L_o and density ρ_o via the relations

$$\tau = (L-L_o)e \ , \quad \pi = (\rho-\rho_o)\dot{u} \ , \tag{2.1}$$

where e denotes the strain tensor and u the displacement. It was shown in [5] that use of the relations (2.1) in the equation of motion leads to the representation for the displacement u in terms of τ, π and the Green function G for the comparison body:

$$u = u_o - S\tau - M\pi \ . \tag{2.2}$$

S,M are certain integral operators whose kernels involve derivatives of G and u_o is the solution of the given boundary value problem for the comparison body. The representation (2.2) can be differentiated to get equations for $\dot{u}$ and e which in turn can be substituted back into (2.1) giving a pair of integral equations for τ and π. Full details are given in [5].

The medium is now assumed to be composed of n different perfectly bonded phases such that the rth phase has tensor of moduli L_r and density ρ_r. The body is also assumed to have been chosen from some sample space of media over which a probability density is defined. A reasonable objective for such media is to try to find the expectation value of the solution, denoted $\langle u \rangle$, which is obtained from (2.2) as

$$\langle u \rangle = u_o - S\langle\tau\rangle - M\langle\pi\rangle \ . \tag{2.3}$$

One route that can now be followed is to take conditional expectations of the governing integral equations to obtain a hierarchy involving various averages of τ and π. This procedure is discussed in [1] and by J.R. Willis in these proceedings. However, it is also shown in [1] that the hierarchy is implied by a stochastic stationary principle; also, substitution of suitable trial fields into the principle leads to equations making optimal use of available statistical information. Substituting the trial fields

$$\tau = \sum_{r=1}^{n} \tau_r(x,t)\, f_r(x) \ , \quad \pi = \sum_{r=1}^{n} \pi_r(x,t)\, f_r(x) \ , \tag{2.4}$$

where τ_r, π_r depend only on x, t and f_r is the indicator function for the event "x is in phase r", into the principle and optimising leads to the pair of operator equations

$$(L_r - L_o)^{-1}\tau_r P_r + \sum_{s=1}^{n}\left[S_x(\tau_s(P_{sr} - P_s P_r)) + M_x(\pi_s(P_{sr} - P_s P_r))\right] = P_r \langle e\rangle, \qquad (2.5)$$

$$(\rho_r - \rho_o)^{-1}\pi_r P_r + \sum_{s=1}^{n}\left[S_t(\tau_s(P_{sr} - P_s P_r)) + M_t(\pi_s(P_{sr} - P_s P_r))\right] = P_r \langle \dot{u}\rangle. \qquad (2.6)$$

Details of the operators appearing in (2.5), (2.6) are given in [5]. In obtaining these equations e_o and $\dot{u}_o$ have been eliminated using the expressions corresponding to (2.3) for <e> and <u̇>; also $P_s(x)$ denotes the probability that x lies in phase s, with $P_{sr}(x,x')$ the corresponding two point probability, and a factor P_r has been introduced for convenience. Equations (2.5) and (2.6) make optimal use of two point probabilities in the sense that any trial fields which depended upon the configuration in a more complicated way than (2.4) would give rise to probabilities involving at least three points.

Equations (2.5), (2.6) and (2.3) afford an approximate description of any boundary value problem for random composites; however, possible plane wave solutions will now be studied. The fields τ_r, π_r are now taken to have the form

$$\tau_r(x,t) = \tau_r \exp\left[-i(\kappa n.x + \omega t)\right], \quad \pi_r(x,t) = \pi_r \exp\left[-i(\kappa n.x + \omega t)\right], \qquad (2.7)$$

where τ_r, π_r on the right hand sides are constants. Boundary conditions are discarded by setting $u_o = 0$ and the body is taken to be infinite and statistically uniform. The direction of propagation of the plane wave, n, and circular frequency ω are assumed to be given, the problem being to find the wave number κ. Substituting (2.7) into (2.3) with $u_o = 0$, the relation

$$\langle u\rangle = -(\tilde{S}\bar{\tau} + \tilde{M}\bar{\pi}) \exp\left[-i(\kappa n.x + \omega t)\right], \qquad (2.8)$$

follows, where $\tilde{S}$, $\tilde{M}$ are the Fourier transforms of the operators S, M evaluated at $(\kappa n, \omega)$ and $\bar{\tau}, \bar{\pi}$ are given by

$$\bar{\tau} = \sum_{r=1}^{n} P_r \tau_r, \quad \bar{\pi} = \sum_{r=1}^{n} P_r \pi_r. \qquad (2.9)$$

If time reduced versions of the operators are used in (2.5), (2.6) the factor $e^{-i\omega t}$ can be suppressed. The operaotrs then follow from the time reduced version of the Green function

$$G(x) = \frac{1}{8\pi^2}\sum_{N=1}^{3}\int_{|\xi|=1} dS\, \frac{U^N(\xi)U^{NT}(\xi)}{\rho_o c_N^2}\left[\delta(\xi.x) + \frac{i\omega}{2c_N} e^{i\omega|\xi.x|/c_N}\right], \qquad (2.10)$$

where the U^N and c_N are eigenvectors and corresponding eigenvalues of the acoustic tensor $L_o(\xi)$ for the comparison material. The advantage of the representation (2.10) is that it is readily decomposed into a static part (with $\omega = 0$) and a dynamic correction. This greatly facilitates the perturbation theory to follow.

3. LOW FREQUENCY DISPERSION RELATIONS

When all the substitutions have been made, equations (2.5), (2.6) become a set of linear, homogeneous, algebraic equations for the constants τ_r, π_r, r=1,2...n, with κ playing the part of an eigenvalue. The assumed statistical homogeneity also implies that $P_{sr} - P_s P_r$ is translation invariant so that the equations only need to be generated at x = 0. A full treatment would require numerical solution, but in the low frequency limit, $\omega a/c \ll 1$, a being a characteristic length scale of

$P_{sr} - P_s P_r$ and c a typical wave speed, further analytic progress is possible. The details will not be given, but M_x, M_t, S_t are all at least of order ω, while to order zero in ω, S_x reduces to the static operator Γ^∞ whose kernel has components $G_{ij,k\ell}$, evaluated at $\omega = 0$ and symmetrized. The right sides of (2.5),(2.6) are kept exactly since the operators appearing in them are homogeneous of degree zero and κ is expected to be of order ω/c. At $x = 0$, and to order zero in ω, equations (2.5), (2.6) now become

$$P_r(L_r - L_o)^{-1}\tau_r + \sum_{s=1}^{n} A_{rs}\tau_s = P_r\langle e\rangle \quad , \tag{3.1}$$

$$P_r(\rho_r - \rho_o)^{-1}\pi_r = -i\omega P_r\langle u\rangle \quad , \tag{3.2}$$

with

$$A_{rs} = \int dx'\Gamma^\infty(x')(P_{sr}(x',0) - P_s P_r) \quad . \tag{3.3}$$

On the right sides of (3.1), (3.2), $\langle e\rangle$ and $\langle u\rangle$ are now given by

$$\langle u\rangle = -(\tilde{S}\bar{\tau} + \tilde{M}\bar{\pi}) \quad , \quad \langle e\rangle = -(\tilde{S}_x\bar{\tau} + M_x\bar{\pi}) \quad ; \tag{3.4}$$

also A_{rs} is symmetric in r, s because Γ^∞ is an even function and by translation invariance $P_{sr}(x',0) = P_{rs}(0,x') = P_{rs}(-x',0)$. The integral defining A_{rs} converges because, for large $|x'|$, Γ^∞ is order $|x'|^{-3}$ and the bracket either tends to zero or has zero mean value. Now, if the composite were subjected to a uniform static mean strain $\langle e\rangle$, equation (3.1) defines exactly the polarizations that would be generated if the static variational principle of Hashin and Shtrikman [8] were used. First, (3.1) is solved to yield

$$\tau_r = S_r\langle e\rangle \quad ; \tag{3.5}$$

then the tensor of overall moduli $\tilde{L}$, defined via $\langle\sigma\rangle = \tilde{L}\langle e\rangle$, is estimated as

$$\tilde{L} = L_o + \sum_{r=1}^{n} P_r S_r \quad . \tag{3.6}$$

Equation (3.1) is now equivalent to

$$\bar{\tau} = \sum_{r=1}^{n} P_r S_r\langle e\rangle = (\tilde{L} - L_o)\langle e\rangle \quad , \tag{3.7}$$

and similarly it can be shown that (3.2) implies

$$\bar{\pi} = (\tilde{\rho} - \rho_o)\langle\dot{u}\rangle \quad , \tag{3.8}$$

if $\tilde{\rho}$ is defined as the mean density. Thus the mean polarizations given by (3.7), (3.8) are the polarizations that would be generated when the mean wave $\langle u\rangle$ propagated freely in a uniform medium having tensor of moduli $\tilde{L}$ and density $\tilde{\rho}$. Hence the equation

$$\left[\kappa^2\tilde{L}(n) - \tilde{\rho}\omega^2 I\right]\langle u\rangle = 0 \quad , \tag{3.9}$$

is obtained which, with (3.2) and (3.5) defines τ_r and π_r.

Corrections to the relation (3.9) which predict both attenuation and dispersion of waves in the composite require further study of the full equations (2.5), (2.6). The lowest order real and imaginary perturbations, of order ω^2 and ω^3 respectively are retained. Then if $\langle u\rangle_N$, κ_N represent an eigenvector and corresponding eigenvalue of (3.9), the lowest order approximation to $\kappa^2 - \kappa_N^2$ is given by

$$\frac{\kappa^2}{\kappa_N^2} - 1 = \frac{1}{\kappa_N}(Q' + iQ) \quad , \tag{3.10}$$

where

$$Q' = \frac{-1}{\kappa\langle u\rangle \tilde{L}(n)\langle u\rangle} \sum_{r,s=1}^{n} \left\{ \tau_r(\kappa^2 A_{rs}^{(\kappa\kappa)} + \omega^2 A_{rs}^{(\omega\omega)})\tau_s + 2\omega\kappa\tau_r B_{rs}\pi_s + \omega^2\pi_r C_{rs}\pi_s \right\}, \tag{3.11}$$

and

$$Q = \frac{-\omega^3}{\kappa\langle u\rangle \tilde{L}(n)\langle u\rangle} \sum_{r,s=1}^{n} \left\{ \tau_r D_{rs}\tau_s + \pi_r E_{rs}\pi_s \right\} . \tag{3.12}$$

For convenience, suffix N has been suppressed on the right sides of (3.11), (3.12) and $\langle u\rangle, \kappa$ take the values $\langle u\rangle_N, \kappa_N$ with τ_r, π_r the associated polarizations obtained from (3.2) and (3.5). The tensors A, B, C, D, E appearing are given in [2]. They contain details of the statistics of the composite through the "structure factors"

$$\Lambda_{sr} = \int dx'(P_{sr} - P_s P_r) , \quad \Lambda'_{sr}(\xi) = \int dx'\delta(\xi.x')(P_{sr} - P_{sr}) , \tag{3.13}$$

the first appearing only in the expression for Q and the second only in that for Q′. Interpreted physically, Λ_{sr} represents P_r times the difference between the expected volume of phase s conditional on the origin being in phase r and the corresponding unconditional expected volume, while Λ'_{sr} has a similar meaning for areas on the plane $\xi.x' = 0$.

This section is concluded by remarking that Q, Q′ are of order ω^4, ω^3 respectively. Positive Q corresponds to attenuation in the direction of propagation and positive Q′ to a reduction of order ω^2 in the phase velocity.

4. APPLICATION TO A MATRIX CONTAINING SPHERES

The results of the preceding section will now be specialized to a composite comprising an isotropic elastic matrix containing an isotropic distribution of identical isotropic elastic spheres each of radius a. The inclusions will be called phase 1 and the matrix phase 2 with volume fractions c_1, c_2 respectively. The number density of inclusions is denoted by n_1, so that

$c_1 = \frac{4}{3}\pi a^3 n_1$.

For such a composite, the tensor of overall moduli $\tilde{L}$ is isotropic and is characterized by bulk and shear moduli $\tilde{\kappa}$, $\tilde{\mu}$, say. The relation (3.9) thus shows that to lowest order the composite supports P waves with speed $\gamma_1 = [(\tilde{\kappa}+4\tilde{\mu}/3)/\tilde{\rho}]^{\frac{1}{2}}$ and S waves with speed $\gamma_2 = \gamma_3 = (\tilde{\mu}/\tilde{\rho})^{\frac{1}{2}}$. The solution of (3.1) was discussed by Willis [9] who obtained for $\tilde{L}$:

$$\tilde{L} = \sum_{r=1}^{2} c_r L_r \left[I + P_o(L_r - L_o)^{-1} \right]^{-1} \left\{ \sum_{r=1}^{2} c_s \left[I + P_o(L_s - L_o) \right]^{-1} \right\}^{-1} , \tag{4.1}$$

where P_o is the integral of $\Gamma^\infty(x)$ over $|x| < a$.

Now, if m is the polarization associated with the wave number $\kappa = \omega/\gamma$,

$$\langle e\rangle_{ij} = -i\kappa\frac{1}{2}(m_i n_j + m_j n_i) = -i\kappa\varepsilon_{ij} , \tag{4.2}$$

which is taken to define ε. The expressions for Q, Q′ can be rearranged so that the polarizations only appear in the combinations $\tau_1 - \tau_2 = -i\kappa\Delta t$, say, and $\pi_1 - \pi_2$. Then from (3.2)

$$\pi_1 - \pi_2 = -i\omega(\rho_1 - \rho_2)m , \tag{4.3}$$

and from [11] it can be shown that

$$\Delta t = \left\{ I + (c_2 L_1 + c_1 L_2 - L_o) P_o \right\}^{-1} (L_1 - L_2)\varepsilon. \tag{4.4}$$

The equations for Q, Q' can now be partially evaluated to get

$$Q = c_1 \Lambda \left(\frac{4}{3} \pi a^3 \right) \kappa^4 \left[\frac{\gamma \Delta t}{\tilde{\rho}} \Delta S_x \Delta t + \frac{\gamma^3}{\tilde{\rho}} (\rho_1 - \rho_2)^2 \frac{1}{12\pi \rho_o \alpha_o^3} \left(\frac{2}{\sigma_o^3} + 1 \right) \right] , \tag{4.5}$$

and

$$Q' = c_1 \Lambda' \left(\frac{4}{5} \pi a^2 \right) \kappa^3 \left\{ \frac{\Delta t}{\tilde{\rho} \gamma^2} (\gamma^2 \Delta S_x' + \Delta \Gamma) \Delta t + \frac{2}{\tilde{\rho}} (\rho_1 - \rho_2) m \Delta S \Delta t \right.$$

$$\left. + \frac{\gamma^2}{\tilde{\rho}} (\rho_1 - \rho_2)^2 \frac{1}{6\pi \rho_o \alpha_o^2} \left(\frac{2}{\sigma_o^2} + 1 \right) \right\} , \tag{4.6}$$

where the comparison material has been assumed isotropic. The constants α_o, σ_o are respectively the P wave speed and ratio of S wave speed to P wave speed of the comparison material. The tensors ΔS_x, $\Delta S_x'$, ΔS are given by

$$\Delta S_x = \left(\frac{1}{12\pi \rho \alpha^5}, \frac{1}{20\pi \rho \alpha^5} \left(\frac{1}{\sigma^5} + \frac{2}{3} \right) \right), \quad \Delta S_x' = \left(\frac{1}{6\pi \rho \alpha^4}, \frac{1}{10\pi \rho \alpha^4} \left(\frac{1}{\sigma^4} + \frac{2}{3} \right) \right) ,$$

$$(\Delta S)_{ipj} = T_{ijpq} n_j , \quad T = \left(\frac{1}{6\pi \rho \alpha^2}, \frac{1}{10\pi \rho \alpha^2} \left(\frac{1}{\sigma^2} + \frac{2}{3} \right) \right), \tag{4.7}$$

where the notation $(3\kappa, 2\mu)$ has been used for isotropic tensors, and $\Delta\Gamma$ by

$$\Delta \Gamma_{ijk\ell} = R_{ijk\ell pq} n_p n_q , \quad R = \lambda' \delta_{ij} \delta_{k\ell} \delta_{pq} + 2\mu' \delta_{pq} I^1_{ijk\ell}$$

$$- \frac{3}{2} \lambda' \left[\delta_{ij} I^1_{k\ell pq} + \delta_{k\ell} I^1_{ijpq} \right] - 6 \mu' I^2_{ijk\ell pq} ,$$

$$\lambda' = \frac{2}{105\pi \rho \alpha^2} \left[\frac{1}{\sigma^2} - 1 \right] , \quad \mu' = \frac{-1}{70\pi \rho \alpha^2} \left[\frac{4}{3} + \frac{1}{\sigma^2} \right] , \tag{4.8}$$

where the notation I^1, I^2 is that of Ogden [10]. In (4.7), (4.8) ρ, α, σ take the values ρ_o, α_o, σ_o. Next, Λ, Λ' are given in terms of $P(x|o)$, the probability density that a sphere is centred at x conditional on a sphere being centred at the origin, as

$$\Lambda = 1 + \int dx (P(x|o) - n_1) , \quad \Lambda' = 1 + \frac{13c_1}{112} + \frac{13\pi a}{6} \int_0^\infty r dr (P(r|o) - n_1) , \tag{4.9}$$

where isotropy and translation invariance of the composite have been used.

Finally, the wave speed in the overall material $\tilde{c}$ is given approximately by

$$\tilde{c}_N \simeq \gamma_N (1 - Q_N'/2\kappa_N) , \tag{4.10}$$

where the dependence of Q' on the wave type has been made explicit, and specific attenuations δ_N are defined by

$$\delta_N = 4\pi Q_N / (\kappa_N^4 a^3) . \tag{4.11}$$

Results are now presented for the glass/epoxy composite discussed by Kinra, Petraitis and Datta [11]. The comparison material is chosen to have the properties of the matrix, the inclusions or self consistently so that $L_o = \tilde{L}$, $\rho_o = \tilde{\rho}$. The first two choices generate the Hashin-Shtrikman lower and upper bounds for the overall moduli $\hat{L}$. For these three choices the lowest order estimates for the wave speeds, γ_N, normalized to the appropriate wave speeds in the matrix are plotted in figures 1 and 2, for P and S waves respectively, against concentration.

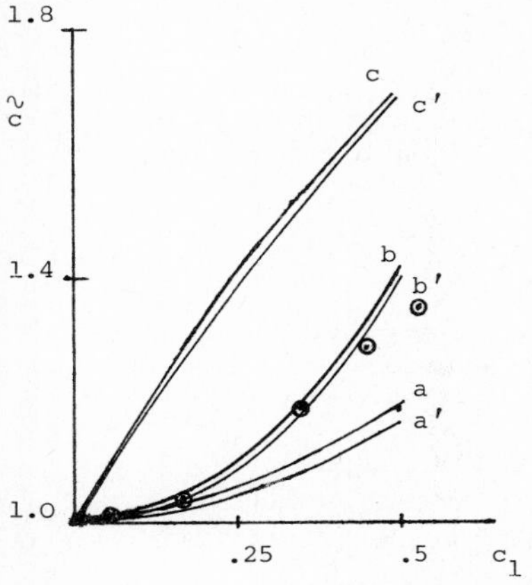

Fig. 1 Normalized P Wave Speeds

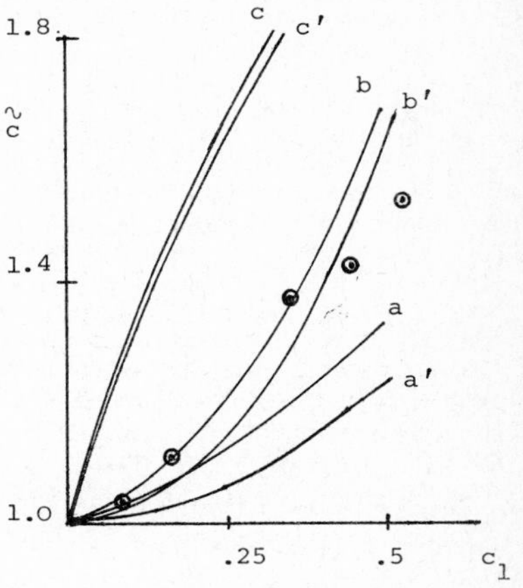

Fig. 2 Normalized S Wave Speeds

Curves labelled a, b, c obtained using matrix, inclusions or overall material respectively as comparison. Curves labelled a', b', c' are the corresponding estimates obtained using (4.10). Points labelled ⊙ are the experimental results taken from [11].

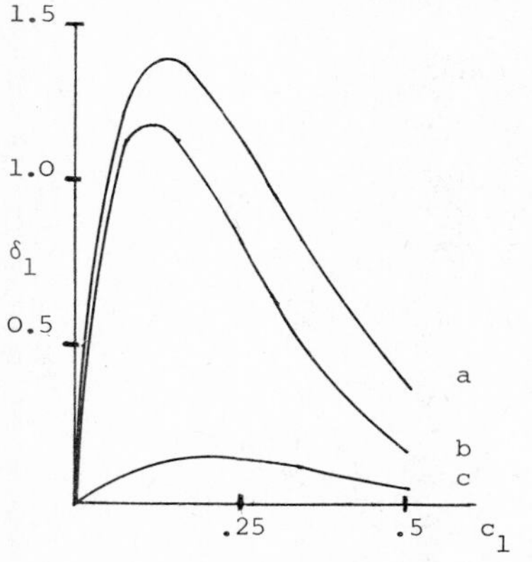

Fig. 3 Specific Attenuation for P Waves

The improved estimate (4.10) requires knowledge of the statistics of the composite through the factor Λ'. This has been evaluated for the approximate distribution of Percus and Yevick [12] and the improved estimates are shown for a frequency of 0.8 mHz for P waves and 1 mHz for S waves. Finally the experimental results reported in [13] are displayed.

It can be seen that predicted phase velocity is sensitive to the choice of comparison material for both P and S waves. However, at least at moderate concentrations, the self consistent estimate seems to agree fairly well with the experimental results.

In figure 3 the specific attenuation for P-waves, δ_1, is shown, again for the Percus-Yevick distribution. As with the estimates for the phase velocity, the three choices of comparison material again give quite different results, although at low concentrations there is fair agreement between the self-consistent estimate and the result obtained using the Hashin-Shtrikman lower bound. All three estimates show the same trend, however: the attenuation rises to a maximum and then decreases. This could be taken to indicate that as the concentration increases the composite becomes more ordered and so the attenuation decreases. The corresponding results for shear waves, although not given, do in fact display the same characteristics.

In conclusion, the sensitivity of the estimates for the phase velocity and attenuation coefficients to the choice of comparison material is noted. However, from the rather limited comparison with experimental data, it might be expected that, at least at low concentrations, choosing the comparison material self consistently will give reasonable first approximations.

1. WILLIS, J.R., Wave Motion Vol. 3, 1981, p.1.
2. TALBOT, D.R.S., and WILLIS, J.R., submitted for publication.
3. LAX, M., Phys. Rev. Vol. 85, 1952, p.621.
4. DATTA, S.K., Continuum Models of Discrete Systems, p.565, 1980. University of Waterloo Press.
5. WILLIS, J.R., J. Mech. Phys. Solids Vol. 28, 1980, p.287.
6. WILLIS, J.R., Ibid. Vol. 28, 1980, p.307.
7. DEVANEY, A.J., J. Math. Phys. Vol. 21, 1980, p. 2603.
8. HASHIN, Z. and SHTRIKMAN, S., J. Mech. Phys. Solids Vol. 10, 1962, p. 335.
9. WILLIS, J.R., Ibid. Vol. 25, 1977, p. 185.
10. OGDEN, R.W., Proc. Camb. Phil. Soc. Vol. 75, 1974, p. 427.
11. KINRA, V.K., PETRAITIS, M.S. and DATTA, S.K., Int. J. Solids Structures Vol. 16, 1980, p. 301.
12. PERCUS, J.R. and YEVICK, G.J., Phys. Rev. Vol. 110, p.1.

Continuum Models of Discrete Systems 4
eds. O. Brulin and R.K.T. Hsieh

CALCULATIONS OF EFFECTIVE PROPERTIES OF REGULAR COMPOSITE MEDIA

L. J. Walpole

School of Mathematics and Physics
University of East Anglia
Norwich
England

We consider composite media consisting of regular densely packed arrays of spherical or circular cylindrical inclusions in an infinite medium. Examples show how variational methods succeed in calculating asymptotic formulae for the effective properties and a comparison with the approaches of Rayleigh (1892) and Keller (1963) is made.

INTRODUCTION

Dispersion composites in which the inclusions and the matrix have greatly differing properties and in which the inclusions are densely packed are important in practice but they have proved to be comparatively difficult to study theoretically even at the level of their 'effective' or 'macroscopic' response. When the inclusions are dispersed in a regular periodic manner a complete analysis is however possible in principle and after the lead given long ago by Lord Rayleigh (1892) a variety of approaches have been proposed in more recent years. Rayleigh considered the steady potential flow (of electricity or heat) through an infinite composite body made up of a connected conducting matrix surrounding a simple rectangular array of either conducting spheres or circular cylinders, for which it suffices to let the direction of the overall flow lie along one of the (either triplet or pair of) directions that fix the rectangular arrangement. The whole composite medium then has an effective conductivity which can be calculated in terms of the fractional volume f occupied by the inclusions and of the individual conductivities of the matrix and the inclusions. Rayleigh was the first to take account of the interaction that occurs between inclusions at all except the most dilute concentrations, but in doing so he felt obliged to sum a non-absolutely convergent series in a specially contrived way that has been justified only recently by McPhedran and McKenzie (1978) and by O'Brien (1979). McPhedran and McKenzie (1978) confirm moreover that a great amount of numerical computation is faced in circumstances where the matrix and the inclusion differ greatly in their conductivities and where the inclusion concentration is increased to the point where neighbouring inclusions make, or nearly make, contact with each other. As an alternative starting point that avoids these particular complexities and offers hope of a simplified asymptotic analysis, we can focus instead on just a single inclusion centred in the basic rectangular domain which has a pair of opposite sides as equipotentials while on its other sides the potential function has a vanishing normal derivative. H. B. Keller and Sachs (1964) solve this mixed boundary-value problem by the numerical method of finite differences for the case of the full range of volume concentrations f of a square lattice of perfectly conducting cylinders in a uniform matrix of conductivity σ_0. When f is brought close to its maximum value of $\pi/4$ they are led to propose that the effective conductivity σ can be calculated by the formula

$$\sigma/\sigma_o = \pi^{3/2}/2(\pi/4 - f)^{\frac{1}{2}} - 1\cdot 95 \qquad (1)$$

which is accurate to within 1% at the higher concentrations (f greater than about 57%). At all the lower concentrations they attribute accuracy to a formula due to Rayleigh (1892). Levine (1966) brings the alternative analytical approach of integral equations to the same two-dimensional boundary-value problem and proceeds far enough through its intricacies to reach agreement with the outcome of Rayleigh and to claim that the method could reproduce and improve the asymptotic result (1). The precise form of the singularity displayed by (1) at the maximum value of f was revealed first by J. B. Keller (1963). He argued further that its reciprocal would apply when the cylinders are non-conducting and later (1964) he proved this theorem to be a special case of one applying when they are finitely conducting. Examples of densely packed, perfectly and finitely conducting spherical inclusions have been subjected by Keller (1963) and by Batchelor and O'Brien (1977), respectively, to asymptotic analyses which have in turn received numerical support in the extensive calculations of McPhedran and McKenzie (1978) and McKenzie, McPhedran and Derrick(1978) for the whole range of volume concentrations. Here we wish to demonstrate by consideration of the example (1) how the use of variational principles offers the advantage of a relatively straightforward rigorous derivation and numerical improvement of existing and hence future asymptotic calculations.

DENSE SQUARE ARRAY OF PERFECTLY CONDUCTING CYLINDRICAL INCLUSIONS

The Boundary-Value Problem

In view of the periodicity of the potential gradient and of the symmetries within the basic square domain ($|x|, |y| < c$) we may follow Keller and Sachs (1964) and Levine (1966) by evaluating the potential function $\phi(x,y)$ within that subdomain D of the x-y plane in which $x^2 + y^2 > a^2$, $0 < y < c$, $0 < x < c$. On the side $x = 0$, $a < y < c$ and also on the adjoining circular part of the boundary the potential is to take a constant value ϕ_1 say while on the side $x = c$, $0 < y < c$ it takes a different constant value ϕ_2 say. Over the remaining parts of the boundary D ($y = 0$, $a < x < c$ and $y = c$, $0 < x < c$) its normal derivative $\phi,_y$ is to vanish. Then the effective conductivity σ of the whole heterogeneous medium is calculated by the expression

$$\sigma = \frac{\sigma_o}{\phi_2 - \phi_1} \int_o^c \phi,_x (c,y)dy$$

which equates the total (thermal or electrical) flux across the side $x = c$, $-c < y < c$ to that which would prevail if a homogeneous medium of conductivity σ were to entirely replace the perfectly conducting inclusions and the matrix of conductivity σ_o.

Variational Procedure

For this mixed boundary-value problem of classical potential theory we can make available a pair of complementary extremum principles which yield (close and possibly coincident) upper and lower bounds on the 'potential energy' and hence on the effective conductivity, once suitable, constrained but optimizable, approximating solutions are constructed. The lower bound principle offers with its fewer constraints a clearer insight into the form of the optimal asymptotic solution. It is derived from the basic inequality

$$p_x^2 + p_y^2 + \phi,_x^2 + \phi,_y^2 \geq 2(p_x\phi,_x + p_y\phi,_y) = 2(p_x\phi),_x + 2(p_y\phi),_y$$

which holds for all vectors (p_x, p_y) that meet the constraint

$$p_{x,x} + p_{y,y} = 0 \quad \text{in} \quad D, \tag{2}$$

since the addition of the positive quantity $(p_x - \phi_{,x})^2 + (p_y - \phi_{,y})^2$ to its right-hand side would restore equality. Next the inequality is integrated over the domain D and some applications of the divergence theorem are made, on the assumption that (p_x, p_y) offers no intruding discontinuities and furthermore that p_y is chosen so as to vanish (like $\phi_{,y}$) on the two sides $y = 0$, $a < x < c$ and $y = c$, $0 < x < c$. Thus after a little further manipulation we conclude that

$$(\phi_2 - \phi_1)^2\sigma/\sigma_o \geq 2(\phi_2 - \phi_1)\int_o^c p_x(c,y)dy - \iint_D (p_x^2 + p_y^2)dx\,dy\,. \tag{3}$$

Now especially in the densely packed circumstances for which the difference $(1 - a/c)$ is very small, it is natural to assume that (as a first approximation to $\phi_{,y}$) p_y keeps a vanishing value everywhere in D and hence that in keeping with the constraint (2) p_x varies only with y in whatever optimal manner is indicated by the extremum principle (3). Thus the choice

$$p_x = (\phi_2 - \phi_1)\,F(y), \quad p_y = 0 \tag{4}$$

makes

$$\sigma/\sigma_o \geq 2\int_o^c F\,dy - \int_o^a \{c - (a^2 - y^2)^{\frac{1}{2}}\}F^2 dy - \int_a^c c\,F^2\,dy\,,$$

after which it is readily seen that the definition

$$F(y) = 1/\{c - (a^2 - y^2)^{\frac{1}{2}}\},\ 0 < y < a; \quad F(y) = 1/c, \quad a < y < c;$$

establishes the sharpest inequality

$$\sigma/\sigma_o \geq \int_o^a \frac{dy}{c - (a^2 - y^2)^{\frac{1}{2}}} + \int_a^c \frac{dy}{c}$$

$$= \frac{2}{(1 - a^2/c^2)^{\frac{1}{2}}}\tan^{-1}\{(1 + \tfrac{a}{c})/(1 - \tfrac{a}{c})\}^{\frac{1}{2}} - \frac{\pi}{2} + \frac{c-a}{c}$$

$$= \{\pi^{3/2}/2(\pi/4 - f)^{\frac{1}{2}}\} - 1 - \frac{\pi}{2} + \ldots\,, \quad \text{if} \quad \frac{\pi}{4} - f \ll 1\,. \tag{5}$$

The manner in which p_x (and hence $\phi_{,x}$) is made large when y is small was perceived by Keller (1963). With the help of this insight, it can now be shown (by an analysis omitted here) how the complementary extremum principle gives an upper bound on σ/σ_o which is dominated by the same leading term as the lower bound is in (5). The asymptotic statement

$$\sigma/\sigma_o \to \pi^{3/2}/2(\pi/4 - f)^{\frac{1}{2}} \quad \text{as} \quad f \to \pi/4$$

is thereby confirmed. Furthermore the accuracy of the upper and lower bounds may be extended by a little further effort. Thus an appropriate generalisation of

(4) is the definition

$$p_x = F(y) + (c - x)^2 g'(y), \qquad p_y = 2(c - x)g(y)$$

in which g is any continuous function of y such that $g(0)$ and $g(c)$ both vanish. The choice which makes

$$g(y) = Ky(a^2 - y^2)/\{c - (a^2 - y^2)^{\frac{1}{2}}\}^2 \qquad \text{for} \quad 0 < y < a$$

and $g(y)$ vanish for $a < y < c$, where K is a disposable parameter, is one that suggests itself by leading to comparatively simple, non-singular integrations. Then by the best remaining choice of $F(y)$ and K the inequality

$$\frac{\sigma}{\sigma_0} \geq \pi^{3/2}/2(\pi/4 - f)^{\frac{1}{2}} + C + \ldots \tag{6}$$

is reached, where

$$C = -1 - \frac{\pi}{2} + 5(8 + 3\pi)^2/3(448 + 117\pi) = -1{\cdot}95032$$

and where the terms neglected in (6) are much smaller than those retained, when f is close to $\pi/4$. The lower bound has thus been brought already to the accuracy of the formula (1) of Keller and Sachs (1964).

REFERENCES

[1] Rayleigh, Lord, On the influence of obstacles arranged in rectangular order upon the properties of a medium, Phil. Mag. 34 (1892) 481-502.
[2] McPhedran, R. C. and McKenzie, O. R., The conductivity of lattices of spheres, I. The simple cubic lattice, Proc. Roy. Soc. A 359 (1978) 45-63.
[3] O'Brien, R. W., A method for the calculation of the effective transport properties of suspensions of interacting particles, J. Fluid Mech. 91 (1979) 17-39.
[4] Keller, H. B. and Sachs, D., Calculations of the conductivity of a medium containing cylindrical inclusions, J. Appl. Phys. 35 (1964) 537-538.
[5] Levine, H., The effective conductivity of a regular composite medium, J. Inst. Maths Applics 2 (1966) 12-28.
[6] Keller, J. B., Conductivity of a medium containing a dense array of perfectly conducting spheres or cylinders or nonconducting cylinders, J. Appl. Phys. 34 (1963) 991-993.
[7] Keller, J. B., A theorem on the conductivity of a composite medium, J. Math. Phys. 5 (1964) 548-549.
[8] Batchelor, G. K. and O'Brien, R. W., Thermal or electrical conduction through a granular material, Proc. Roy. Soc. A 355 (1977) 313-333.
[9] McKenzie, D. R., McPhedran, R. C. and Derrick, G. H., The conductivity of lattices of spheres, II. The body centred and face centred cubic lattices, Proc. Roy. Soc. A 362 (1978) 211-232.

Continuum Models of Discrete Systems 4
eds. O. Brulin and R.K.T. Hsieh

VARIATIONAL PRINCIPLES FOR WAVES IN RANDOM COMPOSITES

J.R. Willis

School of Mathematics
Bath University
Bath BA2 7AY
England

A recently-developed formulation of dynamic problems for inhomogeneous media, in terms of stress and momentum "polarizations" is reviewed and associated "hierarchy equations" are derived when the medium is random. These are then shown to be derivable from variational principles which reduce, in the static limit, either to the classical energy principles or to the Hashin-Shtrikman principle. The dynamical extension of the Hashin-Shtrikman principle facilitates a treatment of wave propagation problems that is "optimal" if only two-point correlations are known: some detailed results will be presented by D.R.S. Talbot. The "classical" principles, on the other hand, provide an "optimal" treatment, if two-and three-point correlations are known. The full implications of the latter treatment have still to be worked out.

1. INTRODUCTION

Waves in randomly inhomogeneous media are usually studied either by generating a formal perturbation series solution, truncating, and taking expectation values, or by generating a hierarchy of equations for conditional expectation values and closing the hierarchy at some stage by appeal to an ad hoc closure assumption. An example of the former approach is contained in the work of Karal and Keller [1], in which they introduced the "method of smoothing" to avoid the appearance of secular terms. The approach is useful if applications to media with small fluctuations in moduli and density are envisaged. An example of the latter approach is contained in the work of Twersky [2], which applies to a composite comprising a matrix containing discrete inclusions. As discussed by Hinch [3], the closure assumptions that are used are valid strictly only at low concentrations of inclusions. Typically, however, composites involve both large variations in moduli and density, and high concentrations of inclusions. Polycrystals, too, provide examples of random media which may display large variations in moduli and do not have a matrix-inclusion structure.

The present work summarises an approach to such problems which employs variational principles. Principles will be displayed which generate the hierarchy equations exactly. Furthermore, substitution of simple configuration-dependent trial fields into the principles and optimising generates "optimal" approximate equations which involve only low-order correlation functions. This, in effect, truncates the hierarchy in a way similar to that obtained from ad hoc closure assumptions, but only approximations that are "optimal" relative to a variational principle are picked out. The equations thus have some status at least, for any type of composite. Three principles are considered which reduce, in the static limit, to the Hashin-Shtrikman principle and to the two classical energy principles, all of which have been used elsewhere to produce bounds on overall moduli.

2. POLARIZATION FORMULATION

Consider a body, which occupies a domain Ω with boundary $\partial\Omega$, composed of "nonlocal" viscoelastic material, for which stress components σ_{ij} are related to strain components e_{ij} through

$$\sigma_{ij}(x,t) = \int_{\Omega} dx' \int dt' L_{ijkl}(x,x',t-t') e_{kl}(x',t'). \tag{2.1}$$

The components L_{ijkl} are generalized functions which will be assumed to have the symmetries

$$L_{ijkl}(x,x',t) = L_{jikl}(x,x',t) = L_{klij}(x',x,t) \tag{2.2}$$

and, in addition,

$$L_{ijkl}(x,x',t) = 0 \quad , \; t < 0. \tag{2.3}$$

Initial value problems, for which $e_{kl}(x,t) = 0$ when $t < 0$, will be considered, so that $e_{kl}(x,t)$, and the displacement field $u_i(x,t)$ from which it is derived, can be considered as a generalized function of t, even though it may be desirable to restrict its behaviour as a function of x, to ensure that the right side of (2.1) is defined. Suffixes will be avoided as far as possible in the sequel and (2.1) will be written in the abbreviated form

$$\sigma = Le. \tag{2.4}$$

In addition to the stress-strain relation (2.4), the material has a momentum-velocity relation

$$p = \rho\dot{u}. \tag{2.5}$$

Here, ρ is a "mass density operator", similar in form to L, except that it has components $\rho_{ij}(x,x',t)$. It was introduced, along with L, by Willis [4], who observed that it could be carried along at no cost and might prove useful for some future application.

The problem now is to solve the equations of motion

$$\operatorname{div}\sigma + f = \dot{p}, \quad x \in \Omega, \tag{2.6}$$

together with appropriate boundary conditions on $\partial\Omega$. Equation (2.6) involves generalized functions, which are zero for $t < 0$. The conventional initial conditions, applied at $t = 0$, are built into the body-force f ; this has been discussed in detail by Willis [5].

The operators L,ρ may vary in a complicated fashion with x,x' and it is likely to be convenient to formulate the problem relative to a "comparison material", with moduli L_0 and density ρ_0 (which may still be operators of the type already described). With this object, polarizations

$$\tau = (L-L_o)e \ , \quad \pi = (\rho-\rho_o)\dot{u} \tag{2.7}$$

are introduced. τ is the stress polarization of Hashin and Shtrikman [6,7,8] and π is the momentum polarization introduced in [4]. Equations (2.7) imply, with (2.4), (2.5)

$$\sigma = L_o e + \tau \ , \quad p = \rho_o \dot{u} + \pi \tag{2.8}$$

which, substituted into (2.6), give

$$\text{div}(L_o e) + (\text{div}\,\tau - \dot{\pi} + f) = (\rho_o \dot{u})^{\cdot} \ . \tag{2.9}$$

Suppose, now that the given boundary value problem can be solved for a body composed of the comparison material, in the sense that the associated Green's function $G(x,x',t)$ is known. It follows, then, that equation (2.9) has solution

$$u = u_o - S\tau - M\pi \ , \tag{2.10}$$

where u_o is the solution of the boundary value problem for the "comparison" body and S,M are operators involving derivatives of G; details are given in [4]. The representation (2.10) may now be differentiated to give expressions for $e,\dot{u}$. Substitution back into (2.7) then gives operator equations

$$(L-L_o)^{-1}\tau + S_x\tau + M_x\pi = e_o \ , \tag{2.11}$$

$$(\rho-\rho_o)^{-1}\pi + S_t\tau + M_t\pi = \dot{u}_o \ . \tag{2.12}$$

Again, the operators are defined explicitly in [4], and L_o,ρ_o are assumed to be chosen in such a way that the inverses $(L-L_o)^{-1},(\rho-\rho_o)^{-1}$ exist.

The formulation (2.11), (2.12) is closely related to a quantum-mechanical formulation used, for example, by Devaney [9], which would take as its starting point the representation

$$u = u_o + G(\text{div}\,\tau - \dot{\pi}) \tag{2.13}$$

(which can be related to (2.10) by integration by parts). The "source" term $(\text{div}\,\tau - \dot{\pi})$ is then represented in the form

$$(\text{div}\,\tau - \dot{\pi}) = Tu_o \tag{2.14}$$

and an operator equation is derived for the "transition operator" T. This equation is closely related to equations (2.11), (2.12) which, however, have the advantage of separating stress and momentum effects.

3. HIERARCHY EQUATIONS FOR A COMPOSITE

Consider now a body which is made up of n different types of material (called phases for convenience of reference), firmly bonded at all interfaces. The rth phase is assumed to be either elastic or viscoelastic, with moduli L_r (which, therefore, are *local* operators, dependent at most upon t) and mass density ρ_r (which is a constant and not an operator). The comparison material, however, may still be defined by nonlocal operators. The indicator function $f_r(x)$ is defined so that

$$\begin{aligned} f_r(x) &= 1 \ , \quad x \in \text{phase } r \\ &= 0 \ , \quad x \notin \text{phase } r \ . \end{aligned} \tag{3.1}$$

If, as will be assumed, the body is random, the probability $P_r(x)$ of finding phase r at x is given as the ensemble average of $f_r(x)$:

$$P_r(x) = \langle f_r(x)\rangle \tag{3.2}$$

and, more generally, the probability of finding simultaneously phase r_1 at x_1, r_2 at x_2 and so on is

$$P_{r_1r_2\ldots}(x_1,x_2\ldots) = \langle f_{r_1}(x_1)f_{r_2}(x_2)\ldots\rangle \ . \tag{3.3}$$

Also, for any function $\phi(x)$ defined over Ω, the conditional expectation $\langle\phi\rangle_{r_1r_2\ldots}$ is defined by

$$\langle\phi\rangle_{r_1r_2\ldots}P_{r_1r_2\ldots}(x_1,x_2\ldots) = \langle\phi(x)f_{r1}(x_1)f_{r2}(x_2)\ldots\rangle \ . \tag{3.4}$$

It will be helpful, for the manipulations to follow, to expand $\phi(x)$ in the form

$$\phi(x) = \sum_{r=1}^{n}\phi(x)f_r(x) \quad , \tag{3.5}$$

which gives, upon substitution into (3.4),

$$\langle\phi\rangle_{r_1r_2\ldots}P_{r_1r_2\ldots}(x_1,x_2\ldots) = \sum_{r=1}^{n}\langle\phi\rangle^r_{rr_1r_2\ldots}P_{rr_1r_2\ldots}(x,x_1,x_2\ldots) \ , \tag{3.6}$$

the superscript r being introduced to indicate that phase r relates to the point x at which ϕ is evaluated.

Equations (2.11), (2.12) apply to any particular realization of the composite. The required hierarchy is obtained by multiplying by $f_r(x)f_{r_1}(x_1)\ldots$ and taking expectations. This gives

$$(L_r-L_o)^{-1}\langle\tau\rangle^r_{rr_1r_2\ldots}P_{rr_1r_2\ldots}$$
$$+\sum_{s=1}^{n}\left\{S_x(\langle\tau\rangle^s_{srr_1r_2\ldots}P_{srr_1r_2\ldots}) + M_x(\langle\pi\rangle^s_{srr_1r_2\ldots}P_{srr_1r_2\ldots})\right\} = e_o, \tag{3.7}$$

with a similar equation obtained from (2.12). The additional suffix s appears with the terms containing the operators because these involve integrals; for instance,

$$S_x\tau = \int_\Omega dx'\int dt'S_x(x,x',t-t')\tau(x',t') \tag{3.8}$$

so that, in line with (3.5), a factor $f_s(x')$ is introduced.

The hierarchy (3.7) (with the other equations obtained from (2.12), is insoluble unless some approximating closure assumption is made. It is possible, in fact, to make closure assumptions that are "optimal", in the sense of making best-possible use of a limited number of joint probabilities $P_{rr_1r_2\ldots}$, relative to a variational principle. With this in mind, variational principles are introduced next.

4. VARIATIONAL PRINCIPLES

Subject to the symmetry restrictions (2.2) applied to L_o, with a similar symmetry assumed for ρ_o, the operators that appear in (2.11), (2.12) have the properties

$$\int_\Omega dx(\tau_1{*}S_x\tau_2) = \int_\Omega dx(S_x\tau_1){*}\tau_2 \ ,$$

$$\int_\Omega dx(\tau{*}\underset{\sim}{M}_x\pi) = \int_\Omega dx(S_t\tau){*}\pi \ , \tag{(4.1) cont.}$$

$$\int_\Omega dx(\pi_1 * M_t \pi_2) = \int_\Omega dx(M_t \pi_1) * \pi_2 \quad , \tag{4.1}$$

where * denotes the operation of time-convolution. It follows immediately that (2.11), (2.12) are equivalent to

$$\delta \mathcal{H}(\tau,\pi) = 0 \, , \tag{4.2}$$

where

$$2\mathcal{H}(\tau,\pi) = \int dx \Big\{ \tau * \Big[2e_o - S_x \tau - M_x \pi - (L-L_o)^{-1} \tau \Big] + \pi * \Big[2\dot{u}_o - S_t \tau - M_t \pi - (\rho - \rho_o)^{-1} \pi \Big] \Big\} \tag{4.3}$$

The static limit of (4.2) is the Hashin-Shtrikman principle [6]. (4.2) was generated in the way outlined here by Willis [10]. Also, in [4], the principle (4.2) was shown to follow from generalized versions of the classical energy principles due, essentially, to Gurtin [11] and Leitman [12]. These were given in [4] just for elastic bodies but they apply unaltered to "nonlocal" bodies of the type considered here. The analogue of the minimum energy principle is

$$\delta \mathcal{F}(u) = 0 \, , \tag{4.4}$$

where

$$2\mathcal{F}(u) = \int_\Omega dx \Big\{ e * Le + \dot{u} * \rho \dot{u} - 2(f-f_o) * u - 2p_o * \dot{u} - 2\sigma_o * e \Big\} \, . \tag{4.5}$$

The variation in (4.4) is taken over fields u(x,t) that satisfy the displacement boundary condition; the fields σ_o, p_o, f_o are related by an equation of motion like (2.6) and σ_o satisfies any given traction boundary conditions. The principle (4.4) follows because, for any σ, p, f satisfying (2.6) and the given traction boundary conditions,

$$\int_\Omega dx \Big\{ (\sigma - \sigma_o) * \delta e + (p - p_o) * \delta \dot{u} - (f - f_o) * \delta u \Big\} = 0 \tag{4.6}$$

for all admissible variations δu.

Dually, an analogue of the complementary energy principle is

$$\delta \mathcal{G}(\sigma,p) = 0 \quad , \tag{4.7}$$

where

$$2\mathcal{G}(\sigma,p) = \int_\Omega dx \Big\{ \sigma * L^{-1} \sigma + p * \rho^{-1} p - 2\sigma * e_o - 2p * \dot{u}_o \Big\} \, , \tag{4.8}$$

the variation being taken over fields that satisfy (2.6) and any given traction boundary conditions. The field u_o, from which e_o and $\dot{u}_o$ are derived, is any displacement field compatible with given displacement boundary conditions.

The trial fields u, σ, p may, in particular, be generated from arbitrarily chosen polarizations τ, π : the displacement u is then given by (2.10) and the associated σ, p are given by (2.8). The relations (2.7) are not satisfied, of course, unless the correct τ, π happen to be chosen. The "comparison material" formulation also generates convenient fields u_o, σ_o, p_o for use in (4.5), (4.8). With these choices, (4.5) gives

$$2\mathcal{F}(u) = \int_\Omega dx\Big\{(S_x\tau + M_x\pi)*L(S_x\tau + M_x\pi) + (S_t\tau + M_t\pi)*\rho(S_t\tau + M_t\pi)$$
$$-2(S_x\tau + M_x\pi)*(L-L_o)e_o - 2(S_t\tau + M_t\pi)*(\rho-\rho_o)\dot{u}_o$$
$$+e_o*(L-2L_o)e_o + \dot{u}_o*(\rho-2\rho_o)\dot{u}_o\Big\} \tag{4.9}$$

Use of the identity (4.6) with $\delta u = S\tau - M\pi$ allows (4.9) to be given in the alternative form

$$2\mathcal{F}(u) = \int_\Omega dx\Big\{(S_x\tau + M_x\pi)*\Big[\tau + (L-L_o)(S_x\tau + M_x\pi - 2e_o)\Big]$$
$$+(S_t\tau + M_t\pi)*\Big[\pi + (\rho-\rho_o)(S_t\tau + M_t\pi - 2\dot{u}_o)\Big]$$
$$+e_o*(L-2L_o)e_o + \dot{u}_o*(\rho-2\rho_o)\dot{u}_o\Big\}. \tag{4.10}$$

Any variation in τ produces an admissible variation in u. Considering such a variation, therefore, (4.3) implies

$$S_x\Big[\tau + (L-L_o)(S_x\tau + M_x\pi)\Big] + M_x(\rho-\rho_o)(S_t\tau + M_t\pi)$$
$$= S_x(L-L_o)e_o + M_x(\rho-\rho_o)\dot{u}_o \tag{4.11}$$

and a corresponding variation with respect to π gives

$$M_t\Big[\pi + (\rho-\rho_o)(S_t\tau + M_t\pi)\Big] + S_t(L-L_o)(S_x\tau + M_x\pi)$$
$$= S_t(L-L_o)e_o + M_t(\rho-\rho_o)\dot{u}_o \quad , \tag{4.12}$$

both of which can be derived from (2.11), (2.12). The variational operator $\mathcal{F}$ has not been expressed in the form (4.10) previously, except that its static limit was derived in [10].

5. STOCHASTIC VARIATIONAL PRINCIPLES AND OPTIMAL APPROXIMATIONS

Consider, in place of (4.2), the variational principle

$$\delta < \mathcal{H}(\tau,\pi) > = 0 \quad , \tag{5.1}$$

where now τ,π are functions defined over the body and the time axis and the sample space. The principle (5.1) produces (2.11), (2.12) directly, if the variations $\delta\tau$, $\delta\pi$ are taken to have "delta-function" dependence on the sample space parameter, so that they generate equations for one particular realization of the body. The principle also produces the hierarchy, the variation

$$\delta\tau = \delta\tau_1(x,t)f_r(x)f_{r_1}(x_1)f_{r_2}(x_2)\ldots \tag{5.2}$$

giving (3.7). An "optimal" approximate description of the waves generated in the body is therefore obtained by seeking stationary points of $<\mathcal{H}>$, when τ,π are restricted to lie in some subspace of configuration-dependent functions. For example, if τ,π are restricted to have the forms

$$\tau = \sum_{r=1}^{n} \tau_r(x,t)f_r(x) \quad , \pi = \sum_{r=1}^{n} \pi_r(x,t)f_r(x) \quad , \tag{5.3}$$

where τ_r, π_r depend only upon x,t, the operator $<\mathcal{H}>$ takes the form

$$<\mathcal{H}> = \frac{1}{2}\int dx\left\{\sum_{r=1}^{n}\tau_r{}^*\left[\left(2e_o-\left(L-L_o\right)^{-1}\tau_r\right)P_r - \sum_{s=1}^{n}\left(S_x\tau_sP_{sr}+M_x\pi_sP_{sr}\right)\right]\right.$$

$$\left.+\sum_{r=1}^{n}\pi_r{}^*\left[\left(2\dot{u}_o-(\rho_r-\rho_o)^{-1}\pi_r\right)P_r - \sum_{s=1}^{n}\left(S_t\tau_sP_{sr}+M_t\pi_sP_{sr}\right)\right]\right\}. \qquad (5.4)$$

Considering variations with respect to τ_r, π_r then produces the equations

$$\left(L_r-L_o\right)^{-1}\tau_rP_r + \sum_{s=1}^{n}\left(S_x\tau_sP_{sr}+M_x\pi_sP_{sr}\right) = e_oP_r\ , \qquad (5.5)$$

$$\left(\rho_r-\rho_o\right)^{-1}\pi_rP_r + \sum_{s=1}^{n}\left(S_t\tau_sP_{sr}+M_x\pi_sP_{sr}\right) = \dot{u}_oP_r\ . \qquad (5.6)$$

Equations (5.5), (5.6) are expressed in terms of the one- and two-point probabilities P_r, P_{sr} and may be regarded as "optimal" in the sense that allowance for the configuration in any way more general than in (5.3) would generate equations involving higher-order probabilities. The next-simplest choice for τ, π would appear to be

$$\tau = \sum_{r=1}^{n}f_r(x)\left\{\tau_r(x,t) + \sum_{s=1}^{n}\int dx'\phi_{rs}(x,x',t)f_s(x')\right\}, \qquad (5.7)$$

with a similar form for π. This would generate, with (5.1), equations involving probabilities for up to four points. Such equations have never been studied except in the static limit: Willis [13,14] used a form rather like (5.7) to estimate the overall moduli to order c^2, for a composite comprising a matrix containing a dilute dispersion of inclusions. Some of the implications of (5.5), (5,6) will be discussed in this Symposium by D.R.S. Talbot.

Finally, the variational statement

$$\delta<\mathcal{F}(u)> = 0 \qquad (5.8)$$

is considered, with u generated by configuration-dependent polarizations τ,π. Taking $\mathcal{F}(u)$ in the form (4.10) and choosing $\delta\tau$ as in (5.2) generates a conditional expectation of (4.11), namely,

$$\sum_{s=1}^{n}\left\{S_x<\tau>^s_{srr_1r_2\ldots}P_{srr_1r_2\ldots} + S_x\left[\left(L_s-L_o\right)\sum_{t=1}^{n}\left(S_x<\tau>^t_{tsrr_1\ldots}P_{tsrr_1\ldots}\right.\right.\right.$$

$$\left.\left.+M_x<\pi>^t_{tsrr_1\ldots}P_{tsrr_1\ldots}\right)\right]$$

$$\left.+M_x\left[\left(\rho_s-\rho_o\right)\sum_{t=1}^{n}\left(S_t<\tau>^t_{tsrr_1\ldots}P_{tsrr_1\ldots}+M_t<\pi>^t_{tsrr_1\ldots}P_{tsrr_1\ldots}\right)\right]\right\}$$

$$= \sum_{s=1}^{n}\left\{S_x\left[\left(L_s-L_o\right)e_oP_{srr_1\ldots}\right]+M_x\left[\left(\rho_s-\rho_o\right)\dot{u}_oP_{srr_1\ldots}\right]\right\}. \qquad (5.9)$$

The variational principle (5.3) thus generates the hierarchy associated with (4.11), (4.12) and so may also be used to derive "optimal" approximations. Substitution of (5.3) into (5.8) and seeking a stationary point gives the equations

$$\sum_{s=1}^{n}\left\{ S_x\left(\tau_s P_{sr}\right) + \sum_{t=1}^{n} S_x\left[\left(L_s - L_o\right)\left(S_x\left(\tau_t P_{tsr}\right) + M_x\left(\pi_t P_{tsr}\right)\right)\right] + \sum_{t=1}^{n} M_x\left[\left(\rho_s-\rho_o\right)\left(S_t\left(\tau_t P_{tsr}\right) + M_t\left(\pi_t P_{tsr}\right)\right)\right]\right\}$$
$$= \sum_{s=1}^{n}\left\{ S_x\left[\left(L_s-L_o\right) e_o P_{sr}\right] + M_x\left[\left(\rho_s-\rho_o\right)\dot{u}_o P_{sr}\right]\right\}, \qquad (5.10)$$

$$\sum_{s=1}^{n}\left\{ M_t\left(\pi_s P_{sr}\right) + \sum_{t=1}^{n} S_t\left[\left(L_s - L_o\right)\left(S_x\left(\tau_t P_{tsr}\right) + M_x\left(\pi_t P_{tsr}\right)\right)\right] + \sum_{t=1}^{n} M_t\left[\left(\rho_s-\rho_c\right)\left(S_t\left(\tau_t P_{tsr}\right) + M_t\left(\pi_t P_{tsr}\right)\right)\right]\right\}$$
$$= \sum_{s=1}^{n}\left\{ S_t\left[\left(L_s-L_o\right) e_o P_{sr}\right] + M_t\left[\left(\rho_s-\rho_o\right)\dot{u}_o P_{sr}\right]\right\}. \qquad (5.11)$$

These make "optimal" use of probabilities involving up to three points. They have not yet been used in the study of dynamic problems but the static limit of (5.10) has been used by Willis [10] to generate bounds of third order for the static overall moduli of a composite. The complementary principle (4.16) could be used to generate an alternative "third-order" description.

REFERENCES

1. KARAL, F.C. and KELLER, J.B., J. Math. Phys. Vol. 5, 1964, p.537.

2. TWERSKY, V., J. Math. Phys. Vol. 18, 1977, p.2468.

3. HINCH, E.J., J. Fluid Mech. Vol. 83, 1977, p.695.

4. WILLIS, J.R., J. Mech. Phys. Solids Vol. 28, 1980, p.287.

5. WILLIS, J.R., Wave Motion Vol. 3, 1981, p.1.

6. HASHIN, Z. and SHTRIKMAN, S., J. Mech. Phys. Solids Vol. 10, 1962, p.335.

7. HASHIN, Z. and SHTRIKMAN, S., J. Mech. Phys. Solids Vol. 10, 1962, p.343.

8. HASHIN, Z. and SHTRIKMAN, S., J. Mech. Phys. Solids Vol. 11, 1963, p.127

9. DEVANEY, A.J., J. Math. Phys. Vol. 21, 1980, p.2603.

10. WILLIS, J.R., Mechanics of Solids, the Rodney Hill Sixtieth Anniversary Volume (H.G. Hopkins and M.J. Sewell, eds.), p.653,1981. Pergamon Press (to appear).

11. GURTIN, M.E., Arch. Rat. Mech. Anal. Vol. 16, 1964, p.34.

12. LEITMAN, M.J., Quart. Appl. Math. Vol. 24, 1966, p.37.

13. WILLIS, J.R., Continuum Models of Discrete Systems (J.W. Provan, ed.), p.185, 1978. University of Waterloo Press.

14. WILLIS, J.R., Variational Methods in Mechanics of Solids (S. Nemat-Nasser,ed.), p.59, 1981. Pergamon Press.

Part 10
ALLOYS, THERMODYNAMICS

Continuum Models of Discrete Systems 4
eds. O. Brulin and R.K.T. Hsieh

A NEW APPROACH DESCRIBING IRREVERSIBLE PROCESSES

K.-H. Anthony

Universität-Gesamthochschule Paderborn
Theoretische Physik
4790 Paderborn, FR Germany

Thermodynamics of irreversible processes can be described by means of the Lagrange-formalism of fields. The general methods are exemplified for heat conduction. From the very beginnung the phenomenological theory is a general dynamics which is not restricted to the neighbourhood of equilibrium. The total information of the system is included in a Lagrange-functional.

1. INTRODUCTION

The Lagrange-formalism of fields is well established for non-dissipative phenomena [1,2]: The total information concerning the processes of a particular system is included in a Lagrange-functional, which is the kernel of Hamilton's variational principle. From this principle the field equations and using Noether's theorem [2] all balance equations and all constitutive equations of the system are derived, the latter two being associated with fundamental invariance properties of the Lagrangian Without going deep into the details and restricting myself to the phenomenon of heat conduction I want to show, that irreversible processes can be described by means of this very straight-forward field theoretical concept, too.1)

As compared with Onsager's approach [3] the Lagrange-formalism as applied to thermodynamics is no extrapolation of thermostatics. (In this context the reader is reminded of the "principle of local equilibrium" which is an ad hoc extrapolation of Gibb's fundamental relation of thermostatics). From the very beginning the Lagrange-formalism describes a general dynamics which must not be restricted to the neighbourhood of equilibrium. As compared with the methods of "rational thermodynamics" [5] the constitutive equations in the Lagrange-formalism are no collection of ad hoc assumptions. They are rather compatible with the existence of a Lagrangian, i.e. they are derived from a Lagrangian. In addition to energy, linear momentum a.s.o., which are traditionally well-known observables in the Lagrange-formalism, it will be shown, that the observable entropy, i.e. its density, flux and production rate, can be defined in quite a natural way. In this context two kinds of observables are distinguished: Entropy belongs to those of the second kind, whereas all traditional observables of the Lagrange-formalism are observables of the first kind.

It should be kept in mind, that the theory to be given here is a phenomenological approach to thermodynamics. In my opinion this new approach is neccessary for further progress in continuum theories of dissipative phenomena. As an example I refer to the dynamics of defects in ordered material structures. The new methods might be an appropriate tool to unify continuum theories of defects and thermodynamics within the scheme of Lagrange-formalism.

In order to keep the presentation of this lecture as vivid as possible the statements of general Lagrange-formalism will be immediately exemplified by the problem of heat conduction.

2. LAGRANGE-FORMALISM OF FIELDS

The processes P of a system are assumed to be characterized by a set of field variables $P = \{\Psi_i(x,t),\ i = 1...N\}$, which are evaluated in space and time.

2.1 ACTION INTEGRAL AND LAGRANGIAN

Lagrange-formalism is based on the *action integral*

$$\mathcal{J} = \int_{t_1}^{t_2} L_t[\Psi_i] = \int_{t_1}^{t_2} \int_{V(t)} \ell\left(x,t;\ \Psi_i(x,t),\ \partial\Psi_i(x,t)\right) d^3x\ dt \tag{1}$$

which is associated with the processes of the system in the time interval $[t_1,t_2]$. Deriving the *Lagrange functional* $L_t[\Psi_i]$ from the *Lagrange density function* ℓ –

$$L_t[\Psi_i] = \int_{V(t)} \ell(\cdots)\, d^3x \quad - \tag{2}$$

I restrict myself to a local theory. In general the volume of a material system takes part in the process, which is indicated by V(t) in eqs. (1,2). *The Lagrange density includes the total information of the system.* It depends on the process variables Ψ_i and on their spatial and time derivatives, both being indicated by $\partial\Psi_i$. Restricting the derivatives to the first order we are dealing with *Lagrange-formalism of the first order*. The Lagrangian may further depend explicitly on the space-time coordinates x and t.

For the *heat conduction* in an undeformable material let me propose the *first order Lagrangian*

$$\begin{aligned}\ell = & -c\,\Psi^*\Psi \\ & - \frac{c}{\omega}\left[\frac{1}{2i}\left(\Psi^*\partial_t\Psi - \Psi\partial_t\Psi^*\right) + \frac{1}{2}\,\frac{\ln\frac{\Psi^*\Psi}{T_0}}{\frac{\Psi^*\Psi}{T_0}}\,\partial_t(\Psi^*\Psi)\right] \\ & + \sum_{\alpha,\beta=1}^{3} \frac{\lambda^{\alpha\beta}}{\omega}\left[\frac{1}{\Psi^*\Psi}\,\frac{1}{2i}\left((\Psi^*\partial_\alpha\Psi)(\Psi^*\partial_\beta\Psi) - (\Psi\partial_\alpha\Psi^*)(\Psi\partial_\beta\Psi^*)\right)\right. \\ & \left. + \frac{T_0}{2(\Psi^*\Psi)^2}\,\partial_\alpha(\Psi^*\Psi)\cdot\partial_\beta(\Psi^*\Psi)\right]\end{aligned} \tag{3}$$
[2]

It is of first degree with respect to time derivatives and of second degree with respect to spatial derivatives. This Lagrangian reproduces completely Onsager's theory of heat conduction as will be shown step by step. It furthermore gives a quite general insight into the structure of Lagrange-formalism as applied to thermodynamics.

The complex scalar field $\Psi(x,t)$ is called

field of thermal excitation.

Ψ and the conjugate complex function Ψ^* enter into ℓ in such a way as to define a real valued Lagrangian. The excitation field is connected with measurement by means of the definition of *absolute temperature :*

$$T(x,t) = \Psi^*(x,t) \cdot \Psi(x,t) \geqq 0 \tag{4}$$

Obviously the *absolut zeropoint of temperature* is involved into the theory from the very beginning. Being positiv definit T vanishes if and only if Ψ vanishes. This corresponds to a qualitativ interpretation of the thermal excitation field: A material is at the absolute zero-temperature at those points (x,t) where no thermal excitation is present and vice versa.

The Lagrangian eq. (3) contains several constants which have the following physical meaning as will subsequently be shown:

c:	Specific heat.
$\lambda^{\alpha\beta}$, $\alpha,\beta=1,2,3$, :	Tensor of heat conductivity.
T_0 :	A reference temperature, which is necessary for dimensional reasons.
ω:	A frequency which is not specified a priori and which is necessary for dimensional reasons, too. (The Lagrangian has the dimension of an energy density).

The last bracket [...] in eq. (3) is symmetric with respect to the indices α and β. Thus from the tensor $\underline{\lambda}$ only the symmetric part is taken into account:

$$\underline{\lambda} = \text{symmetric} \tag{5}$$

This means that *Onsager's reciprocity relations* [3] concerning the coefficients of heat conductivity are involved into the Lagrangian from the very beginning.

Instead of the fields Ψ and Ψ^* we may equally well introduce as independent variables the temperature T and the phase φ:

$$\Psi(x,t) = \sqrt{T(x,t)}\; e^{i\varphi(x,t)} \tag{6}$$

$$\ell = -cT - \frac{c}{\omega}\left[T\partial_t\varphi + \frac{1}{2}\frac{\ln\frac{T}{T_0}}{\frac{T}{T_0}}\partial_t T\right] + \sum_{\alpha,\beta=1}^{3}\frac{\lambda^{\alpha\beta}}{\omega}\left[\partial_\alpha T\,\partial_\beta\varphi + \frac{T_0}{2T^2}\partial_\alpha T\,\partial_\beta T\right] \tag{7}$$

2.2 HAMILTON'S PRINCIPLE, FIELD EQUATIONS

Real processes of the system are distinguished as solutions of Hamilton's variational principle:

$$\mathcal{J} = \text{extremum} \tag{8}$$

by *free variation* of all fields $\Psi_i(x,t)$ keeping fixed the values at the starting and end points of the process:

$$\delta\,\Psi_i(x,t_{1,2}) = 0 \,.$$

Especially variations of the fields at the boundary and of the shape of the boundary are included. Looking at eq. (1) the variation (8) of the action integral has to be performed on the four-dimensional

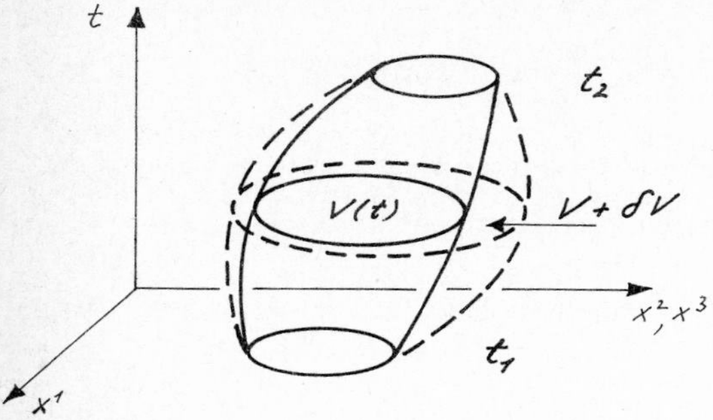

Fig. 1

Range of action integral

world tube in space-time occupied by the material system during its process.

N independent *Euler-Lagrange-field equations* are due to the free variations of the N fields Ψ_i:

$$\partial_t \frac{\partial \ell}{\partial(\partial_t \Psi_i)} + \sum_{\alpha=1}^{3} \partial_\alpha \frac{\partial \ell}{\partial(\partial_\alpha \Psi_i)} - \frac{\partial \ell}{\partial \Psi_i} = 0 \quad , \; i=1,\ldots,N. \tag{9}$$

The variations at the boundary give rise to a

set of boundary conditions. (10)

The fields Ψ and Ψ^* in the Lagrangian eq. (3) are dealt with independently giving rise to two field equations .They are highly non-linear and will not be written down explicitly here. I rather proceed to the variation of the two fields T and φ using the Lagrangian in the form of eq. (7). According to eq. (9) the heat conduction is governed by the
first field equation

$$c\,\partial_t T - \sum_{\alpha,\beta=1}^{3} \lambda^{\alpha\beta} \partial_\alpha \partial_\beta T = 0 \tag{11}$$

and by the *second field equation*

$$c\,\partial_t \varphi + \sum_{\alpha,\beta=1}^{3} \lambda^{\alpha\beta} \partial_\alpha \partial_\beta \varphi = \tag{12}$$
$$= -c\omega - \sum_{\alpha,\beta=1}^{3} \lambda^{\alpha\beta} \left(\frac{T_0}{T^2} \partial_\alpha \partial_\beta T - \frac{T_0}{T^3} \partial_\alpha T \partial_\beta T \right)$$

(Eqs. (11) and (12) are in turn due to variations of φ and T.

The first field equation being completely decoupled from the second one is obviously *Fourier's equation of heat conduction.* Within Lagrange-formalism this equations has to be supplemented by the second field equation which is coupled to the first one and which is non-linear in T.

Variation of φ und T at the surface F of the system, which is assumed to be regid and sufficiently smooth, gives rise to the *boundary conditions*

$$\frac{1}{\omega}\, \vec{n} \cdot (\underline{\lambda} \cdot \nabla T) = 0 \tag{13}$$

and

$$\frac{1}{\omega}\,\vec{n}\cdot(\underline{\lambda}\cdot\nabla\varphi) = 0 \tag{14}$$

Starting with a solution T(x,t) of the first field equation (11), i.e. with the temperature field of a real heat conduction process, it can easily be shown that the second field equation (12) is solved by

$$\varphi = -\omega t + \frac{T_0}{2T(x,t)} \tag{15}$$

for *each* real process. This means, that except for the frequency ω the phase function φ is completely determined by the measured temperature field. As far as the boundary conditions are concerned it will be shown below, that eq. (13) means thermal isolation of the system - as it should be, because in Hamilton's principle eqs. (1,7,8) no physical surrounding of the system is taken into account. Using eq. (15) the condition (14) is compatible with (13).

Eqs. (6,15) define the *thermal excitation wave* associated with heat conduction:

$$\psi(x,t) = \sqrt{T(x,t)}\cdot e^{-i\omega t}\cdot e^{i\frac{T_0}{2T(x,t)}} \tag{16}$$

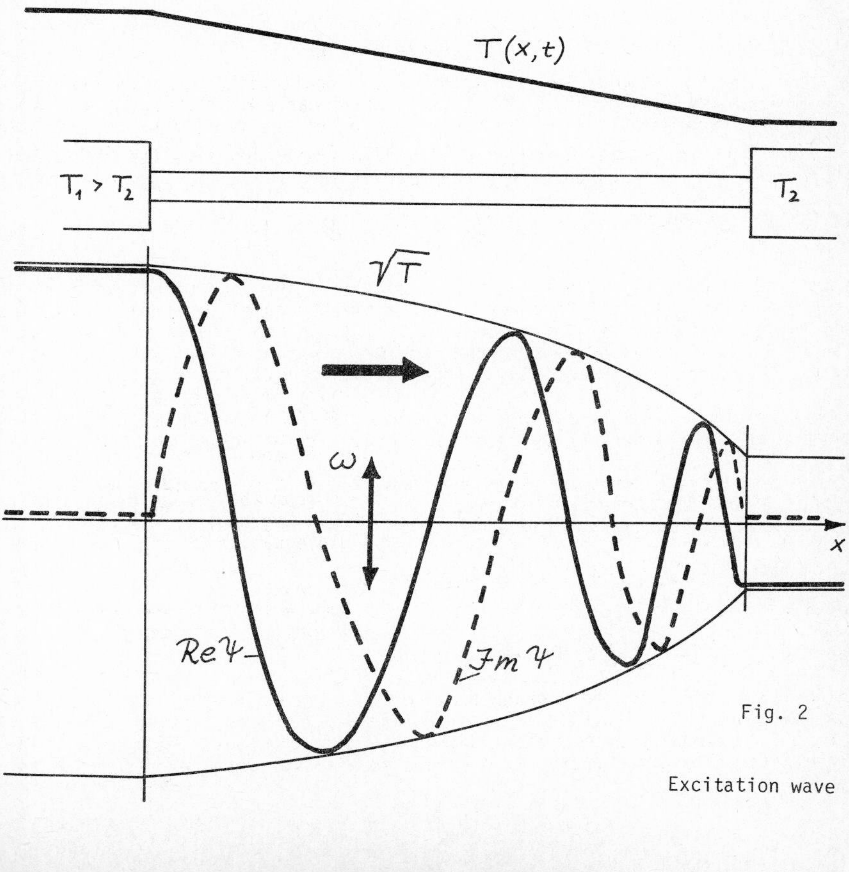

Fig. 2

Excitation wave

For the particular case of a stationary linear temperature gradient between two heat reservoirs Fig. 2 visualizes this wave running down the slope. With decreasing temperature the wave length gets shorter from left to right, whereas the temporal period is constant and is given by the frequency ω, which has to be regarded as a characteristic parameter of the heat conducting material.

Looking at Fig. 2 the question arises if in terms of field theory these oscillations can be regarded as a preformation of those temporal and spatial oscillations which are discussed in synergetics [8]. Of course in pure heat conduction the oscillations of the excitation wave (16) are hidden from the phenomenological point of view. However it might happen that they get obvious in more complex cases when irreversible fluxes of different physical nature are coupled.

In the reference equilibrium state $T(x,t) = T_0$ the excitation wave degenerates into a pure homogeneous oscillation with frequency ω. The Lagrangian (eqs. (3,7)) is normalized in such a way as to vanish at that state.

As compared with a lot of variational principles in Literature the procedure described here is really Hamilton's principle in its true sense. I am dealing with free variations whereas to my best knowledge all preceding variational principles of heat conduction are based on constraint variations and thus cannot be satisfactory [9,10].

2.3 SYSTEM-SYMMETRIES, BALANCE EQUATIONS OF THE FIRST KIND, CONSTITUTIVE EQUATIONS

An operation $\underline{T}_\varepsilon$ of the group of system-symmetries transforms a real process $P = \{\Psi_i(x,t)\}$ into a real process $\bar{P} = \{\bar{\Psi}_i(x,t)\}$:

$$(\bar{x}, \bar{t}, \bar{\Psi}) = \underline{T}_{\varepsilon_1,\dots,\varepsilon_n}(x, t, \Psi) \tag{17}$$

i.e. P and $\bar{P}$ have to fulfill Hamilton's Principle simultaneously. This leads to the sufficient *symmetry criterion* for the Lagrangian

$$\ell(\bar{x}, \bar{t}; \bar{\Psi}, \partial\bar{\Psi}) = \ell(x, t; \Psi, \partial\Psi) + \frac{\partial}{\partial t}F(x, t, \Psi, \varepsilon) \tag{18}$$

which must be identically satisfied in x,t and all Ψ_i for each operation $\underline{T}_\varepsilon$ of the symmetry group. In general $\underline{T}_\varepsilon$ applies to the field variables Ψ_i as well as to the independent space-time variables x and t (eq. (17). In the present context I am dealing with a continuous symmetry group, the group parameters of which are denoted by ε_κ. The term $\frac{\delta}{\delta t}F$ in eq. (18) is due to the restriction $\delta\Psi(x,t_{1,2}) = 0$ in Hamilton's principle. The symmetry criterium requires the existence of a function F depending on the arguments x,t, Ψ_i and the group parameters ε_κ only.

Due to the symmetry criterion *each group parameter* ε_κ *of the symmetry group* (eq. (17)) *defines an observable* ω_κ *of the first kind* which is associated with a *balance equation of the first kind* (Noether):

$$\varepsilon_\kappa \Rightarrow \omega_\kappa \Rightarrow \partial_t a_\kappa + \operatorname{div} \vec{J}_\kappa = 0 \tag{19}$$

The density $a_\kappa = a_\kappa(x,t;\ \Psi,\ \partial\Psi)$ (20)

and the flux $\vec{J}_\kappa = J_\kappa(x, t; \Psi, \partial\Psi)$ (21)

are uniquely defined from the Lagrangian ℓ and thus depend on the same set of variables as ℓ does. Eqs. (20,21) are *constitutive equations* associated with the observable ω_κ. As compared with Onsager's theory, where the densities a_κ are taken from thermostatics and the fluxes $\vec{J}_\kappa$ are related with "driving forces" by ad hoc assumptions, the constitutive equations (20,21) are defined by straightforward methods of Lagrange formalism. They are really compatible with the existence of a Lagrangian ℓ.

Examples for the case of heat conduction:

2.3.1 TIME REVERSAL

The operation

$$\underline{T} : \begin{cases} \bar{t} = -t \\ \bar{x} = x \\ \bar{\Psi} = \Psi \end{cases} \tag{22}$$

being an element of the discrete second order grop {1,T} transforms the real heat conducting process $\Psi(x,t)$ into the process $\bar{\Psi}(x,t) = \Psi(x,-t)$ which with regard to eqs. (3,18) is no longer a real process! (Because of the linear terms $\partial_t\Psi$ and $\partial_t \Psi^*$ in ℓ no function $\mathbf{F}$ exists such that eq. (18) is fulfilled). This means that the *Lagrangian eq. (3)* really *describes irreversible heat conduction.*

2.3.2 TRANSLATION IN TIME

The one-parameter group of translations in time

$$\underline{T}_\varepsilon : \begin{cases} \bar{t} = t + \varepsilon & -\infty < \varepsilon < +\infty \\ \bar{x} = x \\ \bar{\Psi} = \Psi \end{cases} \tag{23}$$

is required to be a symmetry group of each closed physical systems, which means, that the time continuum is homogeneous with respect to such a system. This requirement is an essential part of physical causality.

Once and for all *the translation of time of a closed system is associated with the observable "energy"* and with the *energy balance equation*

$$\partial_t u + \operatorname{div} \underset{(u)}{\vec{J}} = 0 \tag{24}$$

The constitutive equations (20, 21) are now specified as

energy density
$$u = \sum_{i=1}^{N} \frac{\partial \ell}{\partial(\partial_t \Psi_i)} \partial_t \Psi_i - \ell , \tag{25}$$

energy flux
$$\underset{(u)}{\vec{J}} = \sum_{i=1}^{N} \frac{\partial \ell}{\partial(\nabla \Psi_i)} \partial_t \Psi_i \tag{26}$$

Evaluating these statements for the Lagrangian eq. (3) or (7) it is easily shown that ℓ is invariant with respect to time translation eq. (23), i.e. the symmetry criterion eq. (18) is fulfilled with $\mathbf{F} = 0$. The Lagrangian ℓ is thus dealing with heat conduction in a thermally isolated system and eq. (24) has to be looked upon as *the first law of thermodynamics*. Evaluating eqs. (25) and (26) for the Lagrangian eq. (3) or (7) and using eq. (15) I get for the

internal energy
$$u = c \cdot T \tag{27}$$

and for the *energy flux (heat flux)*

$$\underset{u}{\vec{J}} = -\underline{\lambda} \cdot \nabla T, \tag{28}$$

which are the well known constitutive equations for heat conduction in Onsager's theory [3]. From these equations the constants c and λ in ℓ are identified as specific heat and tensor of heat conductivity respectively.

Taking account of eq. (28) the boundary condition eq. (13) forbids heat transfer through the boundary.

By the way it can easily be shown by means of energetical arguments that it is impossible to describe heat conduction by means of Lagrange-formalism using the temperature T as the only variable. Let me assume the existence of a Lagranian

$$\ell = \ell(T, \partial_t T, \nabla T). \tag{29}$$

Then the energy flux eq. (26) reads

$$\vec{J}_{(u)} = \frac{\partial \ell}{\partial(\nabla T)} \partial_t T, \tag{30}$$

which vanishes in the stationary case of Fig. 2. However, there is no doubt, that in this situation energy is flowing from left to right. The contradiction is solved by introducing the complex excitation field Ψ, which retains an appropriate time dependence even in the stationary state (factor $e^{-i\omega t}$ in eq. (16)).

2.3.3 TRANSLATION IN SPACE

As in the case of time translation the 3-parameter group of translations in space

$$T_{\varepsilon^1,\varepsilon^2,\varepsilon^3}: \begin{cases} \bar{t} = t \\ \bar{x}^\alpha = x^\alpha + \varepsilon^\alpha, & \alpha = 1,2,3, \\ \bar{\Psi} = \Psi & -\infty < \varepsilon^\alpha < +\infty \end{cases} \tag{31}$$

is required to be a symmetry group of each closed system. This means homogeneity of space with respect to such a system. *Linear momentum is the vector-valued observable associated once and for all with space translation.* The *balance-equation of linear momentum*

$$\partial_t \vec{p} + \operatorname{div} \underline{\sigma} = 0 \tag{32}$$

is compatible with the constitutive equations of the *density of linear momentum*

$$\vec{p} = -\sum_{i=1}^{N} \frac{\partial \ell}{\partial(\partial_t \Psi_i)} \nabla \Psi \tag{33}$$

and of the *flux of linear momentum (= stress tensor)*

$$\underline{\sigma} = \sigma^\alpha{}_\beta = -\sum_{i=1}^{N} \frac{\partial \ell}{\partial(\partial_\alpha \Psi_i)} \partial_\beta \Psi_i + \ell\, \delta^\alpha_\beta. \tag{34}$$

Evaluating this formalism for heat conduction one finds, that the Lagrangian eq. (3, 7) is invariant with respect to space translations. Thus heat conduction is associated with a balance of linear momentum. Quantities (33) and (34) are found to be

$$\vec{p} = \frac{1}{\omega}\frac{c}{2}\left(\frac{\ln \frac{T}{T_0}}{\frac{T}{T_0}} - \frac{T_0}{T}\right)\nabla T \tag{35}$$

$$\sigma = -\frac{1}{\omega}\frac{c}{2}\left(\quad '' \quad \right)\partial_t T \cdot 1 \tag{36}$$

The bracket being negativ in the neighbourhood of T_0 the density vector $\vec{p}$ of linear momentum is parallel with energy flux. On the other hand momentum flux only occurs in non-stationary processes.

It is a remarkable fact that the unspecified frequency ω of the Lagrangian enters into the definitions of linear momentum flux and density. This offers the possibility of measuring ω by means of experiments taking account of linear momentum exchange.

2.3.4 ROTATION IN SPACE

The 3-parameter rotation group

$$\underline{T}_{\varepsilon_1,\varepsilon_2,\varepsilon_3} : \begin{cases} \bar{t} = t \\ \bar{\vec{x}} = \underline{R}(\varepsilon_1, \varepsilon_2, \varepsilon_3)\cdot\vec{x} \\ \bar{\Psi} = \Psi \end{cases} \tag{37}$$

is associated with isotropy of space with respect to the particular system. In eq. (37) $\underline{R}$ denotes the (3.3)-rotation matrix depending on three Eulerian angles ε_1, ε_2, ε_3. The transformation is already specified for the case of scalar fields, especially for the thermal excitation field Ψ and Ψ^*, which is not functionally affected by the space rotations.

Angular momentum is the observable associated with space rotations. Again the Lagrangian of heat conduction is invariant with respect to the transformations eq. (37). (The tensor $\underline{\lambda}$ which is associated with the material background has to be subjected to the same transformation. See the remarks in Ch. 3.2.) Heat conduction is thus accompanied with a balance of angular momentum and the frequency ω again enters into the definitions of the density and flux of angular momentum, i.e. exchange experiments of angular momentum can be used for measuring ω, too.

2.3.5 GAUGE TRANSFORMATION OF COMPLEX FIELDS

Finally the Lagrangian ℓ of heat conduction is invariant with respect to the gauge group

$$\underline{T}_{\varepsilon} : \begin{cases} \bar{t} = t \\ \bar{x} = x \\ \bar{\Psi} = \Psi\cdot e^{i\Lambda\varepsilon} \\ \bar{\Psi}^* = \Psi^*\cdot e^{-i\Lambda\varepsilon} \end{cases} \tag{38}$$

This group is related with the impossibility of absolutely fixing the phase φ of Ψ by means of experiment. In fact $T = \Psi\Psi^*$ is the measurable quantity, whereas φ is associated e.g. with energy flux leaving undetermined an additiv constant in φ.

Via the symmetry criterion (eq. (18)) *the gauge group* of the thermal excitation field *is associated with a mass like observable*[3] which I call the *"thermal excitation"*. Its balance equation reads

$$\partial_t w + \operatorname{div} \vec{J}_{(w)} = 0 \tag{39}$$

with the *excitation density*

$$w = i\Lambda\left(\frac{\partial \ell}{\partial(\partial_t\Psi)}\Psi - \frac{\partial \ell}{\partial(\partial_t\Psi^*)}\Psi^*\right) \tag{40}$$

and the excitation flux

$$\vec{J}_{(W)} = i\Lambda\left(\frac{\partial \ell}{\partial(\nabla\psi)}\psi - \frac{\partial \ell}{\partial(\nabla\psi^*)}\psi^*\right). \tag{41}$$

Evaluating these quantities for the Lagrangian of heat conduction I get

$$W = \Lambda \cdot c \cdot T \tag{42}$$

and

$$\vec{J}_{(W)} = -\Lambda \cdot \underline{\lambda} \cdot \nabla T. \tag{43}$$

The gauge factor Λ occuring in eqs. (38-43) is not specified within the phenomenological scheme. It characterizes those microscopic thermal excitation processes which phenomenologically give rise to the escitation field Ψ.[3)]

Comparing eqs. (27, 28) with eqs. (42, 43) it seems that the observables "energy" and "thermal excitation" could be identified. However they should carefully be destinguished as qualitatively different observables which belong to time translations and gauge transformations respectively. The coincidence of eqs. (27, 28) and (42, 43) is accidental and is due to the restricted problem of pure heat conduction. If additional degrees of freedom are taken into account both sets of equations will split up.

The observable "thermal excitation" can be regarded as a carrier of energy. To support physical imagination it is provisionally suggested to associate this observable with quantities such as phonon density and phonon flux.

3. SOME ITEMS FROM THE THEORY OF SECOND VARIATION

Only those items are briefly reported which are neccessary for an insight into the definition of entropy using the framework of Lagrange-formalism.

3.1 VARIATIONAL EQUATIONS OF THE FIELD EQUATIONS

Within a one-parameter class C_ε of real processes $\Psi_i(x,t,\varepsilon)$ I define a real process variation

$$\delta\Psi_i = \left.\frac{\partial \Psi_i(x,t,\varepsilon)}{\partial\varepsilon}\right|_{\varepsilon=0} \cdot \delta\varepsilon \tag{44}$$

The set of functions $\Psi_i(x,t,\varepsilon)$, $i = 1,\dots N$, being a solution of the field equations (9) and of the boundary equations (10) for each ε (examples are given below) involves the set of functions

$$\eta_i(x,t) = \left.\frac{\partial \Psi_i(x,t,\varepsilon)}{\partial\varepsilon}\right|_{\varepsilon=0}, \quad i=1,\dots,N, \tag{45}$$

to be a solution of *Jacobi's equations*

$$\partial_t \frac{\partial\Omega}{\partial(\partial_t\eta_i)} + \sum_{\alpha=1}^{3} \partial_\alpha \frac{\partial\Omega}{\partial(\partial_\alpha\eta_i)} - \frac{\partial\Omega}{\partial\eta_i} = 0, \quad i=1,\dots,N, \tag{46}$$

which are the variational equations of the field equations (9). The kernel Ω is defined from the Lagrangian ℓ and depends on the fixed reference process

$$\Psi_i(x,t) = \Psi_i(x,t,\varepsilon=0)\,, \tag{47}$$

on the associated real variation η_i and eventually explicitly on x and t:

$$2\Omega(x,t;\Psi,\partial\Psi;\eta,\partial\eta)$$
$$= \sum_{i,j=1}^{N}\Big[\frac{\partial^2 \ell}{\partial\Psi_i\,\partial\Psi_j}\eta_i\eta_j + \sum_{\alpha=0}^{3} 2\,\frac{\partial^2\ell}{\partial\Psi_i\,\partial(\partial_\alpha\Psi_j)}\eta_i\,\partial_\alpha\eta_j + \sum_{\alpha,\beta=0}^{3}\frac{\partial^2\ell}{\partial(\partial_\alpha\Psi_i)\,\partial(\partial_\beta\Psi_j)}\partial_\alpha\eta_i\,\partial_\beta\eta_j\Big] \tag{48}$$

∂_t is denoted here by ∂_0.

Jacobi's equations are related to investigations of *stability* of the system [6]. They are wellknown in the theory of the second variation of Hamilton's principle [7].

The function 2Ω being homogeneous of second degree with respect to η and $\partial\eta$ can be subjected to Euler's formula. Taking further account of Jacobi's equations (46) it can easily be shown, that the class C_ε of real processes is associated in a natural way with a non-homogeneous balance equation

$$\partial_t s + \operatorname{div} \underset{(s)}{\vec{J}} = \underset{(s)}{\sigma} \tag{49}$$

where s, $\underset{(s)}{\vec{J}}$ and $\underset{(s)}{\sigma}$ are defined by Ω:

density
$$s = \sum_{i=1}^{N}\frac{\partial\Omega}{\partial(\partial_t\eta_i)}\eta_i = s(x,t;\Psi,\partial\Psi;\eta,\partial\eta), \tag{50}$$

flux
$$\underset{(s)}{\vec{J}} = \sum_{i=1}^{N}\frac{\partial\Omega}{\partial(\nabla\eta_i)}\eta_i = \underset{(s)}{\vec{J}}(\quad " \quad), \tag{51}$$

production rate density
$$\underset{(s)}{\sigma} = 2\Omega = \underset{(s)}{\sigma}(\quad " \quad). \tag{52}$$

These three quantities are evaluated for the reference process $\Psi_i(x,t,\varepsilon=0)$ (eq. 47)) and its associated real variation $\eta_i(x,t)$ in the class C_ε (eq.(45)).

3.2 OBSERVABLES OF THE SECOND KIND. THE ENTROPY

Looking at the fundamental invariance transformations (23, 31, 37, 38) each group parameter ε defines a class of real processes in the sense of ch. 3.1. In fact for heat conduction the functions

$$\Psi(x,t,\varepsilon) = T_\varepsilon\Psi(x,t) = \begin{cases} = \Psi(x,t+\varepsilon) & \text{due to eq. (23)}, \quad (53)\\ = \Psi(x^\alpha+\varepsilon^\alpha,t) & \text{due to eq. (31)}, \quad (54)\\ = \Psi(R(\varepsilon_1,\varepsilon_2,\varepsilon_3)\vec{x},t) & \text{due to eq. (37)}, \quad (55)\\ = \Psi(x,t)\cdot e^{i\Lambda\varepsilon} & \text{due to eq. (38)}, \quad (56)\end{cases}$$

define classes of real processes which fulfill the field equations (11, 12) and the boundary conditions (13, 14) for each value of the parameters. (Of course in the case of eq. (55) the conductivity tensor $\underline{\lambda}$ and the surface vector $\vec{n}$ must be submitted to the rotations $\underline{R}$, too, i.e. the class of real processes is defined by transforming the whole physical situation including the material carrier in the case of heat conduction. For details the reader is referred to forthcoming papers.)

Thus, following the formalism of the preceding chapter *each group parameter of the fundamental symmetry groups* of the system (see eqs. 23, 31, 37, 38) *defines* in quite a natural way an additional *observable of the second kind, which is associated with an inhomogeneous balance equation* (49) *and with constitutive equations* (50,51,52). This balance equation might accidentally be empty. However in general it gives an insight into the stability of a process within the classes defined in eqs. (53-56).

In the case of a thermal system *the observable of the second kind associated with the gauge transformation of the thermal excitation field* Ψ (eqs. 38, 56) *is called "entropy"*. Then eq. (49) is the *first part of the second law of thermodynamics* and eqs. (50, 51, 52) give the *constitutive equations for entropy density, entropy flux and entropy production rate* respectively.

Evaluating these expressions for heat conduction and using the functions

$$\eta = i\Lambda\Psi \quad , \quad \eta^* = -i\Lambda\Psi^* , \tag{57}$$

which are due to eqs. (45, 56), I get the wellknown formulae of Onsager's theory [3]:

entropy density $$s = c \ln \frac{T}{T_0} \tag{58}$$

entropy flux $$\vec{J}_{(s)} = -\frac{\underline{\lambda} \cdot \nabla T}{T} \tag{59}$$

entropy production rate $$\sigma_{(s)} = \frac{\nabla T \cdot \underline{\lambda} \cdot \nabla T}{T^2} \tag{60}$$

As far as $$\sigma_{(s)} > 0 \tag{61}$$

is concerned *(second part of the second fundamental law)* this statement is no straightforward outcome of Lagrange-formalism. It seems to distinguish between stable and unstable processes in a particular Lyapunov sense. Investigations are in progress.[4)]

I should finally mention that Jacobi's equations (46) can be associated with a variational principle, too. It is a generalized *principle of least entropy production* which suffers no restrictions as compared with the analogous principle in Onsager's theory [3].[4)]

4. HOW TO GET THE LAGRANGIAN? FINAL REMARKS

I have presented a Lagrange-formalism which completely reproduces Onsager's theory of heat conduction. The question arises, if the newly introduced excitation field Ψ is of real physical relevance as a whole or if it is a mathematical trick only. This question has to be answered in future by experiment. However, even in the case the theory offers a methodical unification of thermodynamics. Especially Hamilton's principle associated with free variation is of some practical use.

Of course the Lagrangian (3) is a very special one which has been constructed in order to reproduce Onsager's theory and to show by this means, that it is possible to include thermodynamics into the Lagrange-formalism. However the Lagrangian can be generalized to include new phenomena.

There are two possibilities to get the Lagrangian:

a. Determination of ℓ for a given set of field equations by means of straightforward mathematics (inverse problem of variation calculus [11]). In a forthcoming paper it will be shown how to get eqs. (3,7) along this line.

b. Starting from an ansatz for ℓ and comparing its consequences with experiment.

Going along the second line let me complete the Lagrangian (7) a little bid by means of a quadratic inertia term (Θ = constant):

$$\begin{aligned} \ell = & -cT \\ & -\frac{c}{\omega}\left[T\partial_t\varphi + \frac{1}{2}\frac{\ln\frac{T}{T_0}}{\frac{T}{T_0}}\partial_t T\right] \\ \Longrightarrow\quad & -\Theta\,\partial_t T\,\partial_t\varphi \\ & +\sum_{\alpha,\beta=1}^{3}\frac{\lambda^{\alpha\beta}}{\omega}\left[\partial_\alpha T\,\partial_\beta\varphi + \frac{T_0}{2T^2}\partial_\alpha T\,\partial_\beta T\right] \end{aligned} \tag{62}$$

The associated first field equation is

$$-\Theta\partial_t^2 T - c\,\partial_t T + \sum_{\alpha,\beta=1}^{3}\lambda^{\alpha\beta}\partial_\alpha\partial_\beta T = 0 \tag{63}$$

As compared with Fourier 's equation (11) it avoids the paradox of an infinite speed of propagation associated with the heat pole.

The singularities of the theory at T = 0 might be cancelled by taking account of a suitable temperature dependence of heat capacity c and conducitivity $\underline{\lambda}$.

REFERENCES

A complete list will be given in forthcoming papers. The following list is restricted to some monographs for general information only.

[1] Corson, E.M.: Introduction to Tensors, Spinors and Relativistic Wave-Equations. (Blackie a. Son, 1957)

[2] Schmutzer, E.: Symmetrien und Erhaltungssätze der Physik. (Akademie-Verlag Berlin, Pergamon Press Oxford, Vieweg Braunschweig, 1972, WTB Nr. 75)

[3] DeGroot, S.R. and Mazur, P.: Non-Equilibrium Thermodynamics. (North Holland Publ. Comp., 1969)

[4] Anthony, K.-H.: Continuum Description of Liquid Crystals, in: Kröner, E. and Anthony, K.-H. (eds.), Continuum Models of Discrete Systems 3 (University of Waterloo Press, 1980)

[5] Truesdell, C.: Rational Thermodynamics. A course of Lectures on Selected Topics. (McGraw-Hill, 1969)

[6] Zubov, V.I.: Methods of A.M.Lyapunov and their Applications. (Noordhoff, Groningen, 1964)

[7] Rund, H.: The Hamilton-Jacobi Theory in the Calculus of Variations. (D. van Nostrand Comp., London, Toronto, NewYork, Princeton, 1966)

[8] Haken, H.: Synergetics (Springer, 1978)

[9] Gyarmati, I.: Non-Equilibrium Thermodynamics, Field Theory and Variational Principles. (Springer, 1970)

[10] Glansdorff, P. and Prigogine, I.: Thermodynamic Theory of Structure, Stability and fluctuations. (John Wiley a. Sons, 1971)

[11] Santilli, R.M.: Foundations of Theoretical Mechanics I. (Springer, 1978)

FOOTNOTES

1) The reader is referred to my paper 4 in the CMDS3-Proceedings, where the Lagrange-formalism is presented in detail for the case of nematic liquid crystals. Concerning thermodynamics I refer to forthcoming publications.

2) $\partial_t = \frac{\partial}{\partial t}$, $\partial_\alpha = \frac{\partial}{\partial x^\alpha}$, $i = \sqrt{-1}$

3) The same scheme is realized in quantum mechanics. The gauge group of the matter field is associated with the observables "particle number", "mass" or "electrical charge" depending on the choise of the gauge factor Λ.

4) The reader is referred to forthcoming papers.

Continuum Models of Discrete Systems 4
eds. O. Brulin and R.K.T. Hsieh

MEMORY ALLOYS - PHENOMENOLOGY AND ERSATZMODEL

Ingo Müller & Krzysztof Wilmanski

Hermann-Föttinger-Institut
FB 9 - Technische Universität Berlin
1 Berlin 12

Institute of Fundamental Technological Research
Polish Academy of Science
00-049 Warsaw

Memory alloys, or pseudoelastic bodies, show a strong dependence of their load-deformation curves on temperature. While at low temperature such an alloy behaves similar to a plastic body, it is non-linear elastic at high temperatures. In-between lies pseudoelasticity because, while the body returns to its original configuration upon unloading, the loading-unloading cycle exhibits a hysteresis. A statistical mechanical Ersatzmodel is described here which simulates all observed phenomena of memory alloys.

1.) PHENOMENOLOGY OF PSEUDOELASTIC BODIES

Typical load-deformation curves of pseudoealstic bodies for different temperatures are shown in Figure 1a. through 1f. in order of increasing temperatures. Except at high temperatures these curves exhibit hysteresis loops in tension as well as in compression.

Low temperatures: In a tensile loading experiment starting at the origin we have an elastic branch first, but when a critical load is reached, the body yields deformation at constant load until it reaches a second elastic branch along which it may be further loaded or unloaded. The removal of the load leads to a residual deformation. An increase in temperature decreases the yield load.

Intermediate temperatures: Again, if the tensile loading starts in the origin, there is a first elastic branch, a yield load and a second elastic branch, but now unloading does not create a residual deformation. Instead, if the load falls below a certain value, the deformation is recovered. This type of behaviour has given rise to the name pseudoelasticity, because like in elasticity the body returns to its original configuration but unlike in elasticity there is a hysteresis. The slope of the initial elastic branch increases with increasing temperature and both the yield limit and the recovery limit also increase; the two grow closer together.

High temperatures: The hysteresis vanishes at high temperatures and the body becomes truely elastic with an increasing elastic modulus as temperature grows.

It is obvious now why pseudoealstic bodies are also called bodies with shape memory. Indeed, if we produce a residual deformation at low temperatures, the body will return to the original configuration at high temperatures. Thus we may say that it "remembers" its original configuration.

Microscopic observations show that the behaviour exhibited in Figure 1 reflects transitions between an austenitic and a martensitic phase of the metallic lattice and between different martensitic twins. At low temperature the body is martensitic with different martensitic twins in coexistence within the hysteresis

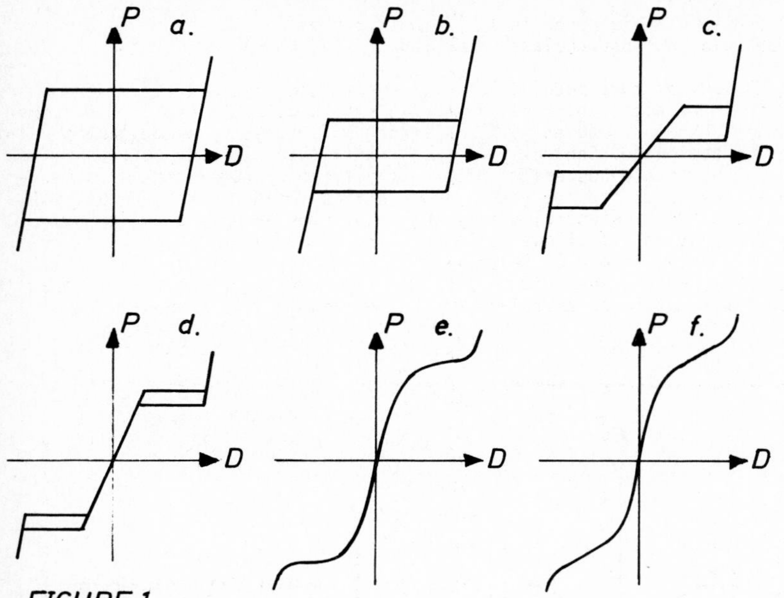

FIGURE 1

loop while on the lateral elastic branches one or the other twin prevails. At intermediate temperatures the body is in the austenitic phase at small loads but it may be forced into one or the other of the martensitic twin phases by tensile or compressive loads, provided these are high enough; within the hysteresis loops austenite co-exists with either one of the martensitic twins. At high temperature there is no well-defined transition between the austenitic and the martensitic phase, rather the transition is gradual.

Knowledge of enough load-deformation curves for different temperatures allows us to construct deformation-temperature curves for any given load. Figure 2 represents such a plot for a load that is bigger than the yield load at a low temperature. We see that this curve exhibits a hysteresis which reflects the hysteresis in the load-deformation diagram.

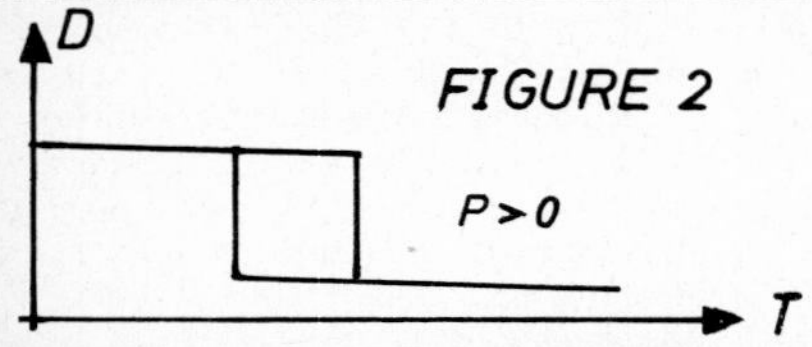

FIGURE 2

Curves like this one are underlying most of the applications of memory alloys. In such applications a loaded sample is alternately heated and cooled and thus it shrinks and expands accordingly.

More detailed information about the phenomenology of pseudoelastic bodies as well as extensive accounts of their crystallography can be found in the literature (see [1] through [4]).

2.) A MODEL FOR PSEUDOELASTICITY

The main purpose of this paper is the formulation of a model that can simulate the observed behaviour of pseudoelastic body.

The basic element of this model is a lattice particle, a small piece of the metallic lattice of the body, which is shown in Figure 3.a. in three different equilibrium configurations denoted by M_+, M_- and A which stands for the two martensitic twins and austenite. Obviously the martensitic particles may be considered as sheared versions of the austenitic one. The shift of the upper edge with respect to the lower one is called the shear length and is denoted by Δ. To each value of Δ there corresponds a potential energy $\phi^{\wedge}(\Delta)$ and Figure 3.a. shows the postulated form of that function. It allows for two lateral stable equilibria at $\Delta = \pm J$, and a metastable one at $\Delta = 0$. Reasons for this choice of the form of $\phi^{\wedge}(\Delta)$ will be given as we proceed. Figure 3.b. gives a schematized picture of the potential which facilitates the calculations that follow below.

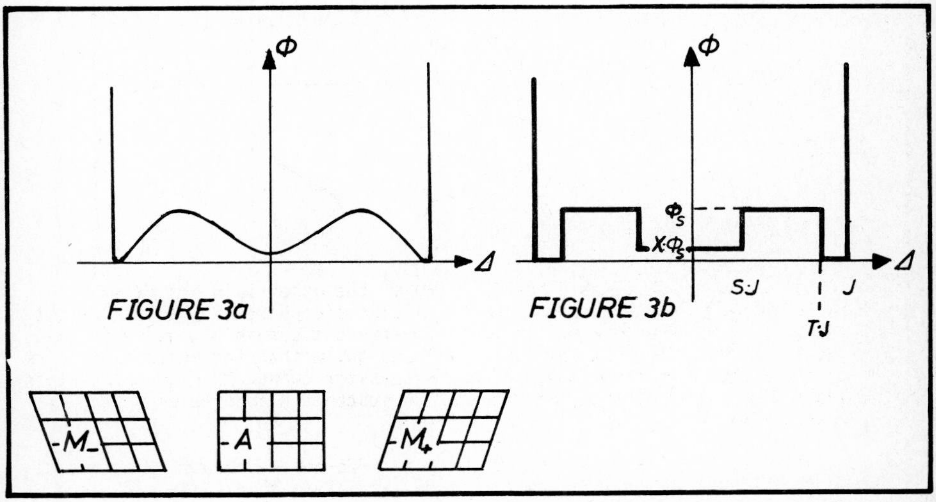

FIGURE 3a FIGURE 3b

To construct the model of the body we arrange the lattice particles in layers and stack these upon each other as shown in Figure 4. The layers are arranged at 45^0 so that under a tensile load P on the body the layers are subject to the shear load $\frac{P}{\sqrt{2}}$.

We proceed to evaluate the qualitative features which, given the arrangement of layers and the form of the potential $\phi^{\wedge}(\Delta)$, the model is expected to have. This discussion will treat the case of low temperature separately from the case of intermediate and high temperature.

3.) THE MODEL AT LOW TEMPERATURE

In a typical virginal configuration after cooling the body from the melt we expect half of the layers to be of type M_- and the other half of type M_+ as shown in Figure 4a. schematically. The reason is this: At high temperature the layers will fluctuate freely about the whole range of Δ, since their mean kinetic energy kT is much bigger than the potential barriers in $\phi(\Delta)$. When the temperature falls, the layers will prefer values of Δ near the minima of $\phi(\Delta)$. At intermediate temperatures a layer with $\Delta \approx 0$ will still be capable of scaling the barrier between configuration A and the configurations $M_\pm$ but, once there, it will not be able to get back, because the minima at $\Delta = \pm J$ are deeper than that at $\Delta = 0$. Thus in the cooling process the layers will become either M_+ or M_-, and roughly there will be half of each as shown in Figure 4a. Indeed, in reality microscopic observations show that in the original state at low temperature the martensitic twins prevail in equal proportions. That observation was the reason for us to postulate deep lateral minima for $\phi(\Delta)$.

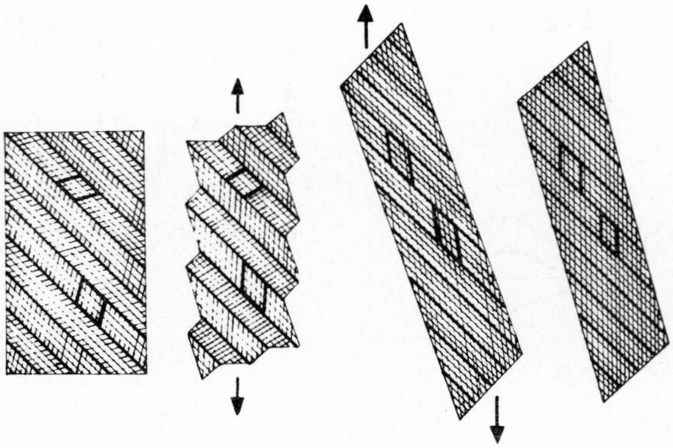

Figure 4b shows what happens to the model when we apply a small tensile load at low temperature. All layers are sheared and the layers M_+ become a little flatter, while layers M_- become steeper. Consequently the length grows by

$$D = \frac{1}{\sqrt{2}} \sum_{i=1}^{N} \Delta_i \,, \tag{3.1}$$

where the summation ranges over all layers. The deformation is elastic, because all Δ_i go back to +J or -J when we remove the load. This behaviour corresponds to the observed virginal curves in Figures 1a and 1b.

The application of the load P changes the potential energy of the lattice layers from $\Phi(\Delta)$ to

$$\Phi(\Delta,P) = \Phi(\Delta) - \frac{P}{\sqrt{2}}\Delta\ , \tag{3.2}$$

because the work of the load must be taken into account. Figure 5a shows such potential energies for various values of P and we conclude that, if the load is increased to a certain value P_Y, it will eliminate the left barrier and the layers M_- will flip into the configuration M_+. This entails a sudden increase of D (see Figure 4c). Thus the model simulates the yield of deformation at a certain load. Subsequent increase of the load will again gradually tilt the lattice layers, which are now all M_+, and this is reflected in the existence of the second elastic Branch in Figures 1a and 1b. Removal of the load brings all particles to the M_+ configuration with $\Delta = +J$, so that there is a residual deformation as indicated in Figure 4d.

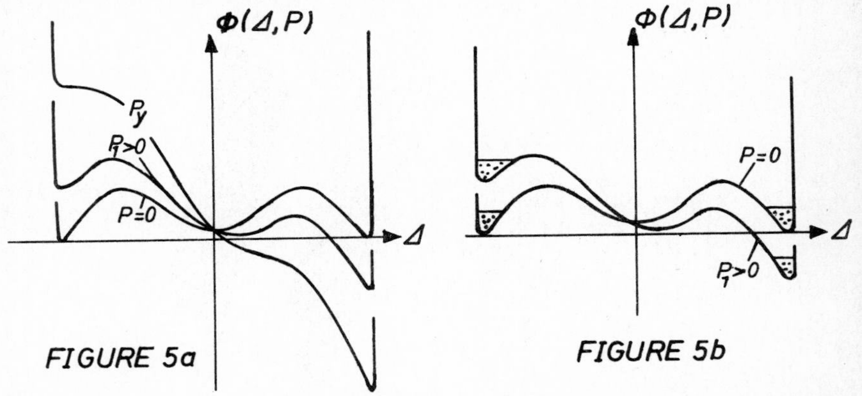

FIGURE 5a FIGURE 5b

If the lattice layers were lying still in their equilibrium positions, as we have tacitly assumed so far, we should experience the yield of deformation at the load which eliminates the left minimum, because only then can the layers flip. This would be so at zero temperature. But even at low temperatures the lattice particels do not lie still, rather they fluctuate about the minima and the fluctuation is stronger the higher the temperature is. The mean energy of this thermal fluctuating motion is of the order of magnitude kT. In Figure 5b. we have indicated this mean energy by the depth of the little pools in the minima. At low temperature, which is the case under consideration, the thermal motion is too slow to overcome high potential barriers, and we say that we have a constrained equilibrium. For instance: Let us consider a situation where the unloaded body contains more layers of type M_- than M_+. Since the potential minima for M_- and M_+

are equally deep in this case, the thermal motion will tend to bring an equal number of layers into each minimum. But, because of the barrier, it will take a very long time to achieve this; and for all practical purposes we may therefore consider the unequal distribution as stationary.

However, when the load has decreased the height of the potential barrier as shown in Figure 5, the layers by their thermal motion may well overcome that barrier. Therefore it occurs that the yield limit is a function of temperature. Indeed, the thermal motion will certainly drive the layers over the barrier when the temperature is so high that the depth of the pool becomes comparable to the height of the barrier. Thus we must expect that the yield load of the model decreases with increasing temperature and this is indeed observed in pseudo-elastic bodies according to the Figures 1a and 1b.

Even at low temperatures it does not make sense to ask for the displacement Δ of a particular lattice layer. But we may ask for the probability p_Δ of finding a layer with the displacement Δ. Statistical mechanics answers that question and sets

$$p_\Delta \sim e^{-\frac{\phi(\Delta,P)}{kT}} .$$

Let x be the fraction of M_+ layers which are those with a Δ in the neighbourhood of the right minimum so that 1-x is the fraction of M_- layers near the left minimum [1]). The probabilities p_Δ of finding a layer with a displacement Δ are

$$p_\Delta = \frac{1}{z}(1-x)\frac{e^{-\frac{\phi(\Delta,P)}{kT}}}{\int_{-J}^{m_L} e^{-\frac{\phi(\Delta,P)}{kT}}\,d\Delta} \quad (\Delta < m_L) \qquad \Bigg\| \qquad p_\Delta = \frac{1}{z}x\frac{e^{-\frac{\phi(\Delta,P)}{kT}}}{\int_{m_R}^{J} e^{-\frac{\phi(\Delta,P)}{kT}}\,d\Delta} \quad (\Delta > m_R) \tag{3.3}$$

for the M_- and M_+ layers respectively. m_L and m_R are the abscissae of the left and right maxima of the curve $\phi(\Delta,P)$. z is the constant of proportionality between $d\Delta$ and the number of positions in $d\Delta$.

The displacement of the body is given by the sum of the expectation values of Δ in the occupied minima formed with the probabilities (3.3)

$$D = \frac{1}{\sqrt{2}}N\left\{(1-x)\frac{\int_{-J}^{m_L}\Delta e^{-\frac{\phi(\Delta,P)}{kT}}\,d\Delta}{\int_{-J}^{m_L} e^{-\frac{\phi(\Delta,P)}{kT}}\,d\Delta} + x\frac{\int_{m_R}^{J}\Delta e^{-\frac{\phi(\Delta,P)}{kT}}\,d\Delta}{\int_{m_R}^{J} e^{-\frac{\phi(\Delta,P)}{kT}}\,d\Delta}\right\}. \tag{3.4}$$

It is useful to introduce dimensionless quantities by the definitions

$$d \equiv \frac{\sqrt{2}D}{NJ}, \quad \delta \equiv \frac{\Delta}{J}, \quad \varphi \equiv \frac{\phi}{\phi_s}, \quad \Theta \equiv \frac{kT}{\phi_s}, \quad p \equiv \frac{J}{\sqrt{2}\phi_s}P, \tag{3.5}$$

because they simplify the expression (3.4). We obtain

$$d(p,\Theta,x) = (1-x)\,\frac{\int_{-1}^{m_L/J} \delta e^{-\frac{\varphi(\delta)-p\delta}{\Theta}}\,d\delta}{\int_{-1}^{m_L/J} e^{-\frac{\varphi(\delta)-p\delta}{\Theta}}\,d\delta} + x\,\frac{\int_{m_R/J}^{1} \delta e^{-\frac{\varphi(\delta)-p\delta}{\Theta}}\,d\delta}{\int_{m_R/J}^{1} e^{-\frac{\varphi(\delta)-p\delta}{\Theta}}\,d\delta} \tag{3.6}$$

d is a function of p,Θ and x whose evaluation requires the knowledge of the explicit form of $\varphi(\delta)$ and a generally difficult integration. Only in the case of zero temperature is the calculation easy, because the probabilities (3.3) are given by δ -function in this case, whose peaks lie at the sites $\delta^{L,R}_{min}$ of the function $\varphi(\delta)$ - pδ in the ranges (-J,m_L) and (m_R,J) respectively. Thus for zero temperature we have

$$d(p,0,x) = (1-x)\,\delta^{L}_{min}(p) + x\,\delta^{R}_{min}(p) \tag{3.7}$$

and the functions $\delta^{L,R}_{min}$ (p) can easily be found from the knowledge of $\varphi(\delta)$ by differentiation: $\frac{d\varphi(\delta)}{d\delta}$ - p = 0. Solving this for δ we obtain $\delta^{L,R}_{min}$(p) and and for a typical potential of the form shown in Figure 3a.) we obtain curves d(p) for different values of x as shown in Figure 6. All of these functions except the one for x = 1 are cut off at the positive yield load as explained before, because the lattice layers will flip at high loads. Analogous arguments hold for the cut-off in compression.

In particular for x = $\frac{1}{2}$, which means that M_+ and M_--layers exist in equal proportion, we obtain the virginal curves shown in Figures 1a. and 1b. If $\Theta \ll 1$, but unequal to zero, calculations indicate that there is little effect of the temperature on the curves of Figure 6 except that the cut-off occurs at a lower load as we have explained above.

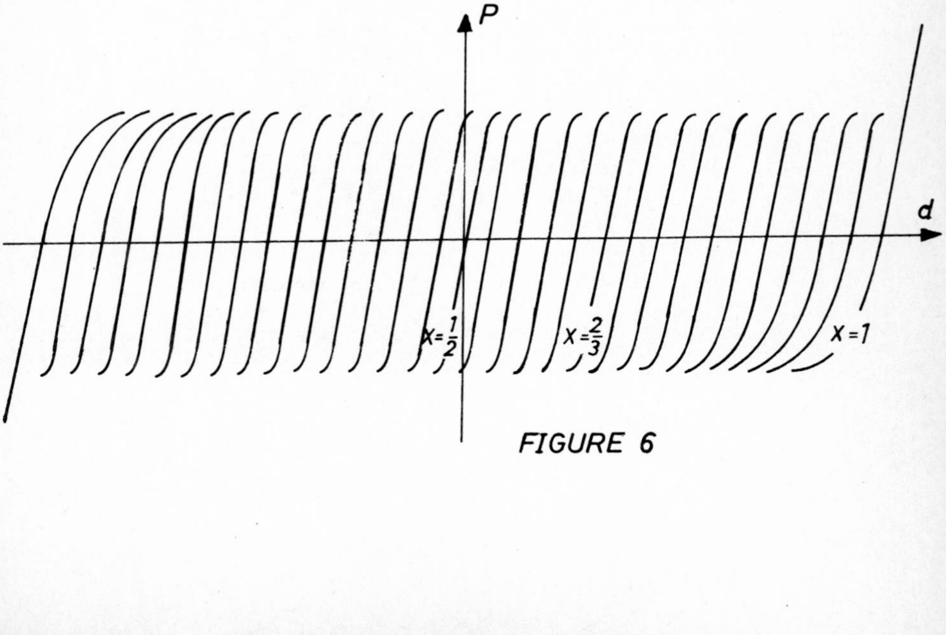

FIGURE 6

We conclude that the model can qualitatively account for the observed low temperature features of a pseudoelastic body.

The observed properties at intermediate and high temperature are not all explainable in so simple terms. At intermediate temperatures the particles will be able to overcome the barriers in the potential $\phi(\Delta)$ even in the unloaded state. Thus they can jump from configuration M^-(say) to A. Under those conditions the macroscopic behaviour reflects a competition between entropy, which tries to distribute the layers uniformly over the range $|\Delta| < J$, and the potential energy, which offers its minima as favourite sites to the particles. For the determination of the outcome of that competition we must rely upon the formal methods of statistical mechanics.

4.) THE MODEL AT INTERMEDIATE AND HIGH TEMPERATURES

a.) Partition Function and Free Energy

The thermodynamic properties of the model are determined by the partition funtion

$$Z(D,T) = \sum_{\Delta_1=-J}^{J} \cdots \sum_{\Delta_N=-J}^{J} e^{-\frac{\sum_{i=1}^{N}\phi(\Delta_i)}{kT}} . \tag{4.1}$$

This sum depends on the deformation D, because the summation is restricted to the values of Δ_i for which $\sum_{i=1}^{N}\Delta_i = \sqrt{2}D$ holds (see 3.1). From the partition function we may calculate the (Helmholtz) free energy

$$\Psi(D,T) = kT \ln Z \tag{4.2}$$

and hence follows the load P as a function of D and T

$$P(D,T) = \frac{\partial \Psi(D,T)}{\partial D} \tag{4.3}$$

This is the function that we wish to determine.[2)]

For analytical convenience we convert the sums in (4.1) into integrals and introduce the δ-function to account for the constraint (3.1). Thus $Z(D,T)$ assumes the form

$$Z(D,T) = z^N \int_{\Delta_1=-J}^{J} \cdots \int_{\Delta_N=-J}^{J} \prod_{i=1}^{N} e^{-\frac{\phi(\Delta_i)}{kT}} \delta\left(D - \frac{1}{\sqrt{2}}\sum_{i=1}^{N}\Delta_i\right) d\Delta_1 \cdots d\Delta_N . \tag{4.4}$$

As before, z is the constant of proportionality between $d\Delta$ and the number of positions on $d\Delta$. An appropriate method for the calculation of this partition function has been described in [5]. The method is akin to the virial expansion of the partition function of a real gas. It proceeds by setting

$$e^{-\frac{\phi(\Delta)}{kT}} = F_0(\Delta) + F_1(\Delta) , \tag{4.5}$$

where

$$F_0(\Delta) = e^{-\frac{\phi_0(\Delta)}{kT}} \quad \text{with} \quad \phi_0(\Delta) = \begin{cases} 0 \\ \infty \end{cases} \text{for} \quad \begin{matrix} |\Delta| \leq J \\ |\Delta| > J \end{matrix} \tag{4.6}$$

and

$$F_1(\Delta) = \begin{cases} e^{-\frac{\phi(\Delta)}{kT}} - 1 \\ 0 \end{cases} \quad \text{for} \quad \begin{matrix} |\Delta| \leq J \\ |\Delta| > J \; . \end{matrix} \tag{4.7}$$

We see that $|F_1(\Delta)| < 1$ holds, so that products of higher order in $F_1(\Delta_i)$ can be neglected. This approximation is obviously better the higher the temperature is, because $|F_1|$ decreases with increasing temperature.

It is fairly easy to calculate the partition function for the simple box-potential $\phi_0(\Delta)$ and the correction by F_1 can be taken into account approximately as explained in [5]. The result is

$$Z(D,T) = (2zJ)^N \left[e^{\beta J \frac{\sqrt{2}D}{NJ}} \left(\frac{1}{J} \int_0^J e^{-\frac{\phi(\Delta)}{kT}} \cosh \beta\Delta \, d\Delta \right] \right]^N \tag{4.8}$$

where β is a function of D whose calculation requires the inversion of the Langevin function $\mathcal{L}(x) \equiv \operatorname{ctgh} x - \frac{1}{x}$. We have

$$\frac{\sqrt{2}D}{NJ} = -\mathcal{L}(\beta J) \, . \tag{4.9}$$

By (4.2) the free energy reads

$$\Psi = -NkT \left[\ln 2zJ + \beta J \frac{\sqrt{2}D}{NJ} + \ln \left(\frac{1}{J} \int_0^J e^{-\frac{\phi(\Delta)}{kT}} \cosh \beta\Delta \, d\Delta \right) \right] . \tag{4.10}$$

b.) Free Energy and Load

In addition to the dimensionless quantities (3.5) we define

$$b \equiv \beta J \quad , \quad \psi \equiv \frac{\Psi + NkT \ln(2zJ)}{N\phi_s} \, , \tag{4.11}$$

because thus the expressions (4.10) and (4.3) for the free energy and the load are much simplified: With the abbreviation

$$H(b,\Theta) \equiv \int_0^1 e^{-\frac{\varphi(\delta)}{\Theta}} \cosh(b\delta) \, d\delta \tag{4.12}$$

one obtains

$$\begin{aligned} \psi(d,\Theta) &= -\Theta \left(bd + \ln H \right) \, , \\ p(d,\Theta) &= -\Theta \left(b + \frac{b^2 \sinh^2 b}{b^2 - \sinh^2 b} \left[d + \frac{H'}{H} \right] \right) \, , \end{aligned} \tag{4.13}$$

and (4.9) assumes the form d = - $\mathcal{L}$(b). H' in (4.13) denotes the derivative of H with respect to b.

For the schematized potential of Figure 3b. the integrals H and H' can easily be calculated and thus we obtain ψ and p as functions of b and Θ as follows

$$\psi(d,\Theta) = -\Theta\left(1 - b\frac{\cosh b}{\sinh b} + \ln\left[(e^{-\frac{X}{\Theta}} - e^{-\frac{1}{\Theta}})\frac{\sinh bS}{b} - (1-e^{-\frac{1}{\Theta}})\frac{\sinh bT}{b} + \frac{\sinh b}{b}\right]\right), \tag{4.14}$$

$$p(d,\Theta) = -\Theta\left(b + \frac{b^2\sinh^2 b}{b^2 - \sinh^2 b}\left\{\frac{1}{b} - \frac{\cosh b}{\sinh b} + \frac{(e^{-\frac{X}{\Theta}} - e^{-\frac{1}{\Theta}})S(\cosh bS - \frac{\sinh bS}{bS}) - (1-e^{-\frac{1}{\Theta}})T(\cosh bT - \frac{\sinh bT}{bT}) + (\cosh b - \frac{\sinh b}{b})}{(e^{-\frac{X}{\Theta}} - e^{-\frac{1}{\Theta}})\sinh bS - (1-e^{-\frac{1}{\Theta}})\sinh bT + \sinh b}\right\}\right)$$

Since there is a unique,and known, relation between b and d, viz. d = - $\mathcal{L}$(b), we may caLculate ψ(d,Θ) and p(d,Θ) from (4.14) for a particular potential $\hat{\phi}(\Delta)$, i.e. for chosen values of X, S and T. The solid and dashed parts of Figures 7 and 8 show the result of that numerical calculation for the following values of the potential function of Figure 3b.

$$X = \frac{1}{4}, \quad S = 0{,}2, \quad T = 0{,}98.$$

c.) Discussion of Load Deformation Curves

The (p,d)-curves of Figure 7 contain parts with negative slope which are thermodynamically unstable. These are dashed in the figure and they correspond to the branches of the (Ψ.d)-curves that have negative curvature (see Figure 8). The model will never be seen to change its state (p,d) along these branches. The solid branches on the other hand are stable and we say that the body is martensitic M_+ or M_-, if its state lies on the left or right solid branch respectively, and that it is austenitic, if the state is on the solid branch in the middle.

For a discussion of the curves of Figure 7 we fix the attention on the curve for Θ = 0,25 and imagine the model to be in the austenitic phase at (p,d) = (0,0) initially. Under progressive loading it will move upward along the curve until the maximum is reached. Upon further loading the model must jump to the right solid branch along the horizontal dotted line. It will thus be in the martensitic M_+ phase and may climb up along the right solid branch upon further loading. If we unload, the model does not leave that branch until the load falls below the value where the (p,d)-curve has a minimum. Then it will move back to the austenitic phase along the lower horizontal dotted line. We conclude that in a loading and unloading experiment the model exhibits a hysteresis like the one observed in memory alloys at intermediate temperatures (see Figure 1c. or 1d.)

It is also seen from Figure 7 that the hysteresis of the model occurs at higher loads when the temperature is raised and that it becomes narrower until finally - above a critical temperature - it vanishes altogether. All this is like in the observed curves of the Figures 1c. through 1f.

It must be remarked here that our calculations show that the hysteresis exhibited by the curves of Figure 7 occurs only, if our postulated potential has a minimum in the middle. A model with only lateral minima does not simulate the observed hysteresis behaviour in the first and third quadrant of the (p,d)-plane. This has been our motivation for chosing a potential of the shape shown in the Figures 3.

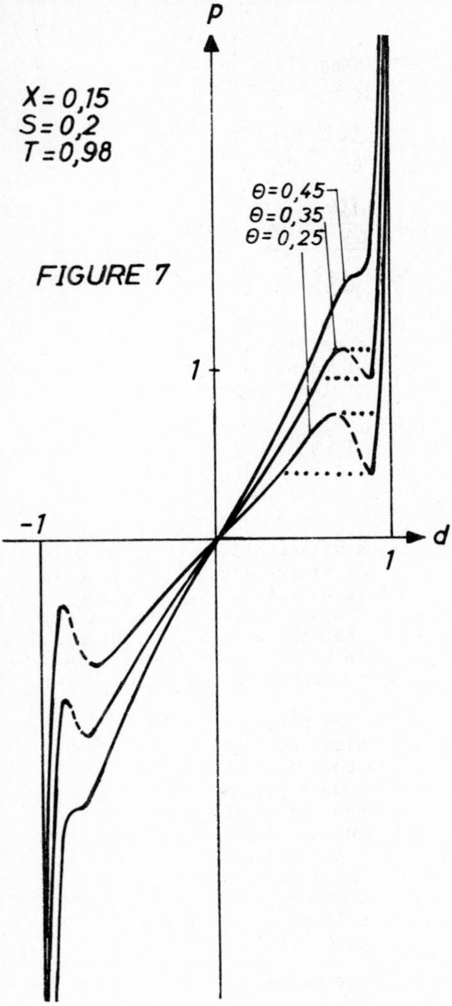

FIGURE 7

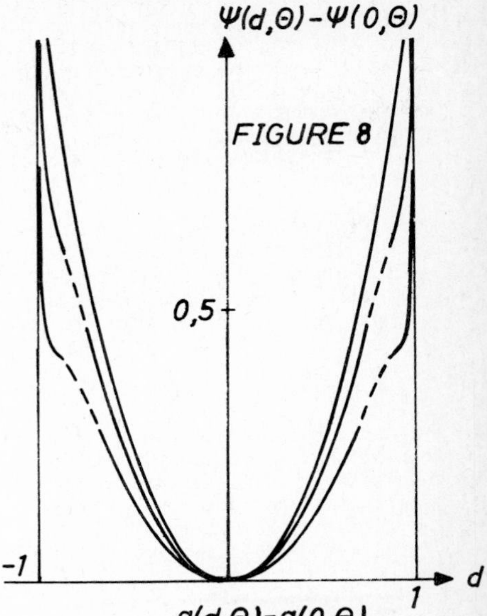

FIGURE 8

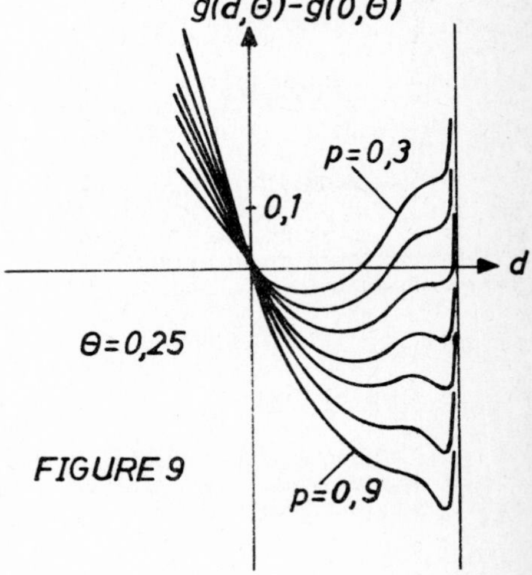

FIGURE 9

As the temperature is lowered there will be some value of temperature below which the approximation described in Section 4.a) is no longer valid. Also at low temperatures, where the layers are constrained to their lateral minima in fixed fractions x and (1-x) respectively, we expect load-deformation curves of the type shown in Figure 6. This theory in its present format is unable to predict at what temperature the transition from the curves of Figure 6 to those of Figure 7 occurs. For that we should need a dynamical theory which would inform us at what temperature the potential barriers can be overcome by sufficiently many layers in a reasonable time.

Nevertheless we conclude, that the present theory, with its low temperature range and its high temperature range can describe load-deformation curves of all types that are observed and reported in Figure 1. Measurement on real materials must be made to adjust the parameters of the model and to determine the point of transition between low temperature and high temperature behaviour.

d.) A Remark on Phase Equilibrium

Load deformation curves of the type shown in Figure 7 are not uncommon in thermodynamics and statistical mechanics. The best-known example is furnished by the isotherms of a van der Waals gas in a pressure-volume diagram. In that case a hysteresis is not observed, rather the gas, instead of following the isotherm until it reaches an extremum proceeds along the horizontal Maxwell-line shown in Figure 10 which makes the two shaded areas equal. Along this line there exists phase equilibrium. Here in a pseudo-elastic body we might also expect that the body should follow a horizontal line like the Maxwell line, but this is not the case. Instead we observe the hysteresis, apparently because the energetic barrier between the two phases is too high, so that equilibrium is constrained.

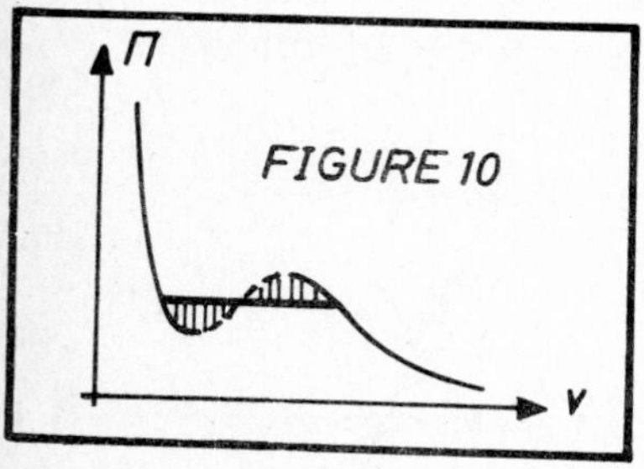

5.) FREE ENTHALPY AND LOAD

The model presents itself from a new and instructive angle when we calculate the Gibbs free energy or free enthalpy G. This function includes the energy PD of deformation by a fixed load P and can be calculated from

$$G(P,T) = -kT \ln \hat{Z}, \tag{5.1}$$

where

$$\hat{Z}(P,T) = z \int_{-\frac{NJ}{\sqrt{2}}}^{\frac{NJ}{\sqrt{2}}} \sum_{\Delta_1=-J}^{J} \cdots \sum_{\Delta_N=-J}^{J} e^{\frac{PD - \sum_{i=1}^{N} \phi(\Delta_i)}{kT}} \, dD \tag{5.2}$$

is the partition function for an ensemble of bodies whose deformation ranges bet-

$-\frac{NJ}{\sqrt{2}}$ and $\frac{NJ}{\sqrt{2}}$, i.e. over all possible deformations. z is the constant of proportionality between dD and the number of positions in dD. Comparison of (5.2) with (4.1) and 4.2) shows that we may write

$$\hat{Z}(P,T) = z \int_{-\frac{NJ}{\sqrt{2}}}^{\frac{NJ}{\sqrt{2}}} e^{\frac{P \cdot D - \Psi(D,T)}{kT}} dD . \tag{5.3}$$

Once G(P,T) is known we may caLculate the deformation D by differentiation of G with respect to P:

$$D = -\frac{\partial G}{\partial P} . \tag{5.4}$$

By use of the dimensionless quantities (3.5) and (4.11) and with

$$g \equiv \frac{G + NkT \ln(2zJ)}{N \emptyset_s} \tag{5.5}$$

we obtain

$$g(p,\Theta) = -\Theta \frac{1}{N} \ln\left(z \frac{NJ}{\sqrt{2}} \int_{-1}^{1} e^{N \frac{pd-\psi}{\Theta}} dd\right).$$

The function ψ - pd in the exponent of the integrand can easily be drawn by adding the function - pd to the energy $\psi(d,\Theta)$ from Figure 8 or equation $(4.14)_2$. Figure 9 shows the result for Θ = 0,25 and a number of values of p, viz.

p = (0,3, 0,4, 0,5, 0,6, 0,686, 0,8, 0,9).

The bigger p is the lower lies the curve g(p,Θ). The interesting point about these curves is the fact that in a certain range of values for p they have two minima and a part of negative curvature in-between. This reflects the stretch of negative curvature exhibited by the curves $\psi(d,\Theta)$ in Figure 8. For p = 0,686 the minima have equal depth.

Even though the minima in the function g (p,Θ) are fairly shallow, the integrand

$$\left[e^{\frac{pd-\psi}{\Theta}}\right]^N$$

of equation (5.6) has sharp peaks at their positions, because we may consider N to be a big number. We conclude from this that, while the ensemble consists of bodies with all possible values of d, these values cluster about one or at most two values of d. Obviously we must interpret this as follows: The cluster around small values of d contains austenitic bodies and the other one at higher values of d consists of martensitic bodies M_+. There is a range of values of p, where both clusters are present so that austenite and martensite may co-exist. This was already clear from Figure 7.

We are now able to construct the form of the free enthalpies as functions of load and temperature for the austenite and the martensites: Because of the steepness and the height of the peaks in the integrand of (5.7), we may replace that integrand by zero except for the value, or values, of deformation where the peaks lie. Thus we obtain

$$g(p,\Theta) = \psi(d_m,\Theta) - p\,d_m \tag{5.8}$$

and Figure 9 shows us d_m as a function of p for $\Theta = 0,25$. Note that, when there are two minima in the curves of Figure 9, the one at the smaller value of d determines g (p,Θ) for the austenite and the one at the bigger d gives us g (p,Θ) for the martensite. Figure 11 shows these functions. The branches corresponding to autenite and the martensitic twins have been denoted by A and $M_\pm$ respectively. The shape of these curves varies with temperature and with increasing Θ the range of overlap decreases until at the critical temperature $\Theta \approx 0,45$ it vanishes altogether.

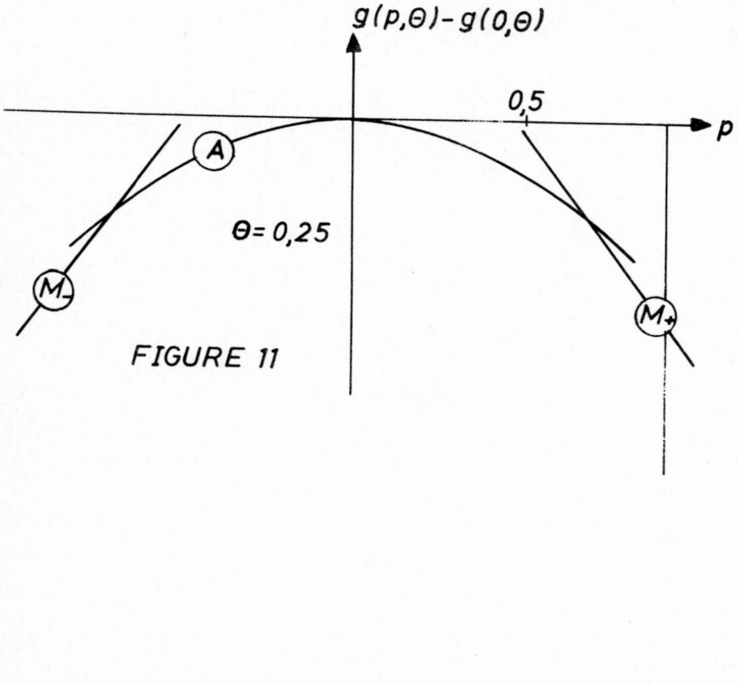

FIGURE 11

[1]) As we have explained earlier, the minimum at $\Delta = 0$ is unoccupied at low temperatures.

[2]) The calculation of P (D,T) is not affected by the kinetic energy of the layers which has therefore been suppressed in (4.1).

[1] Perkins, J., Editor: Shape Memory Effects in Alloys (Plenum Press, New York and London, 1975)

[2] Delaey, L., Krishnan, R.U., Tas, H. and Warlimont, H., Thermoelasticity, Pseudoelasticity and the memory effects associated with martensitc transformations, J. Mat. Sci. 9 (1974) 1511

[3] Jackson, C.M., Wagner, H.J. and Wasilewski, R.J., 55 Nitinol - the alloy with a memory: its physical metallurgy, properties and applications, NASA-SP 5110 (1972)

[4] Bensmann, G., Baumgart, F., Hartwig, J., Haasters, J., Untersuchung der Memory Nickel-Titan und Überlegungen zu ihrer Anwendung im Bereich der Medizin, Technische Mitteilung,Krupp Forschungsbericht Bd. 37 (1979)

[5] Müller, I. and Wilmanski, K., A Model for Phase Transition in Pseudoelastic Bodies, Il Nuovo Cimento 57 (1980)

Continuum Models of Discrete Systems 4
eds. O. Brulin and R.K.T. Hsieh

CONTACT TEMPERATURE, A DYNAMICAL ANALOGUE OF THE THERMOSTATIC TEMPERATURE

W. Muschik

Institut für Theoretische Physik
Technische Universität Berlin, HAD 1
D-1000 Berlin 12
Germany

If systems are far from local equilibrium, the thermostatic temperature cannot be used as a state variable because this temperature is strictly defined only for systems in equilibrium or in local equilibrium. Therefore a dynamical analogue of the thermostatic temperature has to be defined. This is obvious for discrete systems by considering the thermal contact between a non-equilibrium system and a heat reservoir: Analogously to thermostatic temperature contact temperature is defined by the vanishing heat exchange between them.

INTRODUCTION

Thermostatic temperature is defined for discrete thermal unique equilibrium systems being in thermal contact with each other. As it is well known [1] the symmetry and the transitivity of the thermal equilibrium cause a partition of equivalence of all thermal unique equilibrium systems so that all systems of the same class have the same thermostatic temperature. If a system under consideration is not in equilibrium, two possibilities exist: The system may be in local equilibrium [2] and therefore the concept of thermostatic temperature can be applied or the system is far from local equilibrium and thermostatic temperature is not defined. Therefore as in the case of thermostatic entropy which has no natural extension to thermodynamics [3] we have to redefine temperature in non-equilibrium [4,5]. As in the case of equilibrium this definition is naturally given for discrete systems. Therefore to get a field of temperature without using local equilibrium a transition to a field formulation is necessary.

SCHOTTKY SYSTEMS AND DYNAMICAL QUANTITIES

First we consider discrete systems - so called Schottky systems [6] - which interact with their surroundings by exchanging heat, work, and material. For each of these exchange quantities a conjugate dynamical quantity exists which is uniquely defined by the zero of the exchange quantity. This procedure is e.g. well known for the "dynamical pressure": A Schottky system which exchanges neither heat nor material with its surroundings (Figure 1) is in non-equilibrium (NES).

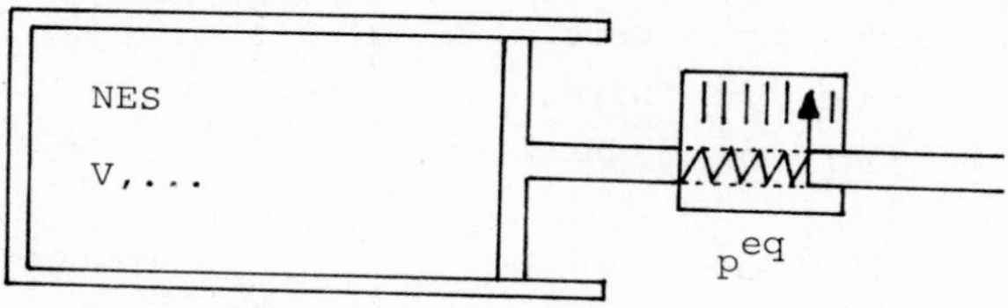

Figure 1
Gas contained in an adiabatic isolating piston and cylinder

The pressure p^{eq} exerted on the surface of the system is measured by a device gauged in equilibrium (length of a spring installed in the piston rod). We now can define the pressure p of NES by the zero of the time derivative of the volume V of the system

$$(p-p^{eq})\dot{V} \geq 0. \tag{1}$$

If in equilibrium the differential of power is given by

$$W^{eq} = \underline{A}^{eq} \cdot \dot{\underline{a}} , \tag{2}$$

where $\underline{a}$ are the work variables and $\underline{A}^{eq}$ the generalized forces (e.g. $\underline{a} \to V$, $\underline{A}^{eq} \to -p^{eq}$) we define the dynamical generalized forces $\underline{A}$ by the same procedure

$$\left.\begin{aligned} &(A_j^{eq} - A_j)\dot{a}_j \geq 0, \\ &\text{for all } j = 1, 2, \ldots, N, \\ &Q = 0, \quad \dot{\underline{n}}^e = \underline{0} , \end{aligned}\right\} \tag{3}$$

(Q = heat exchange, $\dot{\underline{n}}^e$ = change of numbers of moles by external exchange).

In the same way dynamical chemical potentials $\underline{\mu}$ can be defined. Starting from the well known [7] chemical potential in equilibrium $\underline{\mu}^{eq}$ we get for each component k

$$\left.\begin{aligned} &(\mu_k^{eq} - \mu_k)\dot{n}_k^e \geq 0 \\ &\text{for all } k = 1, \ldots, K, \\ &Q = 0, \quad W = 0. \end{aligned}\right\} \tag{4}$$

Of course the dynamical quantities $\underline{A}$ and $\underline{\mu}$ depend on the special chosen contact area between the considered system and its surroundings being in equilibrium.

Up to here two of the three interactions between a Schottky system and its surroundings were used to define dynamical quantities. As discussed below the third, the thermal interaction defines a dynamical analogue of the thermostatic temperature.

CONTACT TEMPERATURE

We now consider two discrete systems contacted by a given diathermal area F being impervious to work and material. To each division of F in non-overlapping parts F_i and to each pair of states (σ, σ') of both the systems instantaneous heat exchanges $Q^{F_i}(\sigma,\sigma')$ exist through the several F_i (Figure 2). A thermal contact between these

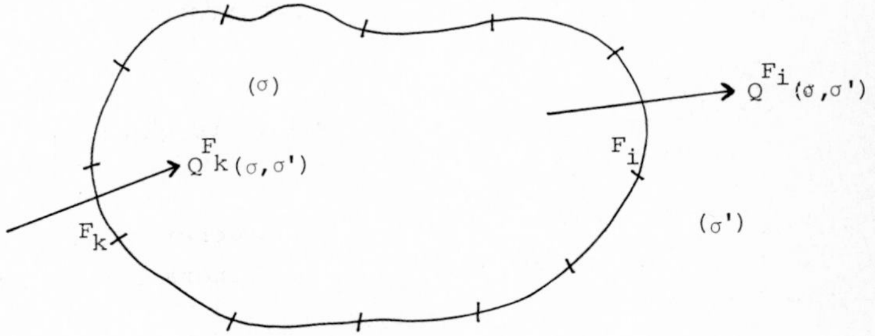

Figure 2

Instantaneous heat exchanges $Q^{F_i}(\sigma,\sigma')$ through partial areas F_i between two discrete systems being in the states σ and σ'.

two systems is defined by the set of all instantaneous heat exchanges

$$\{Q^{F_i}(\sigma,\sigma'),\ \text{for all division of } F = \bigcup_i F_i,\ \text{and all states } \sigma \text{ and } \sigma'\}$$

We now consider a thermal contact between a non-equilibrium system and an equilibrium one having the well defined thermostatic temperature T. Following the general procedure of constructing dynamical quantities discussed above we define the contact temperature Θ of a thermal contact by

$$\left.\begin{array}{l} Q(1/\Theta - 1/T) \geq 0 \\ W = 0,\ \underline{\dot{n}}^e = \underline{0}\ . \end{array}\right\} \tag{5}$$

By this definition the experimental fact is taken into account that exactly one unique thermal homogeneous equilibrium system exists so that the heat exchange through a given contact between it and a non-equilibrium system of given state is zero with change of sign. The contact temperature depends on the state of the non-equilibrium system, on the special choice of the contact area of the thermal contact, but not on its heat conducting properties. Because the contact temperature is defined by the same measuring rules as the thermostatic one we can say that the contact temperature is a dynamical analogue of the thermostatic temperature. As it is easy to be seen in contrast to thermostatic temperature the contact temperature is independent of the internal energy U of the system [8]. This property allows to introduce a state space for non-equilibrium processes in discrete systems called

MINIMAL STATE SPACE

which is the smallest one in which non-equilibrium processes can be described:

$$\left.\begin{array}{l} \mathcal{T}:\ (\underline{a},\ \underline{n},\ U,\ \Theta;\ \underline{\phi}^e)(t), \\ \underline{\phi}^e = (T^e,\ \underline{A}^e,\ \underline{\mu}^e) \end{array}\right\} \tag{6}$$

(t = time, T^e = external thermostatic temperature, $\underline{A}^e$ = external generalized forces, $\underline{\mu}^e$ = external chemical potentials of the guiding surroundings).

The trajectory $\mathcal{T}$ has two characteristic projections onto the sub-space of equilibrium which are called companion processes [9] (Figure 3)

$$\mathcal{T}:(\underline{a},\ \underline{n},\ U,\ \Theta;\ \underline{\phi}^e)(t) \quad \begin{array}{l} \nearrow\ (\underline{a},\ \underline{n},\ U)(t):\ \hat{\mathcal{T}} \\ \searrow\ (\underline{a},\ \underline{n},\ U_\Theta)(t):\ \mathcal{T}^* \end{array} \tag{7}$$

Here U_Θ is the energy belonging to Θ according to the thermostatic caloric equation of state

$$U = f(\underline{a},\ \underline{n},\ T),\quad U_\Theta = f(\underline{a},\ \underline{n},\ \Theta)\ . \tag{8}$$

U_Θ serves only to transform Θ into a measure of energy.

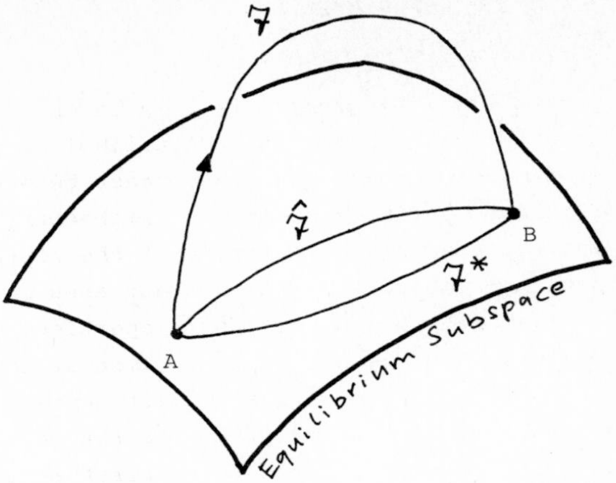

Figure 3
Non-equilibrium process 7 and its companion processes:
U-projection $\hat{7}$ and Θ-projection 7^*

The companion processes are by definition quasi-processes in the equilibrium subspace ($\underline{a}$, $\underline{n}$, U) which is determined by the Zeroth Law. As nearer the non-equilibrium process 7 takes its trajectory to the equilibrium subspace as lower is the difference between U and U_Θ which are equal in equilibrium.

EMBEDDING AXIOM

In non-equilibrium thermodynamics quantities as entropy, enthalpy... have to redefined. If we wish to generalize equilibrium quantities to non-equilibrium, their generalizations must be compatible with the values of these quantities given in equilibrium subspace. Therefore we have to postulate an embedding axiom for a non-equilibrium quantity Z(t) (Figure 3)

$$\left.\begin{array}{l} {}_{7}\!\int_A^B \dot{Z}(t)\,dt = Z_B^{eq} - Z_A^{eq} \\ \\ \text{for each process 7 between} \\ \text{the equilibrium states A und B} \end{array}\right\} \qquad (9)$$

which is immediately obvious. This axiom is one of the two starting points for deriving dissipation inequalities [8,10]. The other one is a suitable formulation of the Second Law for closed discrete systems which can be transformed to a field formulation. Therefore we need an

INTENSIFIED CLAUSIUS' INEQUALITY

because Clausius' inequality usually used

$$\oint \frac{Q(t)}{T(t)}\, dt \le 0 \tag{10}$$

contains the thermostatic temperatures of the guiding reservoirs being in equilibrium. These temperatures can not build into a field formulation because the surroundings of a partial system in the interior of a bigger one consists of non-equilibrium systems. According to (5) we get

$$\oint \frac{Q(t)}{T(t)}\, dt \le \oint \frac{Q(t)}{\Theta(t)}\, dt \tag{11}$$

showing that

$$\oint \frac{Q(t)}{\Theta(t)}\, dt \le 0 \tag{12}$$

is an intensification of (10). To prove (12) we need an additional axiom [11]. This axiom states something about the existence of cyclic processes, if these processes can be simulated by other equivalent ones.

FIELD FORMULATION

To get a field formulation we have to divide the considered system into subsystems. The exchange quantities become surface functions on the surface of the subsystems. The total exchange between the subsystem and its surroundings can be described by surface integrals, whereas the extensive quantities are volume integrals of the subsystem. Therefore we get for the First Law

$$Q(t) = \oint q(\underline{R}, \underline{f}, t)\, df = \int dV(\ldots) \tag{13}$$

and the Second Law

$$\sum_i \frac{Q_i(t)}{\Theta_i(t)} \to \oint \frac{q(\underline{R},\underline{f},t)}{\Theta(\underline{R},\underline{f},t)}\, df = \int dV(\ldots) \tag{14}$$

(q = surface heat density, $\underline{R}$ = point on the surface of the subsystem, $\underline{f}$ = outward unit normal). Using the well known technique that we apply (13) and (14) to arbitrarily small tetrahedrons, we get the existence of the field of the heat flux density $\underline{q} = \underline{q}(\underline{x},t)$ and that of the contact temperature $\Theta = \Theta(\underline{x},t)$.

REFERENCES

[1] Kestin, J., A Course in Thermodynamics (McGraw-Hill, New York, 1979) Vol. I, § 2.3, § 2.4

[2] see [1], § 6.4

[3] Meixner, J., Entropie im Nichtgleichgewicht, Rheol. Acta 7 (1968) 8

[4] Muschik, W., Empirical Foundation and Axiomatic Treatment of Non-Equilibrium Temperature, Arch. Rat. Mech. Anal. 66 (1977) 379-401

[5] Muschik, W. and Brunk, G., A Concept of Non-Equilibrium Temperature, Int. J. Engng. Sci. 15 (1977) 377-389

[6] Schottky, W., Thermodynamik, Die Lehre von den Kreisprozessen, den physikalischen und chemischen Veränderungen und Gleichgewichten (Springer, Berlin, 1929, reprint 1973)

[7] see [1], § 13.6

[8] Muschik, W., Fundamentals of Dissipation Inequalities, I. Discrete Systems, J. Non-Equilib. Thermodyn. 4 (1979) 277-294

[9] Keller, J.U., Ein Beitrag zur Thermodynamik fluider Systeme, Physica 53 (1971) 602

[10] Muschik, W., Fundamentals of Dissipation Inequalities, II. Field Formulation, J. Non-Equilib. Thermodyn. 4 (1979) 377-388

[11] Muschik, W. and Riemann, H., Intensification of Clausius' Inequality, J. Non-Equilib. Thermodyn. 4 (1979) 17-30

Author Index

GL = General Lecturer
SL = Sessional Lecturer